Urban and Community Forestry in the Northeast

Urban and Community Forestry in the Northeast

Second Edition

Edited by

John E. Kuser

State University of New Jersey,
New Brunswick, NJ, USA

Springer

Library of Congress Control Number: 2005937077

ISBN 10-14020-4288-4 (HB)
ISBN 13-978-1-4020-4288-1 (HB)
ISBN 10-1-4020-4289-2 (e-book)
ISBN 13-978-1-4020-4289-8 (e-book)

Printed on acid-free paper.

9 8 7 6 5 4 3 2 1

springer.com

To Dick West and Kim Ching, two of my early forestry mentors

Contributors

Nina L. Bassuk Department of Floriculture and Ornamental Horticulture, Urban Horticulture Institute, Cornell University, Ithaca, NY 14853

Donald F. Blair Sierra Moreno Mercantile Company, Big Pool, Maryland 21711

Edward Cesa US Department of Agriculture, Forest Service, Morgantown, West Virginia 26505

Kathleen E. Clark Endangered and Nongame Species Program, New Jersey Division of Fish, Game, and Wildlife, Woodbine, New Jersey 08270

William R. Comery, C.L.A. Shade Tree and Parks Commission, Borough of Paramus, Paramus, New Jersey 08625

James W. Consolloy Department of Grounds and Building Maintenance, Princeton University, Princeton, New Jersey 08544-5264

Michael V. D'Errico New Jersey Forest Service, Community Forestry Program, Trenton, New Jersey 08625

John F. Dwyer USDA Forest Service, North Central Research Station, Evanston, Illinois 60202-2357

Dr. William F. Elmendorf Assistant Professor, State Extension Specialist, Community Forestry, Penn State School of Forest Resources.

Peter F. Fengler Asplundh Tree Expert Company, Willowgrove, Pennsylvania 19090

Melvin S. Finstein Department of Environmental Sciences, ENRSB, Cook College and the New Jersey Agricultural Experiment Station, Rutgers University, New Brunswick, New Jersey 08901-8551

Bruce R. Fraedrich Bartlett Tree Research Laboratory, Charlotte, North Carolina 28278

Michael Gera Applied Research Technology, Montclair, New Jersey 07042

Henry D. Gerhold School of Forest Resources, Pennsylvania State University, University Park, Pennsylvania 16802-4300

George A Hawver Department of Floriculture and Ornamental Horticulture, Urban Horticulture Institute, Cornell University, Ithaca, New York 14853

James B. Ingram Bartlett Tree Experts, Osterville, Massachusetts 02655-0177

Larry Kuhns Department of Horticulture, Pennsylvania State University, University Park, Pennsylvania 16802-4203

Robert F. Lee GPU Energy Company, Morristown, New Jersey 07960

Edward A. Lempicki New Jersey Forestry Services, Trenton, New Jersey 08625

Victor D. Merullo Merullo, Reister and Swinford, Co., Columbus, Ohio 43206

L. Robert Neville Wolfe Mason Associates, Hampton, New Hampshire 03842

James Nichnadowicz 4-H Agent for Rutgers Cooperative Extension of Union County, New Jersey, Westfield, New Jersey 07009

David J. Nowak USDA Forest Service, Northeastern Research Station, Syracuse, New York 13210

William Porter New Jersey Shade Tree Federation, Cook College, Rutgers University, New Jersey 08901

John Rosenow National Arbor Day Foundation, Nebraska City, Nebraska 68410

Jim Sellmer Department of Horticulture, Pennsylvania State University, University Park, Pennsylvania 16802-4203

E. Thomas Smiley Bartlett Tree Research Laboratory, Charlotte, North Carolina 28278

Deborah Smith-Fiola Rutgers Cooperative Extension of Ocean County, Toms River, New Jersey 08755

Peter F. Strom Department of Environmental Sciences, ENRSB, Cook College and the New Jersey Agricultural Experiment Station, Rutgers University, New Brunswick, New Jersey 08901-8551

Steven Strom Department of Landscape Architecture, Cook College, Rutgers University, New Brunswick, New Jersey 08901-0231

Robert L. Tate Robert L. Tate Associates, Inc., Sacramento, California 95825

Richard S. Wolowicz Public Service Electric and Gas, Newark, New Jersey 07101

Mary Yager National Arbor Day Foundation, Nebraska City, Nebraska 68410

Foreword

During the latter part of 2004, Helen Buitenkamp of Springer Publishing emailed me that the first edition of ***Handbook of Urban and Community Forestry in the Northeast*** is the best volume in its field and inquired whether we'd be interested in compiling a second edition; I replied that we certainly would, and started working on it immediately. We have revised 14 out of 26 chapters in the first edition, and added two new authors. Many things in urban forestry have changed a great deal, while others have not changed at all. Henry Gerhold has written an entirely new Chapter 1 based on a book that he and his graduate student Stacy Franks have written entitled "Our Heritage of Community Trees." Dave Nowak has included the most up-to-date information on the environmental effects of trees in Chapter 2, and Peter Fengler and Tom Smiley have done the same with the diagnosis and treatment of hazard trees in Chapter 17. All told, we have revised or replaced 16 chapters of the original 26; we've kept 10 chapters as originally written, and substituted two entirely new chapters, 1 and 14, respectively.

With the emergence of urban and community forestry as the fastest growing part of our profession in the last several years, the need for a book such as this inevitably developed. The Society of American Foresters' urban forestry working group counts over 40 universities now offering courses in this subject, and the number is growing.

For the last several years, I have coordinated a Continuing Education urban forestry course at Rutgers University for non-matriculated students. Registrants have included arborists, engineers, shade tree commissioners, landscape architects, city foresters, environmental commissioners, park superintendents, and others whose jobs involve care and management of trees. The course was started by Bob Tate in 1980 around a core of managerial subjects such as inventories, budgets, and public relations. After Bob left in 1984 to join Asplundh and later to start his own prosperous business in California, the course languished after it exhausted the local market for those subjects. Bob is now the President of the International Society of Arboriculture.

I revived Bob's course in 1992, adding technical subjects such as tree selection (my own field), planting, pruning, soils, insects, and diseases. Because it is almost impossible for one person to be an expert in all these areas, each lecture was delivered by a practitioner of that particular field. An important part of the course was feedback: each year, registrants were asked what subjects they thought would be meaningful additions to the curriculum for other members of their organizations planning

to attend during the following year. Based on this feedback, new topics and speakers were added, occasionally some were dropped, and thus the current menu of urban forestry topics developed.

During this same time, I also taught an undergraduate course in urban forestry; in looking for a text that covered most of the same topics, as did the Continuing Education course, I could not find one. The only answer was to create one.

What we provide here is a basic framework of information that can be enlarged upon by the reader with respect to any particular specialty. For this reason, reference lists are included at the end of each chapter. City foresters, shade tree commissioners, landscape architects, and others who deal with planning and managing tree populations are required to make decisions that rest on the expertise accumulated in fields as diverse as law, finance, soils, engineering, silviculture, entomology, pathology, plant physiology, and public relations. They must be part scientist, part businessman, part landscape architect, and part politician. All of us involved in urban forestry have our own specialties that we know best and usually two or three others at which we are reasonably competent; this book is designed to help us learn the basics of the parts of urban forestry with which we are least familiar and to provide references for us and our students to use in order to broaden our knowledge as needed.

I broached the idea for this book to our dean, our state forester, and several others, including Harry Wiant, then president-elect of the Society of American Foresters. I offered it as an alternative to one or two very interesting research projects. The answers were unanimous: do the book because it is greatly needed. So here it is. You may notice that our chapter authors do not all agree about everything: tall versus short trees, exactly how to plant a tree, and so forth. But they are all good at what they do, and they verify the old saying that there are more ways than one to skin a cat. Urban Forestry is still young and growing, and as it matures the best ideas will spread.

John E. Kuser

Acknowledgements

Thanks to all our chapter authors, who have taught me much about many parts of urban/community forestry. Special thanks to Jason C. Grabosky, my successor, who has greatly exceeded my achievements in Urban Forestry; Christine Allen for her talented and indefatigable efforts; Eleanor Kuser, for her patience when I bring home piles of paperwork and do not talk much; Jim Consolloy, who knows almost everything about planting and growing trees; Bill Comery, who built Paramus into New Jersey's showcase Tree City and takes time to show it to our urban forestry classes; Les Alpaugh, New Jersey's State Forester, who backed our urban forestry project early on when some were still asking "*urban forestry—what's that?*"; Helen Buitenkamp, who read the first edition and emailed me that it was so good that it was worth a second edition (and not surprisingly, I agreed!), Dave Shaw and Henry Gerhold, who helped us find some of the experts who became chapter authors; and Christine Tizzano who helped Christine Allen with some of the preparation of the chapter revisions.

Contents

Chapter 1
Origins of Urban Forestry
Henry D. Gerhold

1. Introduction 1
   1.1. What Is an Urban Forest? 2
   1.2. Values of Urban Forests 2
2. Design of Urban Landscapes 3
   2.1. Trees Become Urbanized in Europe 4
   2.2. Urban Trees in America 6
   2.3. Choice of Species in Landscape Design 8
3. Arboricultural Practices 10
   3.1. Arboriculture in Europe 10
   3.2. Arboriculture in America 12
4. Urban Forest Management 14
   4.1. Local Authority and Public Involvement in the U.S. 15
   4.2. Municipal Tree Management in North America 15
   4.3. Urban Forest Management in Europe 17
   4.4. U.S. Laws Implementing Urban Forestry 18
5. Professional Organizations 18
6. Research and Education 19
7. Community Trees and Our Future 19

References 20

Chapter 2
Understanding the Benefits and Costs of Urban Forest Ecosystems
David J. Nowak and John F. Dwyer

1. Introduction 25
2. Urban Land in the United States 25
3. Physical / Biological Benefits and Costs of Urban Vegetation 26
   3.1. Urban Atmosphere 27
      3.1.1. Temperature and Microclimate Modifications 27
      3.1.2. Removal of Air Pollutants 28

3.1.3. Emission of Volatile Organic Compounds and Tree Maintenance Emissions ... 31
3.1.4. Net Effects on ozone ... 32
3.1.5. Energy Conservation ... 33
3.2. Urban Hydrology ... 33
3.3. Urban Noise ... 34
3.4. Urban Wildlife and Biodiversity ... 34
3.5 Phytoremediation ... 35
4. Social and Economic Benefits and Costs of Urban Vegetation ... 36
4.1. Benefits to Individuals ... 36
4.2. Benefits to Communities ... 37
4.3. Real Estate Values ... 38
4.4. Tree Value Formulas ... 39
4.5. Other Benefits and Costs of Urban Trees and Forests ... 39
5. Benefit-Cost Analyses ... 40
6. Implications for Planning, Design, and Management ... 41
References ... 41

Chapter 3
Arboriculture Law in the Northeast
Victor D. Merullo

1. General Tree Law ... 47
2. Enabling Statutes for Municipal Commissions to Care for Trees ... 52
References ... 61

Chapter 4
Tree Ordinances
Michael V. D'Errico

1. Introduction ... 63
2. Planning for an Ordinance ... 64
3. Legal Authority ... 64
4. Drafting a Tree Ordinance ... 65
4.1. Outline Community Tree Practices ... 65
4.2. Take a Look at the Tree Resources ... 65
4.3. Identify Needs ... 66
5. Specific Goals ... 66
6. Three Functions of a Successful Ordinance ... 68
7. Ordinance Sections ... 68
8. Linking to Specifications and Standards ... 69
References ... 78

Chapter 5
Using Comprehensive Planning to Conserve the Natural Environment
William F. Elmendorf

1. The Natural Environment Benefits Community 79
2. Community 80
3. Community Development 81
4. Planning 82
5. Comprehensive Planning 83
   5.1. Comprehensive Plans 85
   5.2. Zoning Ordinances 87
   5.3. Subdivision and Land Development Ordinances 87
6. Problems with Comprehensive Planning 88
7. Problems With Planning for the Natural Environment 89
8. Regulatory and Nonregulatory Strategies for Conserving Natural Resources 90
9. Closing Thoughts 94
   9.1. Natural Resource Managers and Planners Working Together 94
   9.2. Strengthen Community Planning 95
   9.3. Excellent Leadership 95
   9.4. A Watershed or Ecosystem Approach 95
   9.5. A Collaborative Process for Conflict Resolution 96
   9.6. A Broader Environmental Ethic 96
References 96

Chapter 6
Urban and Community Forestry: Planning and Design
Steven Strom

1. Planning the Urban Forest 99
   1.1. What is the Urban Forest? 99
   1.2. Who is Responsible? 100
   1.3. What are the Objectives? 100
2. Elements of the Urban Forest 101
   2.1. Nodes 101
      2.1.1. Parks and Open Spaces 101
      2.1.2. Conservation and Natural Areas 101
      2.1.3. Civic and Institutional Facilities 101
      2.1.4. Community Gardens 102
   2.2. Links 102
      2.2.1. Streets 102
      2.2.2. Greenways 103
      2.2.3. Blueways 104
      2.2.4. Highway and Rail Corridors 107
      2.2.5. Visual Linkages 107

3. Design and Urban Forestry ... 107
3.1. Streets ... 108
3.2. Plazas and Squares ... 111
3.3. Parks and Open Spaces ... 111
3.4. Parking Lots ... 111
3.5. Community Gateways ... 113
4. Planning and Design: Future Opportunities ... 115
4.1. Vacant and Underutilized Land and Brownfields ... 115
4.2. Productive Landscapes ... 115
4.3. Highway Interchanges ... 115
5. Conclusion ... 116
References ... 116

Chapter 7
Tree Inventory and Systematic Management
Richard S. Wolowicz and Michael Gera

1. Introduction ... 119
2. Shade Tree Inventory ... 119
3. Inventory and Systematic Management ... 120
4. Benefits of Systematic Management ... 120
5. Different Types of Inventories ... 121
5.1. Random Sample ... 121
5.2. 100% Sample ... 122
6. The Record-Keeping System ... 122
7. Mechanics of Conducting a Tree Inventory ... 123
8. Data Collection ... 124
9. Training Issues ... 126
10. Case Study ... 126
11. Automation ... 127
11.1. Surveying the Marketplace for Hardware, Software, and Training ... 127
11.2. Assessing Internal and External Resources ... 128
11.3. Developing a Project Work Plan ... 129
11.4. Choosing Hardware and Software ... 129
11.5. Review and Implement the Project Work Plan ... 130
11.6. Evaluate the Project Work Plan ... 130
12. Conclusion ... 130
References ... 131

Chapter 8
Urban and Community Forestry Financing and Budgeting
Robert L. Tate

1. Introduction ... 133
1.1. History and Overview ... 133
1.2. Sources of Funding ... 134

1.3. Competition for Budget Funds . . . 134
1.4. Justification for Budget Funds . . . 135
2. Funding Sources . . . 135
2.1. The Property Tax . . . 135
2.1.1. History . . . 135
2.1.2. Limitations on the Property Tax . . . 136
2.1.2.1. Diversification of Municipal and Community Revenues . . . 136
2.2. Aid From Federal and State Governments . . . 137
3. Budgeting . . . 140
3.1. Competing for a Fair Share of Budget Funds with Other Local Government Agencies . . . 140
3.2. Allocation of Budget Funds to Various Tree Management Activities . . . 141
3.3. Establishment of Priorities for Various Tree Management Activities . . . 142
4. Urban and Community Tree Inventories and Their Role in Financing and Budgeting Urban and Community Forestry Activities . . . 142
4.1. Use of the Inventory to Create an Urban Tree Management Program . . 142
4.2. Use of the Inventory to Gain Increased Funding . . . 143
4.3. Use of the Inventory to Insulate against Budget Reductions . . . 143
4.4. Use of the Inventory to Manage the Urban and Community Forest More Efficiently . . . 144
5. Summary . . . 145
References . . . 146

Chapter 9
Community Involvement in Urban Forestry Programs
James Nichnadowicz

1. Introduction . . . 147
2. Public Awareness . . . 147
2.1. Leaflets . . . 148
2.2. Annual Reports . . . 148
2.3. News Releases . . . 149
2.4. Lectures . . . 150
2.5. Television . . . 151
2.6. Videotapes, CD's and DVD's . . . 151
2.7. Comment Forms . . . 151
2.8. Image Check . . . 151
2.8.1. Image Analysis Survey . . . 152
3. Relations with Local Elected Officials . . . 152
4. Involving Volunteers in Your Community Forestry Program . . . 153
4.1. Laying the Foundation for Volunteer Involvement in Your Forestry Program . . . 154
4.2. Analyzing the Job You Need Done Helps Target Recruitment Efforts . . 155
4.3. Applications for Screening and Clarification . . . 155
4.4. Advertising to Find Volunteers . . . 155

4.5. Whom to Accept into Your Program 155
4.6. Benefits of Training Volunteers 157
4.7. Training Volunteers 159
4.8. Financing Volunteer Training and Locating Teachers for Your Training Sessions 159
4.9. Challenges of Volunteer Use in Your Program 159
5. Involving Children in Tree Planting 160
5.1. Tree Seedling Giveaways 161
5.2. Helping Schools Celebrate Arbor Day – The Union County Arbor Day Tree Program 162
5.3. Additional Ideas on How to Involve Children with Trees and Tree Planting 162
6. Conclusion 163
References 163

Chapter 10
Soils: The Key to Successful Establishment of Urban Vegetation
George A. Hawver and Nina L. Bassuk

1. Introduction 165
2. Tree Roots 165
3. Soil Properties 166
3.1. Physical Properties 166
3.1.1. Soil Texture 166
3.1.1.1. Texture Assessment 168
3.1.1.2. Texture Modification 168
3.1.2. Soil Structure 168
3.1.2.1. Bulk Density 169
3.1.2.2. Structure Assessment 169
3.1.2.3. Structure Modification 169
3.1.3. Aeration and Drainage 170
3.1.3.1. Aeration and Drainage Assessment 170
3.1.3.2. Aeration and Drainage Modification 171
3.2. Chemical Properties 171
3.2.1. Nutrients 171
3.2.2. Cation-Exchange Capacity 172
3.2.3. Soil Reaction 172
3.2.4. Soluble Salts 173
3.2.5. Chemical Property Assessment 173
3.2.6. Chemical Property Modification 173
3.3. Biological Properties 173
4. Urban Soils Distinguished from Natural Soils 174
4.1. Destruction of Soil Structure 174
4.2. Reduced Aeration and Drainage 174
4.3. Elevated Soil pH 175
4.4. Additional Characteristics 175

5. Preventing Soil Compaction 175
5.1. Surface Treatments 175
5.2. Choice of Soil Texture 175
5.3. Restricting Access 176
6. Compaction Remediation 176
6.1. Radial Trenching 177
6.2. Mulch Applications 177
6.3. Fracturing with Compressed Air or Water 178
6.4. Mechanical Breakup 178
6.5. Soil Amendments 178
7. Soil Volume 178
7.1. Soil Volume Assessment 179
7.2. Soil Volume Modification 179
8. Soil Specification Summary 180
8.1. Texture 180
8.2. Soil Reaction 180
8.3. Organic Matter 180
8.4. Soluble Salts and Nutrients 180
8.5. Contaminants 181
References 181

Chapter 11
Selecting Trees for Community Landscapes
Henry D. Gerhold and William Porter

1. The Selection Process 184
2. Defining Purposes of Trees in the Landscape 184
3. Evaluating Site Conditions 186
4. Arboricultural Practices 188
5. Selection Criteria 189
6. Using Selection Criteria to Make Choices 190
7. Computer Selection Methods 198
References 198

Chapter 12
Guide to Selecting and Specifying Nursery Stock
Jim Sellmer and Larry Kuhns

1. Introduction 199
2. Selecting Quality Trees 199
2.1. Plant Nomenclature 199
2.2. Nursery Sources 200
2.3. Above Ground Quality 201
2.4. Below Ground Quality 203
2.5. A Summary of Nursery Production Practices Which Affect Post Planting Survival and Growth 205

2.5.1. Propagation or planting practices that lead to deformed root systems . . . 205
2.5.2. Hilling trees or planting too deep in the nursery . . . 205
2.5.3. Postemergence Herbicide Damage – Roundup, Finale, Reward . 206
2.5.4. Poor Fertility . . . 206
2.5.5. Growing on Poor Soils . . . 207
2.5.6. Growing Too Close in the Nursery . . . 207
2.5.7. Insect or Disease Problems . . . 207
2.5.8. Digging an Undersized Ball . . . 207
2.5.9. Plastic Burlap . . . 207
2.5.10. Nylon Twine . . . 208
3. Writing Bid Specifications . . . 208
3.1. Realities . . . 208
3.2. Resources . . . 209
4. Specifying Quality Results . . . 211
4.1. The General Form . . . 211
4.2. Business Section Details . . . 212
4.2.1. Pre-Qualifying the Bidders . . . 212
4.2.2. Bid Proposal Format . . . 212
4.2.3. Awarding of the Contract . . . 213
4.2.4. Expectations, Limitations, and Insurance Requirements . . . 213
4.2.5. Inspection, Completion, Guarantees, and Contacts . . . 213
4.3. The Project Section Details . . . 214
4.3.1. Scope of the Job . . . 214
4.3.2. Listing of Applicable Standards . . . 214
4.3.3. Project Materials . . . 214
4.3.4. Phytopathological Certification . . . 216
4.3.5. Requirements for Selection, Tagging and Inspection of the Plant Materials . . . 216
4.3.6. Procedures for Handling Plant Material . . . 216
4.3.7. Deadlines for Delivery of the Plant Materials . . . 217
4.3.8. Guarantee and Replacement Requirements and Limitations . . . 217
5. Conclusions . . . 218
6. Appendix: Plant Locator Resources on the Internet . . . 218
References . . . 219

Chapter 13
Planting and Maintenance
James W. Consolloy

1. Introduction . . . 221
2. Site Selection . . . 222
2.1. Temperature and Hardiness . . . 222
2.2. Water as a Factor . . . 223
2.3. Light Considerations . . . 223
2.4. Air Quality . . . 223
2.5. Soil Conditions . . . 224

2.6. Biological Factors 224
2.7. Physiological and Morphological Factors 224
3. Site-Specific Factors 224
4. Transportation, Care, and Handling 225
4.1. Inspection 226
5. When and How to Plant 226
5.1. Soils: Native vs. Amended 227
5.2. Planting Hole 227
5.3. Pruning 229
5.4. Fertilizing 229
5.5. Bare-Root Planting 230
5.6. Balled and Burlapped Planting 230
5.7. Container Planting 231
5.8. Large Tree Transplanting 232
6. Staking and Guying 233
7. Mulching 233
8. Watering 234
9. Tree Care and Maintenance 235
References 235

Chapter 14
Pruning
David L. Johnson

1. Introduction 237
2. Trees in an Urban or Suburban Environment 237
3. Reasons for Pruning 238
3.1. Pruning for Form 238
3.2. Pruning for Strength 238
3.3. Pruning for Tree Health 238
3.4. Pruning for Safety 239
3.5. Pruning for Aesthetics 239
3.6. Specialty Pruning 240
4. Seasonal Timing of Pruning 240
4.1. Pruning Evergreens 240
4.2. Pruning Deciduous Trees 241
4.3. Pruning Newly Planted Trees 241
4.4. Pruning Established Young Trees 241
4.5. Pruning Mature Trees 242
5. Proper Pruning Cuts 242
5.1. Types of Pruning Cuts 242
6. General Pruning Practices 244
7. Maintenance Pruning 244
7.1. Crown Cleaning 244
7.2. Crown Thinning 244
7.3. Crown Raising 245
7.4. Crown Reduction 245

7.5. Vista Pruning ... 245
7.6. Crown Restoration ... 245
8. Pruning Specifications ... 245
9. Utility Pruning ... 246
10. Pruning Tools ... 246
11. Treatment of Pruning Wounds ... 246
References ... 247

Chapter 15
Trees, Utilities, and Municipalities
Robert F. Lee and Richard S. Wolowicz

1. Introduction ... 249
2. Background ... 249
3. Distribution and Transmission ... 250
4. Tree and Wire Conflict ... 250
5. Regulations ... 251
6. Myths about Utility Line Clearance Programs ... 252
7. Utilities and Municipalities Working Together ... 254
7.1. Avoiding Municipal – Utility Conflicts ... 256
7.2. Joint Ventures ... 257
8. Summary ... 259
References ... 260

Chapter 16
Tree Roots versus Sidewalks and Sewers
William R. Comery, C.L.A.

1. Introduction ... 261
2. Root Systems ... 262
2.1. Identification of Root Systems ... 262
2.2. How and Why Root Systems Grow ... 262
2.3. What Problems Impede Root System Growth? ... 263
2.4. Common Sidewalk Problems ... 264
2.5. Common Sewer Problems ... 265
2.5.1. Old systems ... 266
2.5.2. Construction techniques ... 266
2.5.3. Unforseen perils ... 266
2.6. What is the Government's Responsibility When Trees Conflict with Sidewalks and Sewers? ... 266
2.6.1. "Do-Nothing": No Tree Requirements ... 267
2.6.2. "Do-Something": Requires Trees but no Tender Local Care (TLC) ... 267
2.6.3. "Do-It-All": Requires Trees and Provides Tender Local Care ... 267
3. Choosing the Right Tree for the Right Spot ... 268
3.1. Tree Species Selection ... 269
3.2. Planning for the Future ... 271

3.2.1. Confined Urban Space . . . 271
3.2.2. Restricted Suburban Space (Within ROW) . . . 271
3.2.3. Unrestricted Suburban Space (Planting Easement) . . . 271
3.3. Proper Planting Location . . . 272
3.4. Proper Construction Practices . . . 272
3.5. Options to Fix Sidewalks . . . 273
3.5.1. Tree Removals . . . 273
3.5.2. Root Cutting and Tree Trimming . . . 273
3.5.3. Contoured Sidewalks . . . 275
3.5.4. Replacement Tree Planting Policy . . . 276

Chapter 17
Hazard Tree Inspection, Evaluation, and Management
E. Thomas Smiley, Bruce R. Fraedrich, and Peter Fengler

1. Introduction . . . 277
2. Methods of Tree Evaluation and Hazard Assessment . . . 278
3. Identification of Hazardous Conditions . . . 278
3.1. Tree Location . . . 279
3.2. Problem Conditions and Locations . . . 280
3.3. Tree Defects . . . 281
3.3.1. Crown . . . 281
3.3.2. Trunk and Scaffold Branches . . . 282
3.3.3. Roots/Root Collar . . . 283
3.3.4. Stumps . . . 283
3.4. Decay . . . 284
3.5. Root Defect Evaluation . . . 285
3.6. Symptoms of Root Problems . . . 285
4. Management of Hazard Trees . . . 287
4.1. Decision Making . . . 287
4.2. Strength Loss Thresholds from Stem Decay . . . 288
4.2.1. Sapwood Decay and Mechanical Damage to the Trunk . . . 290
4.2.2. Roots . . . 290
4.3. Utility Line Clearance and Tree Hazards . . . 291
4.4. Storms and Tree Hazards . . . 292
4.5. Learning More About Predicting Tree Failures . . . 292
5. Summary . . . 293
References . . . 293

Chapter 18
Integrated Pest Management
Deborah Smith-Fiola

1. Introduction . . . 295
1.1. Pest Resistance . . . 296
1.2. Target Pest Resurgence . . . 296

1.3. Secondary Pest Outbreaks 296
1.4. IPM-Multiple Strategies 297
2. Monitoring 298
2.1. Thresholds 301
3. Key Plants and Key Pests 301
4. IPM Pest Control Strategies 303
4.1. Cultural Control 303
4.1.1. Resistant Plant Varieties 304
4.2. Mechanical Control 305
4.3. Biological Control 306
4.4. Microbial and Viral Control 307
4.5. Pesticides and Biorational Pesticides 307
5. Summary 308
References 308
Case Studies
A. Sycamore Anthracnose 310
B. Eastern Tent Caterpillar 311
C. Dutch Elm Disease 313
D. Gypsy Moth 315
E. Banded Ash Clearwing Borer 317
F. Hemlock Woolly Adelgid 319
G. Chestnut Blight 321

Chapter 19
Tree Appraisal: The Goal is Equity
James B. Ingram

1. Introduction 323
2. Appraisal Procedures 325
2.1. Qualifications of the Tree and Plant Appraiser 325
2.2. Diagnostic Instruments and Tools 325
3. Two Basic Methods of Plant Appraisal 326
3.1. Replacement Cost 326
3.2. Trunk Formula 327
4. Factors That Affect Tree Value 327
4.1. Condition 327
4.1.1. Structural Integrity 328
4.1.2. Plant Health 328
4.1.2.1. Disease-Causing Biotic Agents 329
4.1.2.2. Noninfectious Disorders Caused by Abiotic Agents 329
4.1.2.3. Direct Mechanical Injury 330
4.1.3. Calculating the Condition Rating 331
4.1.3.1. The Five Condition Factors 331

4.2. Location . . . 331
4.2.1. Site . . . 333
4.2.2. Contribution . . . 333
4.2.3. Placement . . . 333
4.2.4. Calculating the Location Rating . . . 334
4.3. Species . . . 336
4.4. Size . . . 337
4.4.1. Trunk Area and Adjusted Trunk Area . . . 338
4.4.2. Other Factors in Determining Tree Size . . . 338
4.4.3. Accurate Measurement of Diameter at 4.5 Feet (1.4 m) . . . 339
4.5. Replacement Cost and Basic Price . . . 339
5. Summary of Trunk Formula Method . . . 340
6. Compounded Replacement Cost . . . 342
7. Cost of Repair and Cost of Cure Methods . . . 342
8. Plant Appraisals and Property Worth . . . 343
8.1. Estimating Total Property Value . . . 344
8.2. Contributory Value of Trees, Shrubs, and Other Vegetation . . . 344
8.2.1. Market Evidence . . . 344
8.2.2. Paired Sales Analysis . . . 344
8.3. Arboricultural Appraisal Law and Property Value . . . 345
8.3.1. Measures of Damages . . . 345
8.3.2. Reasonableness Testing . . . 345
8.3.3. Appraisers as Expert Witnesses . . . 346
9. Summary: Role of the Appraiser and the Appraisal Process in Urban Forestry . . . 346
References . . . 347

Chapter 20
Leaf Composting
Peter F. Strom and Melvin S. Finstein

1. Introduction . . . 349
1.1. Background . . . 349
1.2. Current Management Alternatives . . . 350
1.2.1. Source Reduction . . . 350
1.2.1.1. Backyard Composting . . . 350
1.2.1.2. Grass: "Cut It and Leave It" . . . 351
1.2.1.3. Backyard Mulching . . . 351
1.2.2. Farmland "Mulching" . . . 351
1.2.3. Composting . . . 351
2. Process, Product, and Objectives . . . 351
2.1. Composting (the Process) . . . 351
2.2. Compost (the Product) . . . 352
2.3. Composting Objectives . . . 352

3. Underlying Scientific Principles ... 352
3.1. Temperature ... 352
3.2. Oxygen ... 353
3.3. Windrow Size and Turning ... 353
3.4. Water ... 353
3.5. pH ... 354
3.6. Inorganic Nutrients ... 354
3.7. Microorganisms ... 354
3.8. Destruction of Pathogens and Weed Seeds ... 354
3.9. Leaf Type ... 355
3.10. Pregrinding ... 355
4. Facility Siting and Initial Preparation ... 355
4.1. Public Participation ... 355
4.2. Area Requirement ... 356
4.3. Buffer Zone ... 356
4.4. Location ... 357
4.5. Stream Encroachment and Water Pollution ... 357
4.6. Slope and Grading ... 358
4.7. Percolation and Water Table ... 358
4.8. Water Supply ... 359
4.9. Security ... 359
4.10. On-Site Roads ... 359
4.11. Fire and other Safety Considerations ... 359
5. Appropriate Level of Technology ... 359
5.1. Minimal Technology ... 360
5.2. Low-Level Technology ... 360
5.2.1. Site Preparation ... 361
5.2.2. Delivery of Leaves/Staging Area ... 362
5.2.3. Wetting ... 363
5.2.4. Forming Windrows ... 363
5.2.5. Combining Windrows ... 363
5.2.6. Turning Windrows ... 364
5.2.7. Curing ... 364
5.2.8. Shredding or Screening ... 365
5.3. Intermediate-Level Technology ... 366
5.4. High-Level Technology ... 367
6. Management of Other Yard Wastes ... 368
6.1. Grass Clippings ... 369
6.2. Woody Materials ... 369
6.3. Other Organic Materials ... 369
7. Potential Problems and their Solutions ... 369
7.1. Odor ... 369
7.2. Leachate ... 372
7.3. Inadequate Composting Rate ... 373
7.4. Bags ... 373
7.5. Aspergillus fumigatus ... 374

7.6. Toxic Contaminants ... 374
7.7. Other Potential Problems ... 375
8. Use of Leaf and Yard Waste Compost ... 375
References ... 376
Tables
1. Appropriate Leaf Composting Technology ... 356
2. Maximum Initial Site Capacity ... 357
3. Buffer Zone Area Requirements ... 358
4. Generalized Schedule and Requirements for Moderate Sized (15,000 cubic yard) Low Level Technology Leaf Composting ... 362
5. Problems Encountered at Leaf Composting Sites ... 370
Figures
1. Combining two small piles using front-end loader. ... 364
2. Finished compost after shredding. ... 365
3. Example of a turning machine. ... 366
4. Steam released from an oversized pile. ... 371
5. Leachate. ... 373

Chapter 21
Urban Tree Removals
Donald F. Blair

1. Introduction ... 377
1.1. Reasons for Removal ... 378
2. Understanding Tree Failure ... 378
2.1. Biological Failure ... 378
2.2. Mechanical Failure ... 378
3. Removal Options ... 379
3.1. Mechanical Options ... 379
3.1.1. Bulldozer ... 379
3.1.2. Crane ... 379
3.1.3. Auxiliary Equipment ... 380
3.2. Manual Methods ... 380
3.2.1. Felling from the Ground ... 381
3.2.2. Dismantling ... 381
3.2.3. Rigging ... 381
3.3. Safety in Tree Removal ... 381
3.3.1. Site Security ... 382
3.3.1.1. Traffic ... 382
3.3.1.2. Pedestrian ... 382
3.3.1.3. Site ... 382
4. Planning a Removal Operation ... 383
5. Thoughts on Tree Removal ... 383
5.1. Changing cultural practices may reduce the need for future removal ... 383
5.1.1. Species to Site Match ... 383
5.1.2. Improper Pruning ... 385
5.1.3. Good Plant Material ... 385

6. Bid Specifications 389
7. Conclusion 391

Appendix A
Cordage Manufacturers 392
Chainsaw Manufacturers 392
Associations 392
Arborist Supplies 393
Appendix B
Applicable American National Standards 394
Applicable Federal Regulations 394
Appendix C
Tree Removal Work Plan Assessment Form 394
Job Site Safety 395
Removal Plan 396
Appendix D
Recommended Guidelines for Safety Training and Standard Performance for Qualified Personnel 396
References 398

Chapter 22
Recycling Urban Tree Removals
Edward A. Lempicki and Edward Cesa

1. Introduction 399
2. The Market 400
3. Products and Specifications 401
3.1. General Sawlog Specifications 402
3.2. Metal and Other Foreign Material in Street Trees 402
3.3. Proper Log Manufacturing 404
3.4. Log Loading and Unloading Safety Procedures 406
3.5. Log Quantities and Volume Determination 407
3.6. Storage Requirements 409
4. Marketing Merchantable Sawlogs 409
4.1. Commercial Sawmill Operations 409
4.1.1. State Forestry Offices 410
4.1.2. Consulting Foresters 410
4.1.3. Forestry/Wood Products Extension Offices 410
4.1.4. Resource Conservation and Development Program 410
4.2. Local Crafters/Hobbyists/Custom Sawmillers 411
4.2.1. State Forestry Offices 411
4.2.2. Sawmill Equipment Manufacturers 411
4.2.3. Craft Organizations 412
4.3. In-house Markets: Viable Options for Municipalities 412
Appendix: State Forestry Offices 413
References 417

Chapter 23
Tree City USA
John Rosenow and Mary Yager

1. History . . . 419
2. Benefits of Tree City USA . . . 420
3. How to Become a Tree City USA . . . 421
3.1. Standard 1: Tree Board or Department . . . 423
3.2. Standard 2: Tree Care Ordinance . . . 423
3.3. Standard 3: Community Forestry Program with Annual Budget at Least $2 per Capita . . . 423
3.4. Standard 4: Arbor Day Proclamation and Observance . . . 423
4. Tree City USA Growth Award . . . 424
5. Tree Line USA . . . 424
6. Conclusion . . . 425
References . . . 426

Chapter 24
A Case Study of Paramus, New Jersey: A "Tree City USA" Community
William R. Comery, C.L.A.

1. Introduction . . . 429
2. History of Paramus, New Jersey . . . 429
3. Shade Tree Commission Established . . . 430
4. Tree City USA Award . . . 432
5. Tree City USA Growth Award . . . 434
5.1. Category A: Education and Public Relations . . . 435
5.2. Category B: Partnerships . . . 435
5.3. Category C: Planning and Management . . . 435
5.4. Category D: Tree Planting and Maintenance . . . 436

Chapter 25
Attracting and Managing for Wildlife
Kathleen E. Clark

1. Introduction . . . 437
2. Attitudes Toward Wildlife . . . 438
3. Elements of Wildlife Habitat . . . 438
3.1. Cover . . . 438
3.1.1. Shelter and Escape Cover . . . 439
3.1.2. Nesting/Breeding Cover . . . 439
3.2. Food . . . 439
3.3. Water . . . 439
3.4. Arrangement of Elements in the Landscape . . . 440
4. Wildlife in Urban and Suburban Landscapes . . . 440
4.1. Typical Species . . . 440
4.1.1. Urban Landscape . . . 441

4.1.2. Suburban–Residential Landscape 442
4.1.3. Suburban–Farmland Landscape 443
4.2. Benefits 443
4.3. Problems and Nuisance Wildlife 443
5. Managing for Wildlife 444
5.1. Attracting Birds 444
5.1.1. Landscape Elements and Tree–Shrub Species 445
5.2. Attracting Other Wildlife 447
5.2.1. Landscape Elements and Tree–Shrub Species 447
5.3. Managing to Discourage "Problem Wildlife" 448
5.3.1. Landscape Design 448
5.3.2. Mechanical and Chemical Controls 448
References 449

Chapter 26
Managing Urban Ecosystems: A Look to the Future of Urban Forestry
L. Robert Neville

1. Introduction 451
1.1. Sustainable Development 452
1.1.1. Effects of Urban Growth on Natural Systems and Processes 452
1.1.2. Effects of Environmental, Social, and Economic Systems on Urban Decay 453
1.2. Existing Management Responsibilities for Urban Natural Systems 454
1.3. The Need for a Comprehensive Management Approach 454
2. Establishing a Context for Urban Forest Resource Management 455
2.1. The Urban–Rural Continuum 455
2.2. Ecosystem Classification: Natural to Human Dominated 456
2.3. Linking Ecosystem Structure and Function 458
2.4. The Multifunctional Role of Vegetation in Urban Environments 459
3. Expanding the Scope of Urban Forestry 459
3.1. Linking Green Infrastructure to Economic Development 459
3.2. Eleven Steps Toward Managing Urban Ecosystems 460
4. Conclusion 462
4.1. Sustainability as a Central Issue in the 21st Century 462
4.2. The Importance of Context in Natural Resource Decision-Making 462
4.3. Managing Urban Ecosystems: An Expanded Role for Urban Foresters 463
References 464

Index 467

Chapter 1

Origins of Urban Forestry

Henry D. Gerhold

1. Introduction

Trees beautify our towns and cities and enrich our lives in various ways. Most of us admire and enjoy these trees that we encounter everyday, but we largely take them for granted. Yet the heritage of landscape trees in our communities has a fascinating history, one that can deepen our appreciation of the trees and the people who take care of them.

A look back at how urban forestry got started can stimulate an appreciation of our heritage of landscape trees, and thus instill an ethic that fully recognizes their value. The intriguing history of city trees, tree care practices, and community tree programs may inspire us and challenge us to improve public support for urban forestry programs, and thereby contribute to the strategy for keeping urban forests healthy. If people can be challenged to take a fresh, long-term look at trees in their own communities, they may be inspired to persuade community leaders to build a better tree program.

A series of developments in landscape design spanning over 300 years in Europe eventually culminated in what we now consider commonplace—street trees. Primitive pruning, fertilizing, and pest control practices were described in English publications in 1597 and 1662; much later they were improved through research and practical experience in Europe and America.

But it was in America that the urban forestry concept first took root even before the term urban forestry was coined and later developed into a national movement. For example, in 1792 citizens of Philadelphia petitioned to have public trees planted, in recognition of their healthful effects. In 1896, this city hired its first professional arborist with the title Chief Forester—possibly the first urban forester? Other early examples of urban forestry have been documented in Georgia, California, Massachusetts, New York, Michigan, and other places.

Henry D. Gerhold School of Forest Resources, Pennsylvania State University, University Park, Pennsylvania 16802-4300.

Urban and Community Forestry in the Northeast, 2nd ed., edited by, J. E. Kuser.

Only recently, experts have fully understood many benefits of landscape trees (Chapter 2), and how communities can care for them and keep them healthy in stressful urban environments. This field of knowledge, now known as *urban forestry* in North America and in other parts of the world, has gradually emerged mainly from the disciplines of arboriculture, landscape architecture, and forestry, and includes subjects such as horticulture, soil science, plant pathology, entomology, and social sciences.

Before delving into the origins and development of urban forestry, let us consider what urban forests are and why landscape trees are important to people.

1.1. What Is an Urban Forest?

The very idea of an *urban forest* at first seemed strange, even contradictory, when this term was invented. But it has become accepted and even commonplace. *Community forest* and *city green* are related terms with similar meanings. All three refer to the trees, lower vegetation, open green spaces, and associated wildlife within a municipality or adjacent to it.

The term "urban forestry" was coined in 1965 by Prof. Erik Jorgensen at the University of Toronto. This unusual juxtaposition of "urban" and "forestry" arose in searching for an appropriate title of a graduate student's thesis. Jorgensen gave this scholarly definition of urban forestry: "a specialized branch of forestry (that) has as its objective the cultivation and management of trees for their present and potential contributions to the physiological, sociological, and economic well-being of urban society." Furthermore, he believed urban forestry does not deal only "with the city trees or with single tree management, but rather with the tree management in the entire area influenced by the urban population." It was in the United States that urban forestry first took root and developed into a national movement.

In traditional forestry, trees are managed for timber harvests and to provide other goods and services. But in urban forestry trees are cultivated mainly for their aesthetic and environmental qualities, to be "harvested" only when they die or become hazardous. A traditional type of forest set aside for the good of a community sometimes is referred to as a "community forest" and may be considered a part of the larger urban forest that also includes trees within the town. A community forest at Zurich, Switzerland, has provided timber, firewood, recreation, and watershed protection since 853 AD; the oldest community forest in America was established in 1640 at Newington, New Hampshire.

"Community forestry" sometimes is used as a synonym for urban forestry. That is because residents of towns and hamlets usually consider themselves rural, not urban. They think that urban forestry does not apply to them even though it is meant to be inclusive of all municipalities, regardless of size.

1.2. Values of Urban Forests

A well-managed urban forest has numerous advantages, ranging from aesthetic, economic, and health benefits of various kinds to favorable social and psychological influences. Some of these were recognized long ago. For example, a 1700 ordinance in Philadelphia directed that every owner of a house "should plant one or more trees before the door that the town may be well-shaded from the violence of the sun ... and thereby be rendered more healthy." In 1792, the citizens of Philadelphia petitioned the mayor to plant trees in the public squares because "it is an established fact that trees and vegetation ... contribute to ... the increased salubrity of the air." The New York City Commissioner of Health recommended in 1872 that street trees be planted to mitigate the intense heat and thereby diminish the death rate among children.

Recent studies have quantified other kinds of benefits: property values increased by landscaping, energy saved in heating and air conditioning of buildings due to the proper placement of trees, pollutants removed and carbon stored by urban forests, and even improvements in human health. Amazingly, hospitalized people recover faster when trees are in view.

Furthermore, urban forests that are aesthetically pleasing attract people and businesses to communities. Studies indicate that people have an innate desire to be close to the natural environment, and the urban forest can make this connection for people. Urban forests offer places for social gatherings and can lead to a greater sense of community pride. The urban forest plays three main social roles according to H. W. Lawrence: (i) a natural element in a humanized landscape, (ii) an aesthetic object in designed compositions, and (iii) an object expressing attributes of power, such as ownership and accessibility.

The values and purposes that people see in urban trees have changed over the centuries. Already in the 19th century, a psychologist recorded his ideas about dendro–psychoses, proposing that through natural selection in prehistoric times humans had become genetically predisposed to regard trees as indicators of elements that favored survival (i.e., necessities such as food, water, and security). More recently, the biophilia theory has been developed, which maintains that human preferences have been shaped over millennia through gene–environmental coevolution.

Attitudes of people and trends in landscape architecture have been shaped by events, such as the growth in populations, the Renaissance, the Industrial Revolution, the democratization of governments, the installation of utility wires and sewer systems, and the paving of streets and sidewalks. Cultural traditions of various ethnic groups also have affected views of trees in urban landscapes. For example, a Canadian study found that in Toronto people from the British background were most eager to have shade trees, Italian and Portuguese neighborhoods preferred fruit trees and vegetable gardens, and the Chinese least wanted trees on their properties. These kinds of influences may be related to historical developments in the design of urban landscapes.

2. Design of Urban Landscapes

Various ancient civilizations incorporated trees into their landscapes, but little is known about how that happened. Because these trees were transplanted and served specific purposes it follows that the plantings were designed, even long before there were professional landscape architects. The Hanging Gardens of Babylon during the 6th century BC have been cited as the first intentional use of vegetation in an urban setting. But even earlier, royal gardens in Egypt during 1400–1362 BC had rows of sycamores, palms, and pomegranates; Ramses III (1198–1167 BC) had trees planted, some along streets, for promenading and recreation. The Chinese, Greeks, Egyptians, Phoenicians, Persians, and Romans all held trees in high regard and even worshipped them. They used trees for aesthetic purposes in formal gardens, sacred groves, and landscapes that enhanced temples, statues, and buildings. In ancient Greece, trees were planted in plazas to beautify cities and for shade along pathways leading to marketplaces.

2.1. Trees Became Urbanized in Europe

Changing property rights and relationships between people and trees in "the Old World" probably have some lingering influence today. Tribes that overran Europe regarded forests as communal property. City forests came into existence in the 10th century when kings gave land to the cities that emerged from the feudal system. By the end of the Middle Ages, urban areas were well established but with few interior trees or greenspaces. Thereafter, urban design began its long development from the formal gardens of the elite to public greenspaces and street trees, as thinking shifted from survival to aesthetics. Throughout the literature pertaining to urban trees in the Old World, there are indications that city officials ordained trees in cities, designers prepared plans for them, and skilled gardeners tended the trees; but the details and explanations for the most part are obscure or unrecorded.

After the Dark Ages in Europe, say 500 AD, turbulent times caused people to move into fortified cities. Crowding inside the walls led to the demise of open garden areas and trees within the cities. Folks meadows and festival places with trees were common outside the city gates, gathering places for all sorts of activities and celebrations. For example, a public meadow at Nuremberg purchased in 1434, had rows of lindens planted in 1443, and remained in continuous use for more than 500 years. A few special purpose trees were planted near churches or in public squares as witnesses and durable reminders of agreements.

Already in 1486, an Italian city planner wrote about the design of cities, foreshadowing city greens and tree planting that would add beauty and recreational opportunities to the residences and ordinary necessities of life. Gardening activities gradually increased during the next century, mainly by the wealthy and powerful. Some patricians of grand estates showed off their gardens and botanical collections to the public, such as the Villa Borghese in Rome and the estate of Gessner (1516–1565) in Zurich.

Early in the 16th century, European cities had extensive forests outside their fortifications. So nature was always close by, but on land mainly used by nobility for hunting or to produce income during times of war. As cities expanded, more forests were cut for agricultural purposes, less space was available for outdoor recreation, and trespassing by common people caused the nobility to enclose their possessions.

The beginning of the Italian Renaissance in the 16th century, which later spread throughout Europe, brought with it the opportunity to "get back to nature." Garden design flourished under the direction of the nobility and the wealthy, and towns and cities encouraged the introduction of trees into the urban landscape. The first of several distinct landscape elements described by H. W. Lawrence appeared in Italy, the *garden allée*, a tree-lined pathway in a garden generally enclosed in walls. The French modified the Italian garden style by making the allée longer and more prominent. A notable example was the royal gardens of Tuileries, where upper class people would stroll and relax among the shady allées.

During the late 16th century, military engineers constructed ramparts, earthworks, and ditches instead of masonry walls around cities in reacting to new weapons and siege tactics. These new types of fortifications became public open space in some cities. Double or quadruple rows of trees sometimes were planted, creating a new type of allée—the *wall promenade*. Early examples occurred in Antwerp, Amsterdam, Strasbourg, and Lucca. Promenades just outside cities were planted in the 18th century at Stuttgart, Heilbronn, and Vienna.

Closely related to the wall promenade was the *waterside promenade*, which appeared early in the 17th century in the Low Countries. The 1615 "Plan of Three Canals" in Amsterdam was the first recorded use of buildings, traffic, and rows of trees together as a unified spatial form in a city interior rather than in isolation as with allées and wall promenades.

Two new recreational pastimes arose in the 17th century that caused modifications of the garden allée. The game *pallo a maglio* originated in Italy with *malls* consisting of lawn areas and rows of trees to provide shade for spectators. This game resembling croquet became popular throughout western Europe, and many malls were installed near urban centers. In Paris a mall was built in 1599, Berlin built the forerunner of the famous Unter den Linden in 1647, and the Pall Mall in London existed before 1650. As the game went out of fashion, the malls were converted to other uses.

Another development, which also originated in Italy, was carriage riding for pleasure along *cours*. Marie de Medici introduced pleasure riding to France and in 1616 had the Cours de la Reine designed. The Cours resembled the name of a roadway outside of Florence known as the Corso. The Cours de la Reine was open only by invitation and became popular with Parisian high society. Some other cours were accessible to anyone with a carriage. The malls and cours added a new dimension to the European landscapes—greenspaces that were part of the urban landscape outside garden walls or estates. Lawrence considered the cours very important "because it transformed the garden allée into a place for vehicles, albeit one not yet integrated into a city's street system."

The *exterior avenue* was the next landscape form to evolve at about the same time as the cours. Henry IV (1589–1610) had the thoroughfare Cours de Vincennes constructed, bordered by four rows of elms. By the late 17th century, these avenues were built on the outskirts of urban areas as grand entryways to city gates or important buildings, such as a hospital or a chateau. They often would form a large network of avenues, providing spatial organization. The Avenue des Tuileries is a prime French example, and exterior avenues also could be found elsewhere in northwestern Europe. As the cities grew in size, these tree-lined country roads became tree-lined city streets. The Paris example spread to other French cities, including Dijon in 1670, Montpelier in 1688, Nimes in 1730, and Toulouse in 1754. Elements of the Paris Greensystem installed under Louis XIV (1693–1715) were a precursor of similar installations in other European cities.

Open spaces at important intersections of avenues, called *places* in France, became an important part of urban design. The place was inspired by the 16th century Italian *piazza*, or town *square*, which typically was devoid of vegetation. The French *places*, the German *platzen*, and the Spanish *plazas* were usually open to vehicular traffic and integrated with structures as public amenity areas, including trees.

The baroque *boulevard* is the individual form that contributed most to the 19th century boulevard. In 1670, Louis XIV ordered the destruction of walls around Paris. These areas were transformed into broad, elevated, tree-lined promenades that were open to carriages and pedestrians. The transformation of city walls from fortifications to promenades was not prevalent in most of Europe until after the Napoleonic wars in the early 1800s.

As the character of city streets changed, starting late in the 17th century, trees began to be planted along general use streets for shade and ornament rather than for recreational purposes. The old medieval streets that were rough, winding, dark, and narrow became more functional when they were widened, straightened, paved, and lighted. German publications during 1771–1776 gave recommendations for developing lovely cities. Trees were planted to beautify towns and cities, improve health of people, and provide recreational opportunities.

Public perceptions of trees changed in the 19th century, as people came to realize that vegetation was important to a healthy city. Trees were expected to shade the widened sunny streets and were thought to assist in public sanitation by purifying the air of "miasmas," a presumed cause of disease.

2.2. Urban Trees in America

During colonial times in eastern America, the settlers came mainly from towns and cities in England and northwestern Europe where trees were scarce, and were just becoming part of the designed urban landscape. Though landscape trees were not the main concern to settlers struggling to survive in the wilderness, trees were planted along streets and around the "common" pasture in nearly every New England hamlet.

Attitudes of America's founders no doubt had a role in shaping the urban landscapes that developed as settlements grew and a new nation was founded—attitudes about freedom, independence from European domination, and beliefs in private property and free enterprise. Thus, the architecture in the rebellious colonies was far removed from spatial dependence on Old World models.

Tree protection regulations were instituted early on, so people or animals would not destroy the trees, for example in 1637 at Watertown, Massachusetts, at the Boston Common in 1661, and at Newark, New Jersey in 1676. The first public planting for the relief of travelers occurred in "Boston Towne" in 1646. One of these trees, the famous "Liberty Elm," served as a symbol of colonial defiance against the British.

A chronology of landscape trees in Philadelphia is revealing of early tree planting practices and attitudes toward trees. Starting in 1681, William Penn made plans for a "Greene Countrie Towne which will never be burnt and always be wholsome." In 1700, an ordinance required owners of houses to plant trees, but the first organized planting of street trees did not take place until 1750. An act passed in 1782 sponsored by the Philadelphia Contributorship, the first fire insurance company that had been founded by Ben Franklin, required the removal of all trees in streets because they were believed to obstruct passages, interfere with watercourses, and spread fires. By 1787 many trees had been planted, particularly Lombardy poplars and American elms. From 1817 to 1825, more than 60 varieties of trees were planted in Washington Square. In 1896, the city hired its first professional arborist, Chief Forester John C. Lewis; with this title and role, perhaps he could be called the first urban forester in America.

Some other American cities also designed trees into their landscapes early in their history. In 1773 Oglethorpe planned Savannah, Georgia to have 26 squares for public use, and live oaks planted in long avenues were festooned with Spanish moss. On the western frontier, Sacramento was built around existing trees in order to preserve its natural beauty. Major Pierre L'Enfant's design for Washington, DC in 1791 called for trees on both sides of the main avenues.

In 1841, Andrew Jackson Downing, a nurseryman who educated himself to become America's first professional landscape architect, discussed principles for designing with trees. He suggested that villages should form an "Ornamental Tree Society" to "turn dusty lanes and bald highways into alleys and avenues of coolness and verdure." By 1850, five states had ornamental tree societies; later these associations grew to over 1000.

> ... how many towns, how many villages, could we name where rude and uncouth streets bask in the summer heat, and revel in the noontide glare, with scarcely a leaf to shelter or break the painful monotony! ... What must be done in such cases? There must be at least one right-feeling man in every such Sodom. Let him set vigorously at work, and if he cannot induce his neighbors to join him, he must not be disheartened—let him plant and cherish carefully a few trees, if only half a dozen ... (Andrew Jackson Downing, 1847).

At about the same time states changed their fencing laws to require livestock owners to enclose their animals, whereas formerly it had been the responsibility of property owners to exclude livestock if they wished to protect their greenery. Accordingly, state legislation, the spatial reorganization caused by fencing laws, and village improvement associations during 1850–1900 resulted in landscape features that are specifically American ... the tree-lined street with spacious lawns sweeping back to buildings, unencumbered with walls or fences.

The natural, picturesque school of landscape design that originated in Europe and gave a romantic feel for the environment was adopted widely after 1840. In 1857, Frederick Law Olmsted and Calvert Vaux employed this naturalistic style to design New York's Central Park, which had been inspired by Downing's writings. Olmsted and Vaux invented the name "landscape architect" in 1858 to convey their intent to design the landscape, just as an architect designs a building.

Downing, Olmsted, Vaux, and other designers attempted to create landscapes that they believed would promote social progress, such as greater tolerance and appreciation of diverse peoples. They revised gridiron designs commonly employed by real estate developers toward the concept of a new, naturalistic urban landscape that furnished more open space, parks, trees, and living conditions and, thereby, favored family values and a moral society.

Industrial developments during the late 1800s and early 1900s caused grave problems for street trees everywhere. Installation of utility wires, sewer systems, and pavements for roads and sidewalks grew at exponential rates. To cope with these problems, states passed laws that fostered municipal street tree ordinances and shade tree commissions that could preserve and manage their landscape vegetation in keeping with the desires of residents. Shade tree commissions thus were enabled to exert local control over tree planting, maintenance, and the choice of species.

2.3. Choice of Species in Landscape Design

Choosing appropriate species (Chapter 11) is a crucial part of landscape design and also a vital part of arboriculture. Species characteristics must be matched to site conditions so that the trees complement their surroundings and remain healthy. And there are so many species among which to choose!

Voyagers of the seas and the "discovery of the New World" initiated an age of plant exploration and exchange that greatly enriched American and European landscapes. But according to H.-L. Li, "plant dispersion and introduction has been carried on through all ages and in all countries." The first record of foreign plant introduction was into Egypt in 1570 BC. Chinese writings note that in 481 BC disciples of Confucius planted trees from their native regions near and far as memorials at his grave. Though records before the 16th century were scarce, Li believes plant dispersion then was done mainly by merchants, adventurers, and pilgrims.

One of the earliest plant explorers of the New World was John Tradescant II. He returned to England with many tree species. Among them was *Robinia,* named after two gardeners who planted this tree in the royal gardens of French kings. Another well-known tree that has connections with Tradescant is the London plane (*Platanus x acerifolia*). In 1637, he planted the American sycamore (*Platanus occidentalis*) in the same English garden as the oriental plane (*Platanus orientalis*), where the two apparently hybridized naturally. The resulting London plane turned out to be a remarkably strong and vigorous grower that became one of the most widely planted shade trees in Europe and America.

John Bartram's botanical garden, established near Philadelphia in 1728, usually is credited as the first in America and became so famous that Carl Linnaeus called him the greatest natural botanist of the world. However, Christopher Witt actually established the first American botanical garden at Germantown, Pennsylvania in 1708. Bartram introduced more native American trees for landscape use at home and in Europe than any other person of that period. He explored much of eastern North America from Lake Ontario to Florida. Benjamin Franklin, the largest dealer in exchanging plants with Europe in the 18th century, helped Bartram in arranging seed and root exchanges.

Other European plant explorers were also active at this time. The Swedish government in 1748 sent Peter Kalm, a pupil of Linnaeus, to report on the natural resources of America. The King of France sent his botanist André Michaux to America in 1785 to import trees that could replenish French forests. Michaux also sought trees "to enrich the orchard and ornamentals suitable for garden culture." From 1785 until 1807 he and his son discovered plants from southern Florida to Hudson Bay and westward beyond the Mississippi River.

The American plant collector William Hamilton, a cousin of John Bartram, introduced several trees that were widely planted, and some still are. One was the Lombardy poplar (*Populus nigra* "Italica") in 1784, which he distributed to many plant enthusiasts along the Atlantic seaboard. Subsequently, Lombardy poplar fell out of favor because of its disease problems. Hamilton also helped to introduce the Norway maple (*Acer platanoides*), the tree of heaven (*Ailanthus altissima*), and ginkgo (*Ginkgo biloba*).

Peter Collinson gave five ginkgo trees to Hamilton in 1784, after the species had been introduced first to the Netherlands and then in 1754 to England. Engelbert Kaempfer, a German doctor and botanist employed by the East India Company, had discovered this "living fossil" in China in 1690. This ancient species has survived over 200 million years and today is an esteemed street tree that is tolerant of urban conditions and largely free of pests.

Most foreign tree species of today were imported during the colonial period, and these were prominent in nursery catalogs along with many fruit trees and garden plants. Robert Prince established the first commercial nursery in 1737 at Flushing, New York, which became the largest and best known in the continent. The Prince Nursery sold mainly fruit trees at first but by 1771 also shade and ornamental trees. David Landreth established the first seed business in 1784 at Philadelphia and also a nursery. Bernard M'Mahon, an Irishman who began his seed and nursery business in 1802, was entrusted by Thomas Jefferson with seeds brought back by the Lewis and Clark expedition of 1803–1806. The Parsons family operated another early nursery at Flushing from 1840 to 1907, offering a good selection of trees, many from Asia. By 1950 there were 17,400 nurseries in the United States, which annually sold 326,000,000 ornamental woody plants.

Botanical gardens and arboreta, by displaying and distributing trees, have been very influential in the choice of species used in designing urban forests. Duhamel Du Monceau in France assembled more than 1000 tree species including many from the New World, starting in 1750. However, the word "arboretum" did not appear until 1838 in a book by J. C. Louden. The preeminent Royal Botanical Gardens at Kew, which began privately in 1759, were given to England in 1841 and now contain more than 16,000 woody plant species. In America, Joshua and Samuel Pierce started an arboretum in 1798 on a land grant from William Penn, which later was acquired by the DuPont family and became Longwood Gardens in 1921. The founding of other well-known botanical gardens and arboreta in the United States includes the Missouri Botanical Garden at St. Louis in 1859, Arnold Arboretum at Boston in 1872, New York Botanical Garden in 1891, Brooklyn Botanic Garden in New York in 1910, The Morton Arboretum near Chicago in 1922, and the US National Arboretum in Washington, DC in 1927.

Plant explorations proliferated during the 19th and 20th centuries within North America and to Asia and other continents. The most avid explorers included David Fairchild and Frank N. Meyer of the US Department of Agriculture, and Charles Sargent and Ernest "Chinese" Wilson of the Arnold Arboretum. They and others greatly extended the variety of tree species in use today. For example, 379 exotic tree species are now growing in New Jersey and neighboring regions of New York and Pennsylvania, which were not native in 1609 when Henry Hudson sailed to New York; only 70 of these came from other regions of North America.

Besides the exotic species that were introduced by explorers, plantsmen and tree breeders created novel varieties. Over the centuries, plantsmen have sought out trees that were unusual, even bizarre, and multiplied these by budding, grafting, or rooting of cuttings. Weeping varieties especially were sought after for cemeteries and parks. In the 20th century, plant breeders applied the new science of genetics to produce improved varieties through artificial mating and selection. Plantsmen also continued their work of simpler selection and propagation. The resulting named cultivars (*culti*-vated *var*ieties) greatly diversified the kinds of trees available for landscaping. A compilation of 941 validated landscape tree cultivars indicates that 64% of them originated before 1945. Since then more practical characteristics have been incorporated into improved cultivars, such as disease resistance and branching habits that require less pruning.

Designers of urban landscapes, who specified which trees were to be planted and where, depended first on gardeners, and later on horticulturists, arborists, and foresters for the cultivation and care of the trees.

3. Arboricultural Practices

Arboriculture, which deals with the planting, protection, and maintenance of trees and shrubs, is one of the most important components of urban forestry. The selection of appropriate species and cultivars falls within the responsibilities of both designers and arborists. And just as the design of urban landscapes dates back to ancient times, so does arboriculture.

Early civilizations revered trees and went to great lengths for their care because trees were thought to be under the protection of gods. More than 4000 years ago, Egyptians wrote about transplanting trees with a ball of soil and forming a basin around them to retain water. They transported trees as far as 1500 miles by boat. *Enquiry into Plants*, written by Theophrastus (300 BC) in Greece, recommended when transplanting trees to save the entire root system, to use good soil and manure, and to orient the tree in the same direction as in its original location. Theophrastus also advocated plastering of wounds with mud and leaves to prevent drying. Early Romans were keenly interested in trees and appointed foresters to supervise their care and protection. They spread their arboricultural knowledge throughout the Roman Empire as it expanded.

3.1. Arboriculture in Europe

From the Dark Ages until the 15th century, apparently there was little progress in the field of arboriculture. Tree care practices survived this era because faithful monks passed on their knowledge. With the revival of learning in the 16th century, ideas about tree care spread throughout western Europe and England.

Significant developments in tree care practices occurred in England, especially between 1500 and 1800. The term *arborist* first appeared in 1578 in James Lyte's book, *Dodens*. William Lawson published *A New Orchard and Garden* in 1597, which discussed arboricultural practices. He explained how improper pruning would eventually lead to decay. R. Austin in 1657 recommended feeding trees by inserting fertilizer into holes in the soil made with an iron bar and warned about overfertilizing.

The most renowned author of arboriculture in the 17th century was John Evelyn, whose monumental book *Sylva* was published in 1662. Evelyn offered information on transplanting trees, species selection, and diseases and insects. He recommended that "whole towns be planted with trees in even lines in front of the houses, so that they seemed like cities in the woods." Evelyn described transplanting several hundred large trees at Versailles and moving a tree for the first time horizontally rather than vertically in a special rig. He was also the first to advise that a tree cavity should be cleaned of decaying material and treated with oil or tar.

Evelyn also lamented improper pruning practices:

> For tis a misery to see how our fairest Trees are defac'd, and mangl'd by unskilful Wood-men, ... who go always. ... hacking and chopping off all that comes in their way; by which our Trees are made full of knots, boils, cankers, and deform'd branches, to their utter destruction: Good husbands should be asham'd of it .... As much to be reprehended are those who .... so maim the poor branches .... that they leave most of them stubs, and instead of cutting the Arms and Branches close to the boale, hack them off a foot or two from the body of the Tree, by which means they become hollow and rotten .... (John Evelyn, 1662).

The growing popularity of expansive landscaped gardens among the wealthy in the 1700s and 1800s required the services of "estate foresters," especially in England and France. In 1790 William Forsythe, Gardener to the King of England, gave specific instructions in his book on how to fill cavities of trees with a formula consisting of old lime, wood ashes, cow dung, and sand. Others recommended instead coating wounds with tar. William Pontey, in an 1810 treatise on pruning trees, criticized Forsythe and called attention to "the astonishing and successful exertions of simple nature in healing wounds of trees."

The Planter's Guide of Sir Henry Steuart in 1828 gave specifications for successful transplanting. He recommended a root ball one foot in diameter for every inch of trunk diameter, a standard that is still used today. Steuart recognized the need for research and proposed a Society for the Improvement of Arboriculture so that arboriculture could become a separate profession.

Pruning techniques were advanced in the 1800s by foresters, especially in Germany and France. Robert Hartig, the father of forest pathology, conducted research in Germany on pruning, wound healing, and internal decay. The director of parks in Paris, Baron Des Cars, in 1864 devised a practical system of tree pruning and used coal tar as a wound dressing. He advocated cutting branches close to and even with the trunk. The pruning of branches flush with the trunk has persisted until recently, even though in 1756 Büchtung said, "the branch collar should not be injured or removed." The scientific basis of his good advice was verified and widely advocated only recently.

Chemicals for controlling pests were seldom used in arboriculture before 1850, although insecticidal qualities of compounds, such as arsenic and nicotine, were known already in the 1600s. Before 1850, and even much later, insects were controlled mainly by manual and cultural methods that were labor intensive and only partially effective. Some of the pesticides developed toward the end of the 1800s are still in use today, such as Bordeaux mixture and pyrethrums.

3.2. Arboriculture in America

The development of arboriculture in America, from a craft to the profession it has become, started late in the 19th century. Later it became more solidly founded on a scientific basis through research. A series of events stimulated the interests of people in preserving and maintaining healthy landscape trees.

The creation of Arbor Day by J. Sterling Morton in 1872, for the purpose of promoting tree planting, exemplifies the increasing concern about trees in the United States and elsewhere in this period. People were becoming alarmed about extensive logging and wildfires. At the same time, there was a great influx of intriguing new plants from the Orient, and a tremendous expansion of the nursery industry to meet the growing demand. Wealthy owners of elaborately landscaped estates soon recognized the need to employ workers to protect their trees and keep them healthy and to repair those that were injured.

Another reason for a growing public demand for better tree care in America was the disturbance of trees by the installation of utility lines. Since the introduction of the telegraph in 1839, and electricity in 1882, pruning for clearance of overhead wires has shaped and often distorted the appearance of nearby trees. As electrical service expanded into new areas, street trees were regarded as obstructions and dealt with in a reckless manner. Illuminating gas, which is extremely poisonous to trees, often caused injuries and fatalities; in fact, injured trees were used to detect leaks from gas mains nearby.

The defoliation of trees by gypsy moths also had a significant impact on arboriculture. This insect had been imported for hybridization with the silkworm, in the hope of enhancing silk production. A few gypsy moths escaped accidentally in 1869 at Medford, Massachusetts, though the first destructive outbreak did not occur until 1889. Various eradication and control measures were employed, with only limited success, over the next 50 years as the infestation spread first to neighboring states and then to more distant places. DDT was sprayed over three million acres of trees in the 1950s in a massive eradication program in New Jersey, New York, and Pennsylvania that was considered very successful. But by 1970 the hope of eradication was fading, in 1973 DDT was banned, and so control efforts turned to more intensive research on biological control methods.

George Stone taught the first course in arboriculture in 1895 at the Massachusetts State College. Possibly he had been influenced by his visit to Germany with Robert Hartig, the renowned authority on tree diseases. Stone was regarded as an inspiring teacher and a brilliant investigator. He also founded the Massachusetts Tree Wardens and Foresters Association in 1913, the first such technical organization in the United States.

John Davey, an English immigrant who founded the Davey Tree Expert Company in Ohio, published *The Tree Doctor* in 1901 to help educate the public about trees. His company, the first in the United States that was skilled in tree surgery, was incorporated in 1909 but had started nearly 30 years beforehand. The Davey School of Practical Forestry founded in 1906 taught men how to maintain trees and later became the Davey Institute of Tree Surgery.

The F. A. Bartlett Tree Expert Company was a competitor of Davey and the only other large arboricultural company in North America early in the 20th century. Both

Bartlett and Davey pioneered commercial arboriculture, and both had been influenced by Stone's teachings. The Bartlett School of Tree Surgery was established in 1923 at Stamford, Connecticut to train "dendricians," the company's name for tree workers. The Davey and Bartlett schools were the principal source of technically trained arborists for many years, including the first professional tree surgeons from England. The first laboratory designed specifically for shade tree research, the Bartlett Tree Research Laboratories, was formally initiated in 1926.

A third national tree company, the Asplundh Tree Expert Company, started in 1928 and eventually became the largest of the three. Three Asplundh brothers, all college graduates, who catered exclusively to line clearance for utility companies, organized it in Pennsylvania. The corporate image was a fleet of bright orange vehicles and machines. Asplundh may have been the first commercial arborist company to use a power saw in 1935, long before modern, lightweight chain saws had been perfected.

During the early 1900s, the principal tree care work of cavity repairs, pruning, and cabling was broadened to include other practices such as fertilizing, spraying to control insects, and disease treatments. Before 1920 safety was not emphasized, and workers climbed trees with ladders and spurs, not ropes.

A new era in arboriculture began with the Shade Tree Conference in 1924 at Stamford, Connecticut. F. A. Bartlett arranged the program of talks and field trips on topics, including municipal shade tree problems, tree surgery, practical pruning, healing of wounds, fungi, insects, use of lead arsenate, and training of tree experts. There was no mention of arboriculture or arborists, terms with broader meanings that first replaced the terminology of tree surgery in 1928. At the fourth conference in 1928 the National Shade Tree Conference was formally organized, in 1961 it was renamed the International Shade Tree Conference, and in 1975 it became the International Society of Arboriculture.

The Great Depression, which started in 1929, also had a positive influence on arboriculture despite the pervasive unemployment caused by this economic disaster. From 1933 to 1942, young men in the Civilian Conservation Corps and the Works Progress Administration (WPA) engaged in tree planting, control of gypsy moths and diseased elms, and emergency tree clearing after the deadly 1938 hurricane in the Northeast. The WPA also trimmed trees in public parks and surveyed trees in cities. Afterwards, many of these men entered careers in arboriculture and forestry. However, some were not well trained and so created an unfavorable image for their profession.

Dutch elm disease (DED) was detected in North America in 1930 and quickly became the most serious disease of landscape trees. Two female graduate students in the Netherlands identified the causal fungus in 1921 and 1929, and another graduate student proved that elm bark beetles carried the fungal spores to cause new infections. By 1933 a more serious infection area developed around New York City, and by 1942 six million diseased elms had been destroyed. Despite extensive control efforts, the disease spread next to the Midwest, where it wreaked havoc in towns and cities that had elm monocultures along many of their streets. Eventually, DED reached nearly all states and six Canadian provinces. The disease was also exploited to produce some positive effects, including the promotion of professionalism in arboriculture and the support of public officials for tree management programs.

A great many inventions in the 20th century made the work of arborists more effective and efficient. The simple tools of the tree surgeon—axes, hand saws, chisels, ladders, climbing ropes and spurs—were supplemented or replaced by power saws and drills, aerial lifts, wood chippers, and other types of specialized equipment. Later advances in the scientific knowledge related to decay, other tree defects, and pruning practices enabled arborists to treat trees more effectively. The New Tree Biology popularized by Alex Shigo after 1959 helped arborists to understand how trees defend themselves against microorganisms, and thus to improve arboricultural practices.

Arboricultural spraying practices to apply insecticides and fungicides became commonplace during the first half of the 20th century and then fell out of favor. Lead arsenate and DDT, being very effective in controlling gypsy moths and the beetles that spread DED, were often used indiscriminately without regard to hazards to humans and wildlife. But after Rachel Carson in 1962 raised concerns about pesticides in her book, *Silent Spring*, new laws and regulations were passed that limited their use. The Environmental Pesticide Control Act of 1972 established tolerance levels for pesticides and required certification of pesticide applicators. In response to this challenge, arborists and scientists sought more environmental friendly ways of managing insects and diseases. The integrated pest management (IPM) concept was developed in the 1980s to reduce the need for chemicals by utilizing natural checks and balances in the landscape ecosystem, including resistant tree varieties. The IPM evolved into Plant Health Care, a more holistic approach that focuses on the tree instead of the pest and includes the client in managing landscapes.

The number of arborists and related specialists grew at a phenomenal rate, especially after World War II as indicated by membership in the International Society of Arboriculture. The Shade Tree Conference started with about 36 people in 1924; in 1948 the National Shade Tree Conference first exceeded 1000 members; ISA members topped 5000 in 1989 and in 2001 reached 14,836, including 1051 Canadians and 1724 from 37 other countries. These figures, of course, do not include all practicing arborists, urban foresters, and related scientists, whose membership in other organizations partially overlaps with ISA.

These are just the highlights of the developments that shaped the North American arboricultural industry in the past 130 years. Richard Campana's book describes these and others in much greater detail, including fascinating stories about influential personalities.

4. Urban Forest Management

The heritage of placing trees into urban landscapes and of maintaining them through tree care practices began in antiquity, but the management of trees in towns and cities in comprehensive, systematic programs has come about more recently. In a municipal tree management program there are clear lines of authority and defined methods for planting, arboricultural tree care, and removal of undesirable trees. A tree ordinance, tree inventory, management plan, and public involvement are features that assure continuity, effectiveness, and citizen support for the program.

Jorgensen in 1965 crystallized the concept of urban forest management that had been evolving for some time in North America. It was later brought to Europe via Great Britain but had little impact there in the next 10 years. Arboriculture was just beginning to emerge as a distinct profession, and urban forestry was regarded as a threat to its professional status. So the British at first viewed urban forestry with "a mixture of disinterest, misunderstandings and outright opposition," according to M. Johnston. After 1980 other professions, such as planning and landscape architecture, became interested, leading to some of the first urban forestry projects in Great Britain.

Evidence of North American progress can be seen in city statutes and regulations, municipal methods for managing trees, and supportive state and national laws and policies. Furthermore, the growth of professional organizations, research, and education concerned with urban forest management is indicative that the art and science of this field is maturing.

4.1. Local Authority and Public Involvement in the United States

Civic-minded people promoted trees in their communities long ago through volunteerism and legislation. Laws to protect shade trees were enacted as early as 1635 in the "Towne of Boston." Its citizens probably can be credited with the first public shade tree planting in 1646 for the relief of travelers along a highway. A law in Massachusetts that first mentioned trees on public ways was passed in 1786, and the statewide protection of shade trees was legislated in 1830. The Tree Wardens Act of 1899 made mandatory the annual election of a Tree Warden in every Massachusetts town.

Philadelphia also accorded legal status to its trees in colonial times. An ordinance in 1700 directed every owner of a house "should plant one or more trees before the door that the town may be well-shaded from the violence of the sun." The state assembly in 1732 decreed that along with construction of a State House "walks may be laid out, and trees planted, to render the same more beautiful and commodious." The first organized street tree planting in Philadelphia took place in 1750. Fairmount Park officially was made a public park by an 1855 ordinance.

Early legislation about public trees also was enacted in other places. The territory of Michigan specified in 1807 that trees were to be planted along boulevards in Detroit. The city council of Sacramento, California managed its trees in 1853 under an ordinance that directed plantings and removals. The Trees and Parkings Commission in the District of Columbia was organized in 1872. The Minneapolis Park and Recreation Board, founded in 1883, plants and maintains all public trees through its Forestry Division. The New York City Parks Department in 1902 became responsible for all vegetation in parks and other public places; regulations required a permit for planting and prohibited mutilating any tree or shrub. The New Orleans Parking Commission was given control of street trees in 1909.

4.2. Municipal Tree Management in North America

Who are the practitioners of urban forest management? Approximately 90% of urban trees that are on private properties are managed piecemeal by their owners or agents sometimes with the help of commercial arborists, though private landscape trees often are not managed at all. Electric utility companies have limited responsibilities for managing trees along distribution and transmission lines, to provide clearance for safety and uninterrupted service. Municipalities have the authority for coordinated management of the trees within their boundaries, commonly the trees along streets and in parks. In some circumstances municipalities also may regulate private trees, for example to reduce dangers from diseased or hazardous trees.

The municipal tree program may be organized in various ways. Typically, larger cities locate the tree program in a public works or parks department, whereas smaller towns with more limited resources rely on tree commissions and volunteers. A survey of US cities in 1986 found that although 68% had a tree ordinance, only 39% had some type of systematic tree care program. Accordingly, these are the towns and cities that manage their urban forests, either directly or through their policies, and often with the help of volunteers.

A novel way of managing public trees started in 1893 when New Jersey became the first state to empower municipalities through ordinances to appoint shade tree commissioners with the authority to plant, maintain, and protect street trees. Other states soon followed this model. Tree commissions, sometimes called boards or committees, may function in three alternative ways: (i) advisory, (ii) policymaking, or (iii) operational. They are reminiscent of Downing's Ornamental Tree Societies.

> It is quite incredible how much spirit an Ornamental Tree Society, composed at first of a few really zealous arboriculturists, may beget in a country neighborhood. Some men ... are too much occupied with ... more important matters, ever to plant a tree, unsolicited. But these are readily acted upon by society ...Others there are, who ... will not begin to plant trees until it is the fashion to do so. ...And again, others who grudge the trifling cost of putting out a shade-tree, but who will be shamed into it ...by neighbors stimulated into action by the zeal of society. ...In this way, little by little, the Ornamental Tree Society accomplishes its ends. ...We heartily commend, therefore, this Social Planting Reform, to every desolate, leafless, and repulsive town and village in the country. (Andrew Jackson Downing, 1847).

The management of municipal trees, consisting mainly of planning, planting, maintenance, and removal of trees along streets and in parks, should start with an inventory of trees and resources. Landreth (1895), Solotaroff (1911), and Vick (1919) made early recommendations for inventorying and managing city trees. Master street tree plans that specify species to be planted and are based on tree inventories were in vogue especially before computerized inventories and led to management information systems. The major functions of software for managing urban trees are to store tree inventory data, record work on trees, handle service requests, summarize data, generate lists of trees, and map tree locations. A 1973 survey of 166 US cities and 6 in Canada found that only 30% had conducted tree inventories, and fewer than 3% handled data by computers. By 1979 14 cities were using computers and 6 inventory systems had been developed for general usage. In 1986 about 9% of US cities had computerized tree records. However, there was an apparent decline in cities managing their trees systematically, from 56% in 1974 to 50% in 1980 to just 39% in 1986; only 17% had a management plan.

Early in the 20th century, some of the more progressive cities were applying the arboricultural knowledge of that era in their management practices. An article in 1910 claimed that in New York City, "forestry has been developed more extensively than anywhere else in the country"; a graduate of the Yale Forest School was in charge of 150,000 street trees and employed 600 to 800 workmen. In various cities, suitable species and their characteristics were identified. Planting methods included attention to staking, drainage, watering, and protective devices against injuries by people and horses. Common causes of injuries were recognized, including poor soil, dumping of salt used in freezing ice-cream, air pollutants and escaping illuminating gas, oiling of roads, butchering of trees near wires by ruthless linemen, mutilation by horses, street building, and a wide variety of insects and diseases. Tree repair methods included filling of cavities, tree surgery, cabling, and bridge grafting over bark injuries.

In the latter part of the 20th century, municipalities utilized many innovations and improvements in arboricultural methods. Also a broader view of urban forest management developed, probably resulting from the increasing involvement of professional foresters. Consequently, forest management concepts, such as age structure, species diversity, and rotation period, became more common in municipal tree management plans in the 1980s. Terms, such as green infrastructure, land use planning, native versus foreign species, and urban ecosystems, are now discussed, and may further expand the scope and sophistication of urban forest management in the future.

4.3. Urban Forest Management in Europe

Europeans traditionally have regarded the urban forest as being just outside the city, which is different from the American concept. European communities typically own these forests for recreation, timber, game, and watershed protection. Furthermore, there is a tradition in many countries of everyman's right of free access to the forest, regardless of ownership, for recreation though not for hunting. The urban forest in Germany currently is viewed as having three components: (i) a peripheral forest belt that provides spaces for recreation and watershed protection, (ii) radial green arms reaching into the city, and (iii) greenspaces dotting the inner city.

Tree planting along roads and streets became popular in Europe during roughly the same period as in America. For example, in Czechoslovakia Rudolph II ordered the planting of roadside trees in 1740. Maria Theresa broadened the decree in 1752, making local governments responsible for tree planting and maintenance. In the second half of the 19th century, urban planning and services were carried out in a more organized way than before, and city and national governments provided more urban greenspaces to the public. Public beautification committees in Prague undertook the first major reforestation effort between 1897 and 1908. The Dutch forest service established in 1899 was responsible for state owned urban forests. In Great Britain, too, the emphasis was on "urban fringe forestry" around towns, for already by 1900, 80% of the population was living in larger towns and cities.

European city governments exercise more managerial authority over trees than in America with less participation by the public, although there is a considerable variation among countries. C. C. Konijnendijk's recent study of 16 cities in nine countries concluded that forest management plans were common in all cities, but "holistic, structured, and formal urban forest policies are rare."

Formal urban forest management plans and inventories apparently have come into use only recently in Europe, although public trees have been cared for systematically long ago, so the concept cannot be new. Private citizens, public institutions, or city employees manage city greenspaces in Germany. Remote sensing has been used recently to inventory the trees in Freiburg and Munich for planning purposes. In the United Kingdom, legislation has dealt mainly with protecting trees, not managing them. The Town & Country Planning Acts of 1971 and 1990 have empowered local planning authorities to protect amenity trees.

4.4. US Laws Implementing Urban Forestry

In the United States, a growing national interest in urban forestry became evident in 1967, when several agencies appointed committees or task forces. Prior to this, the state of Georgia had started a Metro Forestry program in 1966, and several other states soon followed suit. For example, Kansas funded its Rural Town Forestry Assistance program in 1972, and the Pennsylvania Bureau of Forestry appointed an urban forester in 1973.

Urban forestry was added to broaden the mission of the USDA Forest Service in 1972, and its Urban Forestry Assistance program was first funded modestly in 1978. From the start, the Forest Service has participated actively in partnerships with state agencies, universities, and professional organizations such as American Forests, the International Society of Arboriculture, and the National Arbor Day Foundation. The USDA Cooperative Extension Service also became involved in providing technical assistance in some states.

In 1990, President George H. W. Bush proposed the America the Beautiful program, which was enacted in the 1990 Farm Bill that increased federal funding for urban forestry to $21 million. This created the Urban and Community Forestry Assistance program administered by the Forest Service, which provides education, technical assistance, and grants to municipalities and local groups through state forestry agencies. The National Tree Trust also was formed to stimulate public awareness, volunteerism, and local tree planting. During 2003, the Urban and Community Forestry program of the Forest Service in the 20 northeastern states provided nearly $21 million in grants to states, communities, and special projects. The states in turn assisted nearly 6200 communities with grants and technical information.

5. Professional Organizations

Several professional organizations exchange information about urban forestry and arboriculture and promote their interests through research, education, public relations, and legislative influence. Besides the main North American organizations described here, there are similar organizations in other countries, and also other associations that serve important related fields such as entomology, pathology, genetics, soils, horticulture, and nurseries.

The International Society of Arboriculture, the principal and most comprehensive organization oriented to urban forestry, developed from the Shade Tree Conference founded in 1924. The National Arborist Association that originated in 1938 is a trade organization for commercial arborists and tree experts. The American Forestry Association helped to establish the National Urban and Community Forestry Leaders Council in 1981 and also has sponsored National Urban Forestry Conferences after the first one had been held in 1978. The Society of American Foresters in 1972 formed an Urban Forestry Working Group to provide liaison with other organizations such as the International Society of Arboriculture. The National Arbor Day Foundation, created in 1972, has been a catalyst for community tree care through its Tree City USA program that began in 1976.

6. Research and Education

In the United States, universities have been at the forefront of research into urban forestry topics. Together with federal and state agencies, they have led the world in research into urban tree selection, establishment, maintenance, and management. Research by private industries usually has been product oriented and often has been conducted via grants to universities.

International collaboration with urban forestry research has increased significantly in recent years. A European Union-financed project named COST Action E12 “Urban Forests and Trees” has compiled an overview of European research. Reports provided by 20 countries categorized 404 research projects. The intent of COST E12 is to coordinate urban forestry research, avoid duplication, and improve efficiency. Its leaders believe that there has been better coordination in North America.

Urban forestry curricula in colleges and universities arose gradually in departments of forestry, horticulture, or landscape architecture, as arboriculture and urban forestry were latecomers compared to other priorities in these departments. Bernard Fernow and William Solotaroff, a shade tree commissioner in New Jersey, wrote two of the first texts dealing with landscape tree maintenance in 1910 and 1911. Fernow, educated in Germany, had been the first chief forester of the United States and had presented the first forestry lecture at the Massachusetts Agricultural College in 1887. He also had been a professor or lecturer at Wisconsin University, Cornell University, Pennsylvania State College, and the University of Toronto.

A 1975 survey found that 33 universities had initiated new courses and curricula in urban forestry since 1968, but many of them had no specific offerings in arboriculture or urban forestry at that time. A 1995 survey of 101 institutions that offered courses in arboriculture, urban forestry, and related subjects listed 18 subfields beyond the basic or general education subjects; 50 of these institutions were in the United States.

Urban forestry has come to mean much more than applications of arboriculture and landscape design. Today it includes some aspects not obvious in Jorgensen’s broad definition of 1970, such as municipal and regional planning for the management of natural resources and urban ecosystem management.

7. Community Trees and Our Future

TREES IN OUR COMMUNITIES—what a wonderful, inspirational, invaluable heritage! These trees that enrich our lives in so many ways have endured urban stresses through the ages. For a very long time, they were at the mercy of rulers, the wealthy, and the vagaries of nature. Then a few centuries ago, here and there, landscape designers and gardeners became more active in helping urban trees to thrive. The professions of landscape architecture and arboriculture evolved next, and then urban forestry. Finally, ordinary people, as they gained influence and knowledge about landscape trees, were able to take a shared responsibility with their communities for having trees planted and given proper care.

Along with this growing concern for trees, the knowledge and tools used in tree care practices progressed slowly and precariously for a long time. At first, traditions passed along orally and later were recorded and published, so scientific and practical information about trees could be disseminated more widely, critiqued, and improved. Not until the 20th century did research and education about community trees flourish, leading to more diverse and improved tree varieties, more effective arboricultural practices, and people skilled in using them.

Now we have all the arts, sciences, tree varieties, and professionals needed to landscape our communities, and to understand and appreciate the benefits of urban forests. So why are not more towns and cities tending their trees? Some communities have excellent community tree programs. Other communities have none, their landscapes show the neglect, and the residents may not even realize what they are missing.

Fortunately, ordinary citizens have the persuasive power to get a tree program started or to improve an ineffective program, as has been demonstrated repeatedly. One person can spark this initiative, persuade a few others to help build public support, and convince community leaders to take action. And help is available in the United States from every state forestry agency or extension service, including technical advice.

References

Abbott, R. E., 1977, Commercial arboricultural practices in North America, *J. Arboric.* **3**(8):141–145.

Altman, I., and Zube, E. H., 1989, *Public Places and Spaces*, Plenum Press, New York, 299pp.

Andresen, J., and Williams, B., 1975, Urban forestry education in North America, *J. For.* **73**(12):786–790.

Archibald, P. L., 1973, Urban and community forestry, past and present, in *Proceedings of the National Urban Forestry Conference*, SUNY College of Environmental Science and Forestry, pp. 4–9.

Ball, J., 1999, A brief history of arboricultural pest management, *Arbor. News* **8**(1):29–33.

Bassett, J. R., and Lawrence, W. C., 1975, Status of street tree inventories in the U.S., *J. Arboric.* **1**(3):48–52.

Biles, L. E., and Deneke, F. J., 1982, Urban forestry programs begins fifth year, *J. Arboric.* **8**(6):154–156.

Blair, D., 1999, Borrowed techniques and innovations, *Arbor. News* **8**(4):12–15.

Blair, G. D., 1951, *Tree Clearance for Overhead Lines*, 2nd edn., Electrical Publications, 22 E, Huron St., Chicago, 242pp.

Bradshaw, A., Hunt, B., and Walmsley, T., 1995, *Trees in the Urban Landscapes, Principles and Practice*, E and FN Spon., London, 272pp.

Campana, R. J., 1999, *Arboriculture, History and Development in North America*, Michigan State University Press, East Lansing, 443pp.

Chadwick, L. C., 1970, 3000 years of arboriculture—past, present, and future, in *Proceedings of the 46th International Shade Tree Conference*, pp. 73a–78a.

Clegg, D., 1982, Urban and community forestry—the delivery, in *Proceedings of the Second National Urban Forestry Conference*, pp. 13–17.

Creech, J. L., 1972, Ornamental plant introduction—building on the past, *Arnoldia* **33**(1):13–25.

Crossman, S. S., 1962, History of the association, in *The Golden Year Edition* (F. C. Basile, ed.), Mass. Tree Wardens' and Foresters' Assoc., Haverhill, MA, pp. 14–23.

DeBruin, H. W., 1983, Urban and community forestry as the American Forestry Association sees it, *J. Arboric.* **9**(3):65–66.

DeBruin, H. W., 1982, Urban and community forestry: The time is now, *Am. For.* **88**(3):19–20, 52.
Deneke, F., 1978, Urban forestry education, *J. Arboric.* **4**(7):154–156.
Deneke, F., 1983, Urban and community forestry: Where are we going? *J. Arboric.* **9**(4):99–101.
Domholt, P., 1996, The Minneapolis urban forest, *City Trees* Nov./Dec:12.
Downing, A. J., 1859, *A Treatise on the Theory and Practice of Landscape Gardening*, 6th edn., A. O. Moore, New York, 576pp.
Downing, A. J., 1881, *Rural Essays*, R. Worthington, New York, 557pp.
Dwyer, J., McPherson, E. G., Schroeder, H., and Rowntree, R., 1992, Assessing the benefits and costs of the urban forest, *J. Arboric.* **18**(5):227–234.
Dyke, R., 1989, Community forests: An investment in the future, in *Shading Our Cities* (G. Moll and S. Ebenreck, eds.), Island Press, Washington, DC, pp. 206–210.
Evelyn, J., 1664, *Sylva, or a Discourse of Forest-Trees and the Propagation of Timber in His Majesties Dominions*, Royal Society, London, 210pp.
Felt, E. P., 1942, *Our Shade Trees*, Orange Judd Publication, New York, 316pp.
Fernow, B. E., 1911, *A Brief History of Forestry in Europe, the United States and other Countries*, University Press, Toronto, 516pp.
Fernow, B. E., 1911, *Care of Trees in Lawn, Street, and Park*, Henry Hold, New York, 392pp.
Floyd, D. W., 1992, *Final Report, Urban Forestry Curriculum Development Project*, School of Natural Resources, Ohio State University, 31pp.
Fraser, E. D. G., and Kenney, W. A., 2000, Cultural background and landscape history as factors affecting perceptions of the urban forest, *J. Arboric.* **26**(2):106–113.
Gerhold, H. D., Lacasse, N. L., and Wandell, W. N., 2001, *Landscape Tree Factsheets, Including Evergreens for Screens*, Pennsylvania State University, College of Agricultural Sciences, University Park, 452pp.
Gerhold, H. D., Steiner, K. C., and Sacksteder, C. J., 1987, Management information systems for urban trees, *Boissiera* **38:**29–37.
Gerhold, H. D., and Frank, S. A., 2002, *Our Heritage of Community Trees*, Pennsylvania Urban & Community Forestry Council, 56 E. Main St., Mechanicsburg, PA 17055, 73pp.
Goode, J., 1988, Andrew Jackson Downing on trees, *Arboric. J.* **12**(2):189–194.
Guerin, J. C., 1989., Les arbres a Paris, *Rev. For. Fran.* **XLI** (Num. Sp. 1989):29–44.
Guerin, J. C., 1987, Le patrimoine de la Ville de Paris, *Boissera* **38:**105–119.
Guinaudeau, C., 1989, La gestion des arbres urbains, *Rev. For. Fran.* **XLI** (Num. Sp. 1989):125–126.
Harris, R. W., Clark, J. R., and Matheny, N. P., 1999, *Arboriculture, Integrated Management of Landscape Trees, Shrubs, and Vines*, Prentice-Hall, Upper Saddle River, NJ, 685pp.
Hebert, V. L., 1962, Liberty tree, in *The Golden Year Edition* (F. C. Basile, ed.), Mass. Tree Wardens' and Foresters' Assoc., Haverhill, MA, pp. 53–54.
Hedrick, U. P., 1950, *A History of Horticulture in America to 1860,* Oxford University Press, New York, 551pp.
Hennebo, D., 1979, *Entwicklung des Stattgrüns von der Antike bis in die Zeit des Absolutismus*, Patzer, Hanover, Berlin, 169pp.
Hildebrandt, G., 1986, Inventur städtischer Baumbestände durch Fernerkundung und Photogrammetrie, *Boissera* **38:**49–71.
Hobhouse, P., 1992, *Gardening Through the Ages*, Simon and Schuster, New York, 320pp.
Holscher, C. E., 1970, European experience in integrated management of urban and suburban woodlands, in *Trees and Forests in an Urbanizing Environment, a Monograph*, Univ. Mass. Coop. Ext. Serv., Planning and Resource Development Series No. **17:**133–137.
International Society of Arboriculture, 1995, Catalog of curricula in arboriculture, urban forestry & related areas, *Champaign*, IL.
Irish, C. F., 1932, Highlights in the early history of arboriculture, *Natl. Shade Tree Conf. Proc.* **8:**8–14.
Jackson, J. B., 1969, A new kind of space, *Landscape* **18**(1):33–35.
Johnson, C., 1982, Political & administrative factors in urban-forestry programs, *J. Arboric.* **8**(6):160–163.
Johnson, H., 1973, *The International Book of Trees*, Simon and Schuster, New York, 288pp.
Johnston, M., 1996, A brief history of urban forestry in the United States, *Arboric. J.* **20**(3):257–278.
Johnston, M., 1997, The early development of urban forestry in Britain: Part 1, *J. Arboric.* **21:**107–126.
Jorgensen, E., 1986, Urban forestry in the rearview mirror, *Arboric. J.* **10:**177–190.
Jorgensen, E., 1970, Urban forestry in Canada, *Proc. 46th Intl. Shade Tree Conf.*, 43a–51a.

Kielbaso, J. J., 1978, The urban forestry working group of the Society of American Foresters, *J. Arboric.* **4**(2):43–45.
Kielbaso, J. J., 1990, Trends and issues in city forests, *J. Arboric.* **16**(3):69–76.
Kielbaso, J. J., Beauchamp, B. S., Larison, K. L., and Randall, C. J., 1988, Trends in urban forestry management, *Int. City Manag. Assoc. Baseline Data Rep.* **20**(1):1–17.
Kleinschmit, J., Khurana, K. K., Gerhold, H. D., and Libby, W. J., 1993, Past, present, and anticipated applications of clonal forestry, in *Clonal Forestry II, Conservation and Application*, Springer-Verlag, Berlin, Heidelberg, pp. 9–41.
Koch, J., 2000, The origins of urban forestry, in *Urban and Community Forestry in the Northeast* (J. Kuser, ed.), Kluwer Academic/Plenum Publishers, New York, pp. 1–10.
Konijnendijk, C. C., 1997, *Urban Forestry in the Netherlands: Lessons from the Past*, XI World Forestry Congress, 8pp.
Konijnendijk, C. C., 1997, A short history of urban forestry in Europe, *Arboric. J.* **23**(1):31–39.
Konijnendijk, C. C., 1999, Urban forest policy-making: A comparative study of selected cities in Europe. *Arboric. J.* **23**(1):1–15.
Konijnendijk, C. C., Randrup, T. B., and Nilsson, K., 2000, Urban forestry research in Europe: An overview. *J. Arboric.* **26**(3):152–161.
Kuser, J. E., 2000, *Exotic Trees in New Jersey, Greater Philadelphia, and Metro New York*, Cook College, Rutgers University, 12pp.
Landreth, B., 1895, Trees for city streets, *N.J. Forester* **1**(3):31.
Lawrence, H. W, 1988, Origins of a tree-lined boulevard, *Geogr. Rev.* **78**(4):355–374.
Lawrence, H. W., 1993, The neoclassical origins of modern urban forests. *For. Conserv. Hist.* **37**:26–37.
Lawrence, H. W., 1995, Changing forms and persistent values: Historical perspectives on the urban forest, in *Urban Forest Landscapes: Integrating Multidisciplinary Perspectives* (G. A. Bradley, ed.), University of Washington Press, Seattle, pp. 17–40.
Lewis, C., 1976, Our American heritage: Trees, *J. Arboric.* **2**(10):181–185.
Li, H.-L., 1963, *The Origin and Cultivation of Shade and Ornamental Trees*, University of Pennsylvania Press, Philadelphia, 282pp.
Lobel, D. F., 1983, Managing urban forests using forestry concepts, *J. Arboric.* **9**(3):75–78.
Loeb, R. E., 1992, Will a tree grow in Brooklyn? Developmental trends of the New York City street tree forest, *J. For.* **90**(1):20–24.
Louden, J. C., 1982, *An Encyclopedia of Gardening*, Reprint of the 1835 edition, vols. 1 and 2, Garland Publishing, New York, 1271pp.
Maillet, L., 1989, Approche méthodologique de la gestion de l'arbre en ville, *Revue Forestière Française* **XLI** (Num. Sp. 1989):119–124.
Matthews, W. E., 1976, The history of arboriculture in the United Kingdom, in *Trees and Forests for Human Settlements*, Centre for Urban Forestry Studies, University of Toronto, Canada, pp. 150–157.
McPherson, E. G., 1984, Employer perspectives on arboriculture education, *J. Arboric.* **10**(5):137–142.
McPherson, E. G., and Luttinger, N., 1998, From nature to nurture: The history of Sacramento's urban forest, *J. Arboric.* **24**(2):72–88.
Mell, C. D., 1910, A forester whose field is the city, *Am. For.* **16**:519–523.
Michaux, F. A., 1853, *The North American Sylva*, Translated from the French by Hillhouse, R. P. Smith, Philadelphia, 3 vols., pp. 123, 128, 143.
Miller, R. W., 1997, *Urban Forestry: Planning and Managing Urban Greenspaces*, Prentice-Hall, Upper Saddle River, NJ, 502pp.
M'Mahon, B., 1806, *American Gardener's Calendar*, B. Graves, Philadelphia, 666pp.
Morrison, N. J., and Sydnor, T. D., 2000, Urban tree care in Europe and the United States, *Tree Care Indus.* **11**(2):36–42.
Mumford, L., 1961, *The City in History: Its Origins, Its Transformations, and Its Prospects*, Harcourt, Brace & World, New York, 657pp.
Nadel, I. B., and Oberlander, C. H., 1977, *Trees in the City*, Pergamon Press, New York, 94pp.
Newton, N. T., 1971, *Design on the Land, the Development of Landscape Architecture*, Belknap Press, 714pp.
Nobles, R., 1980, Urban forestry/arboriculture program, *J. Arboric.* **6**(2):53–56.
O'Callaghan, D. P., 1991, Legal protection for trees in Britain and Ireland, *J. Arboric.* **17**(11):306–312.

Pauleit, S., and Duhme, F., 2000, GIS assessment of Munich's urban forest structure for urban planning, *J. Arboric.* **26**(3):133–141.

Pennsylvania Horticultural Society, 1976, *From Seed to Flower: Philadelphia 1681–1876: A Horticultural Point of View*, Pennsylvania Horticultural Society, Philadelphia, 105pp.

Phillips, L. E., 1993, *Urban Trees: A Guide for Selection, Maintenance, and Master Planning*, McGraw-Hill, New York, 273pp.

Prince, W., 1828, *A Short Treatise on Horticulture*, T. and J. Swords, New York, 196pp.

Profous, G. V., and Rowntree, R. A., 1993, The structure and management of the urban forest in Prague, Czechoslovakia. II. History of the Prague greenspace movement, *Arboric. J.* **17**(2):117–130.

Sacksteder, C. J., and Gerhold, H. D., 1979, *A Guide to Urban Tree Inventory Systems*, Sch. For. Resources, Pennsylvania State University, Research Paper 43, 52pp.

Sangster, M., 1993, Urban fringe forestry in Great Britain, *J. Arboric.* **19**(1):51–55.

Schein, R. D., 1993, *Street Trees: A Manual for Municipalities*, Tree Works, State College, Pennsylvania, 375pp.

Schuyler, D., 1986, *The New Urban Landscape*, Johns Hopkins University Press, 237pp.

Shabel, H., 1980, Urban forestry in the Federal Republic of Germany, *J. Arboric.* **6**(11):281–286.

Shafer, E. L., and Moeller,G. H., 1979, Urban forestry: Its scope and complexity, *J. Arboric.* **5**(9):206–209.

Shigo, A. L., 1991, *Modern Arboriculture*, Shigo and Trees, Associates, Durham, NH, 424pp.

Shue, J., 1991, *Framing the Vision: Progressive Reform, Social Value and the Creation of the Modern Treescape*, M.S. Thesis, Pennsylvania State University 219pp.

Solotaroff, W., 1911, *Shade-Trees in Towns and Cities*, Wiley, New York, 287pp.

Spongberg, S. A., 1990, *A Reunion of Trees: The Discovery of Exotic Plants and Their Introduction into North American and European Landscapes*, Harvard University Press, Cambridge, MA, 270pp.

Steuart, Sir Henry, 1848, *The Planter's Guide*, 3rd edn., J. Murray, Edinburgh, 48 + 518pp.

Stone, O. P., 1999, Tree care in North America, *Arbor. News* **8**(2):9, 10.

Tate, R. L., 1984, Municipal tree management in New Jersey, *J. Arboric.* **10**(8):229–233.

Ulrich, R., 1984, View through a window may influence recovery from surgery, *Science* **224:**420, 421.

USDA Forest Service, 1991, America the Beautiful national tree program, FS-499, 8pp.

USDA Forest Service Northeastern Area, 2004, The year in forestry, state and private forestry in the Northeastern Area, NA-IN-03-04, 45pp.

Vick, A. F., 1919, The classification and census of city trees, *Am. City* **20:**368–370.

Weigley, R. (ed.), 1982, *Philadelphia: A 300-year History*, W. W. Norton, New York, 751pp.

Wyman, D., 1973, The history of ornamental horticulture in America, *Arnoldia* **33**(1):97–112.

Zube, E. H. (ed.), 1970, *Landscapes, Selected Writings of J.B. Jackson*, University of Massachusetts Press, Amherst, 160pp.

Zube, E. H., 1971, Trees and woodlands in the design of the urban environment, in *Trees and Forests in an Urbanizing Environment, a Monograph*, Univ. Mass. Coop. Ext. Serv., Planning and Resource Development Series No. **17**:145–150.

Chapter 2

Understanding the Benefits and Costs of Urban Forest Ecosystems

David J. Nowak[1] and John F. Dwyer[2]

1. Introduction

One of the first considerations in developing a strong and comprehensive urban forestry program is determining the desired outcomes from managing vegetation in cities. Urban trees can provide a wide range of benefits to the urban environment and well-being of people. However, there are also a wide range of potential costs and as with all ecosystems, numerous interactions that must be understood if one is to optimize the net benefits from urban vegetation. Inadequate understanding of the wide range of benefits, costs, and expected outcomes of urban vegetation management options, as well as interactions among them, may drastically reduce the contribution of vegetation toward improving urban environments and quality of life.

By altering the type and arrangement of trees in a city (i.e., the urban forest structure), one can affect the city's physical, biological, and socioeconomic environments. Management plans can be developed and implemented to address specific problems within cities. Although trees can provide multiple benefits at one site, not all benefits can necessarily be realized in each location. Individual management plans should focus on optimizing, in a particular area, the mix of benefits that are most important.

2. Urban Land in the United States

The importance of urban forests and their benefits in the United States is increasing because of the expansion of urban land. The percentage of the coterminous land in United States, classified as urban, increased from 2.5% in 1990 to 3.1% in 2000, an area about the size of Vermont and New Hampshire combined. The states with the

David J. Nowak USDA Forest Service, Northeastern Research Station, Syracuse, New York
John F. Dwyer The Morton Arboretum, Lisle, Illinois

Urban and Community Forestry in the Northeast, 2nd ed., edited by, J. E. Kuser.

FIGURE 1. Urban areas in coterminous United States (2000) based on the US Census Bureau Definition of Urban Land.

highest percentage of urban land are New Jersey (36.2%), Rhode Island (35.9 %), Connecticut (35.5%), and Massachusetts (34.2%) with 7 of the top 10, most urbanized states in the Northeast United States (Fig. 1) (Nowak et al., 2005).

The most urbanized regions of the United States are the Northeast (9.7%) and the Southeast (7.5%), with these regions also exhibiting the greatest increase in percentage of urban land between 1990 and 2000 (1.5% and 1.8%, respectively). States with the greatest increase in percentage of urban land between 1990 and 2000 were Rhode Island (5.7%), New Jersey (5.1%), Connecticut (5.0%), Massachusetts (5.0%), Delaware (4.1%), Maryland (3.0%), and Florida (2.5%) (Nowak *et al.*, 2005). Nationally, urban tree cover in the United States averages 27.1%. However, urban tree cover tends to be highest in forested regions (34.4% urban tree cover), followed by grasslands (17.8%), and deserts (9.3%) (Dwyer et al., 2000; Nowak et al., 2001).

Patterns of urban growth reveal that increased growth rates are likely in the future (Nowak *et al.*, 2005). As the Northeast is the most urbanized region of the country and is likely to have some of the greatest increases in urban land growth over the next several decades, understanding the benefits and costs of urban vegetation is paramount to sustain human health and environmental quality in this region.

3. Physical/Biological Benefits and Costs of Urban Vegetation

Through proper planning, design, and management, urban trees can mitigate many of the environmental impacts of urban development by moderating climate, reducing building energy-use and atmospheric carbon dioxide (CO_2), improving air quality, lowering rainfall runoff and flooding, and reducing noise levels. However, inappropriate landscape designs, tree selection, and tree maintenance can increase environmental costs, such as pollen production and chemical emissions from trees and maintenance activities that contribute to air pollution, and also increase building energy-use, waste disposal, infrastructure repair, and water consumption. These potential costs must be weighed against the environmental benefits in developing management programs.

3.1. Urban Atmosphere

Trees influence the urban atmosphere in the following four general, interactive ways that can be remembered by using the word TREE (Nowak, 1995): (1) **T**emperature and microclimatic effects, (2) **R**emoval of air pollutants, (3) **E**mission of volatile organic compounds by trees and emissions due to tree maintenance, and (4) **E**nergy conservation in buildings and consequent effects on emissions from power plants. The cumulative effect of these four factors determines the overall impact of urban trees on the urban atmosphere and particularly air pollution.

3.1.1. Temperature and Microclimatic Modifications

Trees influence climate at a range of scales, from an individual tree to a forest covering an entire metropolitan area. By transpiring water, altering windspeeds, shading surfaces, and modifying the storage and exchanges of heat among urban surfaces, trees affect local climate and thereby influence thermal comfort and air quality. Often, one or more of these microclimatic influences of trees produces an important benefit, while other influences can reduce benefits or increase costs (Heisler *et al.*, 1995).

Trees alter windspeed and direction. Dense tree crowns have a significant impact on wind, but for isolated trees, their influence nearly disappears within a few crown diameters downwind (Heisler *et al.*, 1995). Several trees on a residential lot, in conjunction with trees throughout the neighborhood, reduce windspeed significantly. In a residential neighborhood in central Pennsylvania with 67% tree cover, windspeeds at 2 m above ground level were reduced by 60% in winter and 67% in summer compared to windspeeds in a comparable neighborhood with no trees (Heisler, 1990a).

Trees also have a dramatic influence on incoming solar radiation, and can reduce it by 90% or more (Heisler, 1986). Some of the radiation absorbed by tree canopies leads to the evaporation and transpiration of water from leaves. This evapotranspiration cools tree leaves and the air. Despite large amounts of energy used for evapotranspiration on sunny days, air movement rapidly disperses cooled air, thereby dispersing the overall cooling effect. Under individual and small groups of trees, air temperature at 1.5 m above the ground is usually within 1°C of the air temperatures in an open area (Souch and Souch, 1993). Along with transpirational cooling, tree shade can help cool the local environment by reducing the solar heating of some below-canopy artificial surfaces (e.g., buildings, parking lots). Together these effects can reduce air temperatures by as much as 5°C (Akbari *et al.*, 1992).

Although trees usually contribute to cooler summer air temperatures, their presence can increase air temperatures in some instances (Myrup *et al.*, 1991). In areas with scattered tree canopies, radiation can reach and heat ground surfaces; at the same time, the canopy may reduce atmospheric mixing such that cooler air is prevented from reaching the area. In this case, tree shade and transpiration may not compensate for the increased air temperatures due to reduced mixing (Heisler *et al.*, 1995). Thus, it is important to recognize that it is the combined effects of trees on radiation, wind, and transpirational cooling that affect local air temperatures and climate.

Besides providing for transpirational cooling, the physical mass and thermal/radiative properties of trees can affect other aspects of local meteorology and microclimate, such as ultraviolet radiation loads, relative humidity, turbulence, albedo, and boundary-layer heights (i.e., the height of the layer of atmosphere that, because of turbulence, interacts with the earth's surface on a time scale of a few hours or less (Lenschow, 1986)).

3.1.2. Removal of Air Pollutants

Trees remove gaseous air pollution primarily by uptake through leaf stomata, though some gases are removed by the plant surface (Smith, 1990). Once inside the leaf, gases diffuse into intercellular spaces and may be absorbed by water films to form acids or react with the inner surfaces of leaves. Trees also remove pollution by intercepting airborne particles. Some particles can be absorbed into the tree (Ziegler, 1973; Rolfe, 1974), though most intercepted particles are retained on the plant surface. Often vegetation is only a temporary retention site for atmospheric particles as the intercepted particles may be resuspended to the atmosphere, washed off by rain, or dropped to the ground with leaf and twig fall (Smith, 1990).

Pollution removal by trees in a city varies throughout the year (Fig. 2).

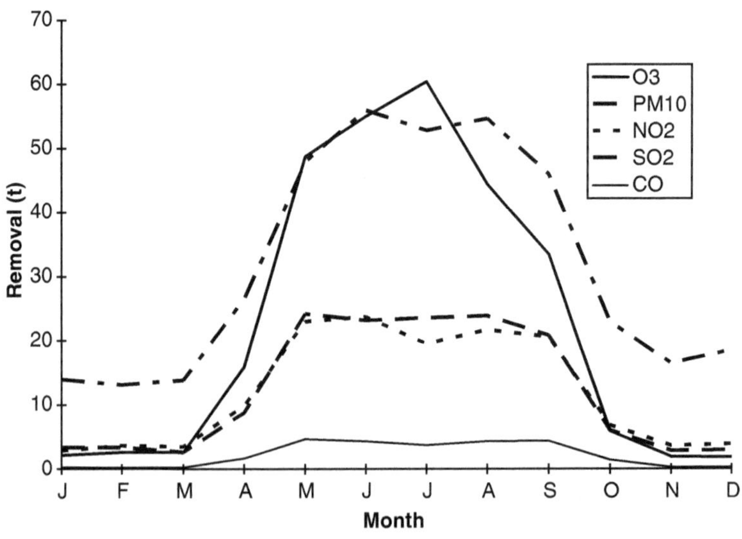

FIGURE 2. Monthly pollution removal by trees (metric tons) in Philadelphia, PA (1994). PM10 = particulate matter <10 microns; O_3 = ozone; NO_2 = nitrogen dioxide; SO_2 = sulfur dioxide; CO = carbon monoxide. PM10 removal assumes 50% resuspension of particles. City area = 350 km^2; tree cover = 21.6%.

Factors that affect pollution removal by trees include the amount of healthy leaf-surface area, concentrations of local pollutants, and local meteorology. Computer simulations using the Urban Forest Effects Model (Nowak and Crane, 2000, Nowak *et al.*, 2002b) with local field data reveal that pollution removal by urban trees in various cities range from 19 metric tons per year in Freehold, New Jersey to over 1500 metric tons per year in Atlanta and New York (Table 1). Pollution removal was typically greatest for ozone, followed by particulate matter less than 10 microns, nitrogen dioxide, sulfur dioxide, and carbon monoxide. Value of pollution removal, based on national median externality values for each pollutant (Murray *et al.*, 1994), ranged from $109,000 in Freehold to $8.3 million in Atlanta.

Average annual pollution removal per square meter of canopy cover was 9.3 g, but ranged between 6.6 g/m^2 in Syracuse and 12.0 g/m^2 in Atlanta (Table 1). The average annual dollar value per hectare of tree cover was $500, but ranged between $378/ha cover in Syracuse and $663/ha cover in Atlanta. As existing canopy cover in cities remove significant amounts of air pollution, increasing tree cover in urban areas will lead to greater pollution removal, as well as reduced air temperatures that can help improve urban air quality.

Average improvement in air quality from pollution removal by trees during the daytime of the in-leaf season among 14 cities (Table 1) was 0.62% for particulate matter less than 10 microns (PM10), 0.61% for ozone (O_3), 0.60% for sulfur dioxide (SO_2), 0.39% for nitrogen dioxide (NO_2), and 0.002% for carbon monoxide (CO). Air quality improvement increases with increased percentage of tree cover and decreased boundary-layer heights. In urban areas (Table 1) with 100% tree cover (i.e., contiguous forest stands), short-term improvements (1 h) in air quality due to pollution removal from trees were as high as 14.9% for SO_2, 14.8% for O_3, 13.6% for PM10, 8.3% for NO_2, and 0.05% for CO. In Chicago in 1991, large, healthy trees—those >77 cm in diameter at breast height (dbh)—removed an estimated 1.4 kg of pollution, about 70 times more pollution than small (<7 cm dbh) trees (Nowak, 1994a).

Trees can also reduce atmospheric CO_2 by directly storing carbon (C) from CO_2 as they grow. Large trees store approximately 3 metric tons of carbon (tC) or 1000 times more carbon than stored by small trees (Nowak, 1994b). Healthy trees continue to sequester additional carbon each year; large, healthy trees sequester about 93 kg C/year vs 1 kg C/year for small trees. Net annual sequestration by trees in the Chicago area (140,600 tC) equals the amount of carbon emitted from transportation in the Chicago area in about 1 week (Nowak, 1994b).

Urban trees in the coterminous United States currently store 700 million metric tons of carbon (335 to 980 million tC; $14,300 million value) with a gross carbon sequestration rate of 22.8 million tC/year (13.7 to 25.9 million tC/year; $460 million/year) (Nowak and Crane, 2002). These results correspond with previous analyses that estimated national carbon storage by urban trees as between 350–750 million tC (Nowak, 1993a) and 600–900 million tC (Nowak, 1994b). Carbon storage by urban trees nationally is only 4.4% of the estimated 15,900 million tC stored in trees in US nonurban forest ecosystems (Birdsey and Heath, 1995). The estimated carbon storage by urban trees in United States is equivalent to the amount of carbon emitted by the US population in about 5.5 months. National annual carbon sequestration by urban trees is equivalent to the US population emissions over a 5-day period (Nowak and Crane, 2002).

Table 1. Total Estimated Pollution Removal (Metric Tons) by Trees During Nonprecipitation Periods (Dry Deposition), and Associated Monetary Value for Various Cities (Pollutant Year = 2000)

City	Pollution removed									
	CO (t)	NO_2 (t)	O_3 (t)	PM10 (t)	SO_2 (t)	Total (t)	Range (t)	g/m^2 cover[a]	$	$/ha cover[b]
New York, NY	67	364	536	354	199	1,521	(619–2,185)	9.1	8,071,000	482
Atlanta, GA	39	181	672	528	89	1,508	(538–2,101)	12.0	8,321,000	663
Baltimore, MD	9	94	223	142	55	522	(183–725)	9.9	2,876,000	545
Philadelphia, PA	10	93	185	194	41	522	(203–742)	9.7	2,826,000	527
San Juan, PR	56	55	161	153	86	511	(222–768)	11.2	2,342,000	511
Washington, DC	18	50	152	107	51	379	(150–568)	8.3	1,956,000	429
Boston, MA	6	48	108	73	23	257	(94–346)	8.1	1,426,000	447
Woodbridge, NJ	6	42	66	62	15	191	(72–267)	10.8	1,037,000	586
San Francisco, CA	7	25	47	42	7	128	(51–195)	9.0	693,000	486
Moorestown, NJ	2	14	43	38	9	107	(41–157)	10.1	576,000	541
Syracuse, NY	2	12	55	23	7	99	(37–134)	6.6	568,000	378
Morgantown, WV	1	5	26	18	9	60	(22–98)	7.5	311,000	387
Jersey City, NJ	2	9	13	9	5	37	(16–56)	8.4	196,000	445
Freehold, NJ	1	3	9	6	1	20	(7–27)	11.4	110,000	632

Estimates are for particulate matter less than 10 microns (PM10), ozone (O_3), nitrogen dioxide (NO_2), sulfur dioxide (SO_2) and carbon monoxide (CO). Pollution removal model methods are described in Nowak *et al.* (1998). Monetary value of pollution removal by trees was estimated using the median externality values for United States for each pollutant. Externality values are: NO_2 = \$6750 t^{-1}, PM10 = \$4500 t^{-1}, SO_2 = \$1650 t^{-1}, and CO = \$950 t^{-1} (Murray *et al.*, 1994). Externality values for O_3 were set to equal the value for NO_2.

[a] Average grams of pollution removal per year per square meter of canopy cover.

[b] Average dollar value of pollution removal per year per hectare of canopy cover.

Table 2. Estimated Carbon Storage, Gross and Net Annual Sequestration, Number of Trees, and Percent Tree Cover for 10 US Cities (Nowak and Crane, 2002)

City	Storage (tC)		Annual sequestration Gross (tC/yr)		Net (tC/yr)		No. of trees (× 10^3)		Tree cover (percent)	
	Total	SE	Total	SE	Total	SE	Total	SE	%	SE
New York, NY	1,225,200	150,500	38,400	4,300	20,800	4,500	5,212	719	20.9	2.0
Atlanta, GA	1,220,200	91,900	42,100	2,800	32,200	4,500	9,415	749	36.7	2.0
Sacramento, CA[a]	1,107,300	532,600	20,200	4,400	na	na	1,733	350	13.0	na
Chicago, IL[b]	854,800	129,100	40,100	4,900	na	na	4,128	634	11.0	0.2
Baltimore, MD	528,700	66,100	14,800	1,700	10,800	1,500	2,835	605	25.2	2.2
Philadelphia, PA	481,000	48,400	14,600	1,500	10,700	1,300	2,113	211	15.7	1.3
Boston, MA	289,800	36,700	9,500	900	6,900	900	1,183	109	22.3	1.8
Syracuse, NY	148,300	16,200	4,700	400	3,500	400	891	125	24.4	2.0
Oakland, CA[c]	145,800	4,900	na	na	na	na	1,588	51	21.0	0.2
Jersey City, NJ	19,300	2,600	800	90	600	100	136	22	11.5	1.2

[a]McPherson (1998).
[b]Nowak (1994b).
[c]Nowak (1993c).
SE = standard error.
na = not analyzed.

Carbon storage within the cities ranges from 1.2 million tC in New York City and Atlanta to 19,300 tC in Jersey City, New Jersey (Table 2).

Urban trees in the North Central, Northeast, South Central and Southeast regions of the United States store and sequester the most amount of carbon, with average carbon storage per hectare greatest in Southeast (31.1 tC/ha), North Central (30.7 tC/ha), Northeast (30.5 tC/ha), and Pacific Northwest (30.2 tC/ha) regions, respectively. The national average urban forest carbon storage density is 25.1 tC/ha as compared to 53.5 tC/ha in forest stands (Nowak and Crane, 2002).

3.1.3. Emission of Volatile Organic Compounds and Tree Maintenance Emissions

Some trees emit into the atmosphere volatile organic compounds (VOCs) such as isoprene and monoterpenes. These compounds are natural chemicals that make up essential oils, resins, and other plant products and may be useful to the tree in attracting pollinators or repelling predators (Kramer and Kozlowski, 1979). Isoprene is also believed to provide thermal protection to plants by helping prevent irreversible leaf damage at high temperatures (Sharkey and Singsaas, 1995). The VOC emissions by trees vary with species, air temperature, and other environmental factors (Tingey *et al.*, 1991; Guenther *et al.*, 1994).

Volatile organic compounds can contribute to the formation of O_3 and CO (Brasseur and Chatfield, 1991). Because the VOC emissions are temperature dependent and trees generally lower air temperatures, it is believed that increased tree cover lowers overall VOC emissions and, consequently, reduces O_3 levels in urban areas. A computer simulation of June 4, 1984 ozone conditions in Atlanta, Georgia revealed that a 20% loss in the area's forest could lead to a 14% increase in O_3 concentrations. Although there were fewer trees to emit VOCs, an increase in Atlanta's air temperatures due to the urban heat island, which occurred concomitantly with the tree loss, increased VOC emissions from the remaining trees and anthropogenic sources and altered O_3 photochemistry such that concentrations of O_3 increased (Cardelino and Chameides, 1990).

A simulation of California's South Coast Air Basin suggested that the impact on air quality from increased urban tree cover might be locally positive or negative. The net basinwide effect of increased urban vegetation is a decrease in O_3 concentrations if the additional trees are low-VOC emitters (Taha, 1996). Examples of low-VOC emitting genera include *Fraxinus* spp., *Ilex* spp., *Malus* spp., *Prunus* spp., *Pyrus* spp., and *Ulmus* spp.; high-VOC emitters include *Eucalyptus* spp., *Quercus* spp., *Platanus* spp., *Populus* spp., *Rhamnus* spp., and *Salix* spp. (Benjamin *et al.*, 1996).

Tree management and maintenance also affects pollutant emissions. The equipment used in many maintenance activities emits pollutants and global gases such as VOCs, CO, CO_2, nitrogen oxides (NO_x), sulfur oxides (SO_x), and particulate matter (US EPA, 1991). Thus, while evaluating the overall net change in air quality due to trees, managers and planners must consider the amount of pollution that results from tree maintenance and management activities. The greater the use of fossil fuels (e.g., from vehicles, chain saws, augers, and chippers) in establishing and maintaining a certain vegetation structure, the longer the trees must live and function to offset the pollutant emissions from vegetation maintenance.

While considering the net effect of tree growth on atmospheric CO_2, managers must also consider that nearly all of the carbon sequestered by trees will be converted back to CO_2 due to decomposition after the tree dies. Hence, the benefits of carbon sequestration will be relatively short-lived if vegetation structure is not sustained. However, if carbon (via fossil-fuel combustion) is used to maintain vegetation structure and health, urban forest ecosystems will eventually become net emitters of carbon unless secondary carbon reductions (e.g., energy conservation) or limiting decomposition via long-term carbon storage (e.g., wood products, landfills) can be accomplished to offset the carbon emissions during maintenance (Nowak *et al.*, 2002c).

Trees in parking lots can also help reduce VOCs emissions by shading parked cars and thereby reducing evaporative emissions from vehicles. Increasing parking lot tree cover from 8% to 50% could reduce Sacramento County, California, light duty vehicle VOC evaporative emission rates by 2% and nitrogen oxide start emissions by <1% (Scott *et al.*, 1999).

3.1.4. Net Effects on Ozone

Besides the studies by Cardelino and Chameides (1990) and Taha (1996), other studies reveal that increased urban tree cover can lead to reduced ozone concentrations. Modeling the effects of increased urban tree cover on ozone concentrations from Washington, DC to central Massachusetts revealed that urban trees generally reduce ozone concentrations in cities. Interactions of the effects of trees on the physical and chemical environment demonstrate that trees can cause changes in pollution removal rates and meteorology, particularly air temperatures, wind fields, and mixing-layer heights, which, in turn, affect ozone concentrations. Changes in urban tree species composition had no detectable effect on ozone concentrations (Nowak *et al.*, 2000). Modeling of the New York City metropolitan area also revealed that increasing tree cover by 10% within urban areas could reduce maximum ozone levels by about 4 ppb, which is about 37% of the amount needed for attainment of the National Ambient Air Quality Standard (Luley and Bond, 2002).

Based on the various research on urban tree effects on ozone, the US Environmental Protection Agency (US EPA) released a guidance document that details how new measures, including "strategic tree planting," can be incorporated in State Implementation Plans as a means to help states meet National Ambient Air Quality Standards (US EPA, 2004).

3.1.5. Energy Conservation

Trees can reduce building heating and cooling energy needs, as well as consequent emissions of air pollutants and CO_2 by power plants, by shading buildings and reducing air temperatures in the summer, and by blocking winds in winter. However, trees that shade buildings in winter can also increase heating needs. Energy conservation from trees varies by regional climate, the size and amount of tree foliage, and the location of trees around buildings. Tree arrangements that save energy provide shade primarily on east and west walls and roofs, and wind protection from the direction of prevailing winter winds. However, wind reduction in the summer can lead to increased energy use for air conditioning, but wind and shade effects combined lead to reduced summer energy use for cooling (Akbari *et al.*, 1992; Heisler, 1990b). Energy use in a house with trees can be 20% to 25% lower per year than that for the same house in an open area (Heisler, 1986). It has been estimated that establishing 100 million mature trees around residences in the United States could save about $2 billion annually in reduced energy costs (Akbari *et al.*, 1988).

Proper tree placement near buildings is critical to maximize energy conservation. For example, it has been estimated that annual costs of air conditioning and heating for a typical residence in Madison, Wisconsin, would increase from $671 for an energy-efficient planting design to $700 for no trees and $769 for trees planted in locations that block winter sunlight and provide little summer shade (McPherson, 1987). In this instance, average annual energy savings with properly placed trees were about 4% more than with no trees and 13% more than with improperly placed trees.

3.2. Urban Hydrology

By intercepting and retaining or slowing the flow of precipitation reaching the ground, trees (in conjunction with soils) can play an important role in urban hydrologic processes. They can reduce the rate and volume of stormwater runoff, flooding damage, stormwater treatment costs, and other problems related to water quality. Estimates of runoff for an intensive storm in Dayton, Ohio, showed that the existing tree canopy (22%) reduced potential runoff by 7% and that a modest increase in canopy cover (to 29%) would reduce runoff by nearly 12% (Sanders, 1986). A study of the Gwynns Falls watershed in Baltimore indicated that heavily forested areas can reduce total runoff by as much as 26% and increase low-flow runoff by up to 13% compared with nontree areas in existing land cover and land-use conditions (Neville, 1996). Further, tree cover over pervious surfaces reduced total runoff by as much as 40%; while tree canopy cover over impervious surfaces had a limited effect on runoff.

In reducing runoff, trees function like retention/detention structures. In many communities, reduced runoff due to rainfall interception can also reduce costs of treating stormwater by decreasing the volume of water handled during periods of peak runoff (Sanders, 1986).

There may also be hydrologic costs associated with urban vegetation, particularly in arid environments where water is increasingly scarce or on reactive clay soils where water uptake by roots may cause localized-soil drying, shrinkage, and cracking. Increased water use in desert regions could alter the local water balance and various ecosystem functions that are tied to the desert water cycle. In addition, annual costs of water for sustaining vegetation can be twice as high as energy savings from shade for tree species that use large amounts of water, e.g., mulberry (McPherson and Dougherty, 1989). However, in Tucson, Arizona, 16% of the annual irrigation requirement of trees was offset by the amount of water conserved at power plants due to energy savings from trees (Dwyer *et al.*, 1992).

3.3. Urban Noise

Field tests have shown that properly designed plantings of trees and shrubs can significantly reduce noise. Leaves and stems reduce transmitted sound primarily by scattering it, while the ground absorbs sound (Aylor, 1972). For optimum noise reduction, trees and shrubs should be planted close to the noise source rather than the receptor area (Cook and Van Haverbeke, 1971). Wide belts (30 m) of tall, dense trees combined with soft ground surfaces can reduce apparent loudness by 50% or more (6 to 10 decibels) (Cook, 1978). For narrow planting spaces (<3 m wide), reductions of 3 to 5 decibels can be achieved with dense belts of vegetation, i.e., one row of shrubs along the road and one row of trees behind it (Reethof and McDaniel, 1978). Buffer plantings in these circumstances typically are more effective in screening views than in reducing noise.

Vegetation can also mask sounds by generating its own noise as wind moves tree leaves or as birds sing in the tree canopy. These sounds may make individuals less aware of offensive noises because people are able to filter unwanted noise while concentrating on more desirable sounds (Robinette, 1972). The perception of sounds by humans is also important. By visually blocking the sound source, vegetation can reduce individuals' perceptions of the amount of noise they actually hear (Anderson *et al.*, 1984). The ultimate effectiveness of plants in moderating noise is determined by the sound itself, the planting configuration used, the proximity of the sound source, receiver, and vegetation, as well as climatic conditions.

3.4. Urban Wildlife and Biodiversity

There are many additional benefits associated with urban vegetation that contribute to the long-term functioning of urban ecosystems and the well-being of urban residents. These include wildlife habitat and enhanced biodiversity. Urban wildlife can provide numerous benefits but also have detrimental effects (VanDruff *et al.*, 1995). Urban wildlife can serve as biological indicators of changes in the health of the environment (e.g., the decline of certain bird populations was traced to pesticides), and can provide economic benefit to individuals and society (VanDruff *et al.*, 1995). For example, bird feeding supports a $170 to $517 million American industry (DeGraff and Payne, 1975; Lyons, 1982).

Surveys have shown that most city dwellers enjoy and appreciate wildlife in their day-to-day lives (Shaw *et al.*, 1985). Among New York State's metropolitan residents, 73% showed an interest in attracting wildlife to their backyard (Brown *et al.*, 1979). Feelings of personal satisfaction from helping wildlife were the most frequently reported reason for feeding wildlife in backyards across America (Yeomans and Barclay, 1981). Detrimental wildlife effects include damage to plants and structures, droppings, threats to pets, annoyance to humans, animal bites, and transmission of diseases (VanDruff *et al.*, 1995).

Urbanization can sometimes lead to the creation and enhancement of animal and plant habitats, which, in turn, usually increases biodiversity. For example, tree species diversity and richness in Oakland, California, increased from an index value of about 1.9 (Shannon–Weiner diversity index value) and 10 species in 1850 to 5.1 and >350 species in 1988 (Nowak, 1993c). However, the introduction of new plant species into urban areas can lead to problems for managers in maintaining native plant structure, as exotic plants can invade and displace native species in forest stands. One example of exotic plant invasion in some areas of the northeastern United States is that by Norway maple (*Acer platanoides* L.) (Nowak and Rowntree, 1990). Also, altering vegetation structure in urban areas can change the prevalence of certain tree insects and diseases (Nowak and McBride, 1992) and could increase the potential for urban wildfires (East Bay Hills Vegetation Management Consortium, 1995).

Urban forests can act as reservoirs for endangered species. For example, 20 threatened or endangered faunal species and 130 plant species are listed for Cook County, the most populated county of the Chicago Metropolitan Area (Howenstine, 1993). In addition, urbanites are increasingly preserving, cultivating, and restoring rare and native species and ecosystems (Howenstine, 1993). A notable example of the involvement of a wide range of individuals and groups in the restoration and management of urban natural areas is the work of the Chicago Region Biodiversity Council, often known as Chicago Wilderness (2005).

Because of increased environmental awareness and concerns about quality of life and sustainability of natural systems, ecological benefits of the urban forest are likely to increase in significance over time (Dwyer *et al.*, 1992).

3.5 Phytoremediation

Trees and other plants show significant potential for remediating brownfields, landfills, and other contaminated sites by absorbing, transforming, and containing a number of contaminants (Westphal and Isebrands, 2001). More information about brownfields and the issues and opportunities that they present can be obtained from USEPA (2000) and De Sousa (2003).

4. Social and Economic Benefits and Costs of Urban Vegetation

In conjunction with the many effects of urban trees on the physical/biological environment, trees and associated forest resources can significantly influence the social and economic environment of a city. These influences range from altered aesthetic surroundings and increased enjoyment with everyday life to improved health and a greater sense of meaningful connection between people and the natural environment. The benefits and costs associated with these influences are highly variable within and among urban areas and often difficult to measure. Nevertheless, they reflect important contributions of trees and forests to the quality of life for urban dwellers.

4.1. Benefits to Individuals

Urban forest environments provide aesthetic surroundings and are among the most important features contributing to the aesthetic quality of residential streets and community parks (Schroeder, 1989). Perceptions of aesthetic quality and personal safety are related to features of the urban forest such as number of trees per acre and viewing distance (Schroeder and Anderson, 1984). Urban trees and forests provide significant emotional and spiritual experiences that are important in people's lives and can foster a strong attachment to particular places and trees (Chenoweth and Gobster, 1990; Dwyer *et al.*, 1991; Schroeder, 1991, 2002, 2004). A wide range of individual benefits has been associated with volunteer tree planting and care (Westphal, 1993). Volunteers continue to play an increasingly important role in urban forestry efforts, such as inventory (Bloniarz and Ryan, 1996), and Sommer (2003) encourages exploration of expanding opportunities for resident involvement in tree planting and care.

Nearby nature, even when viewed from an office window, can provide substantial psychological benefits that affect job satisfaction and a person's well-being (Kaplan, 1993). Reduced stress and improved physical health for urban residents have been associated with the presence of urban trees and forests in a number of environments. Living in a green environment has been associated with a wide range of individual benefits, including improved learning and behavior by children in urban areas (Taylor, Kuo, and Sullivan, 2001a, b; Wells, 2000). Experiences in urban parks have been shown to change moods and reduce stress (Hull, 1992a; Kaplan and Kaplan, 1989), and to provide privacy refuges (Hammitt, 2002). Hospital patients with window views of trees have been shown to recover significantly faster and with fewer complications than the patients without such views (Ulrich, 1984). In addition, tree shade reduces ultraviolet radiation and thus can help reduce health problems associated with increased ultraviolet radiation exposure, e.g., cataracts, skin cancer (Heisler *et al.*, 1995).

Many of the benefits associated with urban trees contribute to improved human health in a wide variety of ways, ranging from improved air quality to reduction of stress and interpersonal conflict. With increased concern over obesity and the need for changing lifestyles (e.g., more exercise) to reduce obesity, trees and forests are receiving increased attention as contributing to a solution. This solution ranges from providing environments that encourage exercise (e.g., playing in well-landscaped parks or walking/running along tree-lined streets and trails) to the actual exercise experienced by the many volunteers who work with trees and associated landscapes (Librett *et al.*, 2005). A comprehensive overview of the relationship of urban design to human health and condition concluded, "There are strong public health arguments for the incorporation of greenery, natural light, and visual and physical access to open space in homes and other buildings (Jackson, 2003)."

Along with the human health benefits, such as those outlined above in this section, some decreases in well-being and increases in health care costs may be associated with urban vegetation. This negative side to urban trees is associated with allergic reactions to plants, pollen, or associated animal and insects, diseases such as Lyme disease that are carried by wildlife, injuries from branch or tree failures, and a fear of trees, forests, and associated environments.

4.2. Benefits to Communities

Urban forests can make important contributions to the economic vitality and character of a city, neighborhood, or subdivision. It is no accident that many cities, towns, and subdivisions are named after trees (e.g., Oakland, Elmhurst, Oak Acres) and that many cities strive to be a "Tree City USA." Often, trees and forests on public lands–and on private lands to some extent–are significant "common property" resources that contribute to the economic vitality of an entire area (Dwyer *et al.*, 1992). The substantial efforts that many communities undertake to develop and enforce local tree ordinances and manage their urban forest resource attest to the substantial return that they expect from these investments.

A stronger sense of community and empowerment of inner-city residents to improve neighborhood conditions can be attributed to involvement in urban forestry efforts (Feldman and Westphal, 1999; Westphal, 1999, 2003). Active involvement in tree-planting programs has been shown to enhance a community's sense of social identity, self-esteem, and territoriality; it teaches residents that they can work together to choose and control the condition of their environment. Planting programs also can project a visible sign of change and provide the impetus for other community renewal and action programs (Feldman and Westphal, 1999; Westphal, 1999, 2003). Several studies have shown that participation in tree-planting programs influences individuals' perceptions of their community (Sommer *et al.*, 1994a, 1994b, 1995, 2003). Conversely, a loss of trees within a community can have a significant psychological effect on residents (Hull, 1992b). A useful framework for considering social benefits of urban and community forestry projects has been developed and illustrated with community examples (Westphal, 2003).

Urban trees and forests can help alleviate some of the hardships of inner-city living, especially for low-income groups (Dwyer *et al.*, 1992). Extensive research in inner-city areas of Chicago suggests that urban trees and forests contribute to stronger ties among neighbors, greater sense of safety and adjustment, more supervision of children in outdoor places, healthier patterns of children's play, more use of neighborhood common spaces, fewer incivilities, fewer property crimes, and fewer violent crimes (Kuo, 2003; Kuo *et al.*, 1998; Kuo and Sullivan, 2001a,b; Sullivan and Kuo, 1996).

While there is sometimes concern over the influence of trees and other vegetation in urban areas on the incidence of crime, research has provided management guidelines that can reduce the fear of crime in urban forest areas (Schroeder and Anderson, 1984; Michael and Hull, 1995).

Consumer behavior has been found to be positively correlated with streetscape greening, suggesting important benefits to commercial establishments and a basis for partnerships with the business community in urban forest planning and management (Wolf, 2003a, 2004). However, improper landscaping of business areas can have a negative impact by blocking business signs and/or reducing the attractiveness of the area.

Regardless of the community benefits derived from urban trees, tree planting and maintenance programs might be perceived by some people as an inappropriate use of resources because of the perception that funds for such efforts could be used to address what they see as more critical urban community problems.

4.3. Real Estate Values

The sales value of real estate reflects the benefits that buyers attach to attributes of the property, including vegetation on and near the property. A survey of sales of single-family homes in Athens, Georgia indicated that landscaping with trees was associated with an increase in sales prices of 3.5% to 4.5% (Anderson and Cordell, 1988). Builders have estimated that homes on wooded lots sell on an average for 7% more than equivalent houses on unwooded lots (Selia and Anderson, 1982, 1984). Research in Baton Rouge, Louisiana indicates that mature trees contributed about 2% of the home market (Dombrow *et al.*, 2000). A recent study in Athens, Georgia indicates that an additional percentage increase in relative tree cover is associated with an increase of $296 in residential value (Sydor *et al.*, 2005). A study of small, urban-wildland interface properties in the Lake Tahoe Basin indicates that forest density and health characteristics contributed between 5% and 20% to property values (Thompson *et al.*, 1999). Shopping centers often landscape their surroundings to attract shoppers, thereby increasing the value of the business and shopping center (Dwyer *et al.*, 1992).

Parks and greenways have been associated with increases in nearby residential property values (Corrill *et al.*, 1978; More *et al.*, 1988; Crompton, 2004). Some of these increased values have been substantial, and it appears that parks with "open space character" add the most to nearby property values. Part of the contribution to the value of residential property is associated with the view from that property. A study of the value of a view in the single-family housing market suggests that a good view adds 8% to the value of a single-family house (Rodriquez and Sirmans, 1994). A premium of 5% to 12% in housing prices in the Netherlands was associated with an attractive landscape view from the property (Luttik, 2000). Although this remains to be investigated, parks also may have a negative impact on local property values if these are perceived as unmaintained or a place where undesirable/criminal activities are concentrated.

Increased real estate values generated by trees also produce direct economic gains to the local community through property taxes. A conservative estimate of a 5% increase in residential property values due to trees converts to $25/year on a tax bill of $500 and is equivalent to $1.5 billion/year based on 62 million single-family homes in the United States (Dwyer *et al.*, 1992). However, from a homeowner's perspective, increased tax expense due to trees is an additional annual cost of owning a home.

4.4. Tree Value Formulas

Various approaches and formulas are used to estimate the value of individual trees (see Chapter 19). One of the most widely used is the Council of Tree and Landscape Appraisers' (2000), *Guide for Plant Appraisal*, which estimates the compensation that landowners should receive for the loss of a tree on their property. For smaller trees, the value is the replacement cost. For larger trees, the formula calculates tree value from measured tree variables and tree assessments by professionals. The species, diameter, location, and condition of the tree are an integral part of the assessment. Because the values estimated with the tree valuation formula are not necessarily tied to the functions that trees perform in the urban environment, they do not relate directly to the values associated with the environmental, social, and economic benefits from trees. An exception is a single study that suggested that the formula produced values that were similar to a tree's contribution to residential property values (Morales *et al.*, 1983).

Compensatory values represent compensation to owners for the loss of an individual tree and can be viewed as the value of the tree as a structural asset. Compensatory value is based on the structure in place as an asset, while the functional value is an annual value based on the various functions of the particular structure. Trees can have both positive (e.g., air pollution removal) and negative functional values (e.g., trees can increase annual building energy use in certain locations). Trees also have various maintenance costs, which are essential for maintaining tree health, human safety, and overall tree functional benefits. Management of urban forests is needed to enhance functional values and improve human health and well-being, and environmental quality in cities. Maximizing net functional benefits of the urban forest will lead to the greatest value to society (Nowak *et al.*, 2002a).

Based on the data from eight cities, overall citywide compensatory values ranged between \$23 and \$64/m^2 (\$2.1–\$5.9/feet^2) of tree cover. However, 75% of the city values were between \$27 and \$39 m^2 (\$2.5–\$3.6/feet^2) of cover. The total compensatory value for the urban forests of the 48 adjacent US states is estimated at \$2.4 trillion or \$630/tree (Nowak *et al.*, 2002a).

Urban forest compensatory values can be used to estimate actual or potential loss due to catastrophic agents. For example, the loss to the urban forest in Oakland, California, due to a large fire in 1991 was estimated at \$26.5 million (Nowak 1993b). Compensatory value of potential loss due to Asian longhorned beetle *(Anoplophora glabripennis)* infestation in various US cities ranges between \$72 million (Jersey City, New Jersey) and \$2.3 billion (New York, New York). The estimated maximum potential national urban impact of infestations by *A. glabripennis* is \$669 billion (Nowak *et al.*, 2001b).

4.5. Other Benefits and Costs of Urban Trees and Forests

The presence of urban trees and forests can make the urban environment a more pleasant place to live, work, and spend leisure time. A study of urbanites that use parks and forest preserves indicated that they were willing to pay extra to have trees and forests in recreation areas (Dwyer *et al.*, 1989). For example, they would be willing to pay an addition $1.60/visit to have a site that was "mostly wooded, some open grassy areas under trees" rather than "mowed grass, very few trees anywhere." The total contribution of trees in urban park and recreation areas to the value of the outdoor leisure and recreation experiences in the United States may exceed $2 billion/year (Dwyer, 1991).

A national survey indicated that drivers prefer trees as a screen of commercial developments along highways (Wolf, 2003b). Reduced driver aggression (Cackowski and Nasar, 2003) and stress recovery (Parsons, *et al.*, 1998) have also been associated with treed thoroughfares. These findings provide the basis for opportunities to incorporate urban forestry into the planning of high-speed urban transportation corridors (Wolf, 2003b).

Urban trees and forests often figure prominently in urban environmental education programs. The high visibility, variability, and complexity of urban forest ecosystems make an outstanding laboratory for environmental education. The lessons learned about forest ecosystems have implications for the management of public and private forest resources far beyond the city boundary (Dwyer and Schroeder, 1994).

Because trees and forests can increase the quality of the urban environment and make spending leisure time there more attractive, there can be a substantial saving in the amount of automobile fuel used because people do not need to drive long distances to reach recreation sites.

At the same time there are direct economic costs associated with urban trees. These include costs of planting, maintenance, management, and removal, as well as costs of damage from falling tree limbs and cracked sidewalks due to tree roots (Dwyer, 1995). However, these costs can be offset by economic benefits generated by trees. For example, homeowners may pay for tree care and driveway repair due to root damage, but receive savings on their utility bill from the energy conserving effects of the trees. At a larger scale, a municipality paying for street and park tree maintenance and management may receive increased tax revenues due to the contribution of trees to property values, and also may achieve savings in storm water management costs due to the influence of trees. Net benefits or costs need to be considered when developing urban vegetation designs or management plans.

5. Benefit–Cost Analyses

The wide range of important benefits and costs that may be associated with managing the urban forest and the significant interactions between the processes that produce important outcomes complicate the analysis of options available to urban forest resource managers. This complexity makes it difficult to predict the influence of trees on the urban environment for various vegetation designs and management options. In many instances, the location of trees with respect to other resources can make a substantial difference in the benefits that they provide, such as with building heating and cooling costs and the management of rights-of-way where improperly placed trees can greatly increase costs. Not all of the benefits are easily translated into monetary terms, and even when they are, it often is difficult to assess the incidence of benefits and costs, i.e., who pays and who gains? Trees planted on a residential property may provide benefits to others in the neighborhood and across the city in terms of aesthetics, reduced air temperatures, and improved air quality. Yet these very trees may present problems for one's neighbor by blocking solar heating through windows in the winter and making it difficult to grow flowers or a vegetable garden in the summer. The management of trees in public areas and rights-of-way often is intertwined with that of other resources, such as park and recreation facilities and programs, streets and roads, utilities, and other aspects of the urban infrastructure. When attempting benefit–cost analyses, one must be aware of these various interconnections, as well as the limitations of the information used in the analyses.

6. Implications for Planning, Design, and Management

It is clear that careful planning and design are critical to increasing the net benefits of trees and forests in urban environments. A change in species or location of trees with respect to each other or buildings and other components of the urban infrastructure can have a major impact on benefits and costs. Similarly, maintenance activities can greatly influence benefits and costs. It often is critical that forest resources are managed in the context of other aspects of the urban structure; including people, buildings, roads and streets, utility rights-of-way, recreation areas, and other open spaces.

Management plans must consider the potential of vegetation to improve individual site conditions or alleviate local problems (e.g., poor air quality, neighborhood revitalization) and design appropriate vegetation structure at the site with consideration of how individual sites interact across the landscape (i.e., the benefits at one site might lead to costs and benefits at other site). Determining the benefits and costs over the urban environment is a complex task that often calls for approaching problems at the landscape level (and sometimes extending beyond the urban system), particularly with respect to aesthetics, meteorology, pest problems, risk of fire, and air quality. Urban landscape designs and management plans must take account of these numerous interactions and the myriad of potential benefits and costs to implement appropriate strategies to maximize the net environmental, social, economic, and human health benefits of urban vegetation. In addition, careful attention must be given to the question of who gains and who pays as a result of forest management efforts across the urban landscape.

References

Akbari, H., Davis, S., Dorsano, S., Huang, J., and Winnett, S., 1992, *Cooling Our Communities: A Guidebook on Tree Planting and Light-colored Surfacing*, US Environmental Protection Agency, Washington, DC.

Akbari, H., Huang, J., Martien, P., Rainier, L., Rosenfeld, A., and Taha, H., 1988, The impact of summer heat islands on cooling energy consumption and CO_2 emissions, in *Proceedings of the 1988 Summer Study in Energy Efficiency in Buildings*, American Council for an Energy-Efficient Economy, Washington, DC.

Anderson, L. M., and Cordell, H. K., 1988, Influence of trees on residential property values in Athens, Georgia (USA): A survey based on actual sales prices, *Landscape Urban Plann.* **15**:153–164.

Anderson, L. M., Mulligan, B. E., and Goodman, L. S., 1984, Effects of vegetation on human response to sound, *J. Arboric.* **10**(2):45–49.

Aylor, D. E., 1972, Noise reduction by vegetation and ground, *J. Acoust. Soc. Am.* **51**(1):197–205.

Benjamin, M. T., Sudol, M., Bloch, L., and Winer, A. M., 1996, Low-emitting urban forests: A taxonomic methodology for assigning isoprene and monoterpene emission rates, *Atmos. Environ.* **30**(9):1437–1452.

Birdsey, R. A., and Heath, L. S., 1995, Carbon changes in US forests, in *Climate Change and the Productivity of America's Forests*, (L. A. Joyce, ed.), Gen. Tech. Rep. RM-271, US Department of Agriculture, Forest Service, Rocky Mountain Research Station, Fort Collins, CO, pp. 56–70.

Bloniarz, D. V., and Ryan, H. D. P., 1996, The use of volunteer initiatives in conducting urban forest inventories, *J. Arboric.* **22**(2):75–82.

Brasseur, G. P., and Chatfield, R. B., 1991, The fate of biogenic trace gases in the atmosphere, in *Trace Gas Emissions by Plants* (T. D. Sharkey, E. A. Holland, and H. A. Mooney, eds.), Academic Press, New York, pp. 1–27.

Brown, T. L., Dawson, C. P., and Miller, R. C., 1979, Interests and attitudes of metropolitan New York residents about wildlife, *Trans. 44th North. Am. Wildl. and Nat. Resour. Conf.* **44:**289–297.

Cackowski, J. M., and Nassar, J. L., 2003, The restorative effects of nature: Implications for driver anger and frustration, *Env. Behav.* **25:**736–751.

Cardelino, C. A., and Chameides, W. L., 1990, Natural hydrocarbons, urbanization, and urban ozone, *J. Geophys. Res.* **95**(D9):13,971–13,979.

Chicago Wilderness website at http://www.chicagowilderness.org, Last accessed May, 2005.

Chenoweth, R. E., and Gobster, P. H., 1990, The nature and ecology of aesthetic experiences in the landscape, *Landscape J.* **9:**1–18.

Cook, D. I., 1978, Trees, solid barriers, and combinations: Alternatives for noise control, in *Proceedings of the National Urban Forestry. Conference* (G. Hopkins, ed.), SUNY College of Environmental Science and Forestry, Syracuse, NY, pp. 330–339.

Cook, D. I., and Van Haverbeke, D. F., 1971, Trees and shrubs for noise abatement, in *Res. Bull.*, vol. 246, *Nebraska Agricultural Experiment Station,* Lincoln.

Corrill, M., Lillydahl, J., and Single, L., 1978, The effects of greenbelts on residential property values: Some findings on the political economy of open space, *Land Econ.* **54:**207–217.

Council of Tree and Landscape Appraisers, 2000, *Guide for Plant Appraisal*, International Society of Arboriculture, Urbana, IL.

Crompton, J. L., 2004, *The Proximate Principle: The Impact of Parks, Open Space and Water Features on Residential Property Values and the Property Tax Base,* National Recreation and Park Association, Ashburn, VA, 184pp. plus appendix.

De Sousa, C. A., 2003, Turning brownfields into green space in the City of Toronto, *Env. Behav.* **63:**181–198.

DeGraff, R. M., and Payne, B. R., 1975, Economic values of nongame birds and some research needs, *Trans. North. Am. Wildl. Nat. Resour. Conf.* **40:**281–287.

Dombrow, J., Rodriquez, M., and Sirmans, C. F., 2000, The market value of mature trees in single family housing markets, *Apprais. J.* 68: 39–43.

Dwyer, J. F., 1991, Economic value of urban trees, in *A National Research Agenda for Urban Forestry in the 1990's*, International Society of Arboriculture, Urbana, IL, pp. 27–32.

Dwyer, J. F., 1995, The significance of trees and their management in built environments, in *Trees and Building Sites: Proceedings of an International Conference held in the Interest of Developing a Scientific Basis for Managing Trees in Proximity to Buildings* (G. Watson, and D. Neely, eds.), International Society of Arboriculture, Savoy, IL, pp. 3–11.

Dwyer, J. F., Nowak, D. J., Noble, M. H., and Sisinni, S. M., 2000, *Assessing our Nation's Urban Forests: Connecting People with Ecosystems in the 21st Century*, Gen. Tech. Rep. PNW-460, US Department of Agriculture, Forest Service, Pacific Northwest Research Station, Portland, OR.

Dwyer, J. F., and Schroeder, H. W., 1994, The human dimensions of urban forestry, *J. For.* **92**(10):12–15.

Dwyer, J. F., Schroeder, H. W., and Gobster, P. H., 1991, The significance of urban trees and forests: Toward a deeper understanding of values, *J. Arboric.* **17:**276–284.

Dwyer, J. F., Schroeder, H. W., Louviere, J. J., and Anderson, D. H., 1989, Urbanites willingness to pay for trees and forests in recreation areas, *J. Arboric.* **15**(10):247–252.

Dwyer, J. F., McPherson, E. G., Schroeder, H. W., and Rowntree, R. A., 1992, Assessing the benefits and costs of the urban forest, *J. Arboric.* **18**(5):227–234.

East Bay Hills Vegetation Management Consortium, 1995, *Fire Hazard Mitigation Program and Fuel Management Plan for the East Bay Hills*, Report prepared by Amphion Environmental, Inc., Oakland, CA.

Feldman, R., and Westphal, L., 1999, Restoring participatory design and planning: Incorporating community empowerment as a tool for social justice, *Places* **12**(2):34–37.

Guenther, A., Zimmerman, P., and Wildermuth, M., 1994, Natural volatile organic compound emission rate estimates for US woodland landscapes, *Atmos. Environ.* **28**(6):1197–1210.

Hammitt, W. E., 2002, Urban forests and parks as privacy refuges, *J. Arboric.* **28**(1):19–26.

Heisler, G. M., 1986, Energy savings with trees, *J. Arboric.* **12**(5):113–125.

Heisler, G. M., 1990a, Mean wind speed below building height in residential neighborhoods with different tree densities, *ASHRAE Trans.* **96**(1):1389–1396.

Heisler, G. M., 1990b, Tree plantings that save energy, in *Proceedings of the 4th Fourth National Urban Forestry Conference* (P. Rodbell, ed.), American Forestry Association, Washington, DC, pp. 58–62.

Heisler, G. M., Grant, R. H., Grimmond, S., and Souch, C., 1995, Urban forests' cooling our communities?, in *Proceedings of the Seventh National Urban Forestry Conference* (C. Kollin, and M. Barratt, eds.), American Forests, Washington, DC, pp. 31–34.

Howenstine, W. L., 1993, Urban forests as part of the whole ecosystem, in *Proceedings of the 6th National Urban Forestry Conference* (C. Kollin, J. Mahon, and L. Frame, eds.), American Forests, Washington, DC, pp. 118–120.

Hull, R. B., 1992a, Brief encounters with urban forests produce moods that matter, *J. Arboric.* **18**(6):322–324.

Hull, R. B., 1992b, How the public values urban forests, *J. Arboric.* **18**(2):98–101.

Jackson, L. E., 2003, The relationship of urban design to human health and condition, *Landscape Urban Plan.* **64:**191–200.

Kaplan, R., 1993, Urban forestry and the workplace, in *Managing Urban and High Use Recreation Settings* (P. H. Gobster, ed.), Gen. Tech. Rep. USDA Forest Service, North Central Forest Experiment Station, NC-163, St. Paul, MN, pp. 41–45.

Kaplan, R., and Kaplan, S., 1989, *The Experience of Nature: A Psychological Approach*, Cambridge University Press, Cambridge, UK.

Kramer, P. J., and Kozlowski, T. T., 1979, *Physiology of Woody Plants*, Academic Press, New York.

Kuo, F. E., 2003, The role of arboriculture in healthy social ecology, *J. Arboric.* **29**(3):148–155.

Kuo, F. E., Bacaicoa, M., and Sullivan, W. C., 1998, Transforming inner-city landscapes: Trees, sense of safety, and preference, *Enviorn. Behav.* **30**(1):28–59.

Kuo, F. E., and Sullivan, W. C., 2001a, Environment and crime in the inner city: Does vegetation reduce crime?, *Enviorn. Behav.* **33**(3):343–365.

Kuo, F. E., and Sullivan, W. C., 2001b, Aggression and violence in the inner city: Impacts of environment via mental fatigue, *Enviorn. Behav.* **33**(4):543–571.

Lenschow, D. H. (ed.), 1986, *Probing the Atmospheric Boundary Layer*, American Meteorological Society, Boston, MA.

Librett, J., Yore, M., Buchner, D. M., and Schmid, T. L., 2005, Take pride in America's health: Volunteering as a gateway to physical activity, *Am J. Health Ed.* **36**(1):8–13.

Luley, C. J., and Bond, J., 2002, *A Plan to Integrate Management of Urban Trees into Air Quality Planning*, Report to Northeast State Foresters Association, Davey Resource Group, Kent, OH, 73pp.

Luttik, J., 2000, The value of trees, water, and open space as reflected by house prices in the Netherlands, *Landscape Urban Plan.* **48:**161–167.

Lyons, J. R., 1982, Nonconsumptive wildlife-associated recreation in the US: Identifying the other constituency, *Trans. North. Am. Wildl. Nat. Resour. Conf.* **47:**677–685.

McPherson, E. G., 1987, *Effects of Vegetation on Building Energy Performance*, Ph.D. Discussion, SUNY College of Environmental Science and Forestry, Syracuse, NY.

McPherson, E. G., and Dougherty, E., 1989, Selecting trees for shade in the Southwest, *J. Arboric.* **15:**35–43.

McPherson, E. G., 1998, Atmospheric carbon dioxide reduction by Sacramento's urban forest, *J. Arboric.* **24**(4):215–223.

Michael, S. E., and Hull, R. B., 1995, Effects of vegetation on crime in urban parks, *Arborist News*, February, p. 45.

Morales, D. J., Micha, R. R., and Weber, R. L., 1983, Two methods of valuating trees on residential sites, *J. Arboric.* **9:**21–24.

More, T. A., Stevens, T., and Allen, P. G., 1988, Valuation of urban parks, *Landscape Urban Plan.* **15:**139–152.

Murray, F. J., Marsh, L., and Bradford, P. A., 1994, *New York State Energy Plan,* Vol. II: *Issue Reports*, New York State Energy Office, Albany.

Myrup, L. O., McGinn, C. E., and Flocchini, R. G., 1991, An analysis of microclimate variation in a suburban environment, in *Seventh Conference of Applied Climatology*, American Meteorological Society, Boston, MA, pp. 172–179.

Neville, L. R., 1996, *Urban Watershed Management: The Role of Vegetation*, Ph.D. Discussions., SUNY College of Environmental Science and Forestry, Syracuse, NY.

Nowak, D. J., 1993a, Atmospheric carbon reduction by urban trees, *J. Environ. Manage.* **37**(3):207–217.

Nowak, D. J., 1993b, Compensatory value of an urban forest: An application of tree-value formula, *J. Arboric.* **19**(3):173–177.

Nowak, D. J., 1993c, Historical vegetation change in Oakland and its implications for urban forest management, *J. Arboric.* **19**(5):313–319.

Nowak, D. J., 1994a, Air pollution removal by Chicago's urban forest, in *Chicago's Urban Forest Ecosystem: Results of the Chicago Urban Forest Climate Project* (E. G. McPherson, D. J. Nowak, and R. A. Rowntree, eds.), Gen. Tech. Rep. NE-186, USDA Forest Service, Northeastern Forest Experiment Station, Radnor, PA, pp. 63–81.

Nowak, D. J., 1994b, Atmospheric carbon dioxide reduction by Chicago's urban forest, in *Chicago's Urban Forest Ecosystem: Results of the Chicago Urban Forest Climate Project* (E. G. McPherson, D. J. Nowak, and R. A. Rowntree, eds.), Gen. Tech. Rep. NE-186, USDA Forest Service, Northeastern Forest Experiment Station, Radnor, PA, pp. 83–94.

Nowak, D. J., 1995, Trees pollute? A "TREE" explains it all, in *Proceedings 7th National Urban Forestry Conference* (C. Kollin, and M. Barratt, eds.), American Forests, Washington, DC, pp. 28–30.

Nowak, D. J., Civerolo, K. L., Rao, S. T., Sistla, S., Luley, C. J., and Crane, D. E., 2000, A modeling study of the impact of urban trees on ozone, *Atmos. Environ.* **34**:1601–1613.

Nowak, D. J., and Crane, D. E., 2000, The Urban Forest Effects (UFORE) Model: Quantifying urban forest structure and functions, in *Proceedings of the Integrated Tools for Natural Resources Inventories in the 21st Century, IUFRO Conference, 16–20 August 1998, Boise, ID* (M. Hansen, and T. Burk eds.), Gen. Tech. Rep.NC-212, US Department of Agriculture, Forest Service, North Central Research Station, St. Paul, MN, pp. 714–720.

Nowak, D. J., and Crane, D. E., 2002, Carbon storage and sequestration by urban trees in the USA, *Environ. Poll.* **116**(3):381–389.

Nowak, D. J., Crane, D. E., and Dwyer, J. F., 2002a, Compensatory value of urban trees in the United States, *J. Arboric.* **28**(4):194–199.

Nowak, D. J., Crane, D. E., Stevens, J. C., and Ibarra, M., 2002b, *Brooklyn's Urban Forest*, Gen. Tech. Rep. NE-290, US Department of Agriculture, Forest Service, Northeastern Research Station, Newtown Square, PA.

Nowak, D. J., and McBride, J. R., 1992, Differences in Monterey pine pest populations in urban and natural forests, *For. Ecol. Manage.* **50**:133–144.

Nowak, D. J., Noble, M. H., Sisinni, S. M., and Dwyer, J. F., 2001a, Assessing the US urban forest resource, *J. For.* **99**(3):37–42.

Nowak, D. J., Pasek, J., Sequeira, R., Crane, D. E., and Mastro, V., 2001b, Potential effect of *Anoplophora glabripennis* (Coleoptera: Cerambycidae) on urban trees in the United States, *J. Econom. Entomol.* **94**(1):116–122.

Nowak, D. J., and Rowntree, R. A., 1990, History and range of Norway maple, *J. Arboric.* **16**(11):291–296.

Nowak, D. J., McHale P. J., Ibarra, M., Crane, D., Stevens, J., and Luley, C., 1998, Modeling the effects of urban vegetation on air pollution, in *Air Pollution Modeling and Its Application XII* (S. Gryning, and N. Chaumerliac, eds.), Plenum Press, New York, pp. 399–407.

Nowak, D. J., Stevens, J. C., Sisinni, S. M., and Luley C. J., 2002c, Effects of urban tree management and species selection on atmospheric carbon dioxide, *J. Arboric.* **28**(3):113–122.

Nowak, D. J., Walton, J. T., Dwyer, J. F., Kaya, L. G., and Myeong, S., 2005, The increasing influence of urban environments on US forest management, *J. Forestry* 103(8): 377–382.

Parsons, R., Tassinary, L. G., Ulrich, R. S., Hebl, M. R., and Grossman-Alexander, M., 1998, The view from the road: Implications for stress recovery and immunization, *J. Enviorn. Psych.* **18**(2):113–140.

Reethof, G., and McDaniel, O. H., 1978, Acoustics and the urban forest, in *Proceedings of the National Urban Forestry Conference* (G. Hopkins, ed.), SUNY College of Environmental Science and Forestry, Syracuse, NY, pp. 321–329.

Robinette, G. O., 1972, *Plants/People/ and Environmental Quality*, USDI National Park Service, Washington, DC.

Rodriquez, M., and Sirmans, C. F., 1994, Quantifying the value of a view in single-family housing markets, *Apprais. J.,* 62(4): 600–603.

Rolfe, G. L., 1974, Lead distribution in tree rings, *For. Sci.* **20**(3):283–286.

Sanders, R. A., 1986, Urban vegetation impacts on the urban hydrology of Dayton Ohio, *Urban Ecol.* **9:**361–376.

Schroeder, H. W., and Anderson, L. M., 1984, Perception of personal safety in urban recreation sites, *J. Leis. Res.* **16**(2):178–194.

Schroeder, H. W., 1989, Environment, behavior, and design research on urban forests, in *Advances in Environment, Behavior, and Design* (E. H. Zube, and G. L. Moore, eds.), Plenum Press, New York, pp. 87–107.

Schroeder, H. W., 1991, Preference and meaning of arboretum landscapes: Combining quantitative and qualitative data, *J. Environ. Psych.* **11:**231–248.

Schroeder, H. W., and Anderson, L. M., 1984, Perception of personal safety in urban recreation sites, *J. Leisure Res.* **16:**178–194.

Schroeder, H. W., 2002, Experiencing nature in special places, *J. For.* **100**(5):8–14.

Schroeder, H. W., 2004. *Special Places in the Lake Calumet Area*, USDA Forest Service North Central Research Station, General Technical Report 249, St Paul MN, 23pp.

Scott, K. I., Simpson, J. R., and McPherson, E. G., 1999, Effects of tree cover on parking lot microclimate and vehicle emissions, *J. Arboric.* **25**(3):129–142.

Selia, A. F., and Anderson, L. M., 1982, Estimating costs of tree preservation on residential lots, *J. Arboric.* **8:**182–185.

Selia, A. F., and Anderson, L. M., 1984, Estimating tree preservation costs on urban residential lots in metropolitan Atlanta, *Georgia For. Res. Pap.* No. 48. MACON, 6A. 6pp.

Sharkey, T. D., and Singsaas, E. L., 1995, Why plants emit isoprene, *Nature* **374**(27 April):769.

Shaw, W. W., Magnum, W. R., and Lyons, J. R., 1985, Residential enjoyment of wildlife resources by Americans, *Leisure Sci.* **7:**361–375.

Smith, W. H., 1990, *Air Pollution and Forests*, Springer-Verlag, New York.

Sommer, R., Learey, F., Summit, J., and Tirell, M., 1994a, Social benefits of resident involvement in tree planting: Compressions with developer-planted trees, *J. Arboric.* **20**(6):323–328.

Sommer, R., Learey, F., Summitt, J., Tirrell, M., 1994b, Social benefits of residential involvement in tree planting, *J. Arboric.* **20**(3):170–175.

Sommer, R., Summitt, J., Learey, R., and Tirrell, M., 1995, Social and educational benefits of a community shade tree program: A replication, *J. Arboric.* **21**(5):260.

Sommer, R., 2003, Trees and human identity, in Identity and the natural environment: The psychological significance of nature (S. Clayton, and S. Opotow, eds.), MIT Press, Cambridge and London, pp. 179–204.

Sullivan, W. C., and Kuo, F. E., 1996, Do trees strengthen urban communities, reduce domestic violence? *Arbor. News* **5**(2):33–34.

Souch, C. A., and Souch, C., 1993, The effect of trees on summertime below canopy urban climates: A case study, Bloomington, Indiana, *J. Arboric.* **19**(5):303–312.

Sydor, T., Bowker, J. M., Newman, D. H., and Cordell, H. K., 2005, *Valuing Trees in a Residential Setting: Revisiting Athens*, Clarke County, Georgia, draft paper, 15pp.

Taha, H., 1996, Modeling impacts of increased urban vegetation on ozone air quality in the South Coast Air Basin, *Atmos. Environ.* **30**(20):3423–3430.

Taylor, A. F., Kuo, F. E., and Sullivan, W. C., 2001a, Coping with ADD: The surprising connection to green play settings, *Enviorn. Behav.* **33**(1):54–77.

Taylor, A. F., Kuo, F. E., and Sullivan, W. C., 2001b, Views of nature and self-discipline: Evidence from inner-city children, *J. Enviorn. Psych.* **21:**49–63.

Thompson, R., Hanna, R., Noel, J., and Piirto, D., 1999, Valuation of tree aesthetics on small urban-interface properties, *J. Arboric.* **25**(5):225–234.

Tingey, D. T., Turner, D. P., and Weber, J. A., 1991, Factors controlling the emissions of monoterpenes and other volatile organic compounds, in *Trace Gas Emissions by Plants* (T. D. Sharkey, E. A. Holland, and H. A. Mooney, eds.), Academic Press, New York, pp. 93–119.

Ulrich, R. S., 1984, View through a window may influence recovery from surgery, *Science,* **224:**420–421.

US Environmental Protection Agency, 1991, Nonroad engine and vehicle emission study report, USEPA Office of Air and Radiation ANR-43. EPA-21A-2001, Washington, DC.

US Environmental Protection Agency, 2004, *Incorporating Emerging and Voluntary Measures in a State Implementation Plan (SIP)*, USEPA Air Quality Strategies and Standards Division, Office of Air Quality Planning and Standards, Research Triangle Park, NC.

US Environmental Protection Agency, 2000, *Introduction to Phytoremedia*tion, US EPA Office of Research and Development, Washington, DC.

VanDruff, L. W., Leedy, D. L., and Stearns, F. W., 1995, Urban wildlife and human well-being, in *Urban Ecology as the Basis of Urban Planning* (H. Sukopp, M. Numata, and A. Huber, eds.), SPB Academic Publishing, Amsterdam, pp. 203–211.

Wells, N. M., 2000, At home with nature: Effects of "greening" on children's cognitive functioning, *Enviorn. Behav.* **32**(5):775–795.

Westphal, L. M., 1993, Why trees? Urban forestry volunteers values and motivations, in *Managing Urban and High Use Recreation Settings* (P. H. Gobster, ed.), Gen. Tech. Rep. NC-163, USDA Forest Service, North Central Forest Experiment Station, St. Paul, MN, pp. 19–23.

Westphal, L. M., 1999, Empowering people through urban greening projects: Does it happen? in *Proceedings of the 1999 National Urban Forestry Conference* (C. Kollin, ed.), American Forests, Washington, D.C, pp. 60–63.

Westphal, L. M., and Isebrands, J. G., 2001, Phytoremediation of Chicago's brownfields: Consideration of ecological approaches to social issues, in *Brownfields 2001 Proceedings*, Chicago IL, http://www.brownfields2001.org/proceedings/BB-11-02.pdf, Last accessed April, 2005.

Westphal, L. M., 2003, Urban greening and social benefits: A study of empowerment outcomes, *J. Arboric.* **29**(3):137–147.

Wolf, K. L., 2003a, Public response to the urban forest in inner-city business districts, *J. Arboric.* **29**(3):117–126.

Wolf, K. L., 2003b, Freeway roadside management: The urban forest beyond the white line, *J. Arboric.* **29**(3):127–136.

Wolf, K. L., 2004, Trees and business district preferences: A case study of Athens, Georgia US, *J. Arboric.* **30**(6):336–346.

Yeomans, J. A., and Barclay, J. S., 1981, Perceptions of residential wildlife programs, *Trans. North. Am. Wildl. Nat. Resour. Conf.* **46**:390–395.

Ziegler, I., 1973, The effect of air–polluting gases on plant metabolism, in *Environmental Quality and Safety,* vol. 2, Academic Press, New York, pp. 182–208.

Chapter 3

Laws and Ordinances

Arboriculture Law in the Northeast

Victor D. Merullo

1. General Tree Law

This section is intended to be a brief overview of some of the major principles of tree law as limited to those states under review. In this section, we will look at the rights and duties of landowners, neighbors, municipalities, and public utilities. We will also discuss the different types of damages and their availability.

Let us begin by looking at the duties of landowners. The landowner is held to duty of care equal to that of the common law negligence standard; if the owner knew or should have known of the danger and failed to correct it, he or she will be held liable. *Ivancic v. Olmstead* (1985) was an action brought against a landowner for injuries to the plaintiff caused by a falling tree branch. In this case, the court said, the plaintiff

> must demonstrate that the defendant's tree was in a condition dangerous to the persons on the adjoining property; that the condition of the tree was a proximate cause of his injury; that defendant realized or in the exercise of reasonable care, should have realized its danger to plaintiff; and that defendant knew of the condition for a period of time prior to plaintiffs' injury to permit defendant in the exercise of reasonable care to have corrected it,

in order to recover from the defendant landowner (*Ivancic v. Olmstead*, 1985, p. 1210). This is a clear statement of the negligence standard by which all landowners are bound.

Urban and rural landowners have different levels of duty with respect to inspection of their land for hazardous trees. Urban landowners have a general duty to inspect their land to be sure that the trees located on it do not become a danger to their neighbors or

Victor D. Merullo Merullo, Reister, and Swinford, Co., Columbus, Ohio 43206.
Urban and Community Forestry in the Northeast, 2nd ed., edited by, J. E. Kuser.

neighbors' land. Rural landowners are not charged with the same duty. The courts have found it unreasonable to require a rural landowner to inspect each tree on his or her property. This process could take many days, depending on the amount of property he or she owned. Courts have simply said that an urban owner has a higher chance of noticing a defective tree in his or her yard than a rural owner whose "yard" consists of many acres, which he or she does not see every day. Urban trees also have a higher probability of causing damage to adjoining property or passing people than do trees in rural areas. But still, it is important to remember, rural or urban, if the owner has knowledge of the defective tree and that tree causes damage to another, the landowner will be liable for those damages. In *Hay v. Norwalk Lodge No. 730 BPOE* (1951), the court held that no duty existed for a rural landowner to inspect the trees on his land that abutted the rural highway; but if that landowner had actual, constructive knowledge of the tree's dangerous condition, he would be liable for damages to the plaintiff (*Hay v. Norwalk Lodge*, 1951). However, in *Toomey v. State of Connecticut* (1994), the Superior Court of Connecticut expanded the duty of a rural landowner. Mrs Toomey brought suit against the State of Connecticut when a tree limb fell on her car while the family was out for a drive in the country. The limb injured her and her daughter and killed her husband and mother. Roadside landowners in the State of Connecticut have long been required to inspect and monitor roadside trees that may pose a threat to passerby. The court held that as the risk of harm increases the duty to inspect a given tree increases. To determine the risk of harm, the court looked at several factors, for example, the amount of traffic on the road; the age, type, location, and health of the tree; the type and extent of possible dangers; and the expense and burden of inspection and maintenance of such trees. The *Toomey* (1994) decision basically expanded the duty of care owed by the rural property owner regarding trees growing on his or her property located adjacent to streets and highways.

Landowners are typically not liable for damages caused by acts of God or natural growth. *Black's Law Dictionary* (1990, p. 33) defines an act of God as "an act occasioned exclusively by the forces of nature without the interference of human agency". Using this definition, the courts have found the landowner liable when the damages could have been prevented with reasonable diligence or were contributed to by human action and thereby not caused solely by the forces of nature. For example, in *Short v. Kerr* (1937), the job foreman was killed when a gust of wind blew the tree he and his workers were uprooting in the wrong direction. The court held the employer liable for the foreman's death because it was not caused by the forces of nature alone but contributed by the human action of uprooting the tree.

Many disputes over trees spring from the question of ownership. Ownership of a tree is generally determined by the position of its trunk. A tree that is completely situated on the grounds of one person is obviously owned by that one person. A tree whose trunk is situated on the boundary of adjoining properties is jointly owned by both landowners. Maintaining such a boundary tree can be tricky. For example, *Cooke v. McShane* (1928), where the court held that a hedge that both landowners accepted as a boundary fence was owned by both adjoining owners and each could trim only his side of the hedge as long as that trimming did not interfere with the other owner's side of the hedge. In short, boundary line trees are split down the middle, and one owner can prune his side of the tree so long as it does not cause damage to or interfere with the adjoining owner's enjoyment of his side of the tree.

A tree whose trunk is situated wholly on the land of one landowner yet whose branches and/or roots invade the land of an adjoining landowner, may cause many problems in the neighborhood. Adjoining landowners have a right to be free from bothersome, encroaching limbs and roots. Most states follow one of the following rules when deciding what the adjoining land owner can do to protect his property from such an invasion (Table 1). First, the Massachusetts rule allows the adjoining landowner to trim roots and limbs off to his property line; this is a self-help theory. In a state that follows the Massachusetts rule, the owner must show that the encroaching limbs and roots are a nuisance before the court will intervene in the situation. Other states follow the restatement rule, taken from the Restatement of Torts 2d § 197, which says:

> (1) One is privileged to enter or remain on land in the possession of another if it is or reasonably appears to be necessary to prevent serious harm to
> (a) the actor, or his land or chattels, or
> (b) the other or a third person, or the land or chattels of either, unless the actor knows or has reason to know that the one for whose benefit he enters is unwilling that he shall take such action.
>
> (2) Where the entry is for the benefit of the actor or a third person, he is subject to liability for any harm done in the exercise of the privilege stated in Subsection (1) to any legally protected interest of the possessor in the land or connected with it, except where the threat of harm of avert which the entry is made caused by the tortious conduct or contributory negligence of the possessor.

Table 1. Applicable State Rules

Massachusetts or common law rule
Connecticut
Illinois
Iowa
Massachusetts
New Jersey
New York
Ohio
Rhode Island
Vermont
Delaware
Michigan
Minnesota
New Hampshire
Maine
Pennsylvania
Maryland
Wisconsin
West Virginia
Restatement rule or Hawaii rule
Indiana
Missouri

Under the restatement rule, any damages caused by artificial conditions may be taken to court. Some states, such as Hawaii, have special rules for these situations. The Hawaii rule states that if the tree poses a danger beyond simply shading or dropping leaves, the adjoining owner may demand that the landowner cut the tree back. If the landowner refuses, the adjoining owner may have the trees trimmed and send the bill to the landowner. In many cases, courts are reluctant to allow recovery by the injured landowner unless that landowner first attempted to use the self-help rule adopted by his state. For example, in *Loggia v. Grobe* (1985), the plaintiff brought an action for damages to his patio, caused by the roots of the neighbor's tree. The court held for the defendant saying that the plaintiff should not be allowed to recover when the intrusion was not intentional, the defendant had no notice of the defect, and the plaintiff had not attempted to use the self-help remedy afforded to him by the state.

Property owners in every state have some right to self-help. The rationale for such rules is to keep neighborhood disputes from clogging up the judicial system with petty grievances. Also, there is a better chance that the neighborhood will remain peaceful if the neighbors settle their differences together, outside of the judicial system. There are some generally recognized limits on the right to self-help. First, the adjoining landowner can only trim up to the property line. Second, if the trimming requires an invasion of the landowner's property, the trimming party must get the landowner's permission to enter the property (unless the encroachment threatens immediate harm to the adjoining landowner). The trimming party must be careful not to damage the tree.

If the tree is damaged, the landowner will then have a cause of action against the trimmer. Generally, if someone cuts down, removes, or hurts a tree without permission, the owner of the tree can sue for at least the amount of the actual loss or replacement value of the tree. In order for the owner to recover, there must be an actual damage to the tree, and the tree must not have been a nuisance or danger to others. Actual damages are measured by the tree's health, not its appearance. The tree does not have to be a work of art after the cutting, as long as that cutting does not endanger the tree's life or its ability to produce fruit. A nuisance is an "activity which arises from unreasonable, unwarranted, unlawful use by a person of his own property, working [causing] obstruction or injury to right of another, or to the public, and producing a such material annoyance, inconvenience and discomfort that law will presume resulting damage" (Black's Law Dictionary, 1990, p. 1065). If the tree was a nuisance, the courts will usually find that the tree needed to be destroyed and therefore no damages will be allowed.

When the damages to person or property are caused by a municipality, governmental immunity may stand in the way of recovery. The actions of the municipality must be analyzed in light of what is called the governmental proprietary distinction. This distinction says that a municipality is not liable for damages when performing a governmental function, unless this immunity is waived by statute. A governmental function is an activity that a municipality performs pursuant to a state requirement. A proprietary function is an activity that can be performed by a private citizen but is performed by the municipality at its discretion. When a municipality performs a proprietary function, it steps into the shoes of a private person in like circumstances (*Toomey v. State of Connecticut*, 1994). In such a situation, the state will be liable if a private person in that same situation would be held liable.

Liability of a public utility company for trimming or cutting a tree is generally determined by a three-part analysis (Merullo and Valentine, 1992). First, the court must determine whether the utility company had authority to trim or remove the trees. This authority must be given by the state. If this authority is found, the second step is to determine whether an easement exists that allows the company the right to enter the landowner's property. If an easement exists, the court must make a determination as to whether the cutting was an undue burden on the owner's land. This decision is generally made by considering the reasonableness and necessity of the cutting (Merullo and Valentine, 1992). This "reasonable and necessary" standard has been widely adopted throughout the states.

Once liability has been found, the question becomes one of damages. There are four major types of damages possible in these types of cases: (1) compensatory damages that cover the actual loss of the trees and other injuries flowing therefrom; (2) statutory damages that are extra damages the cutter is mandated to pay by statute if the cutting is willful or wanton; (3) punitive damages that are awarded when the cutter's conduct was especially outrageous or malicious; and (4) actual damages that are damages awarded in compensation for actual and real loss to the party. Some states have statutes that provide for criminal penalties such as arrest, fines, and even jail time. Damages will generally be awarded for the reasonable cost of replacing the injured trees when the depreciation caused to the property by the injury to the trees is insignificant, when the trees are of significant aesthetic value, or when replacement in kind, size, and condition of the trees is feasible. For more details on the damages allowed by your sate, check your state code (see also Table 2 for a brief overview of some state damages statutes).

Now that we have covered some general principles of this area of law, let us take a look at how some states give their municipalities the ability to create programs for the care and maintenance of their public trees.

Table 2. State Damage Limitations

Double damages
Pennsylvania (18 Penn. Con. Stat. 1107)
Rhode Island (RI Gen. Laws 34-20-1; sometimes tripled)
Wisconsin (WI Stat. Ann. 26.09)
Triple damages
Connecticut (Conn. Gen. Stat. Ann. 52-560)
Delaware (DE Code Ann. 25-1401)
Illinois (IL Stat. Ann. Ch. 96 1/2 § 9402)
Iowa (IA Code Ann. 658. 4)
Maine (14 ME Rev. Stat. Ann. 7552)
Maryland (MD Code Ann. N.R. 5-409)
Minnesota (MN Stat. Ann. 561.04)
Missouri (MO Stat. Ann. 537.340)
New York (Con. Laws NY Ann. R.P.A.P.L. 861)
Ohio (O.R.C. Ann. 901.51)
Vermont (13 VT Stat. Ann. 3606)
West Virginia (WV Code Ann. 61-3-48a)

(*continued*)

Table 2. (*Continued*)

5 × Damages
New Hampshire (NH Rev. Stat. Ann. 539.1)
No statutory damages
Indiana
Fines or jail sentences for damaging a tree
Delaware (DE Code Ann. 7-3301)
Idaho (ID Code Ann. 18-7021)
Massachusetts (MA Gen. Laws Ann. 750.382)
Michigan (MI Comp. Laws Ann. 750.382)
New Jersey (Stat. Ann. 2c.18-5)
Ohio (O.R.C. 901.99)
Rhode Island (RI Gen. Laws 11-44-2)
Virginia (Code of VA 18.2-140)
Wisconsin (WI Stat. Ann. §§ 814.04, 26.05)

2. Enabling Statutes for Municipal Commissions to Care for Trees

Municipalities are generally liable for the care and maintenance of shade trees on public grounds and along roadways. This is a very large task, not easily done without some type of organization devoted strictly to its completion. In this section, we will discuss some state statutes that enable the municipalities within that state to establish such organizations (Table 3). These statutes may allow for the formation of one or more of the following types of organizations: shade tree commissions, urban/community forestry commissions, or the appointment of tree wardens. The organization names and the language of the statutes may vary, but the underlying purpose of each is still the same: the care and maintenance of publicly owned shade trees.

Let us begin with the organizations called shade tree commissions. New Jersey and Pennsylvania are two states that specifically allow for the formation of shade tree commissions. The New Jersey Legislature recently has enacted the New Jersey Shade Tree and Community Forestry Assistance Act. The act states as its purpose the following:

a. Assist local governments and shade tree commissions in establishing and maintaining community forestry programs and in encouraging persons to engage in appropriate and approved practices with respect to tree management and care;
b. Advise local governments and shade tree commissions in the development and coordination of policies, programs and activities for the promotion of community forestry;
c. Provide grants to local governments and shade tree commissions applying for assistance in the development and implementation of a comprehensive community forestry plan approved pursuant to section 7 of P.L. 1996 c. 135 to the extent monies are appropriated or otherwise made available therefore;
d. Educate citizens on the importance of trees and forests and their role in the maintenance of a clean and healthy environment;
e. Provide technical assistance, planning and analysis for projects related to community forestry;
f. Provide training assistance to local governments and shade tree commissions regarding community forestry issues as tree diseases, inspect programs and tree planting and maintenance; and
g. Provide volunteer opportunities for the State's citizens and organizations interested in community forestry activities. (§4 Publ. L. 1996 c. 135)

Table 3. Important Enabling Statutes for Municipalities

Connecticut
- CT ST § 7-131: Municipal forests
- CT ST § 23-58: Tree wardens; appointments; compensation; supervision
- CT ST § 23-59: Powers and duties of tree wardens

Indiana
- IN ST 14-12-3-11: Community forestry grants

Illinois
- IL ST CH 30 § 735/1: Short title (Urban and Community Reforestation Act)
- IL ST CH 30 § 735/3: Administration of the act
- IL ST CH 30 § 735/4: Duties of department

Maine
- ME ST 12 § 8002: Bureau of forestry powers and duties
- ME ST T 30-A § 3261: Conservation Commissions
- ME ST T 30-A § 3282: Appointment and duties of tree wardens
- ME ST T 30-A § 3283: Removal of trees

Massachusetts
- MA ST 41 § 69G: Office of lands and natural resources; establishment; powers and duties; directors and assistants
- MA ST 41 § 106: Appointment term
- MA ST 45 § 5: Powers and duties of boards of park commissioners
- MA ST 87 § 2: Powers of tree wardens
- MA ST 87 § 3: Cutting public trees; public hearing; damages to fee owner
- MA ST 132 § 13: Local superintendents of shade tree management and pest control; appointment; notice; assistants

Maryland
- MD Stat Res § 5-424: Definitions
- MD Stat Res § 5-426: Department to establish and administer program
- MD Stat Res § 5-427: Local programs

New Hampshire
- NH ST § 31:112: Management
- NH ST § 231:139: Tree wardens
- NH ST § 231:141: Acquisition of trees
- NH ST § 231:142: Marketing of trees acquired

New Jersey
- Publ. L. 1996 c. 135: New Jersey Shade Tree and Community Forestry Assistance Act

New York
- NY Village § 17-1732: Trees, shrubs, and shade tree commissions
- NY Envir Conser § 53-0303: Definitions
- NY Envir Conser § 53-0305: Designation of commissioner
- NY Envir Conser § 53-0307: Funding of department; municipal authorization for local urban forestry and arboriculture programs

Ohio
- OH ST § 1503.001: Forest conservation and development
- OH ST § 1503.35: Forestry Development Fund

(*continued*)

Table 3. (*Continued*)

Pennsylvania	
	PA ST 53 P.S. § 38801: Shade Tree Commission
	PA ST 53 P.S. § 38803: Powers may be vested in park commission
	PA ST 53 P.S. § 38804: Powers of commission
	PA ST 53 P.S. § 38805: Report of commission
	PA ST 53 P.S. § 47724: Power of commission
	PA ST 53 P.S. § 53853: Power of commission
	PA ST 53 P.S. § 58024: Hiring employees; legislative power of commission
Rhode Island	
	RI ST § 2-14-2: Appointment by town council or mayor
	RI ST § 2-14-3: Term of office; vacancies
	RI ST § 2-14-4: Deputies; compensation of wardens and deputies
	RI ST § 2-14-5: Trees under control of warden; entry on private property
	RI ST § 2-14-6: Town or city forest areas; supervision by department of environmental management
	RI ST § 2-14-7: Powers and duties of wardens
	RI ST § 2-14-8: Removal and pruning of trees
	RI ST § 2-14-9: Suppression of pests and diseases
	RI ST § 2-14-11: Application to Providence
Wisconsin	
	WI ST 23.097: Urban forestry grants

Some of New Jersey's municipalities already had shade tree commissions, but many had done away with them, fearing added tort liability if they continued to use the organizations. This act is a public endorsement of shade tree commissions, by the state, in an attempt to promote their continued use and reimplementation. This act adds the financial and technical support to the local programs, which will help to shield them from the high risks of liability and the costs of litigation (*§6 Publ. L. 1996 c. 135*).

The act provides for the establishment of a community forestry council headed by the state forester whose duty will be to "establish minimum standards, and provide a training skills and accreditation program, for representatives of local governments and shade tree commissions" (*§6 Publ. L. 1996 c. 135*). This accreditation program will state, "the appropriate and approved methods for planting, protection, care, and management of trees and other related natural resources" (*§7 Publ. L. 1996 c. 135*). These will serve as the minimum requirements with which local governments must comply. In order to have a shade tree commission, the local governments must submit a "comprehensive community forestry plan" for approval by the state forester. On receiving such a plan, the state forester, with advice of the council, will approve the plan if "all required parts of the plan adequately address the needs of the community and tree resource" (*§10 Publ. L. 1996 c. 135*). This state program will be funded in part by existing federal grant money and also by the proceeds of the issuance of a special tree license plate.

Pennsylvania allows its local governments to create similar organizations. Chapter 53 of the Pennsylvania state code deals with the establishment of shade tree commissions. Under Pennsylvania statutes, "Any city may, by ordinance, create a commission to be known as the Shade Tree Commission of such city" (Penn. Stat. 53 PS 38801).

The commission has "exclusive custody and control of the shade trees of the city, and may plan, remove, maintain, and protect shade trees on the streets and sidewalks of the city, as council may direct" (Penn. Stat. 53 PS 38804). The commission also has the power to determine the type, location, and alignment of any tree it plants, the only restriction being that it cannot "prevent necessary or reasonable use of streets, sidewalks, abutting property, or the conduct of business" (Penn. Stat. 53 PS 38804). The commission also may hire any professionals (including a tree warden) and create any regulations it requires to accomplish its duties subject to the approval of the city council (Penn. Stat. 53 PS 38804). Shade tree commissions established by other local governments, for example, boroughs (Penn. Stat. 53 PS 47724), incorporated towns (Penn. Stat. 53 PS 53853), and townships (Penn. Stat. 53 PS 58024), are given generally the same powers as those established by cities. All shade tree commissions must make an annual report on their financial status and transactions at the close of each fiscal year (Penn. Stat. 53 PS 38805).

Any city council that chooses not to establish a shade tree commission pursuant to these statutes may exercise the same powers given to such a commission on its own (Penn. Stat. 53 PS 38801) or through any commission to the care of public parks (Penn. Stat. 53 PS 38803).

Other states provide specifically for the appointment of a tree warden to oversee the care and maintenance of public shade trees. The state of Connecticut allows any town, city, or borough to create a forest or shade tree commission comprising three members to serve for staggered terms of 3 years each, without compensation, which will be required to make annual reports to its governing body. This commission must appoint a tree warden and oversee his work (Conn. Stat. §7-131). These tree wardens will serve for a term of 1 year until their successors are appointed. The warden may "appoint such number of deputies as he deems expedient and he may, at any time, remove them from office" (Conn. Stat. §23-58). The warden and his deputies shall receive reasonable compensation for their services (Conn. Stat. §23-58).

The tree warden's scope of power is described in the following manner:

> The town or borough tree warden shall have the care and control of all trees and shrubs in whole or in part within the limits of any public road or grounds and within the limits of his town or borough, except those along state highways under the control of the Commissioner of Transportation and except those in public parks or grounds which are under the jurisdiction of park commissioners. Such care and control shall extend to such limbs, roots, or parts of trees and shrubs as extend or overhang the limits of any such public road or grounds. (Conn. Stat. §23-59)

The care and control described above includes a wide range of duties. The tree warden is responsible for all funds given for the planting, care, and maintenance of trees, enforcing laws for the preservation of trees, and removing illegal signs or advertisements placed on trees or other objects in any area under his jurisdiction. The tree warden may make regulations for the care and maintenance of trees and shrubs, allowing for a fine not to exceed $90. These regulations are subject to approval by the governing body, must be posted publicly in the town, and given the effect of ordinances. The tree warden may remove or trim any tree within his jurisdiction if he determines it to be contrary to public safety and must submit in writing his determination and approval of the proper fee to be paid to the person performing the task to the governing body. If the tree to be trimmed or removed is not an immediate threat to public safety or "particularly obnoxious as hosts of insects or fungus pests," the tree warden must post a notice of the action he will take in regard to the tree at least 5 days before such action and provide a public hearing if there are any objections. The tree warden must grant his decision within 3 days of the hearing. This decision may be appealed to the superior court for the judicial district. If the decision is appealed, the appeal must be filed within 10 days (Conn. Stat. §23-59).

New Hampshire also provides for the establishment of a forestry committee by a city or town. This committee is to be composed of three to five members serving in staggered terms of 3 years each. This committee shall choose its chairman, serve without compensation, make an annual report to the city or town, and forward a copy of that report to the director of the division of forests and lands, department of resources and economic development. The members shall be appointed by the mayors or aldermen of cities and the selectmen of towns. If the city or town has appointed a tree warden, he or she shall serve on the committee (NH Stat. §31:112).

Candidates for the position of tree warden are nominated by the selectmen or other citizens of a town or city. The director of the Division of Forests and Lands, Department of Resources and Economic Development chooses and appoints a tree warden from those candidates to serve for a term of 1 year until a replacement is appointed. The director maintains the right to remove the tree warden from office at his discretion. The tree warden is allowed reasonable compensation for his services (NH Stat. §231:139).

> It shall be the duty of the tree warden to examine the trees growing within the limits of highways and designate them from time to time such as may be reasonably necessary for the purpose of shade or ornamentation and to acquire them in the name of the municipality as hereinafter provided, if it can be done, either by gift or by purchase if at a fair price and funds either public or private are available. (NH Stat. §231:141)

If the trees needed cannot be acquired by the above process, the tree warden may condemn them upon notice and an offer of adequate compensation to the owner. These condemnation proceedings are appealable by the owner (NH Stat. §231:141). The tree warden, or his agent, also must represent the public interest at any hearing where a public service corporation wishes to cut or remove a tree or has caused damage to such a tree (NH Stat. §231:142).

Rhode Island has enacted statutes that require cities and towns to fill the office of tree warden. The appointment process is very complex under these statutes. The town council or city mayor appoints a tree warden each year. This tree warden is required to be or become a licensed arborist within 6 months of his or her appointment. If, in the city or town, there exists no one licensed or about to be licensed as an arborist, the mayor or town council may appoint a resident of that city or town upon notification and approval of the Director of Environmental Management. No appointment shall take effect until the director has approved and certified the appointee. If that appointee is rejected, the mayor or town council must make a new appointment, subject to the same approval and certification (RI Stat. §2-14-2). Once the tree warden's appointment is approved, he or she shall serve as tree warden until the first of March in the year after his appointment and until the next tree warden is appointed (RI Stat. §2-14-3). The tree warden may appoint any number of deputies the warden deems expedient and may remove them from office at his discretion. Both the warden and the deputies shall receive reasonable compensation for their services as determined by the town council or city council (RI Stat. §2-14-4).

The scope of the tree warden's control as given by Rhode Island statutes is described below:

> The tree warden in a respective town or city shall have the care and control of all trees and shrubs in whole or in part within the limits of any public road or grounds and within the limits of his town or city, except those on the roads under the control of the department of transportation and those in public parks or grounds which are under the jurisdiction and control of the department of environmental management or the park commission of any town or city, provided, however, that the tree warden may assume the care and control of trees or shrubs in any public park if requested in writing by the department of environmental management or the park commission of any city or town. The care and control shall extend to such limbs, roots or parts of trees and shrubs as shall extend or overhang the bounds of any public road or grounds, and the tree warden, or his agent, or an authorized agent of the department of transportation, or an authorized agent of the town or city, may enter upon private property when necessary to exercise care and control. (RI Stat. §2-14-5)

This is very similar to the scope of authority given to the Connecticut tree warden, many states borrow ideas and phrases from each other when creating new legislation.

Under Rhode Island law, the tree warden is responsible for planting and protecting the trees under his or her jurisdiction. The tree warden also must follow the recommendations of the Department of Environmental Management in these areas (RI Stat. §2-14-6). The tree warden shall be responsible for spending the funds set out for the care and maintenance of trees under his jurisdiction, and make and enforce laws for the preservation of such trees subject to the approval of the respective town or city council (RI Stat. §2-14-7). If the tree warden deems it necessary to trim or remove a tree or shrub within his jurisdiction, he may do so at the reasonable expense of the town or city. If the tree is not an "immediate public hazard," the warden must give notice, at least 5 days before the removal, of his intentions. If a citizen raises an objection, the warden must hold a public hearing and within 3 days grant his decision (RI Stat. §2-14-8). The tree warden also must carry out the recommendations of the Department of Environmental Management regarding the suppression of pests and tree diseases (RI Stat. §2-14-9).

Most of these statutes do not apply to the city of Providence. Providence has its own law, entitled "An act providing for the care and preservation of shade trees, and for other purposes, in the city of Providence" (Publ. L. 1925 c. 693). However, the appointment of a tree warden in Providence must be approved by the Department of Environmental Management (RI Stat. §2-14-11).

The State of Maine allows its municipalities a choice concerning the type of organization it uses to oversee the care and protection of its public trees. In Maine, the Bureau of Forestry is given the power to create a community forestry program to assist the municipalities in managing its forest areas (ME. Stat. 12 §8002). Municipalities also may establish conservation commissions for the care and custody of public trees. The municipal officers may appoint three to seven conservation commissioners to serve for staggered terms of 3 years each. Associate, nonvoting commissioners also may be appointed for the same 3-year staggered terms. This commission is mainly in control of acquiring land and trees for parks and directing the money appropriated for these duties (ME. Stat. T. 30-A §3261).

Municipalities who choose not to appoint conservation commissions may appoint one or more tree wardens annually. These wardens are to "have care and control of all public shade trees upon and along the highways and in the parks of the municipality and all streets within any village limits. They shall enforce all laws relating to the preservation of those trees" (ME Stat. T. 30-A §3282). Any shade tree considered public property may be trimmed or removed by the land owner only with the consent of the tree warden or conservation commission. The tree warden and the conservation commission in turn must receive the property owner's permission to remove or trim the trees unless that trimming or removal is necessary to widen or alter roads, ensure safe travel on those roads, or suppress insects or tree pests (ME Stat. T. 30-A §3283).

In the state of Massachusetts, municipalities have choices similar to those in Maine. A municipality may establish an office of lands and natural resources to care for the public lands and trees of that municipality. The city manager or town manager or, if those offices do not exist, the city mayor or town selectmen shall appoint a director of the office (MA Stat. 41 §69G). This director must be "qualified by training and experience in the field of arboriculture and licensed by the pesticide board in the department of environmental protection" (MA Stat. 41 §69G). This director may appoint assistants and control all funds appropriated for the office of lands and natural resources (MA Stat. 41 §69G).

The municipality's alternative choice would be to appoint a tree warden. This tree warden once appointed by the municipal officers, will serve for a term of 3 years. In cities of over 10,000 people, the appointed tree warden must be qualified by training and experience in the field of arboriculture and licensed with the Department of Food and Agriculture (MA Stat. 41 §106). The tree warden shall receive reasonable compensation. The powers of the tree warden are the ability to appoint and remove deputies, spend money appropriated to his office for maintenance and care of trees, and make and enforce regulations for the preservation and care of trees that, once approved by city or town council, will have the effect of city ordinances (MA Stat. 87 §2). The tree warden shall exercise these powers in

> the care and control of all public shade trees, shrubs and growths in the town, except those within a state highway, and those in public parks under the jurisdiction of the park commissioners, and shall have care and control of the latter, if so requested in writing by the park commissioners, and shall enforce all the provisions of law for the preservation of such trees, shrubs and growths. (MA Stat. 87 §2)

Once again, this is very similar wording to the scope of duty-afforded tree wardens in several other states.

The process a tree warden must follow for trimming and removing trees is also similar to that of other states. The warden must give public notice and hold a hearing to discuss any objections. At this hearing, the tree warden and the planning board or city/town council will give their decision, and any person whose property is injured by this action may recover damages from the municipality (MA Stat. 87 §3).

Municipalities in Massachusetts also may create a board of park commissioners for the purpose of laying out, improving, and maintaining public parks (MA Stat. 45 §5), or appoint a superintendent of shade tree management and pest control to serve for a 3-year term (MA Stat. 132 §13). This office has many of the same duties, powers, and requirements as a tree warden.

Some states authorize municipalities to establish local community or urban forestry programs. Most of these local programs are developed under the supervision of a similar program at the state level.

The municipalities of Maryland may choose such an option for the care and maintenance of their trees. The Maryland state code defines "urban and community forestry" as "the management of forest and trees in urban and community areas to improve their health, vigor and composition" (MD Stat. Res. §5-424). The state branch of urban and community forestry was created to provide state or federal financial assistance to county or municipal corporations within the state on a matching basis (MD Stat. Res. §5-426). The state program also provides technical assistance to the local programs in the following ways (MD Stat. Res. §5-426):

1. Street tree inventories
2. Technical evaluation of site development plans
3. Management recommendations for the retention, protection, and utilization of forests and trees in developments
4. Assistance to planning and zoning departments
5. Advice on insect and disease control
6. Advice to counties and municipal corporations in implementing and conducting their own urban and community forestry program

As for local programs, "The governing body of a county or municipal corporation, by appropriate resolution or ordinance, may implement a local urban and community forestry program within its jurisdiction or enter into a cooperative agreement with the [state] department" (MD Stat. Res. §5-427). The local governing body may also hire a director to head its program (MD Stat. Res. §5-426). The program outlined by Maryland statutes seems like a very general framework, inside of which the local government can build its own specialized program with guidance and financial assistance from the larger state program.

In New York, urban forestry plan means

> a comprehensive plan for the selection, establishment, maintenance and management of existing and future trees and associated vegetation in an urban area, which plan shall, to the maximum extent possible, present an analysis of physical, biotic, and potential sites for vegetation, and of improvement needs, opportunities and planned projects within an urban area. (NY Environ. Conserv. §53-0303)

Under New York law, any municipality may establish a local urban forestry or other arboriculture program (NY Environ. Conserv. §53-0307). The chief executive officer of a municipal corporation is authorized to appoint a "municipal forester" whose job is to act as a liaison between the state and local programs (NY Environ. Conserv. §53-0303). The state program, headed by the commissioner, is designed to:

> Provide technical assistance to urban areas for urban forestry and arboriculture, and encourage the participation of municipal foresters, private agencies, organizations, firms, and individual, to furnish urban forestry and arboricultural services.... The commissioner may take steps necessary to facilitate the cooperation of private agencies and organizations with municipalities. (NY Environ. Conserv. §53-0305)

Once again, this is a very loose framework in which the municipalities have some room to build a unique program for their needs.

Small towns and villages in New York do have the option of creating shade tree commissions whose members serve without compensation and whose acts are funded through local taxes. These commissions are created by ordinance of the village board of trustees that also can create ordinances for, "the planting, care, preservation, maintenance, protection and control of shade trees and shrubs upon the public highways or public places of such village, and for the enforcement thereof by fines and penalties" (NY Village §17-1732).

The state of Ohio has established the office of chief of the division of forestry to, among other things, "provide urban forestry assistance to individuals, nonprofit organizations, and political subdivisions to manage their urban forest resource and develop comprehensive tree care programs" (OH Stat. §1503.011). The state also has created a forestry development fund to be used by the chief of the Division of Forestry, with the approval of the Director of Natural Resources, to "make grants for urban and rural forest resource improvement and development projects" (OH Stat. §1503.35). Municipalities may apply for these grants by submitting their proposed plan to the chief. The local program is not limited by strict guidelines.

The State of Illinois has enacted the Urban and Community Forestry Assistance Act, to provide "technical assistance, training, and financial aid to units of local government for the development of plans and implementation of programs for the establishment, management, conservation, and preservation of the urban and community forest" (IL Stat. CH 30 §735/1). This act is to be administered by the Department of Natural Resources (IL Stat. CH 30 §735/3). The department is in charge of promoting "the development of plans and programs for the establishment, management, and conservation of the urban/community forestry within units of local government" (IL Stat. CH 30 §735/4). The department also provides technical assistance, planning, and analysis for the projects of the local governments (IL Stat. CH 30 §735/4).

Indiana also allows municipalities to fund their individual community forestry programs with state grants. In order to receive a grant, several requirements must be met, including the following (IN Stat. 14-12-3-11):

1. The applicant must be a municipal corporation or a nonprofit corporation that has no affiliation with religion.
2. The land involved in the project must be the one owned or controlled by the municipal corporation.
3. The applicant must demonstrate that there are adequate provisions to maintain the completed project.
4. The applicant must demonstrate that the project will be compatible with existing site conditions, including sewers and utility facilities.

Wisconsin is another state that allows its local governments to submit their proposals and possibly receive financial assistance through grants. Cities and villages may be given grants "for up to 50% of the cost of tree management plans, tree inventories, brush residue projects, the development of tree management ordinances, tree disease evaluations, public education concerning trees in urban areas and other tree projects" (WI Stat. 23.097). Once again, the municipality seems to be left to create its own program. Having created this program or plan, the local government may then submit a proposal to the designated office for approval and state assistance.

We have looked at three types of state programs under which local governments can create local programs dedicated to the care and protection of public trees. First, the shade tree commission, which seems to have very rigid guidelines concerning its setup and jurisdiction. Second, the tree warden, who may be a shade tree commission officer, also has very strict guidelines concerning who can be appointed and the duties and jurisdiction of the office. Finally, the urban and community program, simply a state-run program to provide technical and financial assistance to local governments who create their own programs. These three types of programs share similar goals, as evidenced by their use of similar phrases and ideas. These goals are to encourage local governments to provide for the care and maintenance of public shade and ornamental trees and to be sure that those governments have the technical and financial assistance they need to do so.

References

Black's Law Dictionary (6th ed. 1990).
Conn. Stat. §7-131.
Conn. Stat. §23-58.
Conn. Stat. §23-59.
Cooke v. McShane, 1928, 108 Conn. 97, 142 A. 460.
Hay v. Norwalk Lodge No. 730 BPOE, 1951, 92 Ohio App. 14, 109 N.E. 2d 481.
IL Stat. CH 30 §735/1.
IL Stat. CH 30 §735/3.
IL Stat. CH 30 §735/4.
IN Stat. 14-12-3-11.
Ivancic v. Olmstead, 1985, 490 N.Y.S.2d 914, 112 A.D.2d 508.
Loggia v. Grobe, 1985, 128 Misc. 2d 973, 491 N.Y.S.2d 973.
MA Stat. 41 §69G.
MA Stat. 41 §106.
MA Stat. 87 §2.
MA Stat. 87 §3.
MA Stat. 45 §5.
MA Stat. 132 §13.
MD Stat. Res. §5-424.
MD Stat. Res. §5-426.
MD Stat. Res. §5-427.
ME Stat. 12 §8002.
ME Stat. T. 30-A §3261.
ME Stat. T. 30-A §3282.
ME Stat. T. 30-A §3283.
Merullo, V.D., and Valentine, M.J., 1992, *Arboriculture and the Law,* Kowa Graphics, New York, NY.
NH Stat. §31:112.
NH Stat. §231:139.
NH Stat. §231:141.
NH Stat. §231:142.
NY Environ. Conserv. §53-0303.
NY Environ. Conserv. §53-0307.
NY Environ. Conserv. §53-0305.
NY Village §17-1732.
OH Stat. §1503.011.

OH Stat. §1503.35.
Penn. Stat. 53 PS 38801.
Penn. Stat. 53 PS 38804.
Penn. Stat. 53 PS 47724.
Penn. Stat. 53 PS 53853.
Penn. Stat. 53 PS 58024.
Penn. Stat. 53 PS 38805.
Penn. Stat. 53 PS 38803.
Publ. L. 1925 c.693.
Publ. L. No. 1996 c.135.
Restatement of Torts 2d § 197.
RI Stat. §2-14-2.
RI Stat. §2-14-3.
RI Stat. §2-14-4.
RI Stat. §2-14-5.
RI Stat. §2-14-6.
RI Stat. §2-14-7.
RI Stat. §2-14-8.
RI Stat. §2-14-9.
RI Stat. §2-14-11.
Short v. Kerr, 1937, 104 Ind. App. 118, 9 N.E.2d 114.
Toomey v. State of Connecticut, 1994, 1994 Conn. Super. LEXIS 539.
WI Stat. 23.097.

Chapter 4

Laws and Ordinances

Tree Ordinances

Michael V. D'Errico

1. Introduction

Local shade tree and environmental committees can be mired in a seemingly endless stream of site plan reviews. As one of their main activities, these committees review site plans for environmental concerns and proper tree selection, planting, and care. While they may make a number of appropriate comments to their municipalities, committee members realize that their contribution could be minimal at best, and at times possibly ignored. Often these volunteers are powerless to voice their feelings, and even if they do, it is strictly advisory.

To many committee members, their trees represent the bucolic essence that originally attracted them to their community. They feel that rapid development of the forestlands would not only change their area's essential character but would ultimately have an adverse effect on property values and the quality of life enjoyed by their residents. They do not want to stop development, but they want to draw attention to the fact that the community could no longer tolerate the indiscriminate destruction of their woodlands.

The tool to accomplish this goal is the adoption of a local law—a tree preservation ordinance in the instance previously mentioned. This unique ordinance seeks to provide developers with a framework for preserving and restoring as many trees as possible on site, with particular emphasis on larger, older specimens (Mirelli, 1991). A tree preservation ordinance preserves a town's beauty for future generations, while providing an equitable means of recognizing the legitimate needs of development.

Michael V. D'Errico New Jersey Forest Service, Community Forsestry Program, Trenton, New Jersey 08625.

Urban and Community Forestry in the Northeast, 2nd ed., edited by, J. E. Kuser.

2. Planning for an Ordinance

For these concerned community activists, the retention of their tree resource and the proper specifications and standards of restoration through tree planting is a critical need for which an ordinance is established. Each community operates under different policies with different needs and values. Many communities may already have legislation that regulates trees and their management. State statutes differ in the leeway they allow local governments. The ordinances adopted by a city or town reflect this individuality and the values of a community; the values its residents believe are worth protecting to maintain their quality of life and an environment that is both safe and pleasant (Fazio, 1993).

A tree ordinance encourages beautification, protection, preservation, regulation of development, and restoration through planting and management. It also enables citizens to prevent the spread of tree diseases and regulates sidewalk replacement and utility line clearance work. An ordinance may require the use of professionals, specify duties of municipal employees, and help control careless dealing with this important natural resource, such as the preservation of historically significant trees, protecting forested streamside buffers, and managing for a community's total tree canopy cover. There is no limit to the possibilities.

3. Legal Authority

Perhaps the most important purpose of an ordinance is to give the public entity a guiding principle based on law through both responsibility and authority (Grey, 1993). Ordinances often are initiated shortly after a program is established. These ordinances give the program the legal standing it will need as it emerges. The visibility of a local community tree program can be increased by the ordinance. It establishes the program independent of political variation, budget restrictions, or public opinion. The statement of purpose in the ordinance becomes a legal basis for the program.

While an ordinance can be used to "just" establish a tree commission, most ordinances are adopted as part of a legal package for a comprehensive community forestry program. Such a package often provides a framework for dealing of trees with care on publicly owned or controlled property. However constituted, the ordinance should include the following if the primary task is the establishment of a legal authority, commission, or board for tree care (Garcia, 1991).

Statement of establishment.
Number of members and, if applicable, the number of alternate members.
Qualifications of members and a statement of service.
Term of office and statement of volunteering without compensation.
How to fill vacancies?
Duties of officers and members.
Scope of responsibilities.
Measure of accountability, such as an annual report to mayor and council.
Any rules to operate as a legal entity.

Through the establishment of a local tree ordinance, a community has acknowledged the contribution shade trees and community forest make to a city or town and the need to nurture, protect, and manage a dynamic community forest resource. The formation of a Shade Tree Commission or a Tree Board is a necessary step to properly manage this critical resource. A commission has the time to devote that local officials may not have to research grants that help fund tree planting or to start a tree inventory. This group of individuals has the sincere interest in trees and perhaps the professional experience to make knowledgeable decisions. With a Shade Tree Commission in place, "someone" will actively use the ordinance and update needed standards and practices in tree care and is more likely to take initiative on and promote tree-related issues. This is a key opportunity to involve the public in the management of its tree resources.

4. Drafting a Tree Ordinance

The first and most important step in drafting a tree ordinance is for the local group to begin the effort and research the topic (Bomenblit et al., 2002). Before a tree ordinance is written, two questions should be considered—What does your community have and what does your community need? In identifying what a community has as it relates to trees, the Shade Tree Commission can address this by outlining community tree practices and to take a look at the tree resource itself.

4.1. Outline Community Tree Practices

Understanding the current status of the urban and community forest in your neighborhood and how it has been managed in the past is an important factor in establishing a tree ordinance. Relevant data would include items, such as municipal tree care practices (planting, removal, and maintenance strategies), planning regulations, guidelines (permits), activities by municipal departments that impact trees, any current tree survey, and prior ordinances.

Is there an existing ordinance or current municipal legislation that pertains to trees? If these do exist, it will help the group to determine what is needed to bring tree care regulation to the desired level. A question that could be asked—Does another community have an ordinance that can be adapted? Communities should not just copy word-for-word another town's work but should use it only as a guide in developing an ordinance that meets their community's needs.

Municipal records are the most reliable sources for this type of information—the point to identify all activities that affect trees in the community. Any change in the management of the tree should take into consideration other practices and identify all the players involved in the past.

4.2. Take a Look at the Tree Resources

An assessment of the local tree resource provides basic information necessary for making management decisions. These assessments are usually based on some type of inventory method such as a survey. Information that may be helpful when planning your strategy includes the total number of trees classified by species, condition, age, size, and location. Assessing areas with potential future problem situations or already existing problems may help clarify what you are facing in terms of repair and prevention. These include problem situations, such as sidewalk damage, number of hazard tree situations, disease, and pest problems.

Inventories will help your community reach certain goals and improve response to public questions. Once the current condition of the urban and community forest is known, a community will be able to identify its needs and opportunities.

4.3. Identify Needs

With the background information on the status of the tree resource in place, the municipality can now assess the needs of that community.

Biological needs: Those needs that relate to the tree resource itself. These include: increasing species and age diversity to provide long-term stability, providing tree planting that will keep up with urban growth and offset their removal, increasing the canopy effect, spacing in tree-planting sites, and reduced sidewalk damage.

Management needs: The needs of those involved with the care and maintenance of the community forest such as developing adequate long-term planning to ensure the status of community trees, enhancing the use of the limited financial and personal resources, increasing training and educational opportunities for tree program employees and volunteers, and coordinating tree-related activities for town and municipal departments.

Community needs: How the public views and interacts with the community forests and the management of its program? These include increasing public awareness of the values and benefits associated with trees in a community, promoting acceptable private tree care, and seeking conservation of the urban forest.

These needs will vary with each community but are a good example of what concerns the majority.

5. Specific Goals

Specific goals can be created that will address the needs that have been identified in Section 4.3 and should be reasonable within the parameters of the resources available, as well as environmental constraints. If limited resources are an issue, one may need to prioritize the goals of the tree program to get the maximum benefit.

Many tree ordinances rely on voluntary compliance from the pubic so it is crucial that the community values are publicly supported. Involving the public in the creation of these goals is also a way to educate the public about community forestry issues (Sheay, 1992).

The goals that are created should in some way be quantifiable so that the community can monitor their progress.

Establish and maintain maximum tree cover: This will add to the health, safety, and energy conservation in the community by reducing local pollution, rainfall runoff, wind and noise pollution, and providing shade and habitat for wildlife. This could also be viewed as a no-net-loss program that seeks to maintain the current tree cover by replacing trees one for one.

Conservation: As a tree increases in size so do the benefits that it provides the community. Therefore, it is in the best interest of the community to protect this resource. Also remember that tree conservation is important before and after site construction. The ordinance should include maintaining site trees after building is completed. The ordinance should also address stands of trees, not just individuals. You may want to consider natural influences of your community, such as fire threats, wildlife habitat, and flood control.

Maintain healthy trees through cultural practices: Healthy trees are a benefit to the community and public health. Proper and timely pruning can promote tree health and extend the life of a tree. Trees that are not maintained have a higher risk of breakage that can result in property damage and personal injury.

Encouragement of good tree management on private lands: Educational understanding, as well as good public advocacy will let the residents know the laws pertaining to their trees. It is important for the citizens to understand the role their trees play in the urban forests of their town. This can be achieved by forming a Shade Tree Commission made up of citizens from the municipality.

Centralizing tree management under one group or person: Tree-related issues might arise in a number of municipal departments. In many communities street trees are the responsibility of the public works department, while trees located in parks are the responsibility of the parks and recreation department. In order to effectively manage the trees, it is wise to coordinate between the different departments, centralizing direction and technical responsibility, and designating oversight to a tree commission or board.

Promote urban forest care in an efficient and cost-effective manner: A small amount of resources may be all that is allotted to manage the town's trees. Programs could be enhanced by the addition of resources, such as fees and fines, found within some ordinances.

With all of this information in hand, a tree commission can frame out an effective formula for success in enacting an ordinance. This proven approach is one model that may not only lead to the passage of a tree ordinance but also a stronger one at that. The key points are as follows (Anonymous, 1977):

1. Take the initiative. The tree group should take the lead on tree-related issues and promote themselves as such.
2. Follow your convictions; you may be surprised at how many other share them.
3. State your desires simply. Work with your legal staff only after your group has clearly developed goals, objectives, and plans.
4. Solicit the assistance of all community interests and compile information. Their support could prove instrumental in focusing the attention of the local policy makers. Find a policy maker to "carry your torch."
5. Retain the role of the group as coordinator. Be prepared to be both patient and persistent.
6. Address the needs of all parties, revising your draft ordinance to take into account the proper concerns of others. Be prepared to compromise on issues.
7. Formally introduce the proposed ordinance to the administration, expect comments and questions, and have a strategy ready to address concerns.

6. Three Functions of a Successful Ordinance

Ordinances can provide with the overall care of the community tree resource. It is important that the ordinance provides for and addresses adequate management. To do so, the ordinance must do the following three things (Hunt *et al.*, 1982):

1. Provide authority to conduct the tree care program.
2. Define responsibility.
3. Set forth standards for management to ensure public safety and health while caring for the resource.

7. Ordinance Sections

In drafting a tree ordinance, one will find that some sections or provisions are necessary, while others are optional. These optional sections or specific ordinance provisions are built by the group to help in meeting management goals and assure that the document is tailor-made for that specific community. Together, the basic and specific provisions of the ordinance provide a legal key to facilitate effective management of the community tree resource (NJ Forest Service, 2004).

The basic ordinance provisions are a set of elements that are found within most ordinances, tree, and nontree. These elements, while generic in nature, form the foundation on which the ordinance is built.

For example, the key elements of the base element are as follows (Johnson *et al.*, 1988):

1. Title: The ordinance should have a concise title that reflects its purpose.
2. Purpose: A policy statement to set forth the purpose of the ordinance—the "why."
3. Definitions: Defines the terms used in the ordinance.
4. Exception: Defines any exemptions to the ordinance.
5. Compliance: Specifies the authority of the provisions of this ordinance.
6. Establishment or identification of tree entity: Notes either the creation of the tree entity and defines its composition and charges or identifies who in the municipality will be responsible for the ordinance.
7. Enforcement: Who has the enforcement duties of the ordinance and what enforcement measures can be used if necessary?
8. Permits and fees: A fee schedule to adequately cover the administration costs of this ordinance.
9. Appeals: A method for an individual or group to appeal a decision pursuant to the provisions of the ordinance.
10. Violations/penalty: A provision to identify the monetary or imprisonment penalties for any violation of the provisions of the ordinance.

The specific ordinance provisions identify unique sections of the ordinance designed to address the critical issues and needs of the community. It is recommended that the community start by writing in simple language those specific issues that need addressing. From here the simple issues statement is added to the ordinance as one of its elements and the language expanded to reflect the policy of the group.

Some examples of specific provisions are as follows (Borough of Paramus, 1992):

- *Tree topping*: It shall be unlawful for any person and agency to top any trees, either park trees or public property.
- *Use of certified tree expert*: The municipality will use a New Jersey certified tree expert on all contracted tree-planting and tree-pruning projects within the official boundaries of the municipalities.
- *Street tree planting*: An official list adopted by the Shade Tree Commission will constitute acceptable street trees for named municipalities. No species other than those included on the list may be planted without permission from Shade Tree Commission.

The specific provisions are only a few of those that could be incorporated in an ordinance. By carefully considering the needs of the community, those responsible can incorporate into the ordinance specific information suitable for their community.

8. Linking to Specifications and Standards

Planting and maintenance specifications are two examples of detailed arboricultural practices that should be linked to one's ordinance but not necessarily included. It is generally agreed that the standards and specifications section be a separate but companion part of the ordinance. This way, these can be easily changed as new research may direct. Keeping the ordinance brief and workable, while the specifications can be updated separately, thus avoiding the cumbersome process of amending the ordinance.

A good example is the established and accepted pruning standards by the New Jersey Board of Certified Tree Experts. This document spells out the highest standards by which pruning practices are measured. An ordinance can cite this publication but does not have to include the document word for word. If the Board of Certified Tree Experts revises these standards, the ordinance will not need to be changed, thus eliminating unnecessary work.

Ordinances vary in length and complexity, but the key to effectiveness is to write the ordinance simply, clearly, and tailored to the needs of the community. At the end, a tree ordinance is just another tool for proper tree care. Like any tool, it needs to be of high quality, matched properly to the job, and used with skill and care. As previously mentioned, an ordinance can be singular in nature, such as establishment of a tree care program, or encompassing by including provisions to address tree preservation, protection during and after construction, and diseased tree removal. Two examples that follow may be used as a guide in establishing an ordinance in one's community.

BOARD OF SHADE TREE
AND PARK COMMISSIONERS
BOROUGH OF PARAMUS
Ordinance No. ST-1

AN ORDINANCE REGULATING THE PLANTING, CONTROL, PROTECTION AND IMPROVEMENT OF TREES AND SHRUBBERY UPON THE PUBLIC STREETS PARKS AND PUBLIC PLACES IN THE BOROUGH OF PARAMUS, COUNTY OF BERGEN AND STATE OF NEW JERSEY,

BE IT ORDAINED by the Board of Shade Tree and Park Commissioners of the Borough of Paramus, New Jersey as follows:

SECTION 1. DEFINITIONS.

As used in this ordinance, the term hereinafter set forth shall be defined and deemed to have the following meanings:

COMMISSION: The Board of Shade Tree and Park Commissioners of the Borough of Paramus including any of its duly appointed members and any of its duly authorized agents or employees.

PERMITS: Written permission of the Commission (whenever pursuant in any way to any provision of this Ordinance, the Commission shall have occasion to pass upon an application for a permit, it shall, in determining whether or not to issue such permit, take into consideration the nature, species, size, age, and condition of any tree involved: the location thereof in the street or park, the planting, care protection, maintenance or removal procedures involved; the public safety and welfare; and the improvement and advancement of the shade tree plan or program of the Commission.

PERSON: Any individual, firm, partnership, or corporation, or any combination thereof. Where, in the proper context, it is so required, this term may be construed to designate the plural as well as the singular.

STREET: Any road, avenue, street, or highway, dedicated to the public use for street purpose, regardless of whether or not it has been formally accepted by the Borough of Paramus. A street shall be deemed to include all portions lying between the dedicated or established right-of-way lines and/or planting easement thereof, said lines being identical with the front property line of lands abutting the street.

TREE: Any tree, shrub or plant, or any root, branch, flower, or other part thereof that is located on or upon any street or park. Any term or provision of this ordinance that contemplates, directs, regulates, or prohibits the doing of any act may, in applicable case and where the context so requires, be construed to include the causing, allowing, permitting, or suffering of such act to be done by others under the direction, control or supervision of the person charged therewith. Every such act shall be deemed to be within scope of this ordinance, regardless of whether it is a deliberate, intentional or purposeful act, or a careless, negligent, or unintentional act.

BOARD OF SHADE TREE AND PARK COMMISSIONERS BOROUGH OF PARAMUS Ordinance No. ST-1

SECTION 2. PERMITS.

No person shall, without a permit, do any of the following acts:

- Cut, prune, break, injure, remove, disturb, or interfere in any way with any tree;
- Spray with any chemical any tree or near a tree to cause injury or death to said tree;
- Fasten any rope, wire, sign, or other device to a tree or to any guard about such tree;
- Remove or injure any guard or device placed to protect any tree;
- Close or obstruct any open space provided about the base of a tree to permit the access of air, water, and fertilizer, to the roots of such tree.

SECTION 3. PLANTING.

No person shall plant any tree or shrub in any park or street without a permit.

SECTION 4. OBSTRUCTIONS.

(a) No person shall, without a permit, place or maintain in any street or park, any stone, cement, or other sidewalk or any stone, cement, or other substance, which impede the free access of air and water to the roots of any tree.
(b) Where any tree is to be surrounded by pavement of stone, cement, asphalt, or any other substance tending to impede the free access of air and water to the roots of the tree, no portion of such pavement shall be nearer to any portion of the base of the trunk than four (4) feet.

SECTION 5. INJURY.

(a) No person shall place salt, brine, oil, or any other substance injurious to plant growth in any street or park in such a manner as to injure any tree.
(b) No person shall build any fire or station any tar kettle, road roller, fuel oil dispensing truck, or other engine in any street or any other place in such a manner that the heat, vapors, fuel, or fumes therefrom may injure any trees.
(c) Every person, having or maintaining any underground utility lines in any street or park shall maintain such lines in such a manner as will safeguard the trees against any damage therefrom and shall make periodic adjustments whenever necessary to prevent damage to trees.

SECTION 6. PROTECTION.

(a) In the erection, alteration, or repair of any structure or building, the owner, contractor, or other person in charge thereof shall place such guards around all nearby trees as will effectually prevent injury to such trees.
(b) No person shall do any excavating within four (4) feet without a permit.
(c) No person shall use or operate any power shovel, bulldozer or any other implement or tool in such a manner as to damage or destroy any tree.

BOARD OF SHADE TREE
AND PARK COMMISSIONERS
BOROUGH OF PARAMUS
Ordinance NO. ST-1

SECTION 7. WIRES AND UTILITY TRIMMING.

(a) No person shall string any wires in or through a public park or property without a permit.
(b) Every person having or maintaining any electric, telephone, telegraph, cable TV, or other wires running in or through a street or park, shall securely fasten and maintain such wires in such a manner as will safeguard all trees against any damage therefrom and shall make periodic adjustments whenever necessary to prevent damage to all trees.
(c) No person shall, without a permit, attach or fasten any wire, insulator, or other device for holding any wire to any tree.
(d) Whenever the commission shall deem it necessary to prune or remove any tree, any person having a wire running in or through any street or park, shall temporarily remove such wire within 24 h after the service upon the owner of said wire, or his agent, of a written notice to do so.
(e) Any public utility or its agents may upon receiving (written) permission from the Commission at least 72 h prior to the start of work, prune, or remove trees for line clearance of utility wire in nonemergency situations pursuant to a line clearance program.
(f) Any public utility or its agent may undertake emergency tree work to restore electrical service or spot work to prevent interruption of electrical, telephone, telegraph, cable TV, or other wire services. In such an event the utility will notify the commission of said within three (3) business days of its beginning.

SECTION 8. PARKS.

(a) No person shall enter upon any portion of the lawn or ground within a public park or other public place when notified by a sign placed in such a park or public place, or by a guardian of such park or public place, or by a police officer, not to enter upon such lawn or ground.
(b) No person shall leave or deposit any paper or other waste material in any public park except in such receptacle as maybe provided therein for that purpose.
(c) No person shall, except at such time and under such regulations as may be designated by the commission, play any game in public park.
(d) No person shall, without permit, place any booth, stand, or other structure, or station wagon, car, or other vehicle in any public park, except in the area prescribed.
(e) No person shall offer any article for sale, display any advertising device or distribute any commercial circulars or political activity and/or signs in a public park.

SECTION 9. HINDRANCE.

No person shall prevent, delay, or in any manner interfere with Commission or its authorized agents in the performance of their lawful duties.

BOARD OF SHADE TREE AND PARK COMMISSIONERS BOROUGH OF PARAMUS Ordinance No. ST-1

SECTION 10. PENALTIES.

Any person who violates any of the provisions of this Ordinance or who fails to comply with the terms and provisions of any permit issued pursuant hereto shall upon conviction in municipal court thereof, pay a fine of not less than Two Hundred ($200) dollars or to exceed Fifteen Hundred ($1500) dollars at the discretion of the court. Each day that a violation shall continue shall constitute a separate offense.

SECTION 11. RESTITUTION.

In addition to the penalties authorized by Section 10 of this ordinance, the Commission may require a person who removes or otherwise destroys a tree in violation of a municipal ordinance to pay a replacement assessment to the municipality. The replacement assessment shall be the value of the tree as determined by the appraisal of a trained forester or Certified Tree Expert retained by the Commission for that purpose. In lieu of an appraisal, the Commission may adopt a formula and schedule based upon the number of square inches contained in a cross section of the trunk of the tree multiplied by a predetermined value per square inch, not to exceed $27.00 per square inch. The square inch cross section shall be calculated from the diameter at breast height and, if there is a multiple stem tree, then each trunk shall be measured and an average shall be determined for the tree. For the purposes of this Ordinance, "diameter at breast height" shall mean the diameter of the tree taken at a point 41/2 feet above ground level. The commission shall modify the value of the tree based upon its species variety, location, and its condition at the time of removal or destruction.

SECTION 12. INCONSISTENCIES.

All Ordinances, rules and regulation parts of any ordinances, rules and regulations that are inconsistent with any provisions of this ordinance are hereby repealed as to the extent of such inconsistencies.

SECTION 13. EFFECT.

This ordinance shall take effect immediately upon publication thereof after final passage and approval as required by law.

ORIGINALLY PASSED: February 28, 1952.

AMENDED: August 11, 1982; March 11, 1992.

Sample Tree Ordinance for Small to Mid-Size Communities

An ordinance relating to protection of trees on public property that serves the public interest by providing oxygen, stabilization of the soil, prevention of erosion, shelter for wildlife, conservation of energy by providing shade, filtering air, and adding to the beauty of the City of Red Oak, New Jersey; and providing when this ordinance shall take effect.
Whereas, trees provide a setting with a variety of color unsurpassed in shade and hue; and
Whereas, trees are an invaluable psychological counterpart to the man-made urban setting; and
Whereas, trees absorb a high percentage of carbon dioxide and return oxygen, a vital ingredient to life; and
Whereas trees are a valuable asset that can affect an area economically; and
Whereas, the City Commission of the City of Red Oak has determined that the protection of trees on public property within the city of Red Oak, New Jersey is not only desirable but essential to the present and future health, safety, and welfare of all citizens; and
Whereas, the City of Red Oak, New Jersey that trees on public property are valuable contributors to the City's environment;
Now, therefore, be it ordained by the City Commission of the City of Red Oak, New Jersey.

SECTION 1. DEFINITIONS.

Street trees: "Street trees" are herein defined as trees, shrubs, bushes, and all other woody vegetation on land lying between property lines on either side of all streets, avenues, or ways within the city.
Park trees: "Park trees" are herein defined as trees, shrubs, bushes, and all other woody vegetation in public parks having individual names and all areas owned by the city, or to which the public has free access as park.

SECTION 2. CREATION AND ESTABLISHMENT OF CITY TREE BOARD.

There is hereby created and established a City Tree Board for the City of Red Oak, New Jersey which shall consist of five members who are citizens and residents of this city, and who shall be appointed by the City Commission of the City of Red Oak, New Jersey.

SECTION 3. TERM OF OFFICE.

The term of the five persons appointed to the City Tree Board shall be three years except that the term of two members appointed to the first Board shall be for only one year and the term of two members of the first Board shall be for two years. In the event that a vacancy shall occur during the term of any member, their successor shall be appointed for the unexpired portion of that term.

SECTION 4. COMPENSATION.

Members of the Board shall serve without compensation.

SECTION 5. DUTIES AND RESPONSIBILITIES.

(a) It shall be the responsibility of the Board to study, investigate, counsel, develop, and update annually, and administer a plan for the care, preservation, pruning, replanting, removal, or disposition of trees and shrubs in parks, along streets, and in all other public areas. Such a plan will be presented annually to the City Commission and, upon their acceptance and approval, shall constitute official comprehensive tree plan for the City of Red Oak, New Jersey.

(b) The Board, when requested by the City Commission of the City of Red Oak, shall consider, investigate, make findings of fact, and make recommendations upon any special matter or question coming within the scope of its duties.

SECTION 6. OPERATION.

The Board shall choose its own officers, make its own rules and regulations, and keep a journal of its actions. A majority of the members shall be a quorum for the transaction of business.

SECTION 7. STREET TREES TO BE PLANTED.

The following list constitutes the official Street Tree species for Red Oak, New Jersey. No species other than those included in this list may be planted as Street Trees without written permission of the City Tree Board.
Some examples of street trees and their height are given below. The list contains the most popular trees.

Large Trees (Height of 40 feet and Over)
Beech, European (*Fagus sylvatica*)
Beech, Fernleaf (*Fagus sylvatica fastigiata*)
Coffeetree, Kentucky (*Gymnocladus dioicus*)
Elm, Delaware American (*Ulmus americana* "Delaware")
Ginkgo, Magyar Upright (*Ginkgo biloba* 'Magyar')
Ginkgo, Princeton Sentry (*Ginkgo biloba* "Princeton Sentry")
Hackberry (*Celtis occidentalis*)
Honeylocust (*Gleditsia triacanthos*)
Linden, Greenspire Littleaf (*Tilia cordata* "Greenspire")
Katsura Tree (*Cercidiphyllum japonicum*)
Locust, Sunburst (*Gleditsia triac. inermis* "Sunburst")
Maple, Norway (*Acer platanoides*)
Maple, Red (*Acer rubrum*)
Maple, Sugar (*Acer saccharum*)
Oak, Northern Red (*Quercus rubra*)
Oak, Pin (*Quercus palustris*)
Oak, Sawtooth (*Quercus acutissima*)
Oak, White (*Quercus alba*)
Oak, Willow (*Quercus phellos*)
Sweetgum (*Liquidambar styraciflua*)
Zelkova (*Zelkova serrata*)

Medium Trees (30 to 40 feet)
Birch, River (*Betula nigra*)
Chokecherry, Amur (*Prunus maackii*)
Cherry, Kwanzan (*Prunus serrulata* "Kwanzan")
Holly, American (*Ilex opaca*)
Hophornbeam, American (*Ostrya virginiana*)
Pear, Capitol Callery (*Pyrus calleryana* "Capitol")
Yellowwood (*Cladrastis kentukea*)

Small Trees (20 to 30 feet)
Dogwood, Kousa (*Cornus kousa*)
Dogwood, White Flowering (*Cornus florida*)
Japanese Tree Lilac (*Syringa amurensis japonica*)
Maple, Hedge (*Acer campestre*)
Plum, Newport Purpleleaf (*Prunus cerasifera* "Newport")
Redbud, American (*Cercis canadensis*)
Silverbell, Carolina (*Halesia carolina*)

SECTION 8. SPACING.

The spacing of street trees will be in accordance with the three species classes listed in Section 7 of this ordinance, and no trees may be planted closer together that the following: small trees, 30 feet; medium trees, 40 feet; large trees, 50 feet, except in special plantings designed or approved by the Tree Board.

SECTION 9. DISTANCE FROM CURB OR SIDEWALK.

The distance trees may be planted from curb or curb lines and sidewalk will be in accordance with the three species size classes listed in Section 8 of this ordinance. And no trees may be planted closer to any curb or sidewalk than the following: small trees, 2 feet; medium trees, 3 feet; and large trees, 4 feet away.

SECTION 10. DISTANCE FROM STREET CORNER AND FIREPLUGS.

No street trees shall be planted closer than 20 feet of any street corner, measured from the point of nearest intersecting curbs or curb lines. No street tree shall be planted closer than 10 feet to any fireplug.

SECTION 11. PUBLIC TREE CARE.

The City shall have the right to plant, prune, maintain and remove trees, plants, and shrubs within the lines of all streets, alleys, avenues, lanes, squares, and public grounds, as may be necessary to ensure public safety or to preserve or enhance the symmetry and beauty of such public grounds. The City Tree Board may remove or cause to be removed, any tree or part thereof which is in an unsafe condition or which by reason of its nature is injurious to sewers, electric lines, gas lines, water lines, or other public improvements, or is affected with any injurious fungus, insect, or pest. This section does not prohibit the planting of street trees by adjacent property owners providing that the selection and location of said trees is in accordance with Sections 7 through 11 of this ordinance.

SECTION 12. TREE TOPPING.

It shall be unlawful as a normal practice for any person, firm, or city department to *top* any street tree, park tree, or other on public property. Topping is defined as the severe cutting back of limbs to stubs larger than three inches in diameter within the tree's crown to such a degree so as to remove the normal canopy and disfigure the tree. Trees severely damaged by storms or other causes, or certain trees under utility wires or other obstructions where other pruning practices are impractical may be exempt from this ordinance at the determination of the City Tree Board.

SECTION 13. PRUNING: CORNER CLEARANCE.

Every owner of any tree overhanging any street right-of-way within the City shall prune the branches so that such branches shall not obstruct the light from any street lamp or obstruct the view of any street intersection and so that there shall be clear space of eight feet (8″) above the surface of the sidewalk. Said owners shall remove all dead, diseased or dangerous trees, or broken or decayed limbs, which constitute menace to the safety of the public. The City shall have the right to prune any tree or shrub on private property when it interferes with the proper spread of light along the street from a streetlight, or interferes with visibility of any traffic control device or sign.

SECTION 14. DEAD OR DISEASED TREE REMOVAL ON PRIVATE PROPERTY.

The City shall have the right to cause the removal of any dead or diseased trees on private property within the city, when such trees constitute a hazard to life and property, or harbor insects or disease which constitute a potential threat to other trees within the City. The owners shall pay for this at their own expense within 60 days after the date of service of notice. In the event of failure of owners to comply with such provisions, the City shall have the authority to remove such trees and charge the cost of removal on the owner's property tax notice.

SECTION 15. INTERFERENCE WITH CITY TREE BOARD.

It shall be unlawful for any person to prevent, delay, or interfere with the City Tree Board, or any of its agents, or servants, while engaging in and about the planting, cultivating, mulching, pruning, spraying, or removing of any street tree, park trees, or trees on private grounds, as authorized in this ordinance.

SECTION 16. ARBORISTS LICENSE AND BOND.

It shall be unlawful for any person or firm to engage in the business or occupation of pruning, treating, or removing street or park fees within the City without first applying for and procuring a license. The license fee shall be $25 annually in advance, provided, however, that no license shall be required of any public service company or City employee doing such work in the pursuit of their public service endeavors. Before any license shall be issued, each applicant shall first file evidence of possession of liability insurance in the minimum amounts of $25,000 for bodily injury or $10,000 property damage indemnifying the City or any person injured or damaged resulting from the pursuit of such endeavors as herein described.

SECTION 17. REVIEW BY CITY COMMISSION.

The City Commission shall have the right to review the conduct, acts, and decisions of the City Tree Board. Any person may appeal from any ruling or order of the City Tree Board to the City Commission who may hear the matter and make the final decision.

SECTION 18. EXCEPTION.

(a) In the event that any tree shall be determined to be in hazardous or dangerous condition as determined by the City Tree Board so as to endanger health or safety and require immediate removal without delay, verbal authorization for removal may be given by the City.

(b) During the period of emergencies, such as hurricane, windstorm, flood, freeze, or other disasters, the City may waive the requirements of this ordinance. Said waiver must be for a time certain and may not be for an indefinite period.

SECTION 19. PENALTY.

Any person violating any provision of this ordinance shall be upon conviction or plea of guilty, subject to a fine not to exceed $500.00.

References

Anonymous, 1977, An analysis of tree ordinances: The example of New Jersey, *J. Arbor*. **231:**191–197.

Bomenblit, A., Burns, K., D'Errico, M., Ferraro, M., Kanefsky, M., and Schwerd, L., 2002, A *Guide for Writing Tree Ordinances in New Jersey*, NJ Forest Service, Trenton, NJ.

Borough of Paramus, 1992, *An Ordinance to Regulate Planning, Control, Protection and Improvement of Tree-No. St-1*, Board of Shade Tree and Park Commissioners, Borough of Paramus, NJ.

Fazio, J., 1993, *Tree City Bulletin No. 9, How to Write a Municipal Tree Ordinance,* The National Arbor Day Foundation, Nebraska City, NE.

Garcia, G., 1991, Why and how to adopt a tree preservation ordinance, *New Jersey Municipalities Mag.* **16:**116–119.

Grey, G., 1993, *A Handbook for Tree Board Members*, The National Arbor Day Foundation, Nebraska City, NE.

Hunt, R., Adam, D., Roettgeting, B., and Koehler, C., 1982, *Guidelines for Developing Ordinance in Urban Forest*, Cooperative Extension University of California, Davis, CA.

Johnson, C. W., Baker, F. A., and Johnson, W. S., 1988, *Urban and Community Forestry: A Guide for Interior Western US*, USDA Forest Service, Salt Lake City, UT.

Mirelli, J., 1991, South Brunswick's strategy: A stronger tree ordinance, *New Jersey Municipalities Mag.* **16:**125–127.

NJ Forest Service, 1991, *A Model Ordinance for Municipalities, Concerning the Removal, Protection, Planting of Trees during and after Construction*, NJ Forest Service, Trenton, NJ.

NJ Forest Service, 2004, *Tree City USA Program Guidelines*, NJ Forest Service, Trenton, NJ.

Sheay, R. 1992, *Bureau of Forest Management Field Manual*, NJ Forest Service, Trenton, NJ.

Chapter 5

Using Comprehensive Planning to Conserve the Natural Environment

William F. Elmendorf

1. The Natural Environment Benefits Community

Although our relationship with the natural environment has changed as a result of modernization, growth, and development, the natural environment in areas of "open space" continues to provide communities with many economic, environmental, and social benefits that have been well documented (Albrecht, 1993; Center for the Study of Law and Politics, 1991; Dwyer *et al.*, 1991, 2003; Hylton, 1998; McMahon, 2000; National Park Service, 1992; Wolf *et al.*, 2003). In addition to these well-known benefits, the planned conservation of open space helps communities absorb change by creating special places and preserving structured and shared symbols such as historic landscapes (Greider and Garkovich, 1994). By providing a framework for communities, the local autonomy of a community is defined and reinforced through the use of open space, increasing the psychological identification with a locale and making people feel more comfortable and at-home (Bender, 1978; Hawley, 1950; Warren, 1972). Frederick Law Olmstead wrote that parks and open space are benign magnets for social democratization and recreation (Wilson, 1989). The green infrastructure of accessible open space provides opportunities for people to generalize across interest lines and helps create a sense of community—one characterized by shared spatial experiences. It creates an element of social equality because as an equalizer open space offers people the same opportunities regardless of background or social and economic status. The economic value of nature to communities is important. Seemingly ahead of their time, in the Brooklyn Report of 1868, Frederick Law Olmstead and Clarence Vaux wrote that landscape beauty paid enterprising cities directly by raising property values and swelling tax revenues (Wilson, 1989). They argued over 100 years ago that the cost of reserving parks would be more than met by

Dr. William F. Elmendorf Assistant Professor, State Extension Specialist, Community Forestry, Penn State School of Forest Resources.

Urban and Community Forestry in the Northeast, 2nd ed., edited by, J. E. Kuser.

increasing taxable value of real estate and their positive impact on human health and society. Open space improves the quality of life and environment and contributes to economic and social life and is an important factor in the development of healthy and sustainable communities. In order to improve the quality of life and environment in our communities, it is important that the costs and benefits provided by the natural environment through open space are understood and considered through a comprehensive land use planning process. But what is a human community?

2. Community

The notion of community is a fundamental idea with a myriad of definitions surrounding the concept. Some definitions rely on the image of the New England village and express heartfelt sentiments of safety, security, and comfort, including positive visions of the good life (Bender, 1978). Family, supporting institutions (school, church, health care, local government, and financial institutions), social interaction, shared territory, a common life, and collective action are elements found in most social definitions of community (Hillery, 1955; Wilkinson, 1991).

Wilkinson (1991) described three essential properties of community: as a local ecology (meaning an organization of social life that meets daily needs and allows for adaptations to change); as a comprehensive interactional structure, or social whole, that expresses a full round of interest and needs; and as a bond, or field, of local solidarity represented in people acting together to solve common problems. Community is the social field that emerges where people frequently act together in the common concern of life (Kaufman, 1959). This field is locally oriented and is where actions expressing a broad range of views are coordinated. In this field, actions, actors, and associations are oriented toward the general needs and concerns of the community. Community is not a place; rather, it is a place-oriented process. A community is a process of purposive and interrelated actions through which residents express their shared interests in local society (Wilkinson, 1991).

A common life is a complex of interactional relationships, and the foundation of community is social interaction that involves individuals and groups working together in pursuit of generalized and commonly held goals (Wilkinson, 1991). Similarly, using concepts from network theory, Granovetter (1973) has described community as consisting of people in social interaction within a geographic area having one or more common ties. The quality of horizontal social interaction between local people is important, and there are differing types of interactions, including strong ties between family and friends and weak ties between people on a day-to-day basis. The weak ties of day-to-day life help modify strong ties so that they are not stifling and work to bind strong ties into a larger and more cohesive community structure or field (Granovetter, 1973). Also important are the vertical interactions of communities with broader social fields, such as corporations, media, and government agencies, which can provide opportunities for mutual understanding, action, and support. People will continue to live in interaction with each other and with broader social fields and territories. Both the vertical interaction of community with outside institutions and markets and the idea of functionally interdependent local societies are important.

As places experience population growth, change is inevitable. Growth has drastic impacts on a community, including changes to natural, physical, economic, social, historical, political, and spiritual attributes. Change is brought on by social, economic, demographic, and ex-locale forces causing problems and opportunity with both the natural environment and social/economic conditions. In the light of changing conditions, growth decisions can be some of the most baffling, toughest, troubling, and conflicting decisions a community can face. Some communities are better able to deal with change than others—that is, the competent community (Cottrell, 1983). In competent communities, people collaborate effectively in identifying the needs of the community, achieve a working consensus of goals and priorities, agree on ways and means to implement agreed-upon goals, and collaborate effectively on required action. Communities have a choice when faced by growth and change—to try to actively shape the internal and external forces of growth and change, or passively accept unplanned, and often costly, community development.

3. Community Development

There is a basic tension between those whose primary goal is development "in" community and those who prefer development "of" community. Those interested in the latter perspective view community as the quality of relationships among residents of a locality that supports community's well-being (Christenson and Robinson, 1989). The creation and maintenance of activities, organizations, and institutions that strengthen interactional ties among residents is deemed necessary for development of community to occur. Wilkinson (1979) described the community development process as actions ongoing in local social life undertaken with positive regard to community structure. Development of community requires attention to the structure and fabric of social life (Kaufman, 1959; Wilkinson, 1991). In contrast, development in community views the community as a setting in which social, political, and economic development activities occur. The locality is treated like a speculative interest and attention is given to the efficient use and maintenance of productive resources, as well as its adaptability to changes in external markets and environments.

Economic development is an overriding objective in community development. As rural sociologists have discussed, community development, or revitalization, simply cannot begin if it does not start with jobs and income (Wilkinson, 1991), but economic development without development of community can be divisive and exploitative. A developed community is both improved and empowered. Community development, from an interactional perspective, means improving peoples' economic conditions, facilities and services, and empowering the capacity of local people to work together to address their common environmental, social, and political interests. As a process, development means capacity building (e.g., the ability to do work), and community development means building the capacity of people to work together to address their common problems.

In community development, guided by a comprehensive planning process, land is the essential economic resource controlled by a municipality; this control gives a community a substantial ability to direct its own economic viability. Open space, consisting primarily of working and more passive landscapes, can help development in and of communities by increasing and stabilizing economic, social, and environmental elements (Eisner and Gallion, 1993; McHarg, 1992). Furthermore, land use and change is often at the heart of decisions and conflicts facing communities in the light of growth. Planning attempts to deal with change by adapting individuals and organizations investing in development to the values expected by a community; creating the desirable, avoiding undesirable costs (e.g., infrastructure, fire, and flood), and assisting development projects to meet community expectations. A comprehensive planning process is required to actively shape the internal and external forces of change, helping a community to remain vibrant and healthy. In this process, citizen involvement is important. The extent up to which community members are involved and act together in planning for the use and management of land increases the quality of relationships among residents, increase competent community action, and help community occur. A sound, comprehensive planning process is a cornerstone in the development of community and in the integration of the natural environment into community development.

4. Planning

Planning, whether community, financial, or business, has been defined simply as the process for determining appropriate future action (Davidoff and Reiner, 1962). In reality, planning should be an optimizing technique—a system that identifies what plan should yield the best combination of effects and allow for the use of analytical techniques, including cost–benefit analysis, internal rate of return, and net present value to choose among a set of alternative plans. Planning can be used to identify incremental solutions or utopian plans that show end states and courses of action, involving fundamental or structural changes in organizations, social values, or environments (Davidoff and Reiner, 1962). Any possibility of effective planning relies on the assumption that people control their own destiny and have a clear understanding of existing conditions and the velocity and direction of economic, social, and environmental change and other planning variables such as residents' and leaders' values, attitudes, needs, and abilities.

Planning is completed for a number of reasons (Davidoff and Reiner, 1962): (1) To facilitate efficiency and rational action. In a world of scarcity, there is a need to conserve resources through efficient and rational decision-making and action. Resources must be allocated efficiently and optimally, and efficient means reduced costs and destruction. But in a complicated political/power system and with different attitudes, abilities, priorities, and agendas, efficient use can be confused with that which satisfies particular or more powerful actors. (2) To aid in achieving a fully competitive market. Planning would supply little use or worth in an arena where buyers and sellers know fully the relative worth of items and services over time. The planning process collects, analyzes, and publicizes information required to make good decisions. It provides a factual basis that permits the evaluation of various value alternatives and actions. (3) To better deal with change and widening of choices. Given scarcity, planning assists social choices about the allocation of resources. Planning helps allocate scarce resources by answering questions regarding how, when, to whom, to what purpose, and in what combination.

5. Comprehensive Planning

Community planning is defined by the American Planning Association (Eisner and Gallion, 1993; p. 201) as: "a comprehensive, coordinated, and continuing process, the purpose of which is to help public and private decision makers arrive at decisions that promote the common good of society." Planning is an organized approach for solving community problems (Lembeck, 1993). One task of community planning is to provide information to decision makers and the public at large and empower these individuals through education, involvement, and public hearings to take part in community action or agency.

Planning guides growth and change through a number of different functions including: formulation of a vision or plan to guide future development; collection, analysis, and provision of information to allow decision making; identification and consideration of different community alternatives; coordinating local development activities; review and modification of proposed land development; providing information and opportunities for public involvement; and the formulation and update of ordinances. The orientation of planning is to future costs; that is, taking appropriate actions today that will avoid costs in the future. Lembeck (1993) describes four "costs" that may result from not planning for community growth and development: (1) direct mistakes that take municipal revenues to correct (e.g., cost of landslides, fire, and flood); (2) indirect mistakes that cause added municipal expense (e.g., cost of infrastructure, public safety and education); (3) loss of value mistakes that cause public and private property to depreciate (e.g., placement of improper industry in residential area); and (4) failure to secure private investment that would have been made if appropriate actions had been taken (e.g., not providing an attractive place for people and business).

The process used in community planning is a system's technology to make rational decisions that provide positive economic, social, and environmental returns. A sound planning process includes (Eisner and Gallion, 1993; Chaplin and Kaiser, 1979): (1) putting together the planning group by including all those individuals interested, affected and important (2) mutually agreed-upon identification of problems and issues; (3) research and analysis to provide definitive understanding of problems and issues; (4) formulation of goals and objectives to be attained in alleviating problems or resolving issues; (5) development and evaluation of alternative methods (plans and programs) to attain agreed-upon goals and objectives; (6) recommendations of appropriate courses of actions from alternatives; (7) assistance in implementation of approved plans and programs; (8) evaluation of actions taken to implement approved plans and programs in terms of progress toward agreed-upon goals and objectives; and (9) a continuing process of adjusting plans and programs in the light of results of such evaluations or to take into account-changing circumstances.

Municipalities are the creatures of the state that has sovereign power over them. States pass the authority and responsibilities for community planning and land use regulation to local government through state constitutions and enabling legislation. There are different state approaches to planning and land use regulations, with some states providing more authority for land use planning and regulation. In most cases, state-enabling legislation, at a minimum, empowers local government to: (1) plan for and govern development through zoning, subdivision, and other ordinances official maps and reservation and acquisition of land for public purposes; (2) establish planning departments, commissions, committees, and zoning hearing boards; (3) establish transferable development rights, conservation subdivisions, and other innovative processes; and (4) provide for mediation and appeals to courts and penalties.

A distinctive feature of planning in America is the tradition of citizen or volunteer planning commissions. These citizen groups are a by-product of the 19th century civic improvement associations, which were formed to increase the beauty and function of Post-industrial Revolution American cities. Typical duties and powers of a planning commission include: (1) preparation of a comprehensive plan, (2) preparation of official maps, (3) preparation of zoning ordinances, (4) preparation and administration of subdivision and other land development regulations, (5) preparation of environmental studies, (6) preparation of water surveys, (7) promotion of public interest and action in understanding the comprehensive plan and development issues, (8) holding public hearings and testimony, and (9) reviewing and amending zoning, subdivision, and other ordinances and the municipalities comprehensive plan.

In America, states, counties, and municipalities have the power to pass and enforce laws to regulate land use. This power, established and expanded through 200 years of federal and state judicial review and legislation, is termed Police Power. A court case (*Village of Euclid v. Ambler Reality Company* 272, US 365, 1926), helped confirm the democratic foundation of land use regulation as an instrument by which people could attempt to order the destiny of their cities. In this decision, Associate Justice George Sutherland of the US Supreme Court wrote, "Until recently, life was comparatively simple; but with the increase and concentration of population, problems have developed, and constantly are developing, which require additional restrictions in respect to the use and occupation of private lands in communities" (Eisner and Gallion, 1993; p. 113). Land use regulations, such as zoning and subdivision and land development ordinances, are based on the concept of Police Power, which is government's fundamental power to pass laws to promote and protect public health, safety, morals, and the general welfare. The greatest challenge to any land use regulation is that it must effectively demonstrate that the regulation does promote or protect one of these components. The ability of government to regulate the use of private property is associated with the "takings" issue (McElfish, 2004). A "taking" is a problem of government overregulation that goes so far as to result in confiscation of private property without payment of compensation or due process. United States Supreme Court Justice Olivar Wendell Holmes in (*Pennsylvania Coal Company v. Mahon* 260, US 393, 415, 1922) wrote, "The general rule at least is, that while property may be regulated to a certain extent, if regulation goes too far it will be recognized as a taking" (Eisner and Gallion, 1993; p. 113). The beauty of this ruling is the ambiguity surrounding the term "too far." Because of the court's unwillingness to become the "supper legislature," too far continues to be defined on a case-by-case basis. This reinforces the place-based or local nature of land use regulation established by local government through a local democratic process. Planning and land use regulation must balance the investment in and development of private property with reasonable and conforming land use and the health, safety, and welfare of a community.

Planning instruments are developed by, and used in, the planning process to understand and guide community growth and development. In jurisdictions with comprehensive planning, development projects are reviewed in relation to the content of a comprehensive plan, land use regulations or ordinances, and the wishes and desires of leaders and citizens. Comprehensive plans, zoning regulations, and subdivision and land development regulations are fundamental instruments of community planning that can be used to integrate the natural environment into community development (Arendt, 1999; Eisner and Gallion, 1993; Lembeck, 1993; McElfish, 2004).

5.1. Comprehensive Plans

The principal way a planned community deals with future growth and tries to avoid costly mistakes is to prepare a comprehensive plan. The comprehensive plan is a community's attempt to picture future growth and development, outlining what needs to be done, and when and how it will be done in order to create a healthy, well-functioning, and competent community (Lembeck, 1993). The planning process is used to develop this plan, which envisions, identifies, and demonstrates a community's desires (vision, goals, and strategies) for future growth and development. A comprehensive plan is a growth management plan and incorporates community input and background information (such as a natural resource inventory) to provide goals for development, a logical land use map for zoning, a legally recognized basis for all land use ordinances, supporting policy, and maps to guide future placement of roads and other public services. The comprehensive plan and the process from which it evolves (planning process) create community development policy, which essentially are rules that set forth mandatory, optional, and prohibited courses of action in the development of a community or region (Lubka, 1982). A comprehensive plan essentially is a report with maps outlining goals, strategies, standards, and policies toward growth and development. Plans include an inventory and assessment of the economic, social, and environmental states of a community; projection of future population size; and projections of the desired future states of economic activities, traffic, parks, and open space, housing, and other relevant aspects of community.

Lembeck (1993) describes four important parts of a comprehensive plan. (1) *Background information section*: This section provides a general impression of a community by describing history, natural features, population characteristics, community facilities, and other community characteristics. (2) *Resource inventory*: This section describes in detail the economic, social, natural, and cultural resources existing in a community including a map of important locations. (3) *Vision, goals, objectives, policies, and strategies*: This section relates the economic, social, and natural resources of a community to a development vision. A vision statement within the comprehensive plan describes the future community, using such sentiments as well-being, beauty, community image, attractiveness, sustainability, future generations, economic prosperity, environmental quality, and quality of growth and development. Goals are broad statements of intent that define the state of ultimate desired achievement. Goals expressed in a plan can include such concepts as increasing the enjoyment and quality of community life, providing a safe and healthy community environment, and increasing community commerce and business. Objectives describe more specific achievements and requirements necessary to accomplish goals and can become the heading for elements of the comprehensive plan. Within each objective or element, policies and strategies are organized. Often the policies and strategies expressed in the comprehensive plan are later expressed as zoning and other ordinances. (4) *Plan elements*: The following elements, or chapters, are universally prepared as part of a comprehensive plan. (A) The land use element that shows the basic current and expected patterns of land use in the community and how the types of land use (residential, commercial, agricultural, industrial, institutional, and open and natural space) are situated to each other. This element becomes the logical basis for the zoning MAP and regulations. (B) The transportation element, the provision of roads and other infrastructure to move people and goods within and through the community is an extremely important influence on development patterns. This element uses maps to show the desired end states of road and other transportation systems. Transportation planning is not only limited to roadways, but also includes parking, terminal facilities, airports, public transit, bicycle paths, and pedestrian ways. (C) The community facilities and utilities element, which are the public, private, and institutional infrastructure to, serve the community. Development often follows the provision of sewer and water, as well as roads and other infrastructure. This element identifies water, sewer, utilities, and other infrastructure to support and promote future development. Typically this element includes proposed recreation areas, schools, hospitals, churches, police and fire stations, civic buildings, water treatment plants and lines, sewerage treatment facilities and lines, cable TV, recycling centers, libraries, and the like. (D) Conservation element that describes how natural resource will be conserved in development. The first step in acquiring, conserving, and protecting the natural environment, such as riparian areas and open space, is to place the goals, objectives, and strategies of incorporating the natural environment into a community's comprehensive plan. If the natural environment is not considered in the comprehensive plan, it will not be adequately or correctly addressed in zoning, subdivision, or other ordinances, or in capital budgeting used to implement policies expressed within the plan. Furthermore, natural resources will not be adequately considered in the review process for individual developments. A conservation element may be prepared to consider conservation of such things as historic landscapes, forests and woodlots, open space and greenways, sensitive areas, and riparian and ground water resources. (E) Capital budget element that describes how public facilities (such as the purchase of open space) will be funded through bond issue, general taxes, municipal revenues, and partnerships. (F) Plan implementation that describes how objectives and policies will be implemented, including using municipal and private sector resources, developer incentives and other nonregulatory approaches, and land use regulation.

Planning decisions involve the allocation of private and public resources, including the natural environment. To facilitate the incorporation of the natural environment into community development there must be persuasive information placed in a conservation element of a comprehensive plan depicting the natural environment, threats to its sustainability, and the standards and policies required to assure future conservation.

5.2. Zoning Ordinances

In 1916, Edward Bassett, a New York City attorney, defined zoning as, "The regulation by districts under the Police Power of the height, bulk, and use of buildings, and the use of land, and the density of population" (Eisner and Gallion, 1993; p. 113). Under this principle, zoning has survived numerous court challenges of which *Village of Euclid v. Amber Reality Company* (1926), is particularly significant. The accepted origin of state-enabling legislation for zoning is the model Standard Zoning Enabling Act of 1926 published by the United States Department of Commerce (Nolon, 2001). Traditional zoning provides standards for the height of buildings, the percentages of a parcel that may be developed, the density of development, and the use of land (e.g., residential, commercial, industrial, institutional, rural resource, agricultural, forest, open space, special, or other permissible activity such as forestry). Zoning is not planning but is a tool of planning. Zoning is accomplished through a planning process, the development of a comprehensive plan, and enacted by ordinance thorough public hearing and the vote of elected officials. Zoning enacted through a democratic process is a planning implementation tool. One phrase found in many state-enabling statues is that zoning ordinances shall be, "prepared in accordance with a comprehensive plan." Zoning without comprehensive planning will not logically or successfully regulate for the creation of a viable project or community.

Traditional zoning ordinances do not protect any lands beyond those that are wet, steep, or flood prone. Traditional zoning establishes single-lot size minimums on property and the number of lots on a property basically becomes the size of the property divided by the size of the minimum lot size. Thus, with a zoning ordinance that calls for single-family homes on a minimum of 80,000 feet^2 lot sizes, and no conservation of open space, a 6,400,000 feet^2 parcel would be allotted 80 residential lots with no public or continuous open space. Lot lines, streets, and utilities, are laid out on a parcel map without any consideration of natural features. This type of traditional "cookie-cutter" zoning does not encourage the conservation of natural features, such as open space, but rather encourages the maximum number of minimum- or maximum-sized lots. Progressive zoning ordinances are used for the purpose of encouraging innovation and the promotion of flexibility, economy, and ingenuity in development. These ordinances would mandate the conservation of open space and the preservation of riparian areas, steep slopes, and the like.

5.3. Subdivision and Land Development Ordinances

As with traditional zoning ordinances, conventional subdivision ordinances often lead to cookie-cutter development of strips of lots and homes along streets and fitting in as many lots as possible on a piece of property. Subdivision ordinances often do not consider natural resources, but rather only provide standards for subdivision review process and design, such as parceling of land into building sites; a sequence for development-permit process; and required improvements for adequate lots such as utilities, streets, and sewage. To consider natural resource conservation, subdivision ordinances must distinctly address the internal relationship of woodlots, riparian areas, steep slopes, and open space with the design of house lots, streets, and neighborhoods. Subdivision ordinances can be used to mandate developer completion of natural resource inventories and allow for creative development that preserves woodlots, open space, riparian areas, and other natural resources (such as conservation subdivisions), mandatory dedication of open space in development, and in lieu fees to be paid to acquire off-site open space. Subdivision ordinances must work with zoning ordinances. A zoning ordinance might require that 50% of a subdivision over 10 acres be conserved in open space, that a 100 foot buffer be provided for riparian areas, and that allowable density be clustered on smaller lots. The subdivision ordinance would distinctly place standards for the management of storm water, the protection of riparian buffers, and in detail outline a five-step design process for natural resource conservation (1) completion of a natural resources inventory and map to identify primary (e.g., riparian areas, woodlots, and meadows) and secondary (e.g., views, worthy trees, and geological features) conservation features; (2) identification of areas to be preserved; (3) identification of development areas based on cluster density; (4) placement of clustered parcels and lot lines; and (5) placement of streets and other infrastructure (Arendt, 1999).

6. Problems with Comprehensive Planning

Through the planning process, planners have been guiding growth and intervening in community development since the 1893 Chicago's World Fair and the Colombian Exposition gave impetus to planning in America. Natural systems found near growing communities are under continuous and sometimes enormous social, economic, and physical pressures for continued development. In terms of the historical relationship between community development and the natural environment, a question can be posed. In light of modernization, investment, and growth, has land use planning successfully provided guardianship and stewardship to integrate open space into community development, or has it been ignored and disregarded?

While other authors are more optimistic about planning, Warren (1977) is of the opinion that most changes that take place in communities are not the result of purposive planning, but rather are hard-won by some group over hard opposition of another group. He believes that planning promotes an ideology that structurally ineffective services and organizations will become effective only if better organized. Furthermore, he maintains that planning has a number of flaws, including (1) planning discourages major intervention and structural change; (2) planning lulls people into thinking that positive change is being made; (3) planning lulls people into thinking that there will be no negative impacts as a result of growth; and (4) many times planning itself constitutes a biased growth agency. According to McHarg (1992), a progressive and sustainable planning and design philosophy is to design with nature rather than to continually impose unattractive and unsustainable development in the creation of human settlements. Although conceptually acknowledged, McHarg and others conclude that his planning theory continues to be wrongly ignored in ongoing community development.

7. Problems with Planning for the Natural Environment

Jackson (1956) wrote that landscapes are spatial representations of complex social, political, and economic processes, and there is much that can be learned through observing them. As a result of shifting economic and social realities, particularly after World War II, American landscapes have changed radically. The development of rail, air, auto, and other transportation systems; changes in modes of communications and information processing; and the suburbanization of people, commerce, and capital has fueled a different type of development ecology (Lewis, 1994). According to Lewis (1994; p. 17) this "Galactic City" is a new city ecology, "where all the traditional urban elements float in space like stars and planets in a galaxy, but with large spaces between." Elements of this ecology include an internal transportation system dedicated to the automobile, a remarkable degree of internal commercial and industrial clustering along transportation routes, and spreading residential development based on the provision of transportation, water, and sewer. This development ecology is represented by Levittown type subdivisions, the American strip mall, and horizontally spreading urban, suburban, and rural settlements. Furthermore, because of impacts to the interactional patterns of people, some sociologists are of the opinion that modernization and the current settlement pattern of suburbia fuels the eclipse or destruction of a sense of neighborhood and community (Bender, 1978; Hawley, 1950; Warren, 1972). Urban sprawl continues to convert and damage the natural environment and diminish the extent to which communities can be distinguished as identifiable bundles of social interaction.

Given the spreading nature of our current settlement patterns, problems associated with the loss, fragmentation, isolation, and destruction of agricultural lands, forested lands, and watersheds have been described by a number of authors (Bouton *et al.*, 1991; Falk *et al.*, 1992; Harper and Propst, 1990; Lubka, 1982; McMahon, 2000). Negative impacts to the quality and quantity of open space include extinction and removal of species, destruction of ecosystems, destruction and damage to productive forest and agricultural areas, loss of biodiversity, fragmentation of recreational opportunities and wildlife habitat, mass erosion of soil and soil nutrients, and contamination of aquifers. Opportunities for the creation of viable open space systems are compromised or eliminated by unplanned and uncontrolled urban sprawl, which is associated with our society's dependence on the automobile. Planners are of the opinion that from a human standpoint that does not even consider larger requirements for healthy ecosystems, 30 acres of open space should be provided per 1000 people, but existing open space systems have not kept pace with growing populations (Center for the Study of Law and Politics, 1991; Harper and Propst, 1990; Falk *et al.*, 1992). Although land use planners have long promoted open space and greenbelts to be incorporated into community development, the importance and benefits of these natural systems are often ignored because of high-land costs, the profitable nature of development, poor land use planning, speculative or short-term priorities, concerns for private property rights, and a competitive land use arena (Miller, 1988). As time goes on, opportunities for natural resource conservation are eliminated by land conversion, fragmentation, and increasing land costs in a competitive real estate arena.

It is important to consider that natural ecosystems and landscapes are based on natural boundaries, such as floodplains and watersheds, not on jurisdictional boundaries, and the development of high-quality human recreation opportunities and systems such as wildlife habitat requires large-scale or regional planning. This would mean intermunicipal connections and cooperation, something quite rare today because of feuds, power struggles, and competition for tax dollars. The success of past attempts at regional planning to break through structural barriers (such as autonomy and self-sufficiency of local governments long held central to the American individualist paradigm) to bring municipalities together has been questioned. The lack of regional authority to regulate land use and facilitate cooperation and coordination between jurisdictions is seen as a major obstacle in achieving viable large-scale open space because of planning, administrative, and funding concerns.

Because of lack of information on the natural environment, lack of comprehensive planning, inadequate zoning and subdivision ordinances, poor developmental review, lack of partnerships, and the power of speculative interests, the community planning process has had difficulty in the past considering the natural environment, and there appears to be no evidence that such changes will cease in the near future. The creation of larger-scale natural resource systems mandates a single entity dedicated to broad-based and open participation, education and empowerment of residents, setting of mutual goals, working across jurisdictional boundaries, collecting and monitoring data on a large scale, providing adequate funding for resource acquisition, and other elements not always considered in local land use planning at the single municipal scale.

8. Regulatory and Nonregulatory Strategies for Conserving Natural Resources

Objectives for planning, managing, and acquiring open space in developing landscapes can include: preserving the best agricultural lands for agriculture and the best forest lands for forestry; preserving landscape and biodiversity; preserving remnant natural areas; preventing intrusion into sensitive areas by residential and other development; protecting scenic views and historic resources; and preserving riparian areas, slopes, greenbelts, and other special landscapes such as wildlife habitat (Harper and Propst, 1990; Lubka, 1982). Arendt (1994) describes nine "steps" that can be used to plan for the stewardship of open space: (1) an understanding and documentation of community resources; (2) a realistic understanding of the future through a comprehensive plan; (3) reasonable goals for both conservation of the natural environment and economic development; (4) a sound progressive zoning framework; (5) a process for innovative/conservation subdivisions; (6) a good working relationship between municipal officials and landowners; (7) mechanisms for acquiring and managing important lands; (8) a process for multijurisdictional cooperation in planning; and (9) strong leadership and citizen support.

With these steps in mind, there are a number of different regulatory and non-regulatory strategies that can be used to achieve the objectives of incorporating open space preservation and acquisition into community growth and development

(Falk *et al.*, 1992; Harper and Propst, 1990). Although there is a growing and continued debate over governments' authority to limit economic gain from private property (the taking issue), the following strategies are used singularly and in combinations to surmount problems such as limited funding, high-land costs, the competitiveness of the land use arenas, and structural and other barriers to consensus and cooperation.

- *Increased state planning*: In this concept, land use planning and regulation is considered a state responsibility with a broad regional and state perspective. State legislation, such as California Environmental Quality Act would require all municipalities to complete a comprehensive plan and review development proposals in a comprehensive fashion. The state takes a more active role in local planning processes to provide comprehensive and coordinated efforts, as well as monitoring growth and change. State and local comprehensive plans would coordinate land designation and land use. The state would supply increased technical, financial, and educational support and assistance to local agencies (Galvin *et al.*, 2000).
- *Multimunicipal planning and zoning*: Strong regional coordination is crucial for preserving viable open space systems. The implementation of a regional comprehensive plan and zoning ordinance can provide a clearer understanding of how and where the natural environment will be preserved. It can also provide multimunicipal uniformity and consistency in planning for and regulating growth by providing regional planning goals and land use regulations. Also, cooperating municipalities can more effectively address development proposals that have multimunicipal impacts.
- *Natural resource inventory*: Identifying land with important resources and placing this information on a planning tool, such as a geographical information system allows planners to understand existing natural systems, as well as areas of special concern for both conservation and development. In defining the natural environment, the following resource categories can be used (Bouton *et al.*, 1991; Falk *et al.*, 1992): significant wildlife habitats, scenic areas, river and riparian corridors, recreational resources, productive agricultural resources, productive forest resources, woodlands and natural areas, open space, special or unique landscapes, riparian areas and wetlands and their buffers, floodways and floodplains, historical and cultural landscapes, mineral and oil resources, contiguous block, vulnerable landscapes, moderate and steep slopes, viewsheds, and greenways, and corridors for the passage of people, and wildlife and other natural elements.
- *Zoning options*: Zoning options to preserve open space and other natural resources include: (1) existing use zoning that is the legal use of a parcel based on the use, which the parcel is now in. This protects both traditional land uses and the prevailing character of a community and surrounding where the use could be forest or agricultural management areas. (2) Overlay or sensitive area zoning that addresses areas with unique challenges. Additional regulations are "overlaid" on underlying zoning. The use and development of sensitive areas, such as impervious surfaces, wetlands, important habitat, hillsides, riparian

areas, fragile shorelines, and worthy open space is restricted or prohibited. (3) Conservation zoning that sets the groundwork for conservation subdivision by providing for smaller lot sizes, clustered development, and open space protection. A certain percentage of the parçel is conserved in open space and lots are clustered together on a parcel allowing for the protection of open space and other natural and cultural resources on the remaining parcel. (4) Sensitive area ordinances, zoning ordinances are used to regulate and oversee the removal of vegetation, destruction of riparian areas, and limit grading on steep slopes or other areas. They can be used to ensure that sensitive design and grading is accomplished and a reasonable percentage of vegetative cover and natural landforms remain on a property after development. (5) Disturbance standards that place limitations on the percentages of natural resources that can be disturbed by development. (6) Impervious surface standards that place limitations on the percentage of ground that can be covered with asphalt or other impervious surfaces. (7) Natural resource setbacks that place required setbacks for grading and other disturbances from natural resources. (8) Timber harvest ordinances that provide standards for the completion of timber harvesting plans, types of harvesting methods used, protection of riparian and other areas, and construction of roads and landings. (9) Tree preservation ordinances that provide standards for the preservation of trees in development, including the completion of tree inventories, the completion of tree preservation plans, and the use of retaining walls and other engineered devises. (10) Official maps that are a zoning tool enacted by ordinance that show proposed public parks, play grounds, green ways, open space, pedestrian ways and easements, and other reservations. If the owner of the land on an official map applies for a building permit, or submits written notice of intention to subdivide or develop, the municipality has one year to acquire the property shown on the official map.

- *Cluster, open space, or conservation subdivisions (Arendt, 1994, 1999)*: This is a subdivision design that allows for development and community commerce, while maintaining the integrity of worthy farmlands, forests, open space, and habitat (contrasted to traditional segregated, large lot, and dispersed subdivision as development and infrastructures are concentrated). Goals of conservation subdivision design include understanding the natural features of a site, conservation of most sensitive natural areas, such as wetlands, and keeping fragmentation of resources to a minimum by designing individual subdivisions as part of a larger continuous integrated system between subdivisions and municipalities. There is a five-step approach to designing conservation subdivisions (1) development of a context map that identifies the subject property in relationship with surrounding landscapes; (2) identifying primary and secondary conservation areas by site analysis and placing areas, such as woodlots, riparian areas, and landmark trees on a detailed site features map; (3) locating potential house sites on clustered parcels that preserve conservation areas but allow full density; (4) aligning streets in the optimum manner to access homes; and (5) drawing lot lines. This site-specific design process identifies the important natural features of a site and clusters homes and infrastructure to preserve them.

- *Growth areas, growth boundaries, or service boundaries*: Rural resource or other zoning is used to provide a horizontal limit to development of certain types, including in most cases subdivision, commercial, or industrial. These areas are characterized by a firm, long-term (20 years) commitment to compact development within the boundary line or area and rural preservation outside the boundary line. Municipalities must commit themselves to infill housing and cluster development within the area. Furthermore, to limit development outside the growth area, the provision of roads, public water, and public sewer must be restricted.
- *Transferable development rights*: In a comprehensive planning process a municipality identifies areas appropriate for conservation (such as valuable forest or agricultural lands). Conservation areas are classified as sending areas and areas appropriate for development (growth areas) are identified and designated as receiving areas. Development rights of the sending district are purchased from the owner and are transferred to the receiving district, restricting development in a sending area. This tool is often used to transfer development to growth areas.
- *Planned residential development*: A flexible regulatory technique that combines aspects of both zoning and subdivision regulations. It encourages the mixing in the same area of a variety of different and compatible land uses. Development is not limited to residential lands uses, but mixes residences of all types, shopping, offices, and places of entertainment in the same development. It also encourages the incorporation and preservation of natural resources within the development. A key to this development is the preservation of a very large proportion of the site in open space. Perhaps 40%, 50%, or even 60% of the development would be left in open space.
- *Traditional neighborhood development*: Provisions adopted by the municipality for a traditional neighborhood development include the following: (1) the amount, location, and proposed use of common open space to be distributed throughout the development and (2) the location and physical characteristics of the development that is provided for the preservation or enhancing of natural features such as wetlands, ponds, waterways, trees of high quality, wooded areas, and other significant natural features.
- *Development review and regulation*: Dedications from developers are required through conditions placed on the approval of development permits. Dedications to jurisdiction(s), such as land for greenways, open space, and parks, are mandated in the terms of the development approval.
- *Less-than-fee-simple land acquisition*: Land is viewed as a bundle of rights including development, mineral, view, and easement. Less-than-fee-simple is protection of the natural environment by easement, where the easement buyers pay only for land rights that need to be protected (development, access, recreation, river corridor, and view). This concept allows owners to cash in on development rights, while land is preserved and managed as open space. Development rights are 40 to 90 of full market value of property depending on location.
- *Full-fee acquisition*: The total bundle or all land rights are purchased. This method is good for preserving high-quality lands with natural and public values. High-priority acquisitions are identified by natural resource mapping.

- *Tax benefits*: Tax incentives include no capital gains for landowners who sell land rights for conservation purposes to governments and conservation groups. Tax incentives also could be supplied if public agency or conservation groups acquired the right of first refusal or first call from existing property owners. Tax benefits, such as lower property taxes, also can be granted for individuals who hold their land in open space for a period of time under contract with state government.
- *Rolling leases*: If landowners are not interested in selling land rights less-than-fee-simple or fee-simple, rolling leases would allow temporary use and preservation of land rights.
- *Public–private partnerships*: Nonprofit organizations, such as the Santa Monica Mountains Conservancy, can be established to partner with municipal agencies in the acquisition of open space and other natural resources.
- *Nontraditional and multidisciplinary teams*: Include environmental justice and civil rights organizations in land use discussions. Multiple jurisdictions and agencies working together with land conservancies for land acquisition. Cultivating the interest of major corporate and private foundations in land acquisition and educational programs.
- *Empowering the public*: Through public forums, committees, surveys, education, training, and volunteerism, providing citizens abilities and opportunities for participation and involvement.
- *Developer incentives*: Many incentives, such as increased development density, decreased development fees, preferential tax assessments, and preferential permit processing can be used to help motivate developers and property owners to preserve natural landscapes.

In summary, there are five major regulatory and nonregulatory strategies that can be used through planning to help integrate the natural environment into community development—subsidies and incentives, taxation, regulation, public education and information, and public investment. Subsidies and incentives include reduction of permit and development fees, relaxation of development requirements, fast-track permit processing, increases in density and allowing for alternative development design, and transfer of development rights. Taxation can include reducing taxes on agriculture and forestlands held in these uses. Regulation includes zoning, subdivision, and other land use regulations. Public education and information includes planning commissions and ad hoc committees, public hearings, an unbiased media, working with youth and schools, and special celebrations and activities. Public investment includes fee and fee-simple investment funded through general tax revenues, bond issues, or referendum (Lembeck, 1993).

9. Closing Thoughts

9.1. Natural Resource Managers and Planners Working Together

It is clear that both planners and natural resource managers share responsibilities and interests in the same geographic space. Efforts to protect and preserve the natural environment in community development are greatly expanded when resource managers, community planners, and others become involved in joint projects.

9.2. Strengthen Community Planning

In many parts of the country, municipalities have not completed natural resource inventories, practice "negative planning" by reviewing current development proposals with no comprehensive plan to provide a direction, have traditional zoning and subdivision ordinances that prevent conservation of the natural environment, and have difficulty administering and enforcing codes and regulations. Furthermore, there are few examples of working relationships between municipal planning departments, rapid turnover of planning commissioners, leaders with personal rather than public agendas, and planning staff that are marginalized from both leaders and the general public. The success of incorporating viable natural resources into community development depends on the ability of local planning agencies to understand and use the fundamental tools of land use planning and regulation. If local planning agencies can be provided support by state-enabling legislation, grants, mentoring programs, and education to effectively use the fundamental planning tools of subsidies and incentives, taxation, regulation, public acquisition, and education in local jurisdiction, preservation of the natural environment will be more realizable.

9.3. Excellent Leadership

Leadership has been described by Rost (1991; p. 8) as, "an influence relationship among leaders and followers who intend real changes that reflect their mutual purposes." Although important, the values of local autonomy and self-sufficiency can be a weakness in a world of increasing interdependencies. To conserve viable natural systems in community development, local autonomy must be balanced with actions that show concern for others and a commitment to broader social and environmental values. There is a need to provide leaders with education on the values and benefits of the natural environment and planning. Learning activities that help leaders and others understand how public and private interests can be enmeshed in both local and regional interests are needed.

9.4. A Watershed or Ecosystem Approach

The importance of a contextual approach to land use planning that considers ecological, social, and economic elements in development decisions that impact the natural environment has been described by a number of authors (Gramling and Freudenburg, 1992; Freudenburg and Keating, 1985; USDA Forest Service, 1993, 1994). When current suburban and rural land use patterns are considered, many researchers believe that there is a need for more holistic and inclusive process of community planning. Ecosystem-based management is a planning and decision-making process designed to facilitate the integration of natural and social systems into community development. It is intended for larger-scale planning, such as at the watershed scale, and a number of themes revolve around this process, including ecological boundaries, ecological integrity, data collection, monitoring, adaptive management, generalizing leadership, interagency cooperation, organizational change, humanity in nature, and social values.

9.5. A Collaborative Process for Conflict Resolution

Conflict is often at the heart of change and approaches used to manage it. A collaborative approach to conflict resolution (Gray, 1989; Fisher and Ury, 1991) brings people together to listen, define a problem, understand facts, share and understand different values, and develop mutually acceptable solutions, rather than just deciding the winner. It works to separate fact from personal values, beliefs, opinions, and perceptions and clarifies what people can and cannot agree on throughout the process. Disagreement about facts may be resolved by information from an impartial technical expert, but disagreement about values, beliefs, and opinions can be difficult to resolve. Values that represent the worth a person or group places on something are often in the forefront in decision-making processes. Personal and direct explanation (educating/sharing) and acknowledgement (listening/understanding) help to change deeply ingrained values. A collaborative process for conflict resolution includes (1) good communications where importance of listening is emphasized; (2) an inclusive process where diverse viewpoints are brought together and shared; (3) mutual respect for all people at the table; (4) a focus on interest (why people want), not position (what they want); and (5) identification of shared interests and the establishment of mutual criteria based on interests. In this process, positions are moved aside so that mutual interests can be identified. These mutual interests become the criteria for collaboration.

9.6. A Broader Environmental Ethic

Building on Aldo Leopold's conservation ethic, Americans need to broaden their conservation ethic not only to bring balance to the use and conservation of natural resources on public and private forestlands, but also to consider the impact of subdivision and other land uses. The integration of the natural environment into the development of community is crucial. "In wilderness" wrote Henry David Thoreau (Bode, 1979; p. 222) "lies the preservation of the world." Through regional-watershed approaches supported by local comprehensive land use planning, the many human and natural components of landscapes and communities can be identified, understood, and incorporated into community development.

References

Albrecht, J., 1993, *Urban Forestry: A Bibliography*, International Society of Arboriculture, Savoy, IL.

Arendt, R., 1994, *Rural by Design*, American Planning Association, Chicago, IL.

Arendt, R., 1999, *Growing Greener: Putting Conservation into Local Plans and Ordinances*, Island Press, Thousand Oaks, CA.

Bender, T., 1978, *Community and Social Change in America*, Rutgers Press, New Brunswick, NJ.

Bode, C., 1979, *The Portable Thoreau* (C. Bode, ed.), Penguin Books, New York.

Bouton, J., Morion, J., Leary, E., and Sinclair, S., 1991, *Planning for the Future Forest: A Supplement to the Planning Manual for Vermont Municipalities*, University of Vermont Extension Service, University of Vermont, Burlington.

Center for the Study of Law and Politics, 1991, *Building Sustainable Communities: Open Space Preservation and Acquisition*, San Francisco, CA.

Chaplin, F., and Kaiser, E., 1979, *Urban Land Use Planning*, 3rd ed., University of Illinois Press, Urbana.

Christenson, J., and Robinson, J. (eds.), 1989, *Community Development in Perspective*, Iowa State University Press, Ames.

Cottrell, L., 1983, The competent community, in *New Perspectives on the American Community* (R. Warren, and L. Lyon, eds.), Dorsey Press, Homewood, IL, pp. 401–412.

Davidoff, P., and Reiner, T., 1962, Choice theory of planning, *J. Am. Inst. Plann.* May:103–119.

Dwyer, J., Schroeder, H., and Gobster, P., 1991, The significance of urban trees and forests: Towards a deeper understanding of values, *J. Arboric.* **17**(10):276–284.

Dwyer, J., Nowak, D., and Heather Noble, M., 2003, Sustaining urban forests, *J. Arboric.* **29**(1):49–55.

Eisner, S., and Gallion, B., 1993, *The Urban Pattern: City Planning and Design*, 6th ed., Van Nostrand Reinhold, New York.

Falk, L., Harper, S., and Rankin. E., 1992, *The Northern Forest Lands Study of New England and New York*, USDA Forest Service, Rutland, VT.

Fisher, R., and Ury, W., 1991, *Getting to Yes*, Penguin Books, New York.

Freudenburg, W., and Keating, K., 1985, Applying sociology to policy: Social science and the environmental impact statement, *Rural Sociol.* **50**(4):578–605.

Galvin, M., Wilson, B., and Honeczy, M., 2000, Maryland's forest conservation act: A process for urban greenspace protection during the development process, *J. Arboric.* **26**(5):275–280.

Gramling, R., and Freudenburg, W., 1992, Opportunity, threat, development, and adaptation: Toward a comprehensive framework for social impact assessment, *Rural Sociol.* **57**(2):216–234.

Granovetter, M., 1973, The strength of weak lies, *Am. J. Sociol.* **78**(6):1360–1380.

Gray, B., 1989, *Collaborating: Finding Common Ground for Multiparty Problems*, Jossey-Bass, San Francisco.

Greider, T., and Garkovich, E., 1994, Landscapes: The social construction of nature and environment, *Rural Sociol.* **59**(4):578–605.

Harper, J., and Propst, L., 1990, *Creating Successful Communities: A Guidebook to Growth Management Strategies*, Island Press, Washington, D.C.

Hawley, A., 1950, *The Changing Shape of Metropolitan America: Decentralization since 1920*, Free Press, Glenco, IL.

Hillery, G., 1955, Definitions University of California of community: Areas of agreement, *Rural Sociol.*, **20**(2):111–125.

Jackson, J., 1956, *Landscape*, Berkeley, CA.

Hylton, T., 1998, *Save Our Lands, Save Our Towns: A Plan for Pennsylvania*, RB Books, Harrisburg, PA.

Kaufman, H., 1959, Toward an interactional conception of community, *Social Forces* **38**(1):8–17.

Lembeck, S., 1993, Community planning, in *An Ecosystem Approach to Urban and Community Forestry: A Resource Guide* (G. Mason, ed.), USDA State and Private Forestry, Northeastern Region, Radnor, PA, pp. 23–31.

Lewis, P., 1994, *The Urban Invasion of the Rural Northeast*, Pennsylvania State Department of Geography, University Park, PA.

Lubka, J., 1982, *The Role of the Forester in Land-Use Planning*, USDA Forest Service, State and Private Forestry, Northeastern Area, Radnor, PA.

McElfish, J., 2004, *Nature Friendly Ordinances*, Environmental Law Institute, Washington, D.C.

McHarg, I., 1992, *Design with Nature*, Wiley, New York.

McMahon, E., 2000, Green infrastructure, *Plan. Comm. J.* **37**:4–7.

Miller, R., 1988, *Urban Forestry: Planning and Managing Urban Greenspaces*, Prentice-Hall, NJ.

National Park Service, 1992, *Economic Impacts of Protecting Rivers, Trails, and Greenway Corridors, River Trails and Conservation Assistance Program*, National Park Service, US Department of Interior, Washington, D.C.

Nolon, J., 2001, *Well Grounded: Using Local Land Use Authority to Achieve Smart Growth*, Environmental Law Institute, Washington D.C.

Rost, J., 1991, *Leadership for the Twenty-first Century*, Proeger, Westport, CT.

USDA Forest Service, 1993, *An Ecosystem Approach to Urban and Community Forestry: A Resource Guide*, USDA State and Private Forestry, Northeastern Region, Radnor, PA.

USDA Forest Service, 1994, *An Ecosystem-based Approach to Urban and Community Forestry: An Ecosystem Managers Workbook*, USDA Slate and Private Forestry, Northeastern Region Radnor, PA.

Warren, R., 1972, *The Community in America*, 2nd ed., Rand McNally, Chicago.

Warren, R., 1977, *New Perspectives on the American Community: A Book of Readings*, Rand McNally, Chicago.
Wilkinson, K., 1979, Social well-being and community, *J. Comm. Dev. Soc.* **10**(1):5–16.
Wilkinson, K., 1991, *The Community in Rural America*, Greenwood Press, New York.
Wilson, W., 1989, *The City Beautiful Movement*, Johns Hopkins University Press, Baltimore, MD.
Wolf, L., Westphal, L., and Kuo, F., 2003, Social aspects of urban forestry: Special Section, *J. Arboric.* **29**(3):117–155.

Chapter 6

Urban and Community Forestry: Planning and Design

Steven Strom*

1. Planning the Urban Forest

Most communities have not planned their urban forest; it has just happened. This does not mean that communities do not plan. Many communities have open space plans, park and recreation plans, and street tree programs. Very rarely, however, these plans are coordinated or comprehensive with respect to community forest resources.

As with any planning effort, three basic questions must be addressed as part of the process: What is being planned; who is responsible; and what are the objectives? In seeking answers to these questions, the complexity of the urban forest and the practice of urban forestry become readily apparent.

1.1. What Is the Urban Forest?

A clearly articulated operational definition as to what constitutes the urban forest is essential to the planning process. The nature and composition of the urban forest has been described in many ways. The following description is an adaptation of a rather succinct but encompassing definition developed by the Washington State Department of Natural Resources (McFarland, 1994). The urban forest is the land in and around areas of intensive human influence, ranging from small communities to dense urban centers, which is occupied or potentially occupied by trees and associated natural resources. Urban forest land may be planted or unplanted, used or unused, and includes public and private property and street, transportation and utility corridors. This is a compelling definition, because it not only describes what the urban forest is but also what it might become.

Steven Strom Dec'd.

Urban and Community Forestry in the Northeast, 2nd ed., edited by, J. E. Kuser.

In analyzing this definition, the urban forest is not simply street trees and parks. The urban forest is an ecosystem that includes soil, water, animals, utilities, buildings, transportation systems, people, and, of course, vegetation. Vegetation includes all plants—woody and herbaceous—regardless of where they are growing: private yards, parks, school grounds, cemeteries, vacant lots, utility rights-of-way, streets, parking lots, and so forth (Moll *et al.*, 1995).

1.2. Who Is Responsible

In a democratic society, planning should be a participatory process. Ultimately, however, someone needs to assume responsibility. It is perhaps the lack and fragmentation of responsibility that in many cases have hindered a comprehensive approach to urban natural resource planning. The issue of responsibility needs to be addressed on many levels: ownership, investment, management, regulatory authority, and the establishment of policy.

Ownership, because of its implication with respect to control, is a critical point. The extent to which the urban forest is publicly held versus privately owned will directly shape investment and management strategies as well as public policy. Although highly place-dependent, an assumption is that the higher the population density, the greater the percentage of publicly owned urban forest. Support for this assumption may be extrapolated from the Chicago study (McPherson *et al.*, 1994). This study found that street trees accounted for 1 of every 4 trees in 1 to 3 family residential districts in Chicago, 1 of every 10 to trees for residential land in suburban Cook County, and 1 of every 26 trees in the residential areas of the less urbanized DuPage County.

On the basis of the street-tree-to-residential-tree gradient observed in the Chicago study, certain inferences can be made. In dense urban areas, trees within the public realm represent a significant portion of the urban forest. Thus, in these communities it would seem that the health of the urban forest is highly dependent on public commitment and investment. If urban forest resources are to be enhanced, sustained, and integrated into the planning and development process in these dense, urban communities, then "green" infrastructure needs to be elevated to the same level of importance as "gray" infrastructure. Wide-reaching and strongly supported urban forestry programs are a key component in this process.

Conversely, in communities where urban forest resources are predominantly privately owned, regulatory mechanisms, such as landscape and tree ordinances and zoning codes, may play a much larger role in the protection and enhancement of these resources. An interesting finding in a public opinion survey conducted in California (Underhill, 1995) revealed that the amount respondents were willing to pay for urban forestry programs decreased with home ownership. Since ownership tends to increase with decreased density, this would support the need for regulatory techniques. It should be pointed out that regardless of density, communities need to develop a publicly supported urban forestry program and establish appropriate regulatory measures.

1.3. What Are the Objectives?

Objectives can be viewed as to what a community wants and, perhaps more importantly, what it is willing to do to achieve those desires. Objectives may be singular or multiple; however, singular objectives may result in multiple benefits. For example, a community with considerable steep terrain may decide to prevent development on steep, wooded hillsides to control storm water runoff and prevent erosion. However, additional benefits, such as air quality, reduced summer cooling demands, preserved animal habitat, and the establishment of an upland greebelt, could result from this decision.

Needs and desires will vary from community to community or even neighborhood to neighborhood, depending on availability and accessibility to community forest resources and population profile. A method for achieving consensus and guiding decision making is essential to the process (Miller, 1988).

2. Elements of the Urban Forest

For descriptive purposes the elements of the urban forest, primarily within the public realm, can be categorized in two broad groups: nodes and links (Hartman and Strom, 1995). Nodes are spatially defined parcels of land that may serve as focal points, attractions, and/or destinations (Fig. 1). They have the potential to become recognizable and identifiable places within the urban fabric. Links, by their nature, are linear. Links may connect nodes and establish corridors by which we experience and perceive a place as we move through it. They also may preserve and protect natural resource corridors. Links include streets, greenway corridors, stream and river corridors, and abandoned railroad right-of-ways (Fig. 2).

2.1. Nodes

2.1.1. Parks and Open Spaces

These spaces primarily consist of active and passive parks, playgrounds, squares, and plazas (Fig. 3). In dense urban settings these spaces, together with conservation areas, form the backbone of the urban forest system. If properly managed and protected, existing parks and open spaces can be the basis on which a larger urban natural resource system can be established.

2.1.2. Conservation and Natural Areas

Watershed lands, wetlands, coastal marshes, wildlife sanctuaries, and lands with similar functions constitute the major elements of this category. Generally these are sensitive landscapes that have intrinsic environmental value.

2.1.3. Civic and Institutional Facilities

This broad category includes land uses such as churches, cemeteries, schools, college campuses, museums, and other cultural institutions. Arboreta, botanical gardens, and zoological parks may be included here or in either of the two previous categories (Fig. 4).

Vacant land

- Vacant land has no permanent structures and provides a valuable urban land resource
- Vacant land presents a variety of open space and development opportunities including:
  - expansion of natural resource base,
  - expansion of park and open space systems,
  - development of community gardens and working landscapes, and
  - community redevelopment and revitalization.

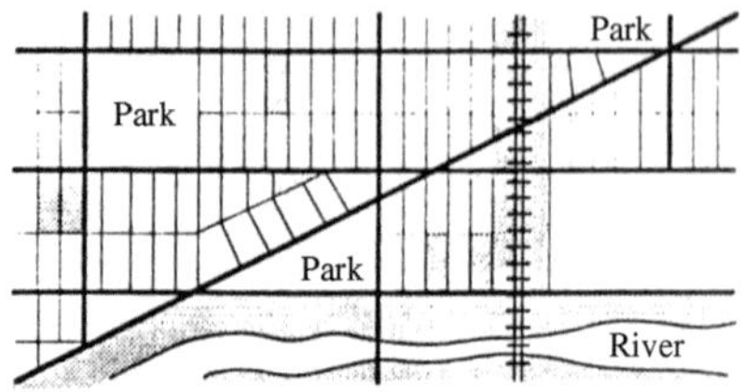

Existing green/open spaces

- Existing urban green/open spaces must be preserved and maintained to have an impact on the future urban landscape.
- Existing green/open spaces of the urban landscape represent a valuable natural resource including wildlife habitat.
- If managed properly and protected, existing green/open spaces can be the basis upon which a larger urban forest can grow.

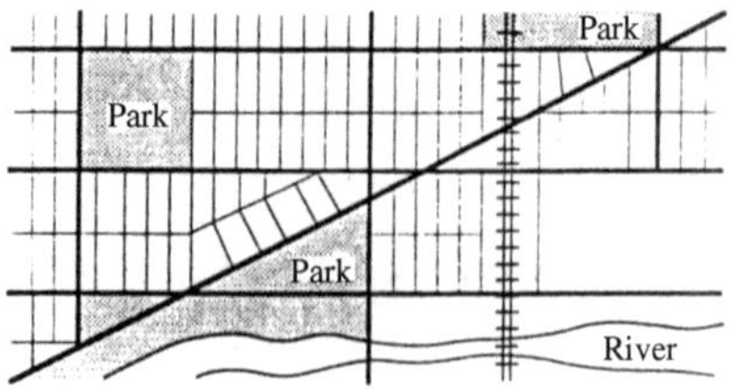

Community gardens

- Community gardens are vacant urban spces that have been turned into horticultural or agricultural oases.
- The gardens are often managed by neighborhood or community groups or associations.
- Community gardens provide a focus for the community and help to foster pride among the citizens who manage and maintain them.
- The gardens present opportunities for environmental education.
- Community gardens are a productive use for vacant land.

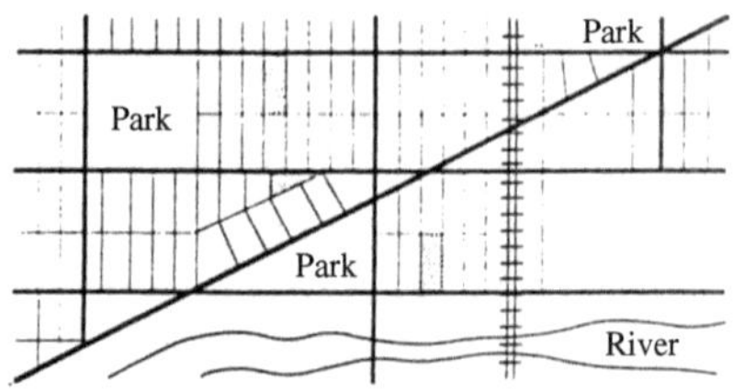

Productive landscapes

- Productive landscapes provide the potential for economic and social benefits.
- Productive landscapes include commercial horticultural gardens and nurseries.
- This type of use provides the opportunity for environmental education and training in the "green industries."
- The potential exists for the generation of revenue for the municipality or organization managing the resource.
- Employment of residents may help to reduce financial burdens on other entities.

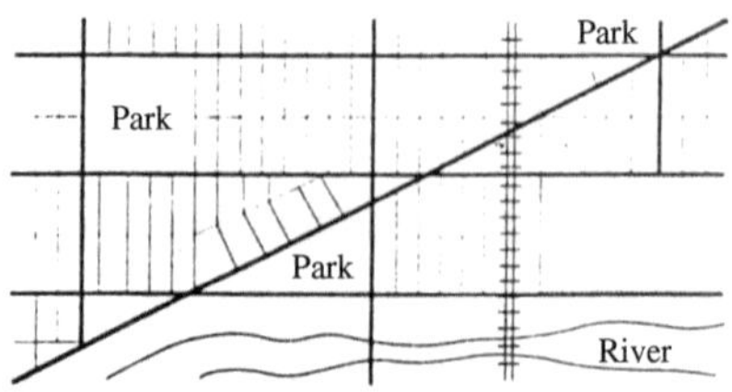

FIGURE 1. Diagram of nodes (source: Newark Urban Forestry Demonstration Project: Findings and Recommendations).

2.1.4. Community Gardens

Community gardens are usually vacant lots that have been turned into horticultural oases. Usually, the gardens are managed by community groups or neighborhood associations. They have the potential to provide a focus for the neighborhood, foster community pride, and present opportunities for environmental education.

2.2. Links

2.2.1. Streets

Streets are the single-most abundant public space within the urban fabric. They are the primary setting for public life, and the network of streets may be thought of

Street corridors

- Streets are an abundant and often underutilized urban forest resource.
- Streets provide existing corridors that can be enhanced to provide a vegetated linkage between nodes.
- Street tree plantings enhance the pedestrian scale, microclimate and aesthetic qualities of the urban environment.
- Street tree plantings enhance the pedestrain experience by providing shade and separation from vehicular traffic.

Greenways

- Greenways are an appropriate use of contiguous and linear areas of vacant land.
- Greenways can provide a vehicular free link between nodes.
- Greenways can be extensively planted with vegetation to become long, narrow parks through the built urban environment.
- Greenways protect and enhance the urban natural resource base.

Blueways (river and stream corridors)

- Blueways are greenways that are directly associated with stream or river corridors.
- Blueways provide a buffer between the built, terrestrial environment, and the aquatic environment.
- The development of blueway networks allow streams and rivers to be maintained and function in a natural state.
- Blueways enhance habitat and water quality.

Visual linkages

- A visual linkage can be created, even when it is not possible to physically link two nodes.
- Visual linkages present the impression of greenness even if the area where one is standing is not green.
- Visual linkages draw people toward green spaces by indicating a green area in the distance.

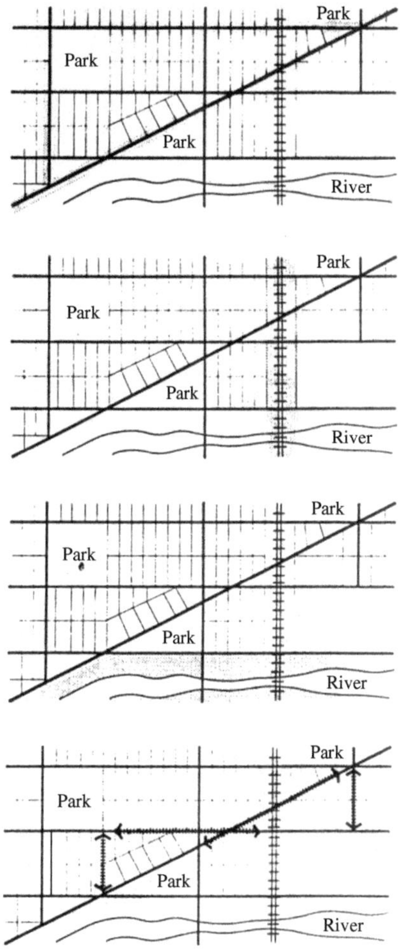

FIGURE 2. Diagram of links (source: Newark Urban Forestry Demonstration Project: Findings and Recommendations).

as the thread that binds together the physical and social fabric of our cities. In many communities, streets are an underutilized urban forest resource. Streets provide existing corridors that can be planted to create vegetated linkages between nodes while enhancing the pedestrian scale, microclimate, and anesthetic quality of the urban environment (Fig. 5).

2.2.2. Greenways

Greenways are an appropriate use of contiguous and linear areas of vacant land such as abandoned rail rights-of-way. On a grander scale, greenways can become greenbelts within metropolitan regions.

(a)

(b)

FIGURE 3. Urban green spaces, such as (a) Central Park in New York and (b) the Boston Common and Public Garden, are important components of the urban forest.

2.2.3. Blueways

Blueways are greenways, which are directly associated with stream and river corridors (Fig. 6). Blueways may perform an important environmental function by buffering the aquatic environment from the built environment, thus preserving or enhancing water quality. The development of blueway networks helps to preserve the natural functions of streams and rivers.

(a)

(b)

FIGURE 4. Public and private institutional facilities including: (a) cemeteries and (b) universities are often overlooked as components of the urban forest.

FIGURE 5. Majestic oaks, with their canopy and column-like trunks, clearly define this street space.

FIGURE 6. The Delaware & Raritan Canal forms a continuous 60 mile blueway/greenway across central New Jersey.

2.2.4. Highway and Rail Corridors

Highway corridors can occupy a significant portion of the urban landscape. Although there may be limitations in terms of safety and access, the potential exists for these landscapes to perform multiple functions. Active rail rights-of-way pose even greater safety, access, and maintenance limitations but should not be overlooked in terms of connecting fragmented landscapes.

2.2.5. Visual Linkages

Where it is not physically possible to link nodes, a perceptual connection may be created by visual linkages. A visual linkage presents the impression of greenness, even if the area where one is standing is not green (Fig. 7).

3. Design and Urban Forestry

The primary underlying principles that guide physical design within an urban forest context are the same as those that would be applied to good urban design. These principles are appropriate to the site, block, and neighborhood scale of development and include spatial organization, unity, and coherence. Trees can be used to create, define, or reinforce spatial structure and/or sequence and to add a humanizing

FIGURE 7. Tree plantings are clustered in groups (rather than in the typical linear manner for street tree plantings) along the downtown pedestrian zone in Munich, Germany. As a result of the cluster locations, tree groupings are always in view to the pedestrian.

scale to the urban environment. Coherence is the principle of making a place readily understandable. Unity serves visually to unify a variety of elements or spaces, often of a disparate nature. At the very detailed level of design, principles, such as rhythm, pattern, and texture, are also applicable. Issues of biological diversity and ecological connectivity also should be addressed; however, these principles need to be planned and managed from a regional perspective. The following sections suggest design considerations for specific components of the urban public landscape.

3.1. Streets

As noted previously, streets constitute the greatest percentage of the public landscape. Without a doubt, they have the greatest impact on public perception and visual quality of a community's residential neighborhoods and commercial shopping districts.

In addition to the environmental and economic benefits, street tree plantings provide spatial enclosure and definition that contribute to the aesthetic quality of the urban environment. Street trees perform functions similar to architecture in which the canopy acts like a ceiling and trunks are analogous to columns. However, for trees to perform these spatial functions effectively, specific design principles must be applied (Strom and Hartman, 1995).

First, large deciduous trees are preferred over small or ornamental trees. Large trees are effective in defining and containing space and are appropriately sized in relationship to pedestrians and the scale of street corridors (Fig. 8). Small trees do not define spaces, interrupt the visual continuity of the street, or permit pedestrians to sit or stroll beneath them easily. Small trees should not be used in deference to overhead

FIGURE 8. London Planetrees are appropriately scaled to the width of the street and center island.

utilities. Examples of large trees and utilities coexisting can be found in almost every community. The Tree Line USA Program, sponsored by the National Arbor Day Foundation, promotes quality tree care and a number of techniques that allow large trees and utility lines to coexist.

Second, trees should be planted close to the curb, preferably in the space between side-walk and street with sufficient soil depth and area. Trees planted close to the curb reduce the apparent width of the street and thus reduce the overall street scale. This placement also separates the pedestrian from vehicular traffic both visually and psychologically. Trees should not be moved back from the curb to accommodate overhead utilities. This approach is similar to moving the walls of a corridor further apart. When this happens, the proportions of the space are lost, as is the sense of enclosure and containment (Fig. 9).

Third, trees should be spaced so that they form a continuous overhead canopy. This continuity provides visual unity that contributes a sense of order and coherence to the urban environment (Fig. 10). Design criteria for streets should establish maximum distances between trunks to ensure a sufficient number of trees to achieve good design rather than a minimum spacing distance between trees (Arnold, 1980). Where it is not possible to create a continuous canopy, trees should be grouped to define their mass and reinforce their ability to define spaces within the urban fabric.

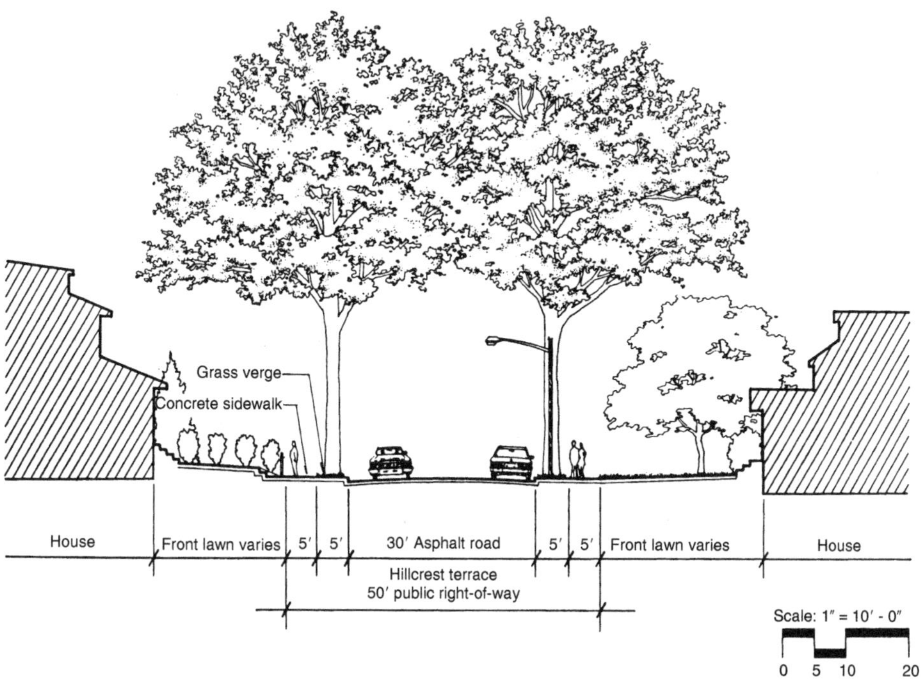

FIGURE 9. Trees placed between sidewalk and street clearly define and separate pedestrian space from vehicular space (source: East Orange Urban Forestry Demonstration Project: Findings and Recommendations).

FIGURE 10. This shows the appropriateness of large canopy trees regardless of the width of the street right-of-way (a), setbacks to buildings (b), or neighborhood density (c).

Finally, sufficient space should be provided to ensure long-term survival. Research has shown that soil volume is critical to the health and survival of street tree plantings. The concept of tree pits has been replaced by continuous trenches with suitable soil mixes and proper aeration and drainage. Urban planting techniques continue to evolve and are addressed in Chapter 13 of this volume and other sections of this text.

3.2. Plazas and Squares

These spaces are usually extensions of the street spaces. Such spaces serve as outdoor rooms, which are most effective when clearly defined by an ordered geometry of tree plantings. Plantings can be used to define spatial limits, create implied walls, or subdivide the space into a series of rooms. In addition to creating voids (i.e., rooms), trees can be used to create volume or mass. In the latter case, a much more architectural type of room is created by the columns of the tree trunks and the ceiling of the canopy. In addition, the mass of the canopy may serve as a boundary to define adjacent spaces (Fig. 11a).

3.3. Parks and Open Spaces

This category is highly diverse and can include large and small passive green spaces, active recreation, and conservation lands. There are no recipes or formulas that can guide the design of these places. Each site must be evaluated within its specific cultural and ecological context to determine its most appropriate function or functions and design responses to accommodate those functions.

Such responses may include ordered geometries, stylized interpretations of natural environments, or ecological restoration. Grids of trees or other geometries may be appropriate for small parks and playgrounds, whereas a large regional park may be an integration of all three approaches.

3.4. Parking Lots

It may be argued that parking lots are incredibly inhospitable environments to plant trees. Parking lots have high surface temperatures, provide little opportunity for proper root aeration and water infiltration, and have degraded water quality and vehicular traffic that can scrape and scar tree trunks. However, the environmental consequences of not planting trees in parking lots are simply not acceptable.

The standards by which parking lots are designed and constructed must be reconsidered so that more hospitable environments can be created for both plants and people. The objectives of revised guidelines should be to reduce the amount of pavement and increase the amount of landscaping. These objectives can be achieved by (1) using realistic rather than inflated parking demand ratios, (2) designing parking stalls and aisles using dimensions that reflect the size of today's automobiles, and (3) increasing requirements for tree plantings and landscaping. Biofilters and bioretention areas should be integrated into most parking lot designs to increase the amount of pervious surface and enhance the quality of storm water runoff.

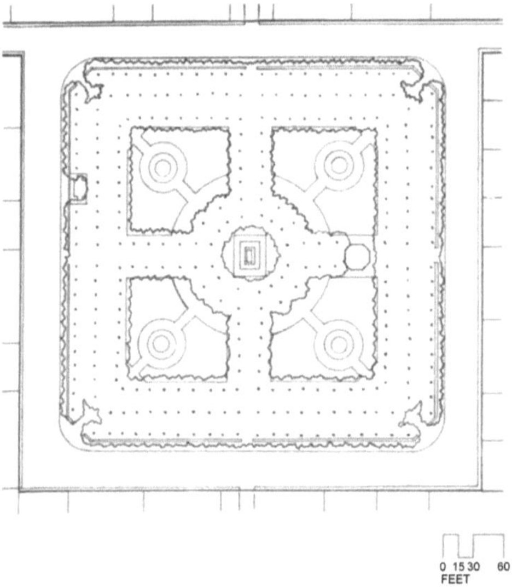

FIGURE 11. The plan (a) of Place des Vosges in Paris is a classic example of a formal, geometrically structured tree planting of a neighborhood square (photo courtesy of John Webster). (b) Louisburg Square in Boston is another excellent example of the importance of small squares to the quality of the urban life.

(a)

(b)

FIGURE 12. The island in this corporate parking lot provides an opportunity for significant tree plantings as well as storm water detention and infiltration.

Surface parking lots represent a significant land use in small- to medium-size cities and for office developments and corporate campuses in suburban areas (Fig. 12). It is essential to look beyond their utilitarian use as temporary storage areas for automobiles and recognize that they represent a significant environmental impact and opportunity.

In designing parking lots it is also important to realize that drivers become pedestrians when they leave their car and the experience from parking to destination should be safe, easily understood for orientation purposes, and comfortable.

Trees planted in parking lots can provide spatial order and human scale and can significantly reduce heat island conditions. The capital investment is relatively small compared to the benefits gained. However, trees cannot continue to be placed in parking lots as a token gesture using improper planting techniques. Under these circumstances tree growth will be minimal and tree mortality will be high.

3.5. Community Gateways

Gateways announce the arrival or entry to a community or place. They can be simple markers, a designed space, or a sequence of spaces. Regardless of the technique, however, gateways set the stage for how a community is perceived and have the potential to become memorable symbols of a place.

(a)

(b)

FIGURE 13. (a) The allée combined with the scale and detailing of the sidewalk create a gateway to New Brunswick, NJ. (b) A mass planting of annuals provides seasonal interest and reinforces the gateway.

Trees may be used effectively to create community gateways. A particular species of tree may be used on major streets leading into a community to announce the passage from one town, or even one neighborhood, to another. A combination of spaces, street tree plantings, and seasonal or special plantings can also be used (Fig. 13). Appropriate scale and spatial structure are important to the design of these spaces so that they do not appear trite or impermanent.

4. Planning and Design: Future Opportunities

4.1. Vacant and Underutilized Land and Brownfields

Vacant and underutilized land has no permanent structures, or the existing structures have become derelict or obsolete. Such places provide a valuable urban land resource for a variety of uses. Brownfields, a term used to describe contaminated sites, is a special category of vacant and underutilized land. In a number of cities there is an abundance of brownfield sites because of the high cost of cleanup, health risks, and general level of uncertainty and anxiety associated with these places. As a result, the supply of brownfield sites appears to be significantly greater than demand (Simons and Iannouly, 1997).

If brownfield sites represent surplus urban land, then these sites should be inventoried and analyzed for the potential of reweaving the natural landscape with the built landscape. Cities should seize the opportunity for reclaiming river and lakefronts, developing and expanding parkland, and creating new urban forests. It is important in this process to realize that formerly developed industrial sites may represent a greater economic benefit to a community as undeveloped or public land.

4.2. Productive Landscapes

Productive landscapes provide the potential for economic and social benefits and are primarily oriented toward urban horticulture. Uses include commercial and community nurseries, greenhouse production, and urban farming. These types of uses provide an opportunity for environmental education, job training in the "green industries," and reconnecting people with their agrarian roots.

For the public sector, the establishment of urban nurseries, tree farms, and greenhouses provides a readily available and regionally acclimated source of plant material. Although such ventures may not be directly profitable for communities because of issues like higher wage and utility costs and investments in programs, such as job training and environmental education, the potential for social benefits and the development of job skills may help to stabilize neighborhoods and relieve fiscal burdens on other social programs.

4.3. Highway Interchanges

Highway interchanges consume a considerable amount of urban and suburban land. The plantings for most of these interchanges have been designed primarily for ease of maintenance and do little to promote ecological diversity. Thus most are large areas of grass allowing for easy mowing (Fig. 14). These interchanges, although highly fragmented landscapes, represent a lost opportunity in developing and expanding the urban forest.

Lessons can be gained from European countries where the availability of land is limited. Perhaps most importantly, interchanges are designed to consume much less land than in the United States, resulting in less pavement and landscape degradation. Oftentimes, land contained by approach and exit ramps perform multiple functions

FIGURE 14. This cloverleaf highway interchange demonstrates a lost opportunity in terms of urban forestry and greater biodiversity 4(/h). photo courtesy of Roy H. Deboer.

such as agricultural production and managed wood lots. Serious consideration must be given to reforesting these landscapes and the potential positive impacts in terms of air quality, microclimate, and storm water runoff that can result.

5. Conclusion

Henry Arnold (1980) said that tree density may be a better indication of the health and liveability of our cities than population density. Whether true or not, this statement underscores the need for a strong vision that integrates the natural and biological environment with the built environment. A continuous network of urban forests from yards, streets, squares, and parks to regional greenbelts, forests, and agricultural lands must become the goal of all future planning and development efforts.

References

Arnold, H., 1980, *Trees in Urban Design*, Van Nostrand Reinhold, New York.

Hartman, J. M., and Strom, S., 1995, *Newark Urban Forestry Demonstration Project: Findings and Recommendations*, Center for Land Planning and Design, New Jersey Agricultural Experiment Station, New Brunswick, NJ.

McFarland, K., 1994, *Community Forestry and Urban Growth*, Washington State Department of Natural Resources, Olympia, WA.

McPherson, E. G., Nowak, D. J., and Rowntree, R. A., 1994, *Chicago's Urban Forest Ecosystem: Results of the Chicago Urban Forest Climate Project*, Northeastern Forest Experiment Station, Radnor, PA.

Miller, R. W., 1988, *Urban Forestry: Planning and Managing Urban Greenspace*, Prentice-Hall, Englewood Cliffs, NJ.

Moll, G., Macie, E., and Neville, R., 1995, Inside ecosystems, *Urban For.* **15**(1):12–15.

Simons, R. A., and Iannone, D. T., 1997, Brownfields: Supply and demand, *Urban Land* **56**(6):36–38.

Strom, S., and Hartman, J. M., 1995, *East Orange Urban Forestry Demonstration Project: Findings and Recommendations*, Center for Land Planning and Design, New Jersey Agricultural Experiment Station, New Brunswick, NJ.

Underhill, K. K., 1995, Public Opinion of Urban Forestry in California (Abstract), http://urbanfor.cagr.calpoly.edu/data/abstracts/OpinionAbst.html.

Chapter 7

Tree Inventory and Systematic Management

Richard S. Wolowicz and Michael Gera

1. Introduction

The question of implementing a shade tree inventory is frequently raised. If used properly, an inventory can be a very productive tool. It also can be a very effective paperweight and dust collector. The purpose of this chapter is to address key elements in organizing a productive shade tree inventory system. The design of the inventory is vital to its usefulness, whether it is a basic card file, organized in a paper binder, or an automated system.

2. Shade Tree Inventory

Let us take the mystery out of the term. The term *inventory* is defined as a listing of goods on hand, a survey of supplies. There is nothing mystical about it; it is a tally of any stock. It is more commonly associated with the retail trades. Inventories play a major role in knowing what supplies to order and when. It is not good business for a vendor to defer restocking of supplies until the last item is sold or when it is in short supply. Inventories come in various shapes and sizes. In fact, all of us do various types of inventory every day. Balancing checkbooks and making change are common examples of the menagerie of inventories people are exposed to. The theory used behind counting supplies can be extended to the natural resource environment. Just like the retail objectives, shade tree inventories assist in determining the supply, condition, and even value of a given tree population.

Richard S. Wolowicz Public Service Electric and Gas, Newark, New Jersey 07101.
Michael Gera Applied Research Technology, Montclair, New Jersey 07042.
Urban and Community Forestry in the Northeast, 2nd ed., edited by, J. E. Kuser.

Look at it from another perspective. Tallying up your natural resources can be thought of as a basic inventory system. Have you ever had to answer a question regarding the number and types of trees that are around your town? Giving a definite answer to the media or to your administrators lends credibility to your program. Shade tree inventories are one of the most valuable tools that shade tree commissioners, arborists, and foresters possess as a source of quantitative and qualitative information. This type of data is a positive armament in an arsenal of tools to manage a shade tree operation and fight for budget dollars. Detailed information is a positive benefit when needed to determine work priorities. It also is invaluable when information must be given to one's manager, the press, a councilperson, or the mayor. You cannot address administrators to request more funding for a special type of program if an estimate of the benefits derived cannot be compared against the expenditures. One is in a better position when a determination can be made based on the knowledge of what resources exist in the field, what your objectives are, and the impact of your actions.

If a town has a major Dutch elm disease problem and wants to treat the healthy trees with a fungicide to prevent the spread of the disease, the number of elm trees located in town and how many are healthy must be identified. It does not add any confidence to your reputation or your program to answer, "There are a lot of 'em in town," when asked by the media or your administrator to justify the cost of the fungicide application against the benefits to the healthy trees. Being able to answer in a quantitative manner will prove more beneficial and provide excellent media opportunities (Gerhold *et al.*, 1987). It also will be critical to those who are involved in the distribution of budget dollars. A program will be respected if a statistical question can be given a quantitative answer. "We have 6500 trees along our street, with 40% of them being American elm trees," sounds professional and irrefutable.

3. Inventory and Systematic Management

Ultimately, directing a shade tree program with a systematic management approach is the utopian goal. Incorporating the inventory into a process that can schedule, organize, sort, and/or identify selected work units and locations is the manner in which an inventoried shade tree population can be used to its fullest potential (Schein, 1993). This propels an inventory to a higher plateau as an active management tool. In this day and age just counting trees and logging the species may not be enough. Being able to use the data from the inventory process and incorporate that into a management plan will pay for itself in a short time. Developing a systematic management operation will help use the information from your inventory data and not just have it collecting dust on a shelf.

4. Benefits of Systematic Management

A functional, user-friendly method to retrieve the inventory data needs to be a high priority in a systematic management approach. It will provide the opportunity to retrieve, sort, search, list, and calculate all different types of data in an efficient manner. Not only it will give accurate field data, but also it will be able to give information on current workload. What was that question, just how many American elm trees are in town? Press a few buttons and you can give that answer, even with their locations!

A systematic management approach also can be used to help obtain future goals and justify management decisions. Do you want to be on a 7-year trim cycle in town? How many trees will have to be trimmed per year? How will it get done? In-house crews or outsourcing? Is there enough budget to handle it? Are there enough resources to remove all of the dangerous trees, thus avoiding the liability issue? Are you losing more trees to mortality than the number being planted? What steps can be taken to turn this around? This information will prove invaluable in justifying budget requests. A systematic approach will provide useful information in managing daily operations, even indicate the type of equipment required, number of crew needed, and provide productivity analysis.

5. Different Types of Inventories

Now is the time to plan it out. Identify what will work for your operation. There is no golden, steadfast rule for this. Inventories are a product of your needs, objectives, resources, and future plans. Look into the various types of inventories that can meet your objectives.

5.1. Random Sample

Data from a 10% sample can be projected to yield approximate totals citywide. This system works well for basic data. In a heavily urbanized city in northern New Jersey, a 10% sample of all the streets was surveyed to estimate the species composition, size, sidewalk condition, and health and mortality of trees in the town easement (Lawrence, 1975). The shade tree department did not posses the resources to do a complete survey. Their objectives were to get approximate results with the basic information. It worked well for them as they planned their budgets and made personnel adjustments to accomplish the goals of their shade tree program.

The sampling of trees is based on a predetermined sampling percentage of a given area such as a town boundary. If there are 240 miles of streets in town, you can randomly select 24 miles to give you a 10% sample. Care must be exerted to maintain the "randomness" of your sample profile. If only predetermined streets will be counted because they are "favorites" or "easy," the data will be biased, and consequently so will be the results. Preplanned approaches to a truly random sample must be used. Determine what the sample proportion will be ahead of time. The methodology cannot be changed as you proceed. Use a random number table to give your grid coordinates, or pick coordinates or street names out of a hat. Also determine what your sample length will be when it is chosen. Determine a specific length that is appropriate to the length of streets in town.

Certainly a percentage greater than 10% can always be surveyed, but sampling less than 10% may yield unreliable data because of not sampling enough of the town to give a true indication of what resources are out there and what conditions they are in. On the other hand, a higher percentage sample may prove useful if more data are being collected and a higher degree of accuracy is desired. It all depends on what information is desired and the resources available to the manager.

5.2. 100% Sample

As the name implies, this is a full-scale inventory, not a portion of the population. The entire system will be at your fingertips with this approach. This still can be a basic survey with only the species and conditions recorded. Or it can be a very thorough, detail-oriented inventory with everything about the tree noted (Schein, 1993). Typical information would include species, size, condition, sidewalk (if any) defects, number of trees on the property, location of any nearby obstructions (street signs, fire hydrants, etc.), presence of overhead power lines, communication cable or street light, pruning requirements, tree limitations, and so forth. It can be extremely thorough, intensely detailed, and very informative. Make sure that you are not afraid to ask all the information that you want to know. Keep your objectives in mind to know what information needs to be collected.

On the positive side, this will be the most reliable and detail-oriented information system you would want (provided the right data are identified to collect). On the negative side, this will take significantly more time and resources. Identify what will be done with the information once it is collected. If you are planning to computerize the data, someone will have to do all the data entry work. If a lot of information will be gathered, more time will be spent inputting the information. This is not necessarily bad; it merely must be put into proper perspective ahead of time—in the planning phase. Merely collecting the field data is only one facet of the inventory process.

Identify your objectives ahead of time. Be realistic with your resources and time allocation. Identify what you wish to accomplish, what will prove most useful for your particular circumstance, and organize a method to collect the necessary information. You will have produced a custom-tailored inventory to fit your needs!

6. The Record-Keeping System

Ultimately, this is the backbone of the inventory system. If you are in the planning stages, now is the time to determine how the information will be updated. Not updating information will yield stale, almost useless data. Conversely, updating field changes, work done, trees planted, and trees removed will provide current, accurate data at your fingertips. This will take a commitment on your part and on the part of the administration. As mentioned earlier, collecting the data is only one part of the task. Using the data intelligently as a useful tool is the art of the process.

Your shade tree inventory will be invaluable if you can get to it. If your inventory will be on a paper system, there will probably be no difficulty in accessing it. It might prove to be a challenge when attempting to update the data. The other extreme is to computerize it. If you go all out and have the information on a computer, will you be able to retrieve it as needed? Will you have to depend on a technician to obtain the basic information you need? If you are in this boat, it is time to learn to do it yourself. The inventory is useful in systematic management if it can provide current information when needed, not making you wait 2 weeks to obtain the data.

Another community dove into an intense, detail-oriented inventory. They computerized all of their data, collected all sorts of information. Unfortunately, this "forward" thinking caused them to wear egg on their face when their record-keeping system was on a mainframe computer, which is so complex no one was able to access it. The town forester had no choice but to depend on the computer staff to run the reports and records as needed, provided they were not too busy with another "higher-priority" job.

Seriously consider the technology to be used in your inventory record-keeping system. It is important that the computer utilized in storing the data is accessible when needed (Miller, 1996). A paper inventory is not conducive to the volume of information that will be produced. Furthermore, if the inventory system will be used to systemize the operation, a computer will be required to manage the database.

7. Mechanics of Conducting a Tree Inventory

The importance of determining and identifying objectives cannot be stressed enough. One of the worst things that can be done is changing the parameters of any job once you have started it. The data collectors will be confused, and you will not be sure (unless you are doing everything yourself) that everyone understands the changes. Consistency is the key. Know what you will be looking for ahead of time. If you desire to know only the type and size of trees in your town, then just gather that information. Standardize the methodology that will be used to measure the trees, and make certain that the collectors are capable of identifying the species correctly while using the proper code abbreviations.

A methodology and standardization system will have to be developed to know how the data will be collected, recorded, and documented (Schein, 1993). It will be most helpful to know what you want to learn and gather the data for it. It will be too late to ask the average diameter of the tree if you have only gathered data about the species and physical condition. Planning is vital.

Consider the future in your planning process. What extra information might prove helpful to you: Plantable locations; Treatable insect or disease locations; Areas where sidewalks are already lifted; Plan on what information will prove useful for today and tomorrow's needs. It will save many hours of extra work and money.

On the other hand, avoid information overload. Inaccurate or sloppy data entered into the database will produce erroneous records. Does it really help your office to know that the first limb of the sugar maple at 353 Main Street's starts at 18 feet above the ground and faces due north, or whether there is a pothole or weed growth in the street adjacent to the tree? Superfluous information will take much more time to gather and demand more office input time, with a greater probability of error. A determination must be made to see if the extra data being collected will be worthwhile.

Here is where a contradiction occurs. Superfluous information regarding potholes or weed growth were cited as examples of having the potential of overburdening the field data collectors and burying you in details. There is a time, however, when gathering such information may prove worthwhile. It becomes a matter of economics and how flexible one can be. An associate had difficulty in financing the field data collection of the town's pending tree inventory. Being aware of the town's curb maintenance program, he approached his Public Works Street Department with an idea. He agreed to note the locations of crumbling curbs during the inventory process, if they assisted him with financing a portion of this project. Needless to say, it was a classic win-win situation.

Gather critical and useful information, keep your objectives in mind, be able to identify dangerous trees and limbs, and note trimming requirements. Prioritize how urgent the trimming is. If you want to know whether there is a potential planting spot, then make provisions to record it as such.

8. Data Collection

The manner in which data will be collected is dependent on the information needed from the inventory process. In a very basic inventory, a windshield survey can provide favorable results (Miller, 1996). One person can drive down a street while a second person counts the trees (safer than one person driving and counting). Bicycles have been used in more congested areas. A creative approach used roller skates/blades to get around town while collecting the data. The basic park and walk (as opposed to park and ride) does have its advantages. Begin on one side of the street, walk to the end (or a designated boundary), and walk back on the other side of the street collecting your data. Utterly simple and efficient. You always wind up back at the starting point to restock with supplies.

Organizing your data is very important. Use acronyms and abbreviations but be consistent with them. One option is to use the rather extensive national coding of tree species found throughout the United States. However, since you will be dealing with a much smaller number of species than those found across America, it probably will be more useful and simpler to develop your own. Keep it simple, alphabetical, and easy to remember (Schein, 1993). Try some of these: POAK for pin oak, ROAK for red oak, RMAP for red maple, and SUGM for sugar maple. Care must be exercised so that no acronym stands for two different trees (with the prior example, a silver maple and sugar maple have the same acronym: SMAP). That is when the acronym changes to include the first three letters of the species: SILM and SUGM. Develop a system that is easy to remember so that field data collectors will not have to look up each individual species or use them incorrectly. Keep things as simple as possible for efficiency and accuracy. Table 1 is a sample of a typical data collection sheet used in the field to collect the identified information. Table 2 shows a key listing for the tally sheet.

Dividing your town into zones can be helpful (Schein, 1993). Consider using existing neighborhood boundaries; this will be an advantage if your data are computerized and you wish to locate certain trees in a certain part of town. For example, if you want to identify all the trees that need immediate pruning in the northward, it can be printed out as such. This will be prudent if you know that a crew will be available in that section of town.

Table 1. Sample of Field Data Sheet

ADRSS	TREE #	SPCS	DBH	CNDTN/ % of dead wood	CVTY/ DSSE/ INSCT PRBM	S/W CND TN	OH WIRES/ OBST- RCTNS	TRIM/ REM PRITY	Notes
120	A	SUGM	24	G/5	None	N/a	No	T3	
124	A	POAK	32	G/10	Basal cavity	Good	No	T2	Small cavity @ base
124	B	NORM	16	P/50	Yes	Good	Yes	R1	decline
125	A	WASH	20	G/5	None	Poor	No	T3	
130	A	AELM	34	D/100	Dead	N/a	Yes	R1	

How will you record data that involve multiple trees on the property? Decide in which direction they will be counted and make a rule on how they will be designated in the field.

Try looking at the front door of the house and designate the first tree on the extreme left as A. The next tree to the right would be B and continue in this fashion to the property line. The same principle works for a corner property. Face the front door and count the A tree as the one on the extreme left side of the property, even if it is around the corner.

Table 2. Key Codes Showing Sample Parameters of Field Data Entries

Species
- AELM: American elm
- NORM: Norway maple
- POAK: Pin oak
- ROAK: Red oak
- SUGM: Sugar maple
- WASH: White ash

Condition
- Dead, poor, fair, or good

S/W Condition
- Good (not lifted)
- Fair (lifted <2 inch)
- Poor (lifted >2 inches)

TRIM/REM priority
- Trimming priority
  - 1 = Immediate
  - 2 = Within 6 months
  - 3 = Long term
- Removal priority
  - 1 = Immediate
  - 2 = Soon, but not emergency

Recording of trees can be aligned with house numbers, block and lot number, or a simple tree count. The easiest would be the house numbers, since the block and lot numbers are not the normal mode of identifying properties along city streets.

As one can imagine, the methodology used in data collection is dependent on information needed as a result of the survey. It is vitally important to organize and standardize your data. Use acronym abbreviations and keys. Make them easy to remember and easy to find. Be consistent throughout.

Questions will arise as to how to get the job done. Think of staffing in terms of supplies and personnel. If all you have to work with the basics, then a pencil and paper system is better than nothing; it is primitive, economical, but limited.

Handheld computers are more "high-tech," more costly, and will require more training, but they will decrease the amount of time needed in the office to log the information, thus saving some dollars at the office-end of the process.

Hiring help may prove to be your most useful strategy. Companies and private consultants can be hired with little need for further training. You are paying for their expertise. Another resource to use would be graduate students, summer interns, or individuals with some tree expertise. Volunteers, both young and old, are also a source of assistance. However, with volunteers, a bit of caution must be exercised. If they are not properly trained, competent to identify species, and monitored sufficiently, the data can be inaccurate. After inventorying a large number of trees, this could prove disastrous if not discovered beforehand during a quality control check.

It would be helpful to make up identification badges for the data collectors. This will help dispel fears or questions raised by property owners seeing unfamiliar people near or on their property with clipboards collecting data. Notify the police of the presence of the data collectors ahead of time. Give the police expected details: who, what, why, where, and when. Also give the police a name and phone number of the municipal contact person who is familiar with the project in case they need to be notified. Press releases for the local paper also will help to lend support to the program and let people know to expect that their neighborhood trees are being inventoried. This will save a lot of the data collector's time in answering potential questions raised by concerned and inquisitive citizens.

9. Training Issues

Once you have gotten this far, decide whether your data collectors will need training (Schein, 1993). Do not forget the office staff. They will be required to decipher the information and should be thoroughly familiar with the design and intent of the process. Make sure that the data collectors know how trees are measured and identified and can handle the parameters they are using to determine a tree's condition.

10. Case Study

A community in New Jersey took their inventory to heart. They spent time in planning a townwide inventory. Then they discussed a computer program to record the inventoried data. Their goal was to get enough information about each tree—its location and condition. They also noted the presence of nearby physical obstructions. They enter the inventory data into the computer and are using the information to schedule work and determine work load. If a resident calls with a question or a problem with a street tree, they can pull up the house address and access the work history on it: what type of tree, where it is on the property, who called about it previously, what work was done by the town, and so forth. The work orders that the town crews are given are identical to the inventory data sheet; thus, the office staff can record work and new plantings or tree removals appropriately, and consequently keep the data in the inventory up-to-date. This is the true art of a shade tree inventory and systematic management operation.

11. Automation

Automating a shade tree inventory requires a combination of administrative and technical skills to make the information useful to a wide variety of constituencies. Several individuals are usually involved in the implementation of an automated shade tree inventory, including the Director of Public Works, the Director of Parks, the forester, information systems personnel, clerical staff, and volunteers. Each individual or group is typically interested in only a subset of the data collected. It is the responsibility of the project manager to clearly define the expected outcomes of the automation project and manage all fiscal and human resources for the successful implementation of an automated shade tree inventory system.

The successful implementation of an automated shade tree inventory is a multiyear process that involves the coordination of resources throughout an organization. Over the past years as the cost of both hardware and software has decreased, more municipal and county Public Works Departments have undertaken the implementation of automated shade tree information systems. Since this project is likely to have a significant impact on the operations of the Shade Tree Department, careful consideration must be given to all facets of this project. The typical shade tree management information system implementation involves six distinct steps, including:

1. Surveying the marketplace for hardware, software, and training
2. Assessing the internal and external resources for the implementation effort
3. Developing a reasonable project work plan
4. Choosing appropriate hardware and software
5. Reviewing and implementing the project work plan
6. Evaluating the overall project implementation

11.1. Surveying the Marketplace for Hardware, Software, and Training

No area has changed more substantially in the last several years than the area of information management technology. Costs have dropped substantially, and the expectations of users concerning the functionality of information systems have increased. For this reason, it is necessary to begin the process of implementing a shade tree inventory by taking a quick survey of the current state of the marketplace. The Director of Public Works or the forester essentially asks several questions during this phase, including:

1. What kind of hardware does the organization that I work for use?
2. Which software programs have my colleagues in other Shade Tree Departments recently implemented?
3. How can I most efficiently train the users of this system?

In most cases, the forester will find that the municipality will have chosen to use standardized IBM-compatible workstations that may be connected to a municipal wide network. While these workstations may vary considerably in their configuration, most will conform to certain standards with respect to brand name and overall system configuration. A quick phone call to several colleagues will reveal who has implemented automated shade tree inventory programs, and several specific questions with regard to software and training will help the forester choose a software and training program that will meet their needs. The key issues to consider when reviewing software are typically:

1. What was the cost of the software?
2. Was the software predesigned, or were you involved in assisting in the design of screens, forms, and reports?
3. How long did it take to fully implement the software?
4. Is the software able to interface with the municipal databases, such as importing data from the Tax Department for establishing property addresses?
5. Was training available on-site, or did staff have to travel to receive training?
6. What was the quality of the training and is follow-up training available?

By answering these questions, the forester will be off to the successful implementation of an automated shade tree inventory.

11.2. Assessing Internal and External Resources

When beginning the process of automating the shade tree inventory, the forester should carefully consider the resources that his organization can bring to bear on the automation effort. Make no mistake; this process requires real human effort and can be easily underestimated. In an effort to fully understand the resources required for successful implementation, the forester should consider both internal and external resources that will be necessary, including:

1. Internal resources such as ongoing clerical support, hardware and software support, and training
2. External resources, including program development, program customization, report writing, and network administration

The successful implementation of an automated shade tree inventory requires a substantial amount of coordinated efforts between internal and external resources. It is not a matter of simply "pushing a button" and having the shade tree inventory completely automated.

11.3. Developing a Project Work Plan

The development of a comprehensive project work plan is the key ingredient in the successful implementation of a shade tree inventory. The typical project work plan contains several elements, including:

1. Developing a statement concerning the overall purpose of the project
2. Establishing a reasonable timeline regarding the project
3. Identifying all of the tasks associated with the project
4. Developing criteria to evaluate the relative success of the implementation

Many project management software packages are currently on the market that can assist the forester with this process. It is most important at this point not to underestimate the work effort required for successful implementation, since once the project work plan is developed, a set of expectations is created throughout the organization concerning the project.

11.4. Choosing Hardware and Software

While many individuals focus on this phase to the exclusion of other tasks, internal organizational decisions and external market forces usually have determined which hardware and software will be used prior to the forester deciding to automate the shade tree inventory. In most organizations, a hardware standard already will have been developed, typically by the municipal manager, finance officer, or director of information systems. While the forester may not fully agree with the hardware platform chosen, it is essential that some level of standardization is established and maintained between the shade tree office and the rest of the organization if the forester expects to receive hardware and software maintenance from the organization.

When choosing hardware, the forester must take care to avoid receiving a personal computer (PC) from another department (typically Finance) that already has outlived its useful life. The typical PC has a useful life of between 3 and 5 years, assuming that it has been adequately maintained. Under ideal circumstances, the Shade Tree Department will receive a new PC with a 3-year on-site warranty. These can be purchased for between $1200 and $1800, depending on features, brand name, and other considerations.

The typical Shade Tree Department will be implementing a software with a database design that runs on an IBM-compatible PC. In most cases, these programs require between 10 and 20 megabytes of hard disk space and between 8 and 16 megabytes of random access memory. Considering the requirements of other software applications in today's world, these are modest requirements indeed. While there are several canned software packages on the market today that have successfully automated the Shade Tree Department, the forester's best strategy when choosing software is to contact his or her colleagues locally to determine which software packages they have used.

Most foresters do not want to "blaze new trails" by implementing untested or unproven software, so they are usually well advised to conduct a brief local information systems inventory among their colleagues to see whether there are any discernible trends in software selection. While process may vary considerable, the typically "canned" software package will range between $1500 and $3000, and comes with 3 to 12 months of software support via telephone. Software development companies have different pricing structures and may or may not include training in the overall price. The forester will do well to carefully consider his options with regard to the purchase of this important component of the overall system. Again, the forester's best option is to briefly discuss with professional colleagues the options which they used, or wished they used, when implementing software.

11.5. Review and Implement the Project Work Plan

Once the resources have been procured, the project work plan has been established, the forester will do well to subject the project work plan to a comprehensive review just prior or to beginning the shade tree inventory implementation phase. The forester must ask some hard questions as following.

1. Have realistic goals been set for the implementation of this system?
2. Are the timetables reasonable, particularly given other demands on our time?
3. Have adequate resources been set aside for this project?
4. Has a local information systems inventory been conducted to see what other foresters have recently implemented?

Once these questions have been answered, the forester then must review the overall project responsibilities with all of the parties involved in the implementation of the shade tree inventory. Such mundane tasks as who will do the data entry, who will back up the system on a routine basis, and who will develop and write reports must be answered. Again, by consulting professional colleagues or talking to software vendors, the forester will begin to develop a clear idea of responsibilities associated with the successful implementation of the shade tree inventory. Unfortunately, there are no "canned" answers to these questions, since each shade tree department operates differently.

11.6. Evaluate the Project Work Plan

Once the project has been successfully implemented, the forester should review the project in detail to determine which aspects of the overall project were successful and which were not. The forester may examine a wide variety of issues during this phase as following.

1. Was the hardware easily installed?
2. Did we receive adequate hardware and software support?
3. Was the training adequate given the complexity of the software?
4. Do we anticipate modifying the software in the near future to accommodate changes in operations?

12. Conclusion

We all will do well to remember that software implementation is not a onetime process. Since we have entered the "Information Age," we are now all linked, for better or worse, to the various information systems that we use as the tools of our trade. Just when we feel that we have mastered one software program, the environment will change or new demands will be placed on our information-gathering and reporting efforts. Once a shade tree inventory has been automated, new demands and expectations are placed on the ability of the Shade Tree Department to quickly report information on a timely basis. The contemporary forester must clearly understand the demands that the Information Age will have on his or her operations.

References

Gerhold, H. D., Steiner, K. C., and Sacksteder, C. J., 1987, Management information systems for urban trees, *J. Arboric.* **13**(10):243–249.

Lawrence III, J. M., 1975, Method for conducting an economical street tree survey, *Shade Tree* **48**(5):45–50.

Miller, R. W., 1996, *Urban Forestry Planning and Managing Urban Green Spaces,* 2nd ed., Prentice-Hall, New York.

Schein, R., 1993, *Street Trees: A Manual for Municipalities,* Penn State University, University Park, PA.

Chapter 8

Urban and Community Forestry: Financing and Budgeting

Robert L. Tate

1. Introduction

Funding is the key component necessary for the development of a comprehensive urban and community forest management program. Optimizing the provisions of the social, aesthetic, and economic benefits available from the urban forest are dependent on an adequate level of funding. Adequate funding allows the resource to be managed in its entirety. The health and survival of the urban forest are directly proportional to the ability of the resource manager to obtain funding.

1.1. History and Overview

According to a 1994 national survey of municipal arborists (Tschantz and Sacamano, 1994), conducted by the Davey Resource Group, during the last decade there has been a significant decrease in municipal funding for urban and community tree management activities. From 1986 to 1994, when adjusted for inflation, the municipal tree management budget decreased nearly 40% from \$4.14 to \$2.49 per capita. It also lost ground as a percentage of the total municipal budget. In 1986, the tree budget was 0.49% of the total municipal budget, while in 1994 it had decreased to 0.31%, a drop of nearly 40%. It is important to note, however, that while the average overall municipal tree management budget as a percentage of total municipal budget in 1994 is less than in 1986, cities in six out of nine population ranges have had increased tree management budgets as a percentage of the total municipal budget (Davey, 1994).

Robert L. Tate Robert L. Tate Associates, Inc., Cohasset, California 95973.
Urban and Community Forestry in the Northeast, 2nd ed., edited by, J. E. Kuser.

1.2. Sources of Funding

The funding picture is also changing. In the past, the municipal general fund, which is primarily financed through the collection of local property taxes, made up the bulk of funds available for tree management. The trend over the last 20 years has been less funding from the general fund. In 1986, 94% of the communities surveyed was partially or mainly funded by the general fund compared to only 67% in 1994 (Tschantz and Sacamano, 1994).

Because of decreasing revenues available to urban and community forestry and other factors, urban forest managers have begun to take the initiative in obtaining funding through outside grants, endowments, special assessments, and a multitude of special use taxes. In 1986, for example, only 1% of the communities surveyed (Tschantz and Sacamano, 1994) received funding via general forestry grants as compared to approximately 30% in 1994. The dependency on the general fund was decreased in favor of nonmunicipal funding sources. During this time period the development of partnerships with public and private groups to facilitate increased funding and support for urban forest management activities increased.

1.3. Competition for Budget Funds

Competing for budget dollars in urban and community tree management programs has never been more difficult. Obtaining funds for tree care is only half of the equation. The other, and probably more difficult part, is retaining them. Getting and keeping budget funds require the urban tree manager to justify, as never before, the existence and importance of the tree resource. Urban and community foresters can almost always convince their peers that there are not enough funds to do the job. Other, more objective officials with a multitude of competing interests have to be convinced that the benefits from the resource justify the costs of funding it. Looking at the cost of the program in light of, as well as the value of, the tree resource through the eyes of local political decision makers is the critical element in adequate funding. These decision makers are faced with major problems that have no easy fixes. Moreover, citizens' concerns about their personal health and welfare are increasing, as well as their general distrust of the political system. Each dollar spent on government services is now examined even more closely in the light of its ability to make the community a better, more desirable place in which to work and live.

Adequate knowledge of the costs and benefits of each element of the tree management program is another important step in successfully competing for budget dollars. People in general want to know how much something is worth and how much it costs; taxpayers are no different. In order to gain a fair share of scarce government money, an urban forester has to know the monetary value of the tree resource derived from the results of a sample or an inventory of the trees in it. The knowledge and communication of the fact that this resource has a dollar value is the first step in capturing the attention of taxpayers and politicians.

After ascertaining the monetary value of the resource, determining the degree to which its value would be impacted when budget funds are reduced or curtailed is paramount. Political decision makers need much more information on how and where each urban forestry dollar is spent and what will be the impact of not spending it. Urban foresters should and can be as successful as many other vocal interest groups that fight for their share of government services. These groups have articulated their positions and informed the politicians what the results of these budget-cutting actions will be. For example, identifying what will happen if the pruning budget is reduced is not saying that fewer trees will be trimmed (that is a given). The key is able to articulate: (1) what will happen to the resource, (2) how much the reduction in funds will cost over the short and long run to it, and (3) to what extent taxpayers will be affected by this action.

1.4. Justification for Budget Funds

The urban tree program has to compare favorably with other city programs. Do the benefits provided by this resource justify the costs spent on it? It is not an easy taks to compare benefits and costs of trees to critical city services such as police and fire. But it has been done successfully at the expense of these services. Cost-efficiency in relation to benefits provided can make the difference. The urban forester, who is principally a good tree technician, may not have the skills needed to compete well for budget dollars. A successful urban tree program is most likely directed by a proactive manager who understands the boundary-spanning activities necessary for the program to survive in the political arena.

The successful competition for budget dollars can take many forms. Pruning three trees today for the same cost needed to prune two trees yesterday is exactly the same as a 50% increase in the pruning budget. If there are not additional budget funds available and even if there are, increasing the amount, quality, and effectiveness of the service may be the only answer for providing the necessary care the resource requires. It is incumbent on an urban and community forester to continually improve the management skills required for doing more with the same, or less.

Realistically, the outlook for massive or even moderate local budget increases is grim. Taxpayers are not in a mood to see their taxes raised. An urban tree manager, who relies on large budget increases in the future to provide necessary services to the urban forest, will be disappointed and unsuccessful as a resource manager.

This is not to say that urban forest managers are unaware of the difficulty of obtaining adequate funding for tree management programs. According to the Tschantz and Sacamano (1994), managers believe that the four greatest challenges to tree management over the next 10 years are general funding, funding for maintenance, public support, and general tree maintenance.

2. Funding Sources

2.1. The Property Tax

2.1.1. History

Although communities in general are relying less on municipal general funds, most still continue to receive funding from this important source, even though diversification of municipal revenues has developed by the formation and use of municipal sales taxes, local income taxes, and many other nonproperty taxes. Municipal general funds are principally financed by the property tax. Originally the property tax was an annual levy on all forms of property such as land, buildings, tangible holdings (such as household goods, machinery in factories, and store inventories), and intangible holdings (such as stocks and bonds). Local assessors determined a value for the various properties and annual rates were levied against assessed valuations (Bromage, 1957).

The property tax now has become basically a tax on land, buildings, and tangible holdings. Income from stocks and bonds is now taxed as income and/or capital gains. The validity of the property tax as a mainstay of municipal revenues does not rest on a person's ability to pay. The owner of a large lot and home may have less income than a more modest homeowner. The property tax has been strongly criticized for its failure to conform to the principle of ability to pay. For example, the large property owner has more value to be protected by fire and police departments and served by the various public works such as road maintenance, parks and recreation, and the like. According to the theory, this property owner's income enables him or her to make a greater contribution to the various municipal services. The property tax is not strictly in accord with benifits because the wealthy large homeowner may require minimum services beyond basic police and fire protection and public works (Bromage, 1957).

In many situations, the assessed value of the homeowner's property has increased more than the property owner's ability to pay. In rapidly escalating real estate markets, for example, retired homeowners living on relatively fixed incomes found themselves having to sell their property because they could not afford to pay the taxes. This and other factors brought about a general dissatisfaction with property taxes and manifested themselves in property tax revolutions such as Proposition 13 in California. This controversial and popular referendum limits the property tax to not more than 1% of the assessed value of a home-owner's property. The tax can only increase by a maximum of 1% per year regardless of the increase in assessed value and can decrease if the assessed value decreases.

2.1.2. Limitations on the Property Tax

During the last 30 years there has been a great diversification of municipal revenues. Federal grants-in-aid, state-collected, locally shared taxes, such as state income taxes, gas taxes, nonproperty taxes, and service charges, have been required to enable municipal budgets to stay balanced (which by law, they have to be). Cities and counties increasingly have had to rely on sources other than property taxes because of state constitutional and statutory limits on the property tax; the failure of assessors to keep assessments proportionate to inflationary trends in real estate; a continued exemption of certain classes of property such as real estate belonging to federal and state government agencies, schools, churches, and charitable organizations; and increasingly popular opposition to any increase in tax rates.

2.1.2.1. Diversification of Municipal and Community Revenues. Because of the limitations of the property tax to adequately supply the needs of the general fund, cities and counties have developed new taxes on sales, income, admissions, utilities, business, gasoline, tobacco, and alcoholic beverages. These new sources of municipal revenue are termed nonproperty taxes. In addition, service charges for sewage disposal and treatment, drinking water hookups, and the like have been growing rapidly to fulfill the need.

One of the most productive nonproperty tax sources for the city has been the municipal income tax. Generally, this tax applies to the gross income of residents, wherever earned, and to nonresident income earned in the city. Some cities levy on the net earnings of unincorporated businesses and professions, others tax the income of corporations (Bromage, 1957).

2.2. Aid from Federal and State Governments

Although most cities and countries are supplementing their property tax revenues through nonproperty taxes, they are widely dependent on aid from federal and state governments. Municipalities have acquired a greater dependency on federal and state grants and state collected, formerly shared revenues. According to a nationwide survey taken in 1994 (Tschantz and Sacamano, 1994), there was a 29% decrease in the number of communities that received funding from municipal general funds since a similar survey was taken in 1986 (Kielbaso, 1988) (see Fig. 1 and Table 1). Conversely, the use of general forestry grants has increased from 1% of the communities responding to nearly 29% during this same period.

State-administered grants for planting, maintenance, and urban forestry management programs are included in this general forestry grants category. Federally

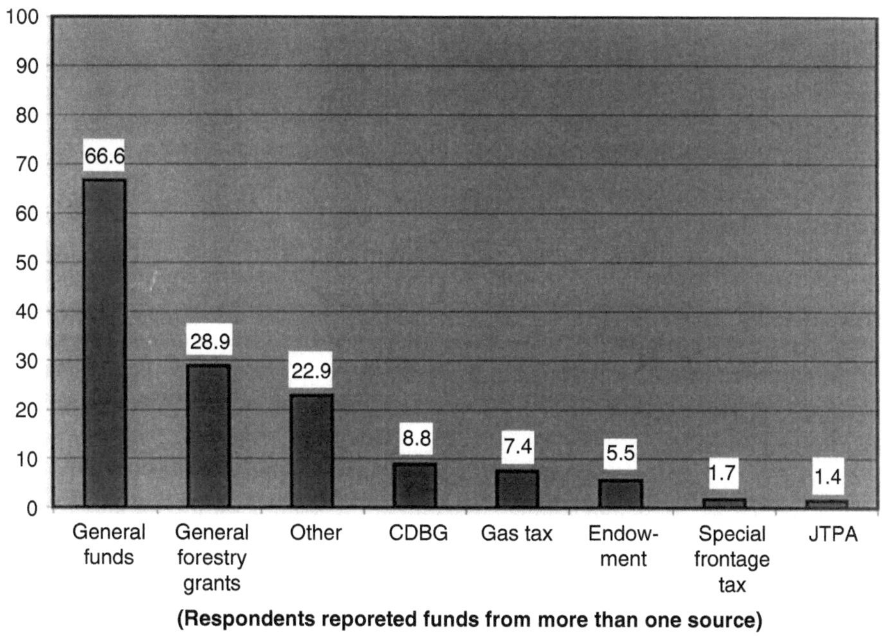

FIGURE 1. Percentage of respondents using various funding sources, from Tschantz and Sacamano (1994).

Table 1. Funding Sources Used for Tree Management Programs[a]

Funding source	Percentage of respondents using funding source	Mean amount of funding used ($)
General funds	66.6	257,155.00
General forestry grants	28.9	27,199.00
Other	22.9	N/A
Community development		
Block grants	8.8	31,389.00
Gas tax	7.4	209,288.00
Endowment	5.5	29,217.00
Special frontage tax	1.7	189,083.00
Job training partnership act	1.4	11,400.00

[a]From Tschantz and Sacamano (1994).

sponsored programs including the Small Business Administration's tree-planting grants and America the Beautiful tree-planting and maintenance activities are two of several examples of federal programs that are considered general forestry grants.

Community Development Block Grants are used by nearly 9% of the communities that responded to this survey (Tschantz and Sacamano, 1994). These federally appropriated allocations are awarded to municipalities that qualify for specific urban development projects. Tree planting, removal, and certain tree maintenance activities that are part of a larger development project can be funded for the duration of the specific activity.

An additional important source of funding for urban forestry activities is the state and federal Cooperative Assistance Program. It offers funding and technical advice to municipalities and was used by nearly 38% of the respondents in the survey. State foresters have detailed information available for those interested in pursuing state and federal Cooperative Assistance for their urban forestry programs.

Because of this increasing dependency on external grants from federal and state agencies, the urban and community forester must have a working knowledge of the grant application process. Obtaining outside funding has always been extremely competitive and will become even more difficult as the competition for these funds increases. Even though urban forest managers appear to be obtaining increasingly more external grants (Tschantz and Sacamano, 1994), many still have not applied for federal funding grants or federal technical assistance grants. In a survey conducted in New Jersey (Tate, 1984), two-thirds of the urban tree managers did not feel that they had sufficient information regarding the design and filing of grant applications and few of the 13 potential sources of funding listed by Unsoeld (1979) were known to the respondents. However, nearly all (97%) would apply if information about the process was made available.

Undoubtedly the best single source of information about federal funding agencies and programs is the Catalog of Federal Domestic Assistance published annually by the Office of Management and Budget. It is available from the Superintendent of Documents in Washington, DC. The catalog describes the federal government's domestic programs and identifies the types of assistance, explains the nature and purpose of the programs, specifies who is eligible, tells what kinds of credentials and documentation are needed to obtain assistance, lists the application and award process, and includes deadlines. Although the catalog provides a tremendous amount of information, direct contact with the target funding agency is usually necessary to acquire additional knowledge to clarify the instructions given in the written guidelines and assess the potential of the target proposal for funding. Additional information needed is the latest update about the particular program of interest, reaction of the agency to specific proposal ideas, lists and costs of projects funded in previous years, and the makeup of the proposal reviewing panel. Telephoning the agency is probably the best way to initially establish contact and gain needed information.

After the urban forester has contacted the potential funding agency, a better informed decision can be made on whether or not to actually write a proposal. This is also the time to contact other urban foresters for information about their success in obtaining funding in similar programs and to determine whether local funds are available to continue the project after external funding ends. At this point the urban and community forester should ascertain whether political decision makers, superiors, and the community are behind the proposal and whether resources are available in the organization to properly complete the project if and after the external funds run out.

Since most federal grants are oriented to construction and development, it is advantageous if the proposal can be subordinate to the main thrust of a larger project. Unsoeld (1979) lists several urban tree projects that probably would not have been funded alone but were funded as parts of larger funding programs, such as federal highway funds, resource conservation and development funds, and community development block grants. In this respect there is an obvious need to develop a close working relationship with engineers, architects, urban planners, and grantspersons who are responsible for preparing and administering grants that may provide ancillary funding for urban and community forestry activities.

If the decision is made to prepare a proposal as a subordinate part of a larger one or to stand on its own, most of the preparation skills needed can be generalized. A good proposal should be well written and organized according to the suggestions of the particular funding agency. In general, good proposals have similar characteristics, which are as follows:

1. The need for the project is clearly demonstrated.
2. Important ideas are highlighted and repeated.
3. Project objectives are given in detail.
4. Collaboration with all interested groups in the project planning stage is evident.
5. Commitment of all involved parties is evident.
6. The uses of the funds are clearly indicated in the proposal.
7. All government procedures have been followed.
8. Directions given in the proposal guidelines have been followed.
9. The proposal is in line with funding agency guidelines.
10. The writing style is clear and concise.

Submitting the proposal is the last important step in the process. After it is written, follow the agency guidelines for submission. Note the date of submission deadlines; they are inflexible. If the proposal is received beyond the stated deadline, it will be rejected and the effort expended will have been a costly exercise. In this respect, after the proposal is sent, it is wise to telephone the agency prior to the deadline to ascertain whether the proposal has been received.

Applying for a federal grant is unquestionably a considerable amount of work. Because of this and other factors mentioned earlier, many urban tree managers who have tried it do not feel the effort justifies the results. While it does take considerable effort and ability, one successful grant may more than justify the effort. Even if the proposal is rejected, learn why. Federal agencies are required to supply information as to the reasons for rejection. If the proposal is appropriate to the agency and was rejected because of specific problems, rewrite it and resubmit it.

Most cities with more than 50,000 population employ grantspersons or a person in a similar capacity to help obtain external funding. Many consultants provide training programs dealing with the process of obtaining grants.

3. Budgeting

3.1. Competing for a Fair Share of Budget Funds with Other Local Governmental Agencies

The purpose of local government is to perform functions and render services that the people in the community expect, require, and demand and that cannot be performed more efficiently by the private sector. The greater the demand and/or the perceived need for a service, the more likely that it will be performed. In order for a service to be rendered, it must have benefits that can be identifiable and quantifiable and be supported by the community. The benefits provided by the urban and community forest have to be compared with other municipal capital expenditures in terms of how these benefits compare with the costs of providing them. Increasingly, more of the benefits provided by the urban and community forest are quantified, yet most are passive and aesthetic. Most citizens would agree that these improve the quality of urban life. However, the electorate, if given a choice, often tends to choose services that are critical over those that provide the amenities of life.

Examples of community services that are critical and essential would be police and fire protection; public works such as clean and abundant drinking water, disposing of sewerage, and passable roads; educational institutions (although funding for schools is becoming more difficult to obtain as the percentage of the population without school-aged children increases); electrical power and communications networks; and a certain level of social services such as minimal medical care for the indigent. Understandably, there must be a minimal level above which these services are maintained or the city and urban area would cease to exist in a manner compatible with continued human habitation.

As budgets shrink in urban and community governments, the essential services are generally supported to a higher degree than nonessential services such as those provided by the urban forest. Therefore, it follows that urban and community forestry activities will require a continuing and increasingly major effort to obtain and retain the funds required to maintain them.

An urban forest management program is truly a long-term consideration. Trees are long-lived and provide the maximum benefits to the community only when they reach a larger size. The urban forestry program must be stable because wild budget fluctuations and reductions on a yearly basis do not allow for consistent planting and maintenance needs. When the program is allowed to fall below a threshold level, a series of events are set into motion. Gaps in stocking occur when planting is curtailed. Decline of tree health and aesthetic values are by-products of a less than stable program. Future maintenance costs increase, and the potential for hazard increases in the future when current funds are reduced. Unfortunately, a true picture of the costs required to maintain a healthy and safe urban forest often is not clearly defined to citizens and their elected officials. As mentioned earlier in this chapter, the need to know the cost of managing, planting, and maintaining an urban forestry program at various levels is critically important.

3.2. Allocation of Budget Funds to Various Tree Management Activities

The four most commonly performed tree management activities in communities surveyed (Tschantz and Sacamano, 1994) were pruning, planting, spraying, and tree and stump removal (see Fig. 2). These activities were performed on only 7.2% of the total publicly owned tree population. This leads to one of two conclusions: either municipal trees do not require maintenance or, because of funding restrictions, the vast majority of trees are not maintained. Most observers would agree that more maintenance is needed on more trees and lack of funding is precluding additional maintenance activities.

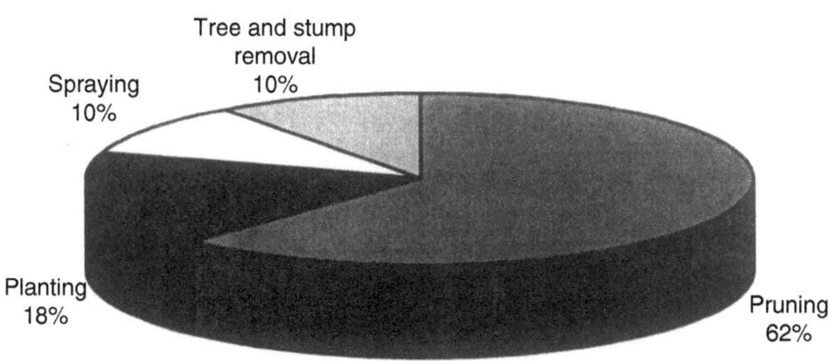

FIGURE 2. Maintenance activity on publicly owned tree population, from Tschantz and Sacamano (1994).

3.3. Establishment of Priorities for Various Tree Management Activities

Which activity should be given the highest priority when applying for tree management funding? Should an urban and community forester consider the health and preservation of the urban forest as the highest priority and only perform maintenance activities that maintain and/or increase the health of the resource? What are these activities? Would planting and replanting enter into this equation? For example, does the urban forester plant trees when there are trees in the existing urban forest that could be made healthier and would live longer by the forester providing an activity, such as fertilization, bracing and cabling, pest control, or girdling root removal, which would be withheld because of the funds expended on planting? Does the urban forester plant trees at the expense of the existing urban forest to create a more desirable urban forest in the future? In the northeastern United States during the height of the Dutch elm disease infection, some communities made the control of Dutch elm disease and the consequent minimal loss of the existing city elms a high priority. Others withheld extensive control measures, removed the elms as they died, and replanted as soon as possible. Controversy and concern over preserving (extensive control) versus conserving (minimal control) the resource raged. There is no clear answer as to which was the better decision.

Does public safety enter into the decision to allocate priorities? Large trees can be dangerous to many persons in the densely populated urban environment. Should the highest priority be only those activities that can increase the safety of the public such as pruning and removal of dead and dangerous trees?

Should politically popular activities, such as planting and replanting, receive the extensive support they are getting today? Money spent on planting programs amounts to nearly one-fifth of the funds allocated to maintenance activities in the urban and community forest. Remember, only 7.2% of the total publicly owned trees had any type of maintenance activity done on them. Most would agree that establishing priorities for tree management activities is a balance between politically popular activities, public safety concerns, and maintaining the health and longevity of the urban forest.

4. Urban and Community Tree Inventories and Their Role in Financing and Budgeting Urban and Community Forestry Activities

4.1. Use of the Inventory to Create an Urban Tree Management Program

Inventory data can be used to demonstrate the need to develop a systematic tree care program and the possible consequences of failing to develop one. The community is ultimately responsible for maintaining the publicly owned urban tree resource and the liability that may result from improperly caring for it. Not having funds to maintain the resource does not absolve a city of an accountability in lawsuits arising from it. Moreover, the cost of a judgment against the community or the defense costs in a lawsuit could conceivably pay for a systematic tree care program for many years (Tate, 1985).

To create an urban and community forest management program in a community that does not have one, data must be arranged and presented to best put forth a message that is concise, straightforward, and graphic. Charts and graphs catch attention and have the ability to impress the need for a program on all who should be concerned but have limited time and interest to spend on interpreting reams of tree-related data. The time for details can come later.

Three major items should be summarized: (1) planting needs, (2) maintenance requirements, and (3) potential hazards to life and property. If a management program is created, data from the inventory can aid in the development of a specific urban tree management plan. While urban and community forest management planning has become commonplace in this country, it has not focused on long-range objectives (Lobel, 1983). The need for a management plan that includes long-range planning is tremendously important. Inventory data can be used to determine the extent, condition, and maintenance needs of the resource and compare these parameters with desired objectives stated in a plan.

4.2. Use of the Inventory to Gain Increased Funding

Increased funding without quantitative information is difficult to secure when competing with other departments that have supporting data (Tate, 1985). Obtaining worthwhile supporting data also lends credibility to budget requests. Police departments, for example, keep crime statistics and usually can make an extremely strong case for the maintenance of their budgets. Obviously, urban tree care does not rank in importance with the protection of citizens from crime, but city trees do improve the general quality of urban life and have continuing maintenance requirements that can be demonstrated by inventory data.

Although increases may be denied, it is important to continue to request what can be justified by an inventory. Political decision makers' and administrators' priorities can and do change after a certain amount of education. Education can be one of the objectives of an inventory, and budget hearings are in many ways akin to educational sessions. Well-presented information that educates and substantiates the need for a budget increase often gains a more favorable response when contrasted to funding requests without proper supporting data. Data presentation for programmatic increases must be concise, straightforward, and oriented to the graphic. The principal purpose is to present a picture of the program, where it is now, and where it has to go to satisfy the stated objectives of a resource management plan.

4.3. Use of the Inventory to Insulate Against Budget Reductions

Each year various city agencies' funding requests are submitted and considered. These are viewed as part of the larger general city budget. Priorities in each community change, sometimes, yearly. Programs are ranked and evaluated for their worth in light of their political and monetary costs and values (Tate, 1985). An urban forestry program that cannot defend itself adequately is in danger of being cut back or eliminated. Obviously, cutbacks result in reduced urban and community forestry maintenance activities. The reductions are not readily apparent to city residents over the short run but may have a tremendous long-term impact on the urban tree resource. Future budget increase do not necessarily reverse the ill effects of the cutbacks.

Unfortunately, many political decision makers do not understand that the existing resource must be maintained at some threshold level. Trees that comprise the resource will continue to grow; limbs and trees will die even though the budget is cut back. Often when systematic tree care is eliminated, tree removal funds seem to be found because of the threat of lawsuits against the city and other factors. Even if the tree program is totally eliminated, tree work will still be done in some reduced format. Operations will probably be less efficient because there is no full-time tree care agency. If the work is done by contract, it will require in-house management and supervision to be effective. More importantly, trees will probably die at greater rates when the tree maintenance budget is reduced.

To insulate and protect against budget cutbacks, inventory data can be used to describe the composition and condition of the resource. If cutbacks will affect planting, pruning, and removal, the presentation of data describing number of planting locations, trees requiring pruning, and dead or dying trees to be removed can be used to demonstrate accomplishments with the existing budget, as compared to what occurs if the budget is reduced. As stated earlier, tree maintenance is needed on a regular and sustained basis, and cutting the budget will not make the need for it go away.

Cutbacks may not directly affect the major service elements, but could impact on minor but critically important expenditure such as fertilizing, pest control, and post-planting care. In these cases data describing vigor conditions, insect and disease extent, and the proportion of newly planted trees to the established tree population can be used to explain the future effects on the resource when these services are reduced.

4.4. Use of the Inventory to Manage the Urban and Community Forest More Efficiently

Efficiency can be improved to some degree in most organizations. Many existing systematic tree care programs are underfunded. Because of intense competition for funds, tree care programs often have extreme difficulty in gaining real budget increases. The additional funds received may not cover yearly cost increases due to inflation. To provide a higher level of service and in some cases just to maintain the same level, given the same budget, a program must become more efficient (Tate, 1993).

More efficient dispatching of work crews can be accomplished if it is known in which area of the city there are large accumulations of trees with similar maintenance needs. For example, trees with low limbs that interfere with pedestrian and/or vehicular traffic should be pruned for clearance. These operations can be easily handled by crews working from the ground without extensive equipment and training. This work can be done during inclement weather, when equipment is being repaired, or as fill-in when key members of a particular crew are absent and other work cannot be done. Minor pruning and lifting can be performed by seasonal workers and temporary employees. Inventory data can be used to determine the number and size of crews needed for a particular tree maintenance operation. For example, the removal of large trees along busy streets requires more personnel and equipment than the removal of smaller trees in residential areas. The sizes, numbers, and locations of trees to be removed in various areas can be provided by inventory data. Then work can be scheduled more effectively.

The purchase of equipment can be made more efficiently by using inventory data. The existing equipment inventory should match today's tree maintenance needs and be the most efficient type for existing and future work. For example, the cost of an aerial life increases proportionally to its working height. If inventory data indicate that most of the street trees needing work are less than 30 feet in height and only one aerial lift can be purchased, buying a lift that can reach 30 feet rather than a taller one is probably a better decision because it is more cost-effective. The small percentage of taller trees may be done by contract, or climbers can be placed into these trees (Tate, 1985).

In planting, species can be better matched with the aid of inventory data to the site. If, for example, one of the inventory variables notes the presence and heights of overhead utility lines, planting locations that are sited under lines can be used to estimate the number of trees needed that are shorter and more compact at maturity. Decisions made about the species mix for future tree purchases will be enhanced. The development of special pruning practices for small trees that will grow into utility lines as they mature can be programmed into the maintenance plan as a result of the inventory.

Properly trained tree crews work more efficiently than untrained crews (Tate, 1981). Inventory data can indicate existing and future tree work. From this, an in-service training program can be designed to adequately prepare the crews for necessary operations. If a large number of potential planting locations exist, for example, and an increased planting program is anticipated, crews can be trained in nursery maintenance practices, proper planting techniques, and postplanting care procedures.

5. Summary

Without adequate funding, the urban tree resource cannot optimize and provide the social, aesthetic, and economic benefits that are available from its citizens of the urban environment. Yet funding has decreased significantly during the last decade for urban and community tree management activities. The sources of funding also are changing from heavy reliance on the general fund, primarily financed by local property taxes, to other sources of income. As the revenue picture continues to get smaller, competition between agencies of local government looms larger, requiring the urban and community forester to become more adept at identifying, obtaining, and retaining inside as well as outside funds for urban forestry management activities. The urban forester must become an expert in utilizing the grant process.

Budgeting or urban and community forest management activities was always an important practice but has become even more so today because of funding reductions. Justification of a budget request requires establishing priorities for each type of management activities and marketing them to local appointed and elected officials who have a role in the budget approving process.

The use of a tree inventory can be critical in the creation of a systematic urban and community forestry program, as well as a tool to justify present funding levels and even gain additional funding. It also can be used to manage the urban tree resource more efficiently. All this can be accomplished only if the inventory is designed as a management tool.

References

Bromage, A. W., 1957, *Municipal Government and Administration*, 2nd ed., Appleton-Century-Crofts, New York.

Kielbaso, J. J., 1988, Trends in urban forestry management, *Baseline Data Rep*. 20 No. 1, 3, 4.

Lobel, D. F., 1983, Managing urban forests using forestry concepts, *J. Arboric*. **9**(3):75–78.

Tate, R. L., 1981, Guidelines for in-service training for urban tree managers. *J. Arboric*. **7**(7):188–190.

Tate, R. L., 1984, Municipal tree management in New Jersey, *J. Arboric*. **10**(8):229–233

Tate, R. L., 1985, Uses of street tree inventory data, *J. Arboric*. **11**(7):210–213.

Tate, R. L., 1993, How to compete for budget dollars by privatizing the tree care operation, *J. Arboric* **19**(1): 44–47.

Tschantz, B. A., and Sacamano, P. L., 1994, *Municipal Tree Management in the United States*, Davey Tree Expert Co, Kent, OH.

Unsoeld, O., 1979, *Federal Assistance for Urban Forestry*. USDA Forest Service, Washington, DC.

Chapter 9

Community Involvement in Urban Forestry Programs

James Nichnadowicz

1. Introduction

This chapter will focus on four topics: (i) How to increase public awareness of your community tree program; (ii) how to involve stakeholders in your program; (iii) using volunteers to increase the size of your community forestry program; and (iv) involving children in urban and community forestry programs.

2. Public Awareness

As an urban or community forester you are responsible for a community's trees. This is what you are compensated to do, and hopefully you are doing a thorough and professional job at it. But no matter how well you do a job, if your supervisors and the other supporters of your program are not aware of it, it will not help you to retain your position or sustain your program. Stakeholders need to know about the high-quality work you are doing.

Stakeholders also need to know what you are doing so that they can refute negative comments they hear about a town's trees, such as: "The town's tree is lifting my sidewalk," "The town's tree is dropping leaves in my house's gutters," "The town's trees get seeds on my car," or The town does not do anything to care for these trees." The only way to refute these claims is to have good public awareness of your activities. The following sections contain several methods that can be used for this purpose.

James Nichnadowicz 4-H Agent for Rutgers Cooperative Research and Extension of Union County, New Jersey, Westfield, New Jersey 07090-1499.

Urban and Community Forestry in the Northeast, 2nd ed., edited by, J. E. Kuser.

2.1. Leaflets

Distributing leaflets about tree care increases people's awareness about your program and how to properly care for trees. One source of arboriculture leaflets is the International Society of Arboriculture. (For address, see "Suggestions for Further Information" at the end of this chapter.)

While such leaflets will increase your client's knowledge about tree care in general, we also want the leaflet to tell the client who you are. Thus, be sure to buy only those leaflets that have room on them for your name, address, and phone number. That way, people not only learn about how to properly plant a tree but also learn about you as well.

Who are you? This is not as simple a question as it seems. If you are not working for yourself, be sure to check with your supervisor before printing your name on any leaflets. Town employees may want to put the mayor and the city council on the leaflet and add their names as a contact for more information. When in doubt, it is better to put as many names on the publication as possible, as long as they have supported the tree program. Sharing the credit helps to develop support for the program in the future.

When these publications generate inquiries for more information, keep a record of these calls so you can develop a mailing list. Citizens who are interested in shade trees in your community may be valuable supporters in the future.

Any frequently visited spot is a good location to distribute these leaflets. These can be displayed for distribution at the town's library, garden centers, and your municipal building. If a display rack is not available for them, you may want to obtain one. Also, the leaflets could be included in a direct mailing. Does your town have a larger mailing that these could be a part of?

2.2. Annual Reports

You have probably seen an annual report from your church, synagogue, college, or from a company that you have invested in. These reports keep you informed about an organization's accomplishments and plans for the future. This is an effective way for key stakeholders in an organization to see all that a group is doing, even though they themselves are involved in only a small part of the organization.

Each year you should do an annual report. It may be only a few pages, but they can be highly effective at educating stakeholders about your program, such as your supervisor, board of directors, or volunteers.

In the annual report, discuss how many trees you planted, what kind they were, what size, their cost, and where you planted them. You can also discuss the number of trees that were removed, pruned, and fertilized. Finally, you may want to talk about any needs you see for the future, such as equipment, supplies, or additional help.

Photographs will help the readers of the annual report to better understand your needs and activities. These pictures may be of your crew or your crew with some of your key stakeholders or of you with an award you received. The photographs can be of many things. For information about taking photographs for publication, see Section 2.3.

A person proficient in desktop publishing will be able to arrange the information and photographs you collected into a well-formed annual report. Be sure to proofread it several times before printing it and share it with your supervisors for review before publishing it. Also, you may want to ask your supervisor or your town mayor to write an introduction for the report.

2.3. News Releases

People learn about their community through the local newspaper. It is usually published once a week and contains information that is not significant nationally or statewide but is important to the local community. These papers are hungry for news, but they do not have enough staff to report on many local stories. This is an opportunity for you.

At least once a month, take the opportunity to write a news release for the local paper. A news release is simply what it says: a release of news (see Fig. 1). With practice, one can be written in an hour or less.

To write a news release you must first think why people read the local paper. They do so because they are interested in local news. Thus, your news release should be about the trees and people in your town, not about a distant forest.

What should you write about? As a general rule, anything that you do that is new is a good topic. New equipment, new personnel, promotions of personnel, training courses you or fellow employees took, number of trees planted, or extent of storm damage to town trees are just some of the possible topics you could write about.

How should you write about the topic you select? In the same way you would tell someone in person about the event. What parts would they be most interested in? Put those parts first. Usually these things are known as the "five magic words of journalism": who, what, when, where, and why. These are the basic things that your story must cover. For example, "Bob Jones of the Westwood Shade Tree Department has just become a certified tree expert. Bob became certified by passing a 100-question test last Thursday." In two sentences, we have covered the five W's. Also, most editors recommend that they should be covered in the first paragraph of your news release. Additional paragraphs can be used after the first one to give more description.

What should your news release look like? It should be typed, double-spaced, and no longer than one page. Above all, it should be accurate. If you are in doubt about some bit of information, double check it or eliminate it. Inaccurate information in an article will generate complaints to you and the newspaper. For instance, if your news release states that "The Mayor's Annual Arbor Day Tree Planting will be at Walnut Park, and the public is invited," but in reality it was at Chestnut Park, you and the paper will get angry calls from misled citizens.

Photographs accompanying your article will not generate complaints but may generate more calls of interest. To have a photograph work to your advantage, submit one that ties into your news release but is catchy. A clear, black-and-white photograph with action in it will reproduce well in the paper and attract readers' interest. For example, if you are covering a tree-planting ceremony, show the people digging the hole, not just standing stiffly around a newly planted tree. Above all, do not submit blurry or dark pictures as these copy poorly. For additional help with writing press releases, see the Public Relations Guide reference at the end of this chapter.

RUTGERS
COOPERATIVE
RESEARCH &
EXTENSION
of Union County

News Release

-300 North Avenue East, Westfield NJ 07090-1499

-Immediate Release/all zones

Contact: -James Nichnadowicz, Union County 4-H Agent
Telephone: -908-654-9854, extension 2236

MASTER TREE STEWARDS TEACH COUNTY YOUTH

Union County—Congratulations go out to Edi Merkel and Joan Laezza of Cranford, Joanne Kavinski of Fanwood, Maureen Kelley of Rahway, Joanne Lindquist of Linden, Al Greenberg of Murray Hill, Lorraine Shively and George Coyne of Plainfield, Karen Kotvas of Colonia, Mary Barth of Bloomfield, Marianne Lang and Shirley Bonilla of South Orange, and Audrey Walther of Montclair for teaching 2500 Union County fourth and fifth graders about trees this spring.

All of the above are 4-H Master Tree Stewards. Prior to teaching children about trees, they completed a 30-h training program sponsored by the Rutgers Cooperative Research & Extension Program of Union County.

For more information about the 4-H Master Tree Steward Program, which is free of charge and supported by the Union County Board of Chosen Freeholders and the State of New Jersey Forestry Department, call the Union County 4-H Agent, James Nichnadowicz, at 908-654-9854, at extension 2236.

Rutgers Cooperative Research & Extension provides information and educational services to all people without regard to race, color, national origin, gender, religion, age, disability, political beliefs, sexual orientation, or marital or family status.

FIGURE 1. Sample publicity release.

2.4. Lectures

While photographs are educational, lecturing can be as well. Contact your local Rotary, Kiwanis Club, Gardening Club, and so on and offer to do a free 30-min lecture about the care of trees. People are interested in tree care because they know it can save them money. Also, lectures accompanied by a slide show or Powerpoint© presentation are very effective. People retain much more information when they can see what you are talking about. You can prepare your own show by taking photographic slides to use in your presentation. If you do not want to prepare on your own, you may be able to borrow a slide show from your local forestry office or Cooperative Research & Extension Office.

If you are anxious about speaking in public, you can boost your confidence by watching an experienced person. From such an experience, you can develop an outline for your own talk.

2.5. Television

Once you have presented your lecture several times, contact your township or local cable television station. You can appear on their station. They may not want to telecast your lecture, but they may want to interview you instead. This is fine as long as you can meet with the station interviewer ahead of time to go over the questions they will be asking you. This eliminates "surprise questions" that could be difficult to answer.

2.6. Videotapes, CDs, and DVDs

Television shows can answer many "people questions" about you and your organization, but they are only aired a few times. To keep a video performance in front of the public for a longer time, consider making a videotape or a CD about tree care and the tree care services your department offers. Often this can be done with the assistance of the local township or high school television production studio.

2.7. Comment Forms

Although videotapes can reach a large audience, it is also important to have ways that inform individuals about what you have done. For instance, when you plant a tree in front of someone's house, be sure to leave a comment form. This asks them some questions about the quality of your work. It should have a stamp on it and your address so that they can easily send it back to you. From this form you can discover how people perceive the quality of your work. You could ask, "Was the tree planted at the proper depth?" "Was it put in straight?" "Was the dirt replaced correctly?" Questions also could be left when a tree pruning has been done.

2.8. Image Check

Using the previously discussed ideas will greatly increase the public's awareness of you and your tree program. With this, new awareness will come with increased observation. People will be watching to see if you and your work are of professional quality. To be sure they are, complete the Image Analysis Survey in Section 2.8.1. It is based on a survey that was developed by Rutgers Cooperative Research and Extension but has been modified for relevance.

2.8.1. Image Analysis Survey

Circle the number that indicates where you stand:

Not responsive to client needs	1	2	3	4	5	Responsive to client needs
Impersonal client service	1	2	3	4	5	Friendly client service
Poor quality service	1	2	3	4	5	Top quality service
Poor organizational credibility	1	2	3	4	5	Good organizational credibility
Poor value to local elected officials	1	2	3	4	5	Good value to local elected officials
Unknown to local officials	1	2	3	4	5	Well known to local officials
Poor source of information to the media	1	2	3	4	5	Good source of information to the media
Dirty trucks and equipment	1	2	3	4	5	Super clean trucks and equipment
Dirty, ununiformed employees	1	2	3	4	5	Super clean, uniformed employees

If your image and service to the community is positive, then making the public aware of what you do should be very successful. If your image is not positive, work on that before attempting anything else. Public awareness cannot correct a flawed program, but many a good program has lost support because no one knew about it.

3. Relations with Local Elected Officials

In a community forestry program, you must always keep in mind who is incharge. If you work for a government agency, it may be the public works supervisor or the mayor. If it is a nonprofit group, such as Save the Windsor Oaks, it may be a board of directors. Whoever is your boss, it is important that you clear all your activities through him or her before acting. They will have a better understanding of the big picture and can help you avoid costly mistakes.

For instance, you may want to plant a row of shade trees on Maple Street. When you check with the boss, she informs you that Maple Street is going to be widened and the place you selected for your trees is going to be paved over. In this case, not checking with the boss could have resulted in much wasted time and effort.

Once established in our jobs, we may want to redirect our efforts to address new tree-related issues. Our community may need a shade tree survey or an increased removal program. How do we go about implementing such new ideas and programs?

First, we must define a need for the new program. For instance, a shade tree inventory has never been done in our town. Such a survey would help us to direct our efforts in tree planting, care, and maintenance. The only obstacle to doing the survey is where to get the money to pay for it.

Now is the time to meet with the people who control your community's finances. See whether they have the funds or they know of other sources from where the funds could be obtained. Try to show them how the idea would save the town money in the long run and would improve the appearance of the town dramatically. They probably will agree with your idea but may not have the money to do it. The best thing to do at this point is to bring the issue up at each year's budget meeting. After several years, they may finally agree to do the survey because they have enough money to do so.

Not all issues with elected officials or bosses will revolve around money. They may revolve instead around the removal of trees for development. If you are a public employee, your role in such removal may be limited. But if you are with a nonprofit group, your objections to removal of trees can be heard. For instance, if you want to save a wooded area from development, put together a rationale for doing so, and then visits the elected official. For more information of this nature, be sure to obtain a copy of the article, "Working Effectively with Government" by Marcia D. Bansley and Edith Makra (address in Section 7). Their four-page article contains a wealth of information on how to influence and build rapport with elected officials.

Finally, elected officials, mayors, councilmen, and so forth need publicity to get reelected. In light of this, they love publicity associated with positive events such as Arbor Day tree planting and memorial tree planting. Be sure to invite them to speak at events of this nature. Not only will they appreciate the effort that you made, but they will remember, and thus give you a chance to be heard about another important tree-related issue.

4. Involving Volunteers in Your Community Forestry Program

As a community or urban forester, it would be prudent to involve volunteers in your program for the following reasons:

1. Volunteers who are committed to the mission of your program can provide skills you do not have. While you may have expertise in arboriculture, they may have knowledge of computers, advertising, writing, politics, and more. All these skills can be used to greatly enhance your town's tree program.
2. Volunteers can provide you with different perspectives and contacts. Some volunteers will know elected officials they can talk to about supporting your program. Other volunteers will have wonderful advices on keeping a volunteer program going because they have run one themselves.
3. Good volunteers are a committed and kind audience. They applaud when you succeed and help to pick up the pieces when things go wrong. It is disappointing to have a great tree program for your town, but few people know about it. When you have good volunteers that have had input into your program (i.e., they have helped pick the types of trees to plant), they notice and applaud a job well done.
4. Enthusiastic volunteers can grow to be involved in more complex volunteer jobs. As new volunteers gain experience, they can organize Arbor Day fairs, develop a speakers' bureau, and even go out to the schools to teach children about trees.
5. Finally, and most obvious, volunteers can help you to expand your tree program. To illustrate this, I relate to you my own work in a field closely related to forestry: home horticulture.

When I first started working for Rutgers Cooperative Research and Extension, my job was to teach and solve problem in the area of home horticulture. I taught classes to the gardeners of Union County, New Jersey about lawn care, vegetable growing, and care for houseplants. I also served as "plant doctor," diagnosing the ills of hundreds of sick plants each year.

I never thought about involving volunteers in my job when I first began working at Rutgers in 1983. Rather, I perceived that I was here to serve the public, not to have them serve me. Also, I was too busy familiarizing myself with the position to have time to involve others in it. How can you teach something to others when you do not know it all that well yourself?

By 1986, however, I was more adept and knew what areas needed improvement. I wanted to try some new ideas and programs, such as starting a demonstration garden, where I could do trials of vegetables and flowers. To do this, I would have to get help from people who could take over the established programs. This would allow me time to work on new ones. Since there was no money available to hire help, I needed to find volunteers. I found them.

By the end of 1987, I had a group of 30 active volunteers. This group enabled me to greatly expand the services my program offered. We increased the hours we were open by 100%. We had a garden fair that attracted over 1000 people and raised $1000. A tomato variety trial was run that produced over 5000 pounds of tomatoes for the needy. Finally, we started a horticulture therapy program for the residents of a nearby nursing home.

When you first start working in community forestry, your experience will no doubt be very similar to mine with home horticulture. You will start out trying to do the best you can by yourself, but as times goes on, you will need to involve others in your program so that you can initiate new programs.

4.1. Laying the Foundation for Volunteer Involvement in Your Forestry Program

Before involving volunteers, you have to decide what they are to do. To decide, compare an outstanding shade tree program in a neighboring town with yours. This will help you to locate the areas of your program that need work. Volunteer involvement will be of great service to you in these areas.

For example, the neighboring town has trees in every available planting site. Your town does not. You have identified a need. To meet this need, you could recruit volunteers who would raise funds to plant trees in your town. Or, you could recruit volunteers to do a shade tree inventory. An inventory would document in great detail your town's need for additional trees.

Once you have documented your town's needs, you are ready to design a volunteer program that addresses them. Designing a volunteer program is much like creating a business. You need to decide what your output will be and then how to staff your project to accomplish this.

For instance, if your output was that every block in your community would have its trees inventoried, you would need to decide how many volunteers you need to do this and what skills are needed. To answer these questions, inventory a few blocks by yourself. From this, you will know how long it takes to inventory a tree and the skills you need to do so.

This process of a pilot test for a program to determine volunteer needs and skills before recruiting volunteers will assist with the design of any community forestry program involving volunteers. Before recruiting volunteers, submit your program plan and a description of volunteer activities to your town's legal department and risk management department. They can analyze them for potential hazards and for uninsured activities. Finally, present your plan to your supervisor for approval.

4.2. Analyzing the Job You Need Done Helps Target Recruitment Efforts

Now that you have identified a need for volunteers, you need to structure their participation. A first step toward doing so is figuring out when the volunteers are needed. Do you need them in the daytime, nighttime, spring, fall, or all seasons of the year? Also, does your project require special skills such as knowledge of computers? Does the project need people to lift heavy objects, walk long distances, have their own transportation, or a driver's license? The answers to these questions will help you to target your recruiting efforts.

For example, I needed volunteers who could work between the hours of 9:00 a.m. and 4:00 p.m. and were interested in gardening and enjoyed working with the public. I decided to target my recruiting efforts toward retirees, especially retirees who belonged to garden clubs and visited local garden centers.

4.3. Applications for Screening and Clarification

To make certain that people understand what your program is about and what it requires, have them fill out an application before they can be accepted into the program (see Figs. 2 and 3). Applications help potential volunteers to really question if they can commit to the program.

4.4. Advertising to Find Volunteers

Now that you have figured out who could volunteer for you and have drawn up an application, how do you make potential volunteers aware of your opportunity? The answer: advertising through a variety of media like press releases to your local paper, television, and radio stations; flyers to all local garden clubs and libraries; and appearances on your local cable television. Speak to local organizations such as the Lions Club, Rotary, and local gardening clubs. Perhaps these groups could become cosponsors of your project. For more information about how to write press releases and get publicity for a new program, see Section 2.

4.5. Who to Accept into Your Program

Your recruitment efforts have been a success. People are applying to be in your program efforts everyday. Review an application as soon as you receive it. Potential volunteers want to know quickly if they have been accepted into the program in less

SAMPLE VOLUNTEER APPLICATION

APPLICATION FOR MASTER TREE STEWARD PROGRAM

NAME: ____

ADDRESS: ____________________________ **ZIP:** ____________

PHONE NUMBER: (DAY) ______________ **(EVENING)** __________

1. Are you available to attend Master Tree Steward classes on Wednesdays, September 6 to December 6, 2006, from 1 p.m. to 3 p.m. and field trips? _______YES _______NO

2. Are you available from 12 noon to 3 p.m., every other Wednesday, January–June 2007, to teach youth?

3. What do you enjoy about working with youth?

____

____

____

4. Experience working with youth (if any)?

____

____

5. Additional information: (if you respond "YES" to any of the following, please explain below.):

a. Do you use or sell illegal drugs? _______YES _______NO
b. Have you ever been convicted of a criminal offense? _______YES _______NO
c. Have you ever been charged with child neglect or abuse? _______YES _______NO
d. Has your driver's license ever been suspended or revoked? _______YES _______NO
e. Other than the above, is there any fact or circumstance involving you or your background that would call into question your being entrusted with the supervision, guidance, and care of young people? _______YES _______NO

6. I am available to attend all Master Tree Steward training classes (up to three absences can be made up) and will return 30 h of volunteer time teaching youth through the 4-H Program. The coordinator of the program will provide training, teaching materials, and supplies.

Signed: ____________________________________ Date: ________________

IF YOU ARE ACCEPTED, YOU WILL BE NOTIFIED BY AUGUST 25.

FIGURE 2. Sample volunteer application.

than 2 weeks. Accept everyone into the program who meets the qualifications. Once you have accepted a person, give them a call. They will be happy to hear directly from you. Follow up with a letter confirming when and where the training program starts.

SAMPLE PROGRAM DESCRIPTION

Fun and Lots of Learning Await You in the Award-Winning* 4-H Master Tree Steward Program

The Master Tree Steward Program is an enjoyable educational program. It teaches concerned adults about the nature and life of trees. Learn how trees are born, grow, and reproduce; how to plant trees, their importance to our environment, and fun and easy ways to share this knowledge of trees with children. Homework is easy reading and a simple project. There are no exams or quizzes. Once trained, you and the other volunteers will be part of a **Tree Team**. Together, you will teach groups of interested and attentive schoolchildren about trees.

Your commitment is short-term, but impact is LONG-TERM

Classes will meet every Wednesday, 1 p.m. to 3 p.m., from September 6 to December 6, 2006. Monthly field trips will be from 10 a.m. to 2:30 p.m. While perfect attendance is desired, up to three absences can be made up.

In return for this training, **provided at no cost to you**, you will be asked to volunteer 30 h of time. We will need you to be available every other Wednesday, from January to June of 2007, from 12 noon to 3 p.m. to team-teach about trees. All teaching materials will be supplied to you.

Convenient, accessible, safe location for classes

Classes will be held at the 4-H office, 300 North Avenue East, in Westfield. Secure and safe parking is available.

It's easy to apply

Just fill out the attached application form and return no later than August 18. The class is filled on a first-come, first-serve basis.

For more information

Contact Union County 4-H Agent, James Nichnadowicz, at 908-654-9854, extension 2236, Fax 908-654-9818, email to nichnadowicz@aesop.rutgers.edu, or write to the RCRE 4-H Office, 300 North Avenue East, in Westfield, NJ 07090-1499.

* Winner of the 1996 Green Community Achievement Award; an award presented by the New Jersey Forest Service for an outstanding educational program in the area of community forestry.

Rutgers Cooperative Research & Extension Programs are open to all people without regard to race, color, national origin, gender, religion, age, disability, political beliefs, sexual orientation, or marital or family status.

FIGURE 3. Sample program description.

4.6. Benefits of Training Volunteers

Training people before they volunteer for your program has many advantages. First, training is an incentive for some people to volunteer. When surveyed as to why people volunteer, many give the answer, "Because I wanted to learn new skills." In fact, a survey done in 1990 of Cooperative Research & Extension master gardener volunteers showed that over half the volunteers in that program volunteered because of a desire to learn new skills (Simonson and Pals, 1990). In essence, you are trading your training for their help.

Second, when part of the training is testing, you can be sure that they are competent. For instance, someone wants to help prune the new trees you transplant out into the community. You might be nervous about letting them do so, but if you test them after training, you need not fear this. You have shown them the way it is done, and a test will show what they learned.

Third, training allows you to bond with your volunteers. Through informal chatting and intentional activities you will get to know each other better. Getting to know your volunteers can eliminate many problems later. When difficulties arise, it is much easier to resolve them with friends.

To help turn a group of strangers into friends, each training session should have a time for informal chatting. (Our training sessions had a break for refreshments. We supplied the hot drinks, and each volunteer took a turn bringing a dessert.) At the first training session, set aside some time to play a game that will put people at ease and allow them to get to know one another. (For a sample "Get to Know You Game," see Fig. 4.)

PEOPLE SCAVENGER HUNT

Directions: Fill out the form below without listing anyone's name more than once. When you have filled in all the blanks, raise your hand.

	Name	Answer
1. Someone who likes ice cream	________	________
2. Someone who is a twin	________	________
3. Someone who plays a musical instrument	________	________
4. Someone who likes to read	________	________
5. Someone who speaks a foreign language	________	________
6. Someone who likes to cook	________	________
7. Someone who likes to play sports	________	________
8. Someone who has a dog	________	________
9. Someone who has a pet	________	________
10. Someone who collects things	________	________

FIGURE 4. People scavenger hunt.

4.7. Training Volunteers

First, be sure that you are thoroughly competent in the job that you need the volunteers to do. Competence can be achieved through additional training and, of course, practice.

Second, train them through a variety of methods. This is important, for it is well documented that we remember:

20% of what we read
20% of what we hear
30% of what we see
50% of what we see and hear
70% of what we see, hear, and discuss
90% of what we see, hear, discuss, and practice.

Thus, train your volunteers by using pictures, drawings, or diagrams. Demonstrate what needs to be done, give them reading materials, discuss the topic with them, observe their practice, and assist when needed.

Videotapes, fact sheets, leaflets, and other educational materials about trees are available from your local Cooperative Research and Extension service, the US Department of Agriculture Forest Service, your state forest service, and the International Society of Arboriculture.

When volunteer training is complete, recognize the volunteers' efforts with a certificate and a ceremony. Our ceremony not only features certificates but includes a potluck luncheon as well. Also, those with perfect attendance receive an award.

4.8. Financing Volunteer Training and Locating Teachers for Your Training Sessions

Some other questions to consider before training volunteers are: Where will you obtain the funds (about $30 per person) to purchase the books they will need for study? If the funds are not available through your organization's budget or grants, you may want to run the class with a textbook as an option and give handouts in class instead, or use the textbook as a further reference that people can buy on their own.

Who will teach the training classes? Contact your local Cooperative Research and Extension office. You can find the phone number of your local Cooperative Research and Extension office by looking under "county government" in the blue pages of your telephone directory. The number may be listed under Cooperative Research and Extension or under the name of your state's land grant college. If the local Extension agent cannot help you, ask in a neighboring county. Your state forestry service and the forestry department at your state land grant university will also be able to help you. Finally, after watching these professionals teach, you may be able to offer the classes yourself, the next time they are needed.

4.9. Challenges of Volunteer Use in Your Program

Below are some of the special challenges that working with volunteers bring. These should not frighten you away from starting a program. Rather, they are mentioned so that if they do arise, you can deal with them quickly and effectively.

1. You will recruit a few volunteers who are unhappy people. They will often wear a frown throughout your classes and may make negative comments. These unhappy folk can be a big challenge. How to deal with them? Do not let them squelch your enthusiasm. The other 20 people in the training are positive. Focus on *their* enthusiasm instead.

Meet privately with each of the unhappy volunteers to see if the cause of their unhappiness is something in the program. You may discover that you are misinterpreting their facial gestures. More than likely, their attitude is not caused by the program, but factors outside of it. These you can do little about, but they will feel better knowing that you took the time to listen to them.

Should you eliminate these volunteers from your program? No, as long as they are respectful of the program's objectives, other volunteers, and they perform their duties satisfactorily, they should not be terminated. Focus on their positive qualities and appreciate their willingness to volunteer.

2. Realize that people have different personalities and ways of relating to each other. Your challenge is to work with everyone fairly. You may be a reserved person, but your volunteer is assertive. As long as the person is respectfully assertive, polite, and does not make unreasonable demands, this is something you have to learn to tolerate. Remember to make all the opportunities available to everyone in the class.

3. While all volunteers have different personalities, what sustains a volunteer program is uniformity. To achieve the harmony of uniformity we need to let volunteers know that there are rules for attendance, performance, and dress. People should be made aware of these rules when they first consider volunteering, before time is invested in a training program.

4. Make sure your volunteers understand that you are depending on them to do the job they agreed to do. As one 4-H agent said to her volunteers, "Nothing will kill our program quicker than people who say `yes' to a job but do not follow through with it." Have a reserve volunteer in case one calls in sick. If you need one volunteer to do a job (such as plant a tree), get two. If both show up, great! The work will take them only half as long.

5. In addition to training, active volunteers need to be managed. This involves observing them, recognizing their efforts, and providing training as needed.

6. Volunteer training can be inefficient if the volunteers do not stay involved with the program. As a general rule, volunteers should give at least as much time volunteering as you spent on training them. Your challenge is to get volunteers to stay on past their initial competency, so they can become experts. Recognition and continuous training help to do this.

5. Involving Children in Tree Planting

In a study done by Washington State University in the year 2000, urban residents were asked, "Do you like the trees in your neighborhood?" (Lohr and Pearson-Mims, 2002). The residents who answered "yes" were then asked "Why?" They reported that, as children, they were exposed to nature (through hikes in parks and woods), as well as were involved in gardening (growing a flower garden, vegetables, and planting a tree seedling). These things helped them to develop an appreciation for trees. As the survey showed, this appreciation did not end in childhood. It continued on into adult life.

We as urban and community foresters *want* citizens who like the trees in their neighborhoods. It is much easier to convince a homeowner who appreciates trees to plant on their property. Union County Shade Tree Advisory Board member, Nicholas Polanin, often laments about the homeowner who does not like trees. "There are some homeowners who just do not want us to plant a tree in front of their houses. I try and try, but they can not see the beauty of the tree. They are all caught up in the potential problems and efforts that a tree requires" (Nicholas Polanin, Union County, New Jersey, and Shade Tree Advisory Board Member).

The study claims that seeing past the potential problems is a vision that needs to be developed in childhood. It goes on to say that this vision is developed by hiking, planting trees, and visiting parks. While hiking and visiting parks are wonderful, I will focus here on planting trees because it is more in the field of a community forester. The rest of this chapter will contain plans on how to involve children in tree planting.

5.1. Tree Seedling Giveaways

In many towns schools give free tree seedlings to the students. On a spring day, the children will be presented with a small pine or spruce seedling. This little 6-inch tall piece of green makes a nice gift for the children. They are excited about receiving it and taking it home to show their parents. They are also inexpensive for the school to buy. Some state forestry departments even provide the seedlings for free or at low cost.

Education is an element often lacking in this seedling giveaway. The children may receive a brief handout with directions on how to plant a tree, but it is a rare occurrence for them to watch someone plant a tree and provide formal instruction. Perhaps this is because community foresters either lack the time or are not confident enough to teach. Take heart. There are ways to overcome these obstacles.

If time is the issue, consider training other adults who can then train the students directly. For example, in 90 min you could train 30 teachers on how to plant a tree seedling. They in turn could each instruct 30 children. If the teachers did not have time for the training, you could train parents who would then teach the students. As an incentive, I would offer to provide participating schools with tree seedlings at a discounted rate or at no charge. With all the issues and projects that schools are expected to deal with, tree planting may get forgotten without an incentive.

The teachers' training session should include possible planting locations and a discussion of how big the tree will ultimately grow to be. Train them how to plant a seedling, the importance of watering, how to remove it from its tube or pot, the dimensions of the hole, staking the tree, etc. Once this instruction is complete, provide time for practical application by allowing each person to plant their own seedling. You should observe them, so you can give encouragement and correction as needed.

Another element lacking in tree seedling giveaways is follow-up. The children are given the seedlings to take home and no one reminds them to plant it. Many of the tree seedlings die a slow death; starved for sunlight on a shady windowsill. Community foresters can prevent this waste. A news release in the local newspapers reminding the children to plant the trees will help. Also, send an advertisement to your town's community access television station.

If the schools in your community do not give away tree seedlings, you could offer to buy them in bulk for your school district. Repack them in individual bags to make them portable for each student. You do not have the help to repack them? Call on local community garden clubs or your high school's environmental club for help. If this is not feasible, provide the schools with a list of where they can buy tree seedlings.

5.2. Helping Schools Celebrate Arbor Day—The Union County Arbor Day Tree Program

Arbor Day is a great opportunity to encourage kids to be aware of trees. Unfortunately, many children are not celebrating it; either because their schools lack the funds to buy a tree or they are not aware of it. To deal with both of these issues, the Union County, New Jersey, Shade Tree Advisory Board created the Union County Arbor Day Tree Program.

The Arbor Day Tree Program provides schools with large trees free of charge. Schools are given container-grown, 4- to 5-foot tall trees in 5-gallon pots. We use container-grown trees because they can be planted well into spring. (Arbor Day is the last Friday in April in New Jersey.) Bare root trees would be more economical, but the time of year that they need to be planted is much sooner than Arbor Day. Also, we have found that schools prefer smaller flowering trees. When we offered a choice in the past between a Red Oak and a Korean Flowering Dogwood, 85% of the schools chose the dogwood. We do not use balled and burlapped trees. They are too heavy for grammar school students to move.

5.3. Additional Ideas on How to Involve Children with Trees and Tree Planting

Many towns have carnivals or festivals in the spring. These events are ideal times to give out tree seedlings. This is also a great time to distribute literature on tree care and selection. Consider giving out literature on how to involve children with tree planting and gardening. This information is available from your Cooperative Research and Extension service, your state forestry department, and the International Society of Arboriculture.

Another educational tool is a poster or poetry contest for children. In the process of creating the poem or poster, the necessary research and introspection help them to develop an appreciation for the wonder of trees. For these contests to be a success, you need to have a prize of sufficient interest for the children to participate. Survey some children to find out what is of relevance to them. In my research, toys and appliances, such as DVD players and televisions, attract more attention than saving bonds. To encourage cooperation, you may also want to offer a prize to participating schools, such as a special tree for Arbor Day.

6. Conclusion

This chapter has touched briefly on the subjects of public relations, working with elected officials, and volunteer use in community forestry programs. For more training in volunteer use, enroll in a volunteer management class at your university. If such a course is not offered, a class in business management would be useful. Classes in public relations also are offered at most universities. If these are too lengthy, contact your local adult school or university short-course department. They may offer 1- or 2-day courses covering these subjects.

Also, volunteer for a local organization that has a reputation for being well run. It could be a student group on campus, your local Cooperative Research & Extension 4-H program, a local religious group, or community forestry program.

References

Interaction by Design: *Bringing People and Plants Together for Health and Well Being: An International Symposium*, Iowa State Press, Ames, IA.

Lohr, V. I., and Pearson-Mims, C. H., 2002, *Childhood Contact With Nature Influenced Adult Attitudes and Actions Toward Trees and Gardening* (C. A. Shoemaker, ed.), pp. 267–277.

Simonson, D. L., and Pals, D. A. 1990, Master gardeners: Views from the cabbage patch, *J. Ext.* **28:**35–36.

Suggestions for Further Information

Alliance for Community Trees, 201 Lathropway, Sacramento, CA 95815.

Bansley, M., and Makra, E., *Working Effectively with Government*, Trees Atlanta, 96 Poplar St. NW, Atlanta, GA 30303.

International Society of Arboriculture, PO Box 3129, Champagne, IL 61826-3129.

Gardening Activities for Youth, 4 pages, a Rutgers Cooperative Research & Extension Fact Sheet by James Nichnadowicz and Lydia Blalock, FS#003., published September 2004. Available at the Rutgers Cooperative Research and Extension website free of charge: www.rutgers.edu. Click on <publications>.

Still, D. T., and Gerhold, H. D., 1997, *Community Forestry Volunteers: Review of Practical Studies*, Pennsylvania State Cooperative Extension, Pennsylvania Urban and Community Forestry Council, 56 East Main St., Mechanicsburg, PA 17055.

Public Relations Guide—subtitle: *"How to Win Support for a Good Cause"* from the P&G Corporation, 37 pages, undated, P&G Educational Services, PO Box 599, Cincinnati, OH 45201-0599, email: pgfund.im@pg.com, website: www.pg.com

Chapter 10

Soils: The Key to Successful Establishment of Urban Vegetation

George A. Hawver and Nina L. Bassuk

1. Introduction

It has been estimated that 80% of urban-vegetation problems are attributable to the soil environment (Patterson *et al.*, 1980). Foresters and horticulturists are facing the question of how to improve woody plant growth under these circumstances. In many cases, the solution is to modify rooting environments. This requires gaining a basic understanding of soil properties, acknowledging the difficulties associated with urban soils, and conducting proper site assessments prior to planting.

2. Tree Roots

An appropriate first consideration is of the part of the tree that interacts most intimately with soil—the roots. Tree roots obtain water, minerals, and oxygen from the soil. These nutrients, in turn, are used by the plant to synthesize carbohydrates and energy through the processes of photosynthesis and respiration. They also synthesize hormones that trigger growth in the entire tree. The soil also provides support and anchorage to root systems.

Tree roots exist predominantly in the top 30 cm (12 inches) of soil and usually do not extend deeper than 1 or 2 m (40 to 79 inches) (Perry, 1982). Roots grow in the top layers primarily because of their requirement for oxygen and a permeable soil; oxygen concentrations are greatest near the soil surface where the soil is less dense. These top layers are also favorable for growth because of the organic matter and nutrients that are usually present.

George A. Hawver and Nina L. Bassuk Department of Horticulture, Cornell University, Ithaca, NY
Urban and Community Forestry in the Northeast, 2nd ed., edited by, J. E. Kuser.

In the absence of restriction, roots will often extend horizontally to occupy an area four to seven times greater than the tree's projected crown area. Roots are opportunistic and will extend into zones where resources are most plentiful and conditions are most favorable for growth (Perry, 1982). Regard for a tree's "natural" root distribution helps to underscore how susceptible roots are to disturbances made at the soil surface, even at points far beyond the projected crown area.

3. Soil Properties

It is necessary to gain a basic knowledge of soil properties in order to fully understand the problems encountered with urban soils. Many of these difficulties, including compaction and impeded drainage, are a function of soil's physical properties. Soil's chemical and biological properties also are significantly impacted in urban forests and warrant consideration.

3.1. Physical Properties

Soil's physical properties include texture and structure. These properties, in turn, influence soil pore space, aeration, infiltration of water and drainage capacity, susceptibility to compaction, nutrient-holding capacity, temperature, and color.

3.1.1. Soil Texture

Soil is a mixture of particles of different sizes. These sizes are summarized in Table 1. The primary particle-size classes are sand, silt, and clay.

Sand particles are between 0.05 and 2 mm in diameter and are relatively rounded or cubelike in shape. This shape allows large pores to exist between particles. Soils that contain a high proportion of large sand particles often exhibit favorable aeration and drainage because air and water can readily move through the large pore spaces.

Silt particles are between 0.002 and 0.05 mm in diameter and diverse in shape. Silt is intermediate between sand and clay in terms of its properties.

Table 1. Particle-Size Classes

Particle-size class	USDA diameter range (mm)
Gravel	6.7–2.0
Very coarse sand	2–1
Coarse sand	1–0.5
Medium sand	0.5–0.25
Fine sand	0
Very fine sand	0
Silt	0
Clay	<0.002

USDA—United States Department of Agriculture.

Clay particles are less than 0.002 mm in diameter and tend to have a relatively flat, or wafer-like shape. The pores between clay particles are very small. Predominance of these small pores contributes to clay's ability to hold water very tightly (Brady, 1990); soils dominated by clay particles typically exhibit poor aeration and reduced water movement.

Due to its small particle size, clay has a very large surface area per unit weight. This large surface area and the presence of negative surface charges enable clay particles to hold or "adsorb" significant quantities of water and mineral ions. Clay contributes substantially to soil fertility.

The term texture refers to the relative proportions of particle size classes (sand, silt, and clay) contained in a soil; the names of the dominant particle sizes in a soil are used to name its texture. For example, a soil dominated by particles of 0.05 to 2-mm diameter is called "sandy," a soil dominated by particles less than 0.002-mm diameter is called "clayey." Many combinations of particle sizes are possible.

The term loam refers to a soil texture that contains moderate amounts of sand, silt, and clay. A loam soil has characteristics that are intermediate to those of sand, silt, and clay. These soils often have relatively favorable water retention and nutrient-holding capacities. Depending on which particle size dominates a loam, it can be classified as a sandy loam, silt loam, or clay loam. Further and increasingly specific classifications can be made (Brady, 1990; Craul, 1992).

Texture influences a soil's structure, porosity, water relations, and chemical properties (Craul, 1992). By identifying a soil's texture, inferences can be made about its characteristics and its capacity to support satisfactory plant growth. Tables 2 and 3 summarize the characteristics of some of the most common textures. The

Table 2. General Characteristics of Common Soil Textures

Soil texture	Susceptibility to compaction	Nutrient-holding capacity	Approximate available water (as percentage of total soil volume)
Sand	Very little	Limited	7
Loamy sand	Limited	Limited	10
Sandy loam	Limited to moderate	Moderate	12
Loam	Moderate	Moderate to substantial	16
Silt loam	Substantial	Substantial	20
Clay loam	Substantial	Substantial	12
Clay	Substantial	Substantial	11

Table 3. General Characteristics of Sandy and Clayey Soils

Soil characteristic	Sandy soils	Clayey soils
Soil surface area	Less	Greater
Pore space	Less	Greater
Aggregation potential	Less	Greater
Bulk density	Higher	Lower
Particle density	Same	Same

ability to predict these qualities is valuable to urban foresters when evaluating a soil's texture and making decisions regarding plant selection and the necessity of soil modification.

3.1.1.1. Texture Assessment. A soil's texture can be identified in a laboratory by performing a particle size analysis on a soil sample. This will indicate the proportions of sand, silt, and clay particles that are present.

Assessment of texture also can be made in the field via the texture-by-feel method. This technique involves moistening a small amount of soil and rubbing it between the fingers and thumb so that a ribbon begins to form. The length of the ribbon that can be produced before it breaks in two from its weight indicates the soil's textural class. Ribbon formation is difficult with sandy soil but enhanced with increasing clay content. This technique can be quite reliable after gaining experience with it. A more detailed description of this method can be found in other publications (Trowbridge and Bassuk, 2004).

3.1.1.2. Texture Modification. To significantly change a soil's texture, it may be necessary to add large quantities of soil from a different textural class. Depending on the scale and budget of the planting, it may be more practical to replace the existing soil with a soil of desired texture. Selection of texture is discussed in Sections 3.1.3.1 and 5.2.

3.1.2. Soil Structure

Soil structure refers to the way individual soil particles aggregate or clump together. The size and proportion of the pores existing between these aggregates determines the likelihood that air, water, and roots can move through the soil. Basically, there are two types of pore spaces. These are called micropores and macropores.

Micropores are smaller than approximately 0.06-mm diameter. Due to their size, they retain water very tightly. Air movement within these pores is quite slow. Clayey soils tend to have a high proportion of micropores (Brady, 1990).

Macropores are larger than approximately 0.06 mm in diameter (Brady, 1990) and very important in terms of soil structure. Macropores are occupied by air or water; soils that contain a high proportion of macropores allow air, water, and roots to move easily through the soil. Sandy soils often have a high proportion of macropores.

A soil that is well aggregated contains many macropores and is said to have favorable structure. Structure is enhanced by the presence of organic matter or clay. These act as binding agents and encourage particles to clump together; simultaneously, pores are formed between the clumps. The organic matter found in the top layers of natural, undisturbed soils contributes to the favorable structure that is typical of those environments. A primary constraint of urban soils is the shortage of macropore space that results when soil becomes compacted.

Ideally, 50% (approx.) of a soil's total volume is made up of pore spaces. Of this, half of the pore volume contains air and half contains water. About 5% of the ideal soil's total volume is made up of organic matter and 45% is composed of mineral particles (Patterson, 1977; Patterson *et al.*, 1980; Brady, 1990).

3.1.2.1. Bulk Density. Bulk density is a measurement that indicates the volumes of solids and pore spaces contained in a given sample. This index is calculated by dividing a soil samples oven-dried weight by its volume (in g/cm^3 or mg/m^3).

Soils that contain a high proportion of pore spaces and low proportion of solids exhibit relatively low-bulk densities. These soils will likely demonstrate favorable aeration, water movement, and root penetration. A soil must have both macropores and micropores to achieve a balance between water-holding capacity (micropores) and aeration (macropores). If a soil's pore space is altered, its bulk density is likewise affected. Soils that have been compacted exhibit high-bulk density values.

Normal or ideal bulk densities can range from approximately 1.30 to 1.40 mg/m^3 (Patterson *et al.*, 1980). When interpreting bulk density, it is necessary to consider soil texture. Relatively high-bulk densities are most detrimental in clayey soils; in these situations, bulk densities of approximately 1.45 mg/m^3 can inhibit root elongation. In sandy soils, however, bulk densities may reach 1.70 mg/m^3 before reducing root elongation (Trowbridge and Bassuk, 2004). Bulk density typically increases with depth in a soil profile.

3.1.2.2. Structure Assessment. Assessment of a soil's structure and degree of compaction can be made by measuring bulk density. This requires obtaining a soil sample in its undisturbed field state so that existing pores remain intact. Several samples should be taken, depending on the size of the site being considered.

A commonly used method of evaluating bulk density—the core sampling method requires that a soil sample be oven dried for 24 h at 105°C and then weighed. The sample's weight is divided by its volume to obtain the bulk density (in g/cm^3 or mg/m^3). The core sampling method is simple and can be performed relatively quickly, but it loses accuracy if used on stony soils or on soils that are very dry or wet. Other methods for estimating bulk density include the soil clod method, volume excavation methods, and nuclear densitometry. These techniques are discussed in detail in other publications (Lichter and Costello, 1994; Blake and Hartge, 1986). Another indication of soil structure can be obtained by using a penetrometer to measure soil strength. Strength, measured in units of pressure (megapascals—MPa), indicates soil's ability to resist penetration.

Strength is influenced by soil's moisture content, texture, and particle mineralogy. In general, wet soils and clayey soils have relatively low strength; conversely, dry soils and sandy soils typically exhibit greater strength (Brady, 1990). Since moisture content and texture vary at different sites, comparisons of soil strength at different locations should be made with caution.

Research to determine the level of resistance that restricts root growth of woody plants is limited. It has been suggested, though, that when a penetrometer with a standard 30°-cone tip is used, woody root growth may be severely restricted when soil strength reaches approximately 2.3 MPa (Day and Bassuk, 1994).

3.1.2.3. Structure Modification. The structure modification that is most often needed in urban situations involves increasing the number of macropores in compacted soils. Modifications for this situation are suggested in Section 6. Structure modification is also discussed in Section 3.1.3.2, as it relates to aeration and drainage.

3.1.3. Aeration and Drainage

Aeration is the exchange of gases, including oxygen and carbon dioxide, which occurs between the soil and the atmosphere. Oxygen concentrations are typically high in the surface layers of natural soils where many macropores exist.

As noted previously, macropores are occupied by air and water. When soil moisture content increases, as from irrigation or a rainstorm, water displaces the air from the macropores. When water is removed from soil via drainage, evaporation, or transpiration, it is replaced by air.

If a soil has few macropores or if its macropores are often occupied by water (i.e., saturated), then gaseous exchange will be impeded. Replenishment of soil oxygen from the atmosphere will be slow, and carbon dioxide, which is produced by the respiring roots and soil organisms, will accumulate. Oxygen deficiencies are significant because roots require oxygen for the process of respiration, which results in energy production for nutrient and water absorption (Brady, 1990; Craul, 1992). An excess of carbon dioxide is significant because high concentrations of this gas can be toxic to plant processes (Brady, 1990).

Favorable soil aeration benefits soil organisms as well as plants. When gaseous oxygen is present, aerobic organisms are actively decomposing organic matter. When gaseous oxygen is very low or absent, the rate of breakdown of organic matter is substantially reduced (Brady, 1990; Craul, 1992).

3.1.3.1. Aeration and Drainage Assessment. Aeration and drainage can be assessed in the field by observing soil color in the top 30 to 40 cm of the soil profile. Soil color is influenced by the oxidation states of elements, including iron and manganese. In well-aerated, well-drained soils, these elements will exist predominately in oxidized states; these soils will often be yellowish-brown or reddish-brown in color.

When a soil drains slowly and thus is poorly aerated, the oxygen levels will be relatively low. Under these conditions, termed anaerobic, elements will exist primarily in their reduced states; gray or blue may be seen in the soil sample.

If both oxidizing and reducing conditions are present in the soil, then a mixture of grays, blues, and yellowish- or reddish-browns may be observed in the profile. This condition is called mottling (Brady, 1990; Craul, 1992).

Another indication of unfavorable drainage and aeration is a foul odor resembling vinegar or rotten eggs. This smell often accompanies anaerobic conditions.

Assessment of drainage also can be made via a field percolation test. Tests should be conducted at several locations on a site. This test is performed by digging a hole of approximately 30 cm (12 inches) depth and filling it with water. After all of the water has drained, the hole should be refilled with water, and the water's depth should be measured. After 15 min, the depth of water remaining should be measured, and the drainage rate can be calculated in centimeters per hour. It is important that the soil surrounding the hole is very wet to minimize the lateral movement of water.

Ideally, the percolation rate should be approximately 3 to 15 cm (1 to 6 inches)/h. If the hole drains more slowly than 3 cm/h, then the soil probably contains a relatively small proportion of macropores. This soil may be compacted or may have a high-clay content. Another possibility is that the soil may be saturated, as from a high-water table. If the hole drains more quickly than 15 cm/h, it contains a high proportion of macropores and is likely to contain a large portion of sand or gravel, or have a good structure.

3.1.3.2. Aeration and Drainage Modification. In urban situations, if a soil drains slowly and is poorly aerated, it is likely that the soil is either compacted or has a high clay content. If the soil is compacted, then modifications to improve structure are warranted.

If the soil is very clayey, then the most effective plan may be to replace it with a soil texture that contains more macropores. A sandy loam is often a good choice for soil replacement. Selection of texture is discussed further in Section 5.2. Recent research suggests that compacted clayey soils may be modified with the addition of 50% well-composted organic matter (1:1 by volume) (Rivenshield, 2003).

Slow drainage and poor aeration will also occur in landscapes where a high-water table exists. Modifications of soil structure are not likely to improve this situation. A technique that can be used on sites with high-water tables and with clayey soils involves creating a raised planting bed, or berm above the unfavorable landscape. This technique is discussed in Section 7.2.

If a soil is composed primarily of coarse sand particles, it may drain excessively. In this scenario, water-holding capacity can be increased by amending the soil. By uniformly amending the soil with up to 20% organic matter (by volume), such as sphagnum peat, the situation can be improved (Urban and Craul, 1993).

Precautions can be taken prior to planting if future drainage problems are anticipated. One option is to install subsurface drainage in tree pits. This requires placing a perforated drainage pipe so that it slopes from the bottom of the pit toward an outlet such as a storm drain. A layer of geotextile over the pipe end that intersects the pit will discourage clogging. A uniform, well-drained soil should be used. Variations of this technique are discussed in other publications (Craul, 1992).

A common fallacy is the idea that placement of a gravel layer at the bottom of a tree pit will improve drainage. In fact, an abrupt textural change will only exacerbate the drainage problem.

3.2. Chemical Properties

Soil chemical considerations include nutrient content, cation-exchange capacity, and soil reaction. Accumulation of soluble salts, a common problem in urban areas, will also be considered.

3.2.1. Nutrients

There are 16 essential elements that are required by plants to complete their life cycle. Of these, carbon, hydrogen, and oxygen are obtained primarily from air and the soil solution (i.e., water in the soil). The rest of the essential elements are obtained from soil solids, or minerals.

Essential elements can be classified as macronutrients, which plants require in large amounts, and micronutrients, which are needed only in small quantities. The macro- and micronutrients are listed in Table 4.

Table 4. Essential Elements[a]

Macronutrients	Micronutrients
Carbon (C)	Boron (B)
Hydrogen (H)	Chlorine (Cl)
Oxygen (O)	Copper (Cu)
Nitrogen (N)	Iron (Fe)
Phosphorus (P)	Manganese (Mn)
Potassium (K)	Zinc (Zn)
Calcium (Ca)	Molybdenum (Mo)
Magnesium (Mg)	
Sulfur (S)	

[a]From Epstein (1972).

Although a given element may exist in a soil, it may not be in a form that is available for uptake by plant roots. Available elements can exist in a dissolved form in the soil solution or are held on the surfaces of particles. Surfaces of clay and organic matter are reservoirs for large quantities of nutrient elements. The supply of elements increases as organic matter is decomposed and as rocks and soil particles weather (Craul, 1992; Harris, 1992).

3.2.2. Cation-Exchange Capacity

Particles of organic matter and clay have surface charges that attract positively charged nutrient ions, or "cations." Cation exchange occurs as a cation is released from the surface of a particle in exchange for a cation from the soil solution.

Soils, rich in organic matter and clay, exhibit high capacities for cation exchange and holding nutrients. In contrast, sandy soils and soils containing little organic matter have reduced cation-exchange capacities and are relatively less fertile. Nutrient-holding capacities for some common soil textures are summarized in Table 2.

3.2.3. Soil Reaction

Soil reaction is expressed as pH and indicates the relative proportion of hydrogen ions (H^+) and hydroxyl ions (OH^-) in a soil. The pH is defined as the log of the reciprocal of the hydrogen ion concentration. A one-unit change in pH indicates a 10-fold change in hydrogen and hydroxyl ion concentrations.

A soil with a pH of 7 has a neutral reaction. Larger pH values indicate an increased proportion of hydroxyl ions; these soils are termed alkaline (or basic). A pH value below 7 indicates a relatively large concentration of hydrogen ions; these soils are said to be acidic.

Soil reaction is correlated with the solubility and thus availability of nutrient elements. A near-neutral pH of approximately 6.5 to 7 favors solubility of many of these elements. In contrast, soils that are quite alkaline or acidic will tend to have a reduced reservoir of available elements. A neutral soil pH also favors activity of microorganisms (Craul, 1992; Harris, 1992).

3.2.4. Soluble Salts

Applications of deicing salts to roadways and sidewalks can lead to accumulation of soluble salts in adjacent soils. High concentrations of salts (>1.000 ppm) can reduce a plant's ability to take up water from the soil, resulting in a droughtlike experience.

When sodium chloride salts dissolve in water, the sodium and chloride ions dissociate. Excess sodium in the soil can reduce the availability of nutrient elements. The chloride ions can be absorbed by plant roots and may accumulate to toxic levels in actively growing parts of the plant. This buildup can result in a marginal leaf scorch (Hudler, 1980).

3.2.5. Chemical Property Assessment

Laboratory analysis can be used to characterize soil pH and identify the concentrations of nutrient elements, organic matter, and soluble salts. Methods also exist to measure soil pH and soluble salts in the field. Based on this assessment, vegetation can be selected, which is appropriate for the existing soil conditions, or modifications can be made.

3.2.6. Chemical Property Modification

A soil's chemical properties are generally easier to modify than its physical properties. Fertility status can be altered by adding fertilizers and modifying the pH. Application rates are dependent on the desired change in pH and the soil's texture and organic matter content (Harris, 1992).

Although soil pH can be raised by applying limestone or lowered by applying sulfur compounds, in a perennial landscape it is difficult to make a permanent change in pH over a large area. It is generally more practical and successful to select plants that will tolerate the existing pH ranther than trying to change it.

Soluble salts will be leached from the soil during snowmelt and via rainwater. This process can be facilitated by flushing root zones with tap water. As a preventative measure, plantings could be situated so that exposure to deicing salts is minimized.

3.3. Biological Properties

Soil organisms significantly influence how favorable a soil is for plant growth. These organisms include earthworms, bacteria, actinomycetes, protozoa, nematodes, fungi, and mycorrhizae.

Soil organisms help to decompose organic matter into simplified nutrients, which can then be used by plants. Some organisms can fix nitrogen from the air. Others improve soil structure and aeration through burrowing and mixing actions; simultaneously, residues are mechanically incorporated into the soil. These activities in turn influence a soil's chemical and physical properties.

4. Urban Soils Distinguished from Natural Soils

Natural soils are dynamic systems that are synthesized in profile over time. Typically, these soils are formed as a gradual continuum over the landscape. Natural soils are composed of mineral and organic components, and their formation is influenced by factors, including climate, living organisms, soil parent material, and topography.

In contrast, urban soil formation is influenced primarily by human activity. The term urban soil is applicable to soils outside the perimeter of urban and suburban areas; the prerequisite for calling a soil "urban" is that it has been substantially impacted or disturbed by human activities.

Human impact includes moving and mixing of soils during landscape construction, traversing the soil surface, and adding materials. Added materials include plastics, glass, paper, construction debris, fertilizers, sewage, deicing salts, and atmospheric pollutants. As a result of these and other activities, the physical, chemical, and biological properties of urban soils differ substantially from those of natural soils. The net effect of these modifications is a rooting environment that is often detrimental to tree growth.

4.1. Destruction of Soil Structure

The most widespread and harmful of human activities may be the impact that occurs when soil structure is destroyed and soil is compacted. Structure is damaged when macropore space is reduced by way of applied forces from foot and vehicular traffic, heavy machinery, and vibrations caused by traffic or machinery (Craul, 1992). Structure is also damaged when soil is moved or mixed; as the existing soil matrix is destroyed, particles become rearranged. Small particles may translocate downward and fill in the spaces between the large particles, resulting in decreased macropore space. Soil structure can also be impaired from applications of deicing salts. These salts result in dispersion of small particles, which can fill in pore spaces (Craul, 1992).

Loss of structure results in soil compaction, increased bulk density, reduced capacity for aeration and water infiltration, and increased mechanical impedance to root penetration (Craul, 1992). Each of these effects is related and detrimental to plant growth and survival. Compaction has been shown to affect reductions in root elongation (Veihmeyer and Hendrickson, 1948; Simmons and Pope, 1987; Gilman *et al.*, 1987), shoot growth, and available rooting volume (Day and Bassuk, 1994).

4.2. Reduced Aeration and Drainage

Urban soils commonly exhibit impeded aeration and drainage as a consequence of soil compaction. Soil volumes that are restricted by the presence of below-ground structures may also exhibit poor aeration and drainage. This occurs because the vertical and horizontal movement of air and water is obstructed. This situation may result where trees are planted into confining pits, or where there are discontinuities in the soil because of below-ground utilities, foundations, buried pavement, hardpans, or because of abrupt changes in soil texture (Craul, 1992).

4.3. Elevated Soil pH

The soil pH in urban areas is often relatively high. This is a consequence of increased calcium concentrations that result as concrete surfaces and building materials are weathered. Applications of calcium or sodium chloride deicing salts also can increase a soil's pH. High pH levels (7.5 to 8.2) can be problematic by limiting the availability of iron, manganese, and zinc.

4.4. Additional Characteristics

In addition to the qualities previously mentioned, Craul (1992) has noted that urban soils are characterized by substantial vertical and spatial variability, modified temperature regimens, interrupted nutrient cycling, modified soil organism activity, and the presence of an often-hydrophobic surface crust. Patterson *et al.* (1980) have observed that plant roots exhibit nonuniform growth patterns in these soils.

5. Preventing Soil Compaction

Compacted soil is often cited as a primary constraint to tree growth in urban areas (Patterson *et al.*, 1980; Patterson, 1977; Craul, 1992; Stone, 1994). After a soil's macropore space has been reduced, it is difficult to repair the damage. Remediation techniques to lessen the negative effect of a compacted soil are often temporary and do not always produce satisfactory results. Ideally, plans for compaction prevention should be made during the landscape design phase. There are several techniques to be considered.

5.1. Surface Treatments

Application of a mulch layer will help to prevent soil compaction. The effect of this layer is to redistribute applied force from foot and vehicular traffic or construction machinery over a wider area. The depth of mulch required depends on expected site activities.

To help protect soil from heavy equipment and traffic during site construction, a layer of at least 15 cm (6 inches) of a coarse wood chip mulch or a 10-cm (4-inches) layer of gravel (3/4 inch crushed) may be required. After landscape installation is complete, the layer may need to be at least partially removed to avoid anaerobic conditions (Lichter and Lindsey, 1994). If the intent is instead to reduce compactive damage from moderate levels of foot traffic, then application of a coarse-textured mulch of approximately 10-cm depth is appropriate.

5.2. Choice of Soil Texture

Some soil textures are better able to resist compaction than others. A soil that contains a narrow distribution of coarse sand particles, such as a "gap-graded" sandy loam, is preferable. It is important to specify soils with few fine to very fine sands (0.05

to 0.25-mm range), as these can contribute significantly to future compaction. Ideally, 60% to 75% of the sand particles should be greater than 0.25 mm in diameter (Urban and Craul, 1993). The similarity of shape and size of these particles allows for maximal particle-to-particle contact that results in a relatively rigid soil matrix. It is important to recognize that while this texture tends to resist compaction, its water-holding capacity is relatively low (Craul, 1992). Supplemental irrigation in the mulch layer may be needed to meet plant moisture requirements.

In contrast, a soil containing a wide distribution of particle sizes, including portions of sand, silt, and clay, will be very susceptible to compaction. This includes well-graded sandy loams as well as finer textured soils. In this situation, the smaller particles can translocate downward through the profile and rearrangement and fill the pore spaces between the larger particles. These soils have the potential to compact even in the absence of applied forces (Spomer, 1983; Craul, 1992) and should be avoided.

In situations where whole sidewalks or streets are being installed concurrently with tree planting, use of a structural soil under the road and sidewalk is recommended. This medium consists of a gravel matrix that contains suspended soil with a hydrogel tackifier in its pores. The structural soil is rigid enough to safely bear loads required by engineering standards for roadways and sidewalks, yet this medium can be used by tree roots (Grabosky and Bassuk, 1995; Grabosky *et al.*, 2001).

5.3. Restricting Access

Wet soils are especially susceptible to compaction because of the lubricating effects of water and the ability of small particles to translocate into macropore space. If possible, it is best to suspend activities involving heavy traffic on a site until the soil is dry (Craul, 1992).

Prevention of compaction around existing trees during construction is possible given proper planning and commitment to preserve the trees. It is crucial to recognize that, barring restriction, roots often extend to occupy a zone four to seven times greater than the tree's projected crown area (Perry, 1982). Placement of sturdy fencing around protected root zones will help to prevent construction vehicles, equipment, and personnel from damaging the soil structure and the existing tree roots (Anella, 1996; Lichter and Lindsey, 1994).

6. Compaction Remediation

Compaction remediation refers to the improvement of soil structure so that the proportion of macropores is increased and bulk density is reduced. In nature, favorable soil structure is produced over the course of many years via influences, including climate and soil organisms (Perry, 1982; Craul, 1992). It is difficult for humans to create favorable soil structure after soil has been compacted. Nevertheless, the attempt must be made if satisfactory plant growth is to occur.

Techniques for compaction remediation can be considered in terms of whether they are feasible for sites that have existing trees and shrubs. This vegetation may have been planted initially into compacted soil, or else the soil may have been compacted since the time of planting. Though these trees may continue to live, they are likely to be growing very slowly or are in a gradual decline. The techniques that are currently available for use in this circumstance have varying levels of applicability and effectiveness; these include radial trenching, mulch applications, and injections of compressed air or water. For situations where a compacted site has yet to be planted, the techniques that can be used include mechanical break up of the soil and incorporation of soil amendments.

6.1. Radial Trenching

This technique requires using a trencher or small backhoe and shovel to create two or more trenches that radiate outward from the tree trunk. The trench dimension is approximately 36 cm (14 inches) in width by 1.5 m or more (5 feet) in length. To facilitate drainage, the depth should grade from approximately 30 cm (12 inches) at the end nearest the tree to a depth of 60 cm (24 inches). Trenches should be oriented so that the tree's major root axes are avoided and impeded root growth is intersected near the trunk.

The soil should be excavated, thoroughly loosened with a shovel, amended with organic matter, and replaced into the trenches. These zones exhibit reduced bulk density, which will allow roots to extend into them and gain increased access to water and nutrients.

In order to protect the trenches from recompaction, a 10-cm (4 inch) surface layer of coarse wood chip mulch is highly recommended. Pedestrian and vehicular traffic should be restricted from these zones.

Variations of this technique have been used successfully by several researchers (Smiley, 1997). Increases in root (Day *et al.*, 1995; Watson *et al.*, 1996) and shoot growth (Day *et al.*, 1995) have been reported within 2 years after trench installation. Trenches also could be installed at the time of initial planting if the landscape is compacted. Recent research has shown that a minimum of 25% to 33% organic amendment (by volume) in a compacted sandy loam soil and 50% (by volume) in a clayey soil will reduce recompaction of the soil and provide for a more favorable rooting environment by reducing soil-bulk density below root restricting thresholds (Rivenshield, 2003).

6.2. Mulch Applications

Another option is to apply and maintain a layer of organic mulch on the compacted surface. The mulch should be approximately 10 cm in depth and extend as far out as the drip line, if possible. This method of improving soil structure is quite slow, but over a period of years it will help to develop more favorable surface soil structure. Roots are likely to grow into the mulch layer as well. Reductions in surface bulk densities and increases in porosity and root development have been reported (Watson *et al.*, 1996; Himelick and Watson, 1990).

6.3. Fracturing with Compressed Air or Water

Some techniques use equipment that creates fractures in the soil by injecting compressed air or water. These methods have generally not proved to be very effective at uniformly reducing surface bulk densities. In fine-textured soils, the changes in soil structure often are limited to the cracks formed at the site of injection (Smiley, 1997; Craul, 1992; Watson *et al.*, 1996).

6.4. Mechanical Breakup

For situations where a compacted site has yet to be planted, mechanical breakup techniques can be used. These techniques, including subsoiling and deep plowing, typically require large machinery that would not be practical for use in settings with existing trees or where above ground space would limit maneuverability. Unfortunately, the effects of subsoiling and tilling are often short-lived (Day and Bassuk, 1994; Craul, 1992).

Breakup also could be accomplished with a backhoe, which might be more maneuverable on confining sites. In situations where the soil is severely compacted or the texture is unfavorable and likely to self-compact later, a better option may be to amend the soil as in Section 6.1 or replace the soil to 18 inches with a better soil.

6.5. Soil Amendments

There is at least one report that low-density soil amendments can be rotary tilled into compacted soil with successful results. In that case, sintered fly ash (20% by volume) and expanded slate (33%) were incorporated on separate plots of a high-traffic playground and picnic area. Even with continued visitor use, bulk densities were reduced for at least 4 years (Patterson, 1977). However, most recommendations for inorganic amendments suggest at least 75% addition to make a meaningful change.

Additions of at least 25% of well-composted organic matter can be added to a compacted sandy loam or 50% of the same added to a loam to clayey textured soil and mixed to a depth of 18 inches to reduce compacted soil bulk density. It is important to amend a site, not just a hole and provide a way for excess water to drain away from the root zone if this is to be successful. As this technique effectively reduces bulk density but rarely achieves perfect drainage, a selection of plants that will tolerate less than perfect drainage should be considered in this situation (Rivenshield, 2003).

7. Soil Volume

In the absence of restriction, tree roots will extend horizontally far beyond the tree's drip line. However, urban trees are often subjected to confined soil volumes. Small tree pits, containers, compacted soil, hardpans, and the presence of belowground structures physically limit rooting space; under these circumstances, root extension is mechanically impeded. Another limitation of soil volume occurs in situations where the soil water table is high.

In addition to experiencing root restriction, urban trees often suffer from water stress (Krizek and Dubik, 1987). Water deficits can occur because soil acts as a water reservoir, and in the absence of frequent irrigation a reduction in soil volume effectively reduces the size of the reservoir (Spomer, 1980). Water deficits are enhanced by the high-atmospheric evaporative demands that are typical of urban environments (Lindsey and Bassuk, 1991).

If the available rooting volume is assessed prior to planting, then vegetation appropriate for the volume can be selected. Studies of whole-tree water requirements led a general recommendation that 2 cubic feet of soil volume be provided for every square foot of a tree's anticipated crown projection). Crown projection refers to the area of the circle formed by the tree's drip line. A circle's area is calculated by squaring the radius of the circle, and then multiplying that value by 3.14 (i.e., area of a circle = $(r^2 \cdot \Pi)$). In this case, the circle's radius is the distance from the tree's trunk to the tree's drip line. This recommendation is reasonable for most of the United States except the desert southwest. For specific information about different climates, the original paper by Lindsey and Bassuk (1991) should be consulted.

After identifying a site's available soil volume, estimates of mature crown projection for different tree species should be obtained. Based on the soil volume available and the recommended guideline, a species can be selected that is appropriate for the volume.

When using this guideline, soil depth that exceeds approximately 91 cm (3 feet) should not be considered usable, as most roots will not grow below this level. A 10-cm (4 inch) layer of mulch should be applied to the soil surface.

7.1. Soil Volume Assessment

A penetrometer can be used to assess available rooting depth. When penetration is no longer possible or the pressure required to penetrate the profile exceeds about 2.3 MPa, then the usable depth has been reached. In some cases, available depth can be estimated by consulting a site construction diagram to identify locations of below-ground utilities. The soil depth is multiplied by the length and width of the site to calculate the available soil volume.

7.2. Soil Volume Modification

In situations where soil volume is inadequate, one option is to construct a berm, or a raised planting bed on top of the soil surface. This technique can be used where the soil is compacted or clayey, or where a high-water table exists.

A raised 90-cm (3 feet) bed of well-structured, noncompacted sandy loam can be constructed. For shrubs, the berm width should be no less than 61 cm (2 feet). For trees, the width should be at least 2 to 4.5 m (6 to 15 feet). The edges of the berm should slope down to the original soil level and a mulch layer should be applied (Lieberman and Weir, 1986). Underdrainage can be installed at the interface between berm and original soil (Craul, 1992). If this technique is to be successful, noncompacted, well-structured soil should be used to create the berm.

8. Soil-Specification Summary

Thoughtful soil specifications will help to ensure the success of planting and reduce the need for future soil modifications. Simply specifying that "topsoil" be used guarantees very little, as there is no standard, quantifiable definition of topsoil. Soil characteristics that should be specified include texture, soil reaction, content of organic matter, soluble salts, and contaminants. Suggested specifications for these characteristics are summarized as follows:

8.1. Texture

For sites where the potential for soil compaction exists, a good choice is a soil that contains a narrow distribution of medium to coarse sand particles such as a gap-graded sandy loam. An example of one specification for a sandy loam texture is shown in Table 5.

8.2. Soil Reaction

Although a moderate soil pH of approximately 6 to 7 is often suitable, pH should be specified based on the requirements of the intended vegetation. Certain plants, such as red maple, will require a more acidic pH. Other plants, including lindens, will tolerate a more alkaline pH. More information about plant preferences for pH may be found in "*Recommended Urban Trees: Site Assessment and Tree Selection for Stress Tolerance*" by Bassuk *et al.* (2003).

8.3. Organic Matter

Soils are commonly specified to contain from 3% to 5% organic matter. Somewhat more than this may be required if the site is expected to be droughty and in situations where irrigation will not be provided.

8.4. Soluble Salts and Nutrients

A soil sample should be submitted to a laboratory that has experience in interpreting results from that particular location. Soluble salt content should be specified

Table 5. A Sample Soil Specification for Sandy Loam

Particle-size class	Particle-size distribution (%)
Gravel (>2 mm)	<10
Coarse to medium sand (1–0.25 mm)	65–75
Fine sand (0.25–0.10 mm)	<10
Silt (0.05–0.002 mm)	10–20
Clay (<0.002 mm)	5–15

to be less than 1000 ppm (Craul, 1992). Fertilizer applications are warranted only if a soil test identifies nutrient deficiencies.

8.5. Contaminants

At certain concentrations, various soil contaminants can be dangerous to plants and animals. Contaminants, including heavy metals, pesticides, herbicides, and industrial wastes, should not exceed levels deemed acceptable by the Environmental Protection Agency (Craul, 1992). Debris, such as wood, glass, and plastics, should be screened out of the soil.

References

Anella, L. B., 1996, Tree-friendly construction planning, *Am. Nurseryman* **183**(5):52–55.

Bassuk, N., *Recommended Urban Trees: Site Assessment and Tree Selection for Stress Tolerance*, October, 2003, Department of Horticulture, 133pp., http://www.hort.cornell.edu/uhi/outreach/recurbtree/index.html

Blake, G. R., and Hartge, K. H., 1986, Bulk density, in *Methods of Soil Analysis*, Part 1: *Physical and Mineralogical Methods* (A. Klute, ed.), American Society of Agronomy, Soil Science Society of America, Madison, WI, pp. 363–375.

Brady, N. C., 1990, *The Nature and Properties of Soils*, 10th ed., Macmillan, New York.

Craul, P. J., 1992, *Urban Soil in Landscape Design*, Wiley, New York. pp. 85–117.

Day, S. D., and Bassuk, N. L., 1994, A review of the effects of soil compaction and amelioration treatments on landscape trees, *J. Arboric.* **20**(1):9–17.

Day, S. D., Bassuk, N. L., and van Es, H., 1995, Effects of four compaction remediation methods for landscape trees on soil aeration, mechanical impedance, and tree establishment, *J. Environ. Hart.* **13**(2):64–71.

Epstein, E., 1972, *Mineral Nutrition of Plants: Principles and Perspectives*, Wiley, New York. pp. 41–63.

Gilman, E. F., Leone, I. A., and Flower, F. B., 1987, Effect of soil compaction and oxygen content on vertical and horizontal root distribution, *J. Environ. Hart.* **5**(l):33–36.

Grabosky, J., and Bassuk, N., 1995, A new urban tree soil to safely increase rooting volumes under sidewalks, *J. Arboric.* **21**(4):187–201.

Grabosky, J., Bassuk, N. L., Irwin, L., and Van Es, H., 2001, Shoot and root growth of three tree species in sidewalks under pavement to increase street tree rooting volumes, *J. Environ. Hort.* **19**(4):206–211.

Harris, R. W., 1992, *Arboriculture: Integrated Management of Landscape Trees, Shrubs, and Vines*, 2nd ed., Prentice-Hall, Englewood Cliffs, NJ.

Himelick, E. B., and Watson, G. W., 1990, Reduction of oak chlorosis with wood chip mulch treatments, *J. Arboric.* **16**(10):275–278.

Hudler, G. W. (Comell Cooperative Extension), 1980, *Salt Injury to Roadside Plants*, Information Bulletin No. 113169, Plant Sciences/Plant Pathology, Cornell University, Ithaca, NY.

Krizek, D. T., and Dubik, S. P., 1987, Influence of water stress and restricted root volume on growth and development of urban trees, *J. Arboric.* **13**(2):47–53.

Lichter, J. M., and Costello, L. R., 1994, An evaluation of volume excavation and core sampling techniques for measuring soil bulk density, *J. Arboric.* **20**(3):161–164.

Lichter, J. M., and Lindsey, P. A., 1994, The use of surface treatments for the prevention of soil compaction during site construction, *J. Arboric.* **20**(4):205–209.

Lieberman, A. S., and Weir, R. (Cornell Cooperative Extension), 1986, *Suggested Practices for Planting and Maintaining Trees and Shrubs, Information Bulletin No. 1B24*, Floriculture and Ornamental Horticulture, Cornell University, Ithaca, NY.

Lindsey, P., and Bassuk, N., 1991, Specifying soil volumes to meet the water needs of mature urban street trees and tree containers, *J. Arboric.* **17**(6):141–148.

Patterson, J. C., 1977, Soil compaction-effects on urban vegetation, *J. Arboric.* **3**(9):161–167.

Patterson, J. C., Murray, J. J., and Short, J. R., 1980, The impact of urban soils on vegetation, in *Proceedings of the Third Conference of the Metropolitan Tree Improvement Alliance (METRIA)* (unpublished). Rutgers University.

Perry, T. O., 1982, The ecology of tree roots and the practical significance thereof, *J. Arboric.* **8**(8):197–211.

Rivenshield, A., 2003, *The Effects of Organic Amendments on Tree Growth in Compacted Soil*, PhD Dissertation, Cornell University, Ithaca, NY.

Simmons, G. L., and Pope, P. E., 1987, Influence of soil compaction and vesicular-arbuscular mycorrhizae on root growth of yellow poplar and sweet gum seedlings, *Can. J. For. Res.* **17**:970–975.

Smiley, T., 1997, Treating soil compaction near trees, *Grounds Maintenance* **32**(l):6–12.

Spomer, L. A., 1980, Container soil water relations: Production, maintenance, and transplanting, *J. Arboric.* **6**(12):315–320.

Spomer, L. A., 1983, Physical amendment of landscape soils, *J. Environ. Hort.* **1**(3):77–80.

Stone, H., 1994, Stalking the silent killer, *Arbor. Age* **14**(10):34–37.

Urban, J., and Craul, P., 1993, Check list for the design and specification of soils, in *The Proceedings of the ASLA Conference* (unpublished).

Veihmeyer, F. J., and Hendrickson, A. H., 1948, Soil density and root penetration, *Soil Sci.* **65**:487–493.

Watson, G., Kelsey, P., and Woodtli, K., 1996, Replacing soil in the root zone of mature trees for better growth, *J. Arboric.* **22**(4):167–172.

Chapter 11

Selecting Trees for Community Landscapes

Henry D. Gerhold and William Porter

1. Introduction

Selecting the most appropriate trees for various kinds of planting sites is crucial for the success of a municipal tree program. A properly selected tree will appreciate in value for a long time and will be cherished by many people. Conversely, an improper choice might soon become unhealthy or die. Even worse, problems could appear years later when a remedy will be difficult and costly. Trees that lift sidewalks, conflict with wires (see Chapter 15, this volume), or interfere with traffic certainly are selection mistakes that can be avoided.

So, in selecting trees for your community, take this responsibility seriously. You want every tree to remain healthy and serve its intended purpose, and that means for many decades. Recall all the building that occurred in the past 50 years, the unusual weather events, the diseases and insects that trees were exposed to, and then imagine what can occur in the future. With this long-range view in mind, it is imperative to consider how the selected tree will react to changeable environments and changes in the buildings and pavements around it, and how large it will grow in its lifetime. Also think about how much care a tree will receive, or how little. And keep in mind the people for whom you are selecting this tree—the immediate neighbors, others who will walk or drive past, and the tax-payers who bear the expenses and should expect to receive its benefits.

Selection of trees for a municipality may seem to be a daunting task to someone who is inexperienced. A new tree commission member might be baffled by the complexities of coping with many tree characteristics, site conditions, and diverse desires of people. The designer of a community project, though familiar with tree selection, could be preoccupied with aesthetic issues without giving adequate attention to

---

Henry D. Gerhold School of Forest Resources, Pennsylvania State University, University Park, Pennsylvania 16802-4300. **William Porter** New Jersey Shade Tree Federation, Cook College, Rutgers University, New Brunswick, New Jersey 08901.

Urban and Community Forestry in the Northeast, 2nd ed., edited by, J. E. Kuser.

biological or managerial concerns. Even an experienced, conscientious city forester will acknowledge that an occasional mistake will be made. Nevertheless, a novice should realize that it is possible to make sound choices and avoid serious mistakes by consulting technical information, using good judgment, and understanding the tree selection process (Schein, 1993).

1. The Selection Process

A logical five-step process for selecting the variety of tree to be planted provides an organized way of dealing with the various types of information needed for making wise choices.

1. Define the purpose of the tree in the landscape
2. Evaluate site conditions that will affect the choice
3. Consider arboricultural practices that can impact the tree
4. Develop selection criteria based on purpose, site, and managerial impacts
5. Match characteristics of candidate trees to the criteria to identify suitable varieties, leading to the final choice

Anyone who is experienced in selecting tree species and varieties for planting automatically follows this thought process, perhaps not consciously. But some people overemphasize one or two steps and neglect others. Some simply choose favorite trees. Thus, their decisions may satisfy their own desires or biases but not necessarily the greater good of the whole community.

The steps in the selection process are addressed in the following sections. A thoughtful, thorough analysis in matching species characteristics to selection criteria is more important than the sequence in which the steps are considered.

In this sorting process a long list of candidate species is reduced to a small number that meet important constraints. The number of alternative species from which the final choice can be made will depend on the severity of these constraints. For example, where the plant environment is ideal for growth and health, the art of landscape design can be practiced with great latitude; whereas on stressful sites the good choices are severely limited.

2. Defining Purposes of Trees in the Landscape

A clear concept of the aesthetic and functional purposes that a tree will serve helps to place its role into proper perspective. The mental exercise of defining its purpose safeguards against overlooking something important and reminds us that this publicly owned tree is valuable to the entire community, as well as its immediate neighborhood.

Each tree has its role and is also part of a landscape. Landscape architects have developed an approach of designing with trees that is both artistic and functional (Zion, 1995). An understanding of landscape design is helpful in defining the purpose of a tree in the context of the landscape around it.

Ideally, a community should have a landscape plan that provides guiding principles for the selection of trees. Many communities do not have a tangible landscape plan, in which case it would be wise to think about some analogous mental guidelines. The plan (or mental guidelines) should describe the existing landscape, comment on its historical context, and offer a vision of its appearance and sustainability in the future.

To implement the vision, the plan should state specific objectives and goals to be attained according to a schedule. Typically the community landscape plan is organized by districts—residential, business, commercial, and so forth. These districts in turn imply who the stakeholders are, that is, those people who take an interest in the trees there. The plan should describe the kinds of landscapes desired in each district and indicate the types of trees that would be appropriate for planting. The landscapes may be formal and uniform or informal with diverse sizes, shapes, textures, and colors. Decisions about particular species for individual streets or properties come later and should be made only after site conditions have been evaluated.

The landscape plan also can comment on communitywide issues that affect tree selection. One important issue is the desirability of diversity in the tree population. Diversity in species composition confers protection against catastrophic losses caused by a disease or insect infestation. Diversity also adds visual interest and can provide food and shelter for more kinds of wildlife. But the use of single species along a street, or even one clone, should not be prohibited. Sufficient diversity within a community can be achieved while accommodating uniformity of plantings in local settings where formality is desirable for architectural reasons.

Another issue concerns preferences for or against native trees versus nonindigenous species. Such a preference may be valid for historical or design reasons but less for biological concerns. Most planting sites in communities are quite different from natural forests to which trees are genetically adapted. No tree species is truly native to urbanized sites. But certain species can tolerate compacted soils and pollutants better than others, and some of these are native to distant, analogous climatic regions. Therefore, some nonindigenous species can withstand urban stresses better than some "native species," and these should not be excluded just because they are classified as exotic.

The local landscape into which a tree is to be planted helps to define its purpose, within the broader context of the community plan. This question may be a good way to start—What will this tree contribute to the landscape in the immediate neighborhood and for whom? Of course, the tree will be just one element in the community landscape, which also consists of other vegetation, buildings, transportation corridors, and furnishings, as well as vehicles and people. Any or all of these elements, plants, architecture, utilities, vehicles, and people, can influence the role of a particular tree. Their influences already have been expressed in older sections of town, where replacement of a tree that has been lost can either renew its purpose or modify the landscape gradually.

The opportunity to create a new landscape is restricted to situations where many trees can be planted such as housing developments on nonforested land, large commercial projects, or street renovation projects. Here the landscape designer has great freedom in defining the purpose of trees, which may differ among individuals or groups of trees to be planted. For example, one or more of the following purposes may apply (see also Chapters 2 and 6, this volume).

- Complement the attractiveness of buildings
- Accent a view or frame a vista
- Provide shade for pedestrians
- Screen unsightly places
- Create a sense of formality or informality along streets
- Shade buildings or shield them against wind
- Express feelings of nostalgia or pride in the neighborhood

Note that each purpose indicates what a tree is to do for people, at least by implication.

The aesthetic and functional purposes of a tree indicate some of the characteristics that are wanted at a particular site. Another important step is to evaluate site conditions that impose limitations upon the choice.

3. Evaluating Site Conditions

Site conditions impose several kinds of constraints in selecting specific trees that are in accord with the landscape plan. Those conditions that affect the choice can only be identified by a thorough inspection of the site and evaluated by a thoughtful analysis of their impacts. The main considerations in evaluating sites are: (1) available space related to tree size; (2) climate and soil related to species adaptation; (3) exposure to pathogens and other causes of injuries; and (4) the anticipated level of tree care. Some of these considerations are specific to the planting site, and others may pertain to the entire municipality.

Moving the exact planting spot a short distance should always be considered, as an alternative location sometimes can dramatically improve the outcome. For example, moving the planting spot of a street tree away from curbside to beyond a sidewalk might provide a tree that is large at maturity instead of a small one, with less exposure to injuries and greater space for normal growth.

Mature tree size is a dominant consideration that comes into play repeatedly in the selection process. If a large tree is specified in an architectural design, then provisions must be made that assure its biological health and longevity. Or if that is impossible, the design must be revised to accommodate a smaller tree.

Adequate soil space for the root system is essential, to support the tree during high winds and supply water needed for good health. The most common barriers to root growth are streets and other paved surfaces, curbs, sidewalks, and underground structures such as sewers or basements.

Aboveground constraints include buildings, utility wires, and space needed for clearance of vehicles, pedestrians, lights, and signs. The crown of a tree may be pruned to circumvent some of these conflicts with the normal growth of branches. But the preferred solution is to select a tree that fits into the available space with as little modification as possible. Estimates of mature sizes of trees are available in references, such as Dirr (1990), Gerhold *et al.* (1993), and Wandell (1994); nursery catalogs commonly give height and spread at an assumed age, such as 25 (not always stated), so these usually are smaller than mature dimensions. Because of regional variation in the size that species reach, visits to local arboreta or mature plantings may be useful.

Climatic effects on species selection should be viewed as extremes that can occur infrequently over the many decades of a tree's life span and yet cause serious injuries or death. Winter hardiness is the most obvious climatic concern, one that is easily handled through hardiness zone maps (Fig. 1) and hardiness classifications of species

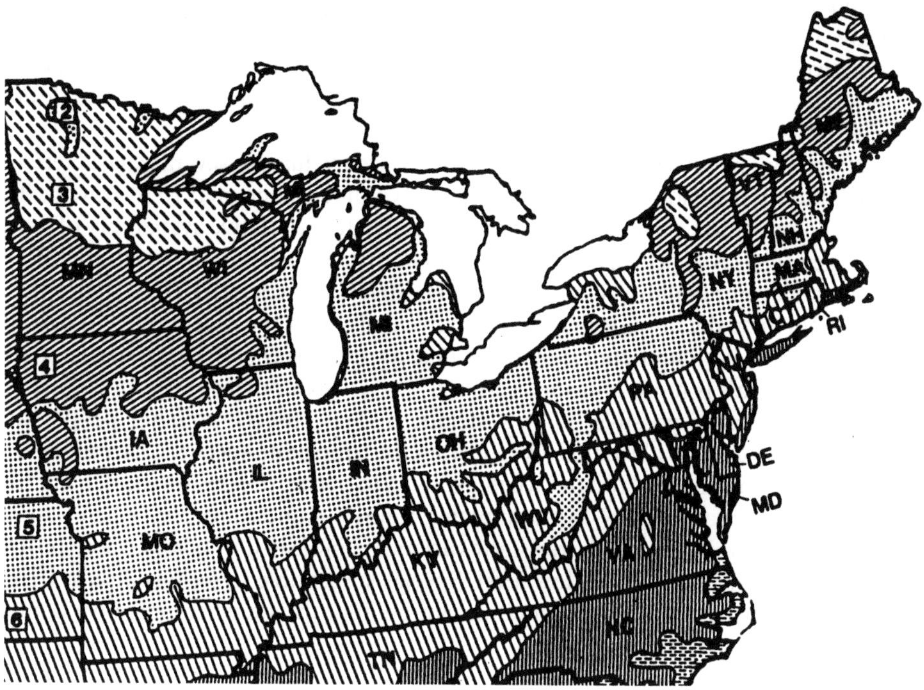

Range of average annual minimum temperatures for each zone

	Fahrenheit	Celsius
ZONE 1	Below −50	−46
ZONE 2	−50 to −40	−46 to −40
ZONE 3	−40 to −30	−40 to −34
ZONE 4	−30 to −20	−34 to −29
ZONE 5	−20 to −10	−29 to −23
ZONE 6	−10 to 0	−23 to −18
ZONE 7	0 to 10	−18 to −12
ZONE 8	10 to 20	−12 to −7
ZONE 9	20 to 30	−7 to −1
ZONE 10	30 to 40	−1 to 4
ZONE 11	Above 40	Above 4

FIGURE 1. United States Department of Agriculture (USDA) plant hardiness zone map.

given in technical references and nursery catalogs. The likelihood of damage caused by climatic events, such as droughts, flooding, or ice storms, may be serious in some geographic locations but relatively unimportant in others. Besides broad climatic influences, the microclimate at each planting site should be evaluated. Reflected light from buildings and wind tunnel effects may aggravate stresses on trees; elsewhere structures may ameliorate temperatures and shelter plants.

Physical and chemical properties of the soil are important too and often interact with climatic variables. Soil analyses are useful for this reason but not always practical or necessary. If a tree is to be planted in native soil, the necessary information can be obtained readily from soil survey publications and observations at the site that indicate any adverse modifications such as compaction or chemical contamination. Alternatively, if good soil is not present at the planting site, it is usually advisable to bring in a generous quantity of ordinary topsoil that is appropriate for the desired species.

When soil survey information or soil tests are evaluated, several characteristics deserve attention besides adequate depth and volume for normal root growth. The physical composition, structure, and drainage indicate the relative availability of water and oxygen to trees, especially during droughts or when soils have been saturated. The relative acidity or alkalinity, expressed as pH, affects some species more than others. Soil nutrients usually are not limiting in native soils, with the exception of a few soil types. The most common contaminant is deicing salt, especially along major streets and highways; salt is more injurious to some species than others.

Other causes of possible injuries include disease, insects, animals, people, and vehicles. These can be alleviated by selecting pest-resistant species or varieties and those that are considered "tough" when exposed to urban stresses or vandalism.

The managerial strategy for minimizing injuries comes into play when making selections. The choice of tough, resistant varieties is indicated in places where protection of plant health will be at a low level. Species susceptible to serious diseases, such as Dutch elm disease of American elms, can be appropriate where effective control measures will be assured.

4. Arboricultural Practices

An assessment of tree care practices that can be anticipated also is a vital part of selection. Those that especially impact selection are measures that promote the health of recently planted trees, protective measures against diseases and insects, and pruning that develops the sound structure of branches and trunk.

Regardless of the level of tree care, trees that can tolerate stresses, resist pests, and develop a strong trunk and crown are always preferable. They remain healthier, cost less to maintain, and reduce the need for pesticides. Such low-maintenance varieties always should be selected for places where trees will receive little care after planting.

In evaluating the level of tree care practices that can be expected, one should start with those just before, during, and shortly after planting. Some species are more difficult to transplant successfully, so these should be avoided if conditions associated with transportation, temporary storage, planting, staking, watering, mulching, and weed control are less than ideal.

If trees are to be planted in the autumn, certain species that are more likely to suffer from harsh winters should be excluded. Lists of such trees are available in various references and nursery catalogs, though experts may differ about their validity.

No species is entirely free of disease and insect problems, but considerations related to selection come into play mainly for the most serious ones such as Dutch elm disease and several diseases of crabapples. Some diseases and insects occur sporadically and injuries vary in severity, often influenced by weather and geographic region. Local conditions may play a role, for example, defoliation of lindens by Japanese beetles is severe only in places where turf enables them to multiply. Despite such variability, each disease or insect can be characterized as to host range and seriousness for particular tree species over their life spans.

For each tree species being considered for selection, the seriousness of any pest problems together with the cost, effectiveness, and pesticide risks of control measures will be pertinent. If integrated pest management is being practiced, both resistant varieties and more susceptible ones can be appropriate for selection.

Pruning practices are the most costly part of tree maintenance, and therefore variation in pruning needs among species is very important to consider in selection. Branching habit varies greatly, so varieties have been developed that have distinctive branching patterns: ascending, shortened, regularly spaced, strong crotches, and others that reduce or eliminate the need for pruning. Such varieties should be favored especially where a good, regular pruning program does not exist.

5. Selection Criteria

Taken together, the aesthetic and functional purposes of a tree, the site conditions that will affect its health, and the arboricultural practices that will affect its growth are the basis for selection criteria that will narrow the choices. Those criteria that ought to be considered for a particular location should be identified during the initial stages of the selection process, as well as the relative importance of each selection criterion. Usually many criteria are applicable, and some of these are complicated to apply. So it is useful to express criteria by subdividing each into a small number of several categories. Examples are given in Table 1.

Each criterion will either eliminate certain species entirely or indicate the desirability of a species relative to others that are being considered. For example, the hardiness rating will exclude all species that require a warmer climate for survival. In contrast, a preferred red flower color may merely favor varieties that are cerise over those whose flowers are pink.

Thus, it is clear that various selection criteria differ in the degree to which they influence a decision. Furthermore, the influence of some criteria varies among different planting sites.

Table 1. Categories Within Selection Criteria Based on Characteristics of Species and Varieties

Selection criteria	Categories
Height at maturity	Under 25 feet, 25 to 45 feet, over 45 feet
Crown width at maturity	Under 15 feet, 15 to 30 feet, over 30 feet
Crown shape	Columnar, conical, ovoid, globe, and vase
Hardiness in winter	Hardy to zone 3, 4, 5, 6, 7, or 8
Light requirement	Full sun, sun or shade, prefers shade
Soil reaction	Acid required, mildly acid to neutral, wide tolerance of acid to alkaline
Soil structure	High, medium, or low tolerance of dense, droughty, or poorly drained soil
Urban stresses	High, medium, or low tolerance of salt, fumes, or other city stresses
Transplant concerns	Normal to slow recovery, or avoid fall planting
Serious diseases	Low, moderate, or high risk
Serious insect injuries	Low, moderate, or high risk
Hazards	Storm breakage, heaved walks, clogged pipes, and thorns
Nuisances	Surface roots, odors, debris from twigs, bark, flowers, or fruit
Flower color	Red, pink, orange, yellow, or white
Fruit color	Red, orange, yellow, purple, or blue
Colorful foliage	Colorful in spring, summer, or fall
Texture	Leaves and branches fine, medium, or coarse

6. Using Selection Criteria to Make Choices

The final stage in the selection process is to determine which species and varieties have characteristics that best match the selection criteria according to their relative importance. The reliability with which these characteristics can be predicted deserves attention too, not just the degree of correspondence with the criteria.

The predictability in turn depends on the genetic makeup that is variable among trees of a species grown from seed. Trees that are propagated clonally (by grafting, budding, rooting cuttings, or tissue culture) remain genetically identical, so their characteristics can be predicted with greater confidence. That is why named cultivars (a contraction of cultivated varieties), which are vegetatively propagated, have become so popular.

Some characteristics are under strong genetic control, such as the color of flowers and fruit, whereas others, such as growth rate, are more variable and subject to environmental influences. Crabapples grown from seed exhibit many colors of flowers and fruit, but "Adams" crabapple always has pink flowers and dark red fruit, and the "Harvest Gold" cultivar always has white blossoms and golden fruit. "Goldspire" sugar maple always has a very narrow crown, though its growth rate may vary with site conditions.

If uniform autumn coloration of a row of sugar maples is desired, then a single cultivar should be planted. Alternatively, if various shades of yellow, orange, and red are preferred, then trees propagated from seed should be obtained. And because sugar maples are rather sensitive to site conditions, the seeds should have come from the same hardiness zone so that they will be reasonably well adapted to the planting site. Unfortunately, most nurseries do not keep track of seed origins, so the only practical recourse may be to procure planting stock from a nursery in the same hardiness zone or a more northerly one. With these genetic influences on selection decisions in mind, the final part of the selection process can proceed.

There are several ways of matching tree characteristics with selection criteria. One methodical procedure is to use lists of trees (Table 2, for example), each one organized by a single characteristic. The most efficient way is to start with a characteristic that is especially important or limiting, for example, a mature height less than

Table 2. Trees for Northeastern States

Species/cultivar	Width, feet	Common name	Zones
A. Street trees[a]			
Tall: 45 feet and up			
Acer platanoides "Cleveland"	30–35	Cleveland Norway maple	4–8
Acer platanoides "Columnare"	15–20	Columnar Norway maple	4–8
Acer platanoides "Crimson King"	25–35	Crimson King Norway Maple	4–8
Acer platanoides "Emerald Queen"	50–60	Emerald Queen Norway Maple	4–8
Acer platanoides "Summershade"	50–65	Summershade Norway Maple	4–8
Acer pseudoplatanus	40–50	Sycamore maple	4–8
Acer rubrum	40–60	Red maple	3–9
Acer rubrum "Bowhall"	20–30	Bowhall red maple	4–8
Acer rubrum "October Glory"	40–50	October Glory red maple	4–8
Acer rubrum "Red Sunset"	40–50	Red Sunset red maple	4–8
Acer saccharum	40–70	Sugar maple	3–7
Acer saccharum "Bonfire"	40–60	Bonfire sugar maple	4–7
Acer saccharum "Commemoration"	30–35	Commemoration sugar maple	4–7
Acer saccharum "Goldspire"	12–15	Goldspire sugar maple	4–7
Acer saccharum "Green Mountain"	40–60	Green Mountain sugar maple	4–7
Acer saccharum "Legacy"	30–35	Legacy sugar maple	4–7
Acer X freemani "Armstrong"	20–25	Armstrong maple	4–7
Acer X freemani "Autumn Blaze"	40–50	Autumn Blaze maple	4–7
Acer X freemani "Celebration"	25–40	Celebration maple	4–7
Acer X freemani "Scarlet Sentinel"	25–35	Scarlet Sentinel maple	4–7
Betula nigra	30–50	River birch	4–9
Betula nigra "Heritage"	30–35	Heritage river birch	4–9
Celtis laevigata "All Seasons"	30–40	All Seasons hackberry	5–9
Celtis laevigata x "Magnifica"	50–60	Magnifica hackberry	5–9
Celtis occidentalis "Prairie Pride"	40–50	Prairie Pride hackberry	4–9
Corylus colurna	25–35	Turkish Filbert	4–7
Eucommia ulmoides	40–60	Hardy rubber tree	5–7
Fagus sylvatica	40–70	European beech	4–7
Fagus sylvatica "Fastigiata"	10–20	Pyramidal beech	4–7
Fagus sylvatica aspenifolia	35–60	Fernleaf beech	4–7
Fraxinus americana "Autumn Applause"	45–55	Autumn Applause white ash	3–8
Fraxinus americana "Autumn Purple"	45–55	Autumn Purple white ash	3–8
Fraxinus americana "Greenspire"	20–30	Greenspire white ash	5–8
Fraxinus americana "Rosehill"	35–45	Rosehill white ash	5–8
Fraxinus pennsylvanica "Newport"	40–50	Newport green ash	3–9
Fraxinus pennsylvanica "Patmore"	40–50	Patmore green ash	3–9
Fraxinus pennsylvanica "Summit"	45–55	Summit green ash	3–9

(*continued*)

Table 2. (*Continued*)

Species/cultivar	Width	Common name	Zones
Fraxinus pennsylvanica "Urbanite"	35–45	Urbanite green ash	3–9
Ginkgo biloba "Magyar"	35–40	Magyar ginkgo	4–8
Ginkgo biloba "Princeton Sentry"	25–30	Princeton Sentry ginkgo	4–8
Gleditsia triacanthos "Continental"	50–60	Continental honeylocust	4–9
Gleditsia triacanthos "Shademaster"	50–60	Shademaster honeylocust	4–9
Gleditsia triacanthos "Skyline"	55–65	Skyline honeylocust	4–9
Gleditsia triacanthos "Sunburst"	35–50	Sunburst honeylocust	4–9
Gymnocladus dioicus	45–65	Kentucky coffee tree	4–8
Liquidambar styraciflua	40–65	Sweetgum	5–9
Liquidambar styraciflua "Rotundiloba"	40–50	Rotundiloba sweetgum (seedless)	6–9
Liriodendron tulipifera "Fastigiatum"	25–35	Fastigiate tuliptree	5–9
Magnolia acuminata	30–55	Cucumber tree	4–8
Metasequoia glyptostroboides	25–40	Dawn redwood	5–8
Nyssa sylvatica	30–45	Black gum, tupelo	3–9
Platanus X acerifolia "Bloodgood"	50–65	Bloodgood London planetree	5–9
Quercus acutissima	35–55	Sawtooth oak	5–9
Quercus alba	55–85	White oak	3–9
Quercus bicolor	50–80	Swamp white oak	3–8
Quercus coccinea	40–60	Scarlet oak	4–9
Quercus imbricaria	40–70	Shingle oak	4–8
Quercus macrocarpa	60–90	Bur oak	2–8
Quercus palustris	40–55	Pin oak	5–8
Quercus palustris "Green Pillar"	25–35	Green Pillar pin oak	4–8
Quercus phellos	45–60	Willow oak	6–9
Quercus robur	50–70	English oak	4–8
Quercus robur "Fastigiata"	15–20	Pyramidal English oak	5–8
Quercus robur "Skymaster"	25–30	Skymaster English oak	5–8
Quercus robur "Skyrocket"	14–17	Syrocket English oak	5–8
Quercus rubra	50–70	Red oak	4–8
Styphnolobium japonicum "Princeton Upright" (*Sophora japonica*)	25–35	Princeton Upright scholartree	6–8
Styphnolobium japonicum "Regent"	30–40	Regent scholartree	4–8
Taxodium ascendens "Prairie Sentinel"	15–25	Prairie Sentinel pond cypress	5–9
Taxodium distichum	20–45	Baldcypress	5–9
Taxodium distichum "Shawnee Brave"	20–30	Shawnee Brave baldcypress	5–9
Tilia americana "Redmond"	30–45	Redmond American linden	3–7
Tilia cordata "Corinthian"	15–25	Corinthian littleleaf linden	3–7
Tilia cordata "Fairview"	35–50	Fairview littleleaf linden	3–7
Tilia cordata "Glenleven"	35–50	Glenleven littleleaf linden	3–7
Tilia cordata "Greenspire"	35–50	Greenspire littleleaf linden	3–9
Tilia euchlora	35–50-	Crimean linden	4–7
Tilia tomentosa "Green Mountain"	45–55	Green Mountain silver linden	4–8
Tilia tomentosa "Sterling"	45–55	Sterling silver linden	4–7
Ulmus americana "New Harmony"	60–90?	New Harmony American elm	3–9
Ulmus americana "Princeton"	60–90	Princeton American elm	3–9
Ulmus americana "Valley Forge"[b]	60–90?	Valley Forge American elm	3–9
Ulmus parvifolia "Dynasty"	50–60	Dynasty Chinese elm	5–9
Ulmus parvifolia "Pathfinder"	40–50	Pathfinder Chinese elm	5–9

Table 2. (*Continued*)

Species/cultivar	Width	Common name	Zones
Zelkova serrata "Green Vase"	50–55	Green Vase zelkova	5–8
Zelkova serrata "Halka"	35–45	Halka zelkova	5–8
Zelkova serrata "Village Green"	50–60	Village Green zelkova	5–8
Medium: 25 to 45 feet			
Acer campestre	25–40	Hedge maple	5–8
Acer campestre "Queen Elizabeth"	25–35	Queen Elizabeth hedge maple	6–7
Acer campestre "Schichtel's Upright"	20–30	Schichtel's Upright hedge maple	4–7
Carpinus betulus "Fastigiata"	20–30	Pyramidal European hornbeam	3–8
Cladrastis kentukea	40–55	Yellowwood	4–8
Halesia carolina	30–40	Carolina silverbell	4–8
Koelreuteria paniculata	25–40	Goldenrain tree	5–9
Koelreuteria paniculata "Fastigiata"	15–25	Fastigiate goldenrain tree	5–9
Koelreuteria paniculata "Rose Lantern"	25–30	Rose Lantern goldenrain tree	5–9
Magnolia X "Elizabeth"	20–35	Elizabeth magnolia	5–8
Ostrya virginiana	20–30	American hophornbeam	4–8
Phellodendron amurense "Macho"	30–40	Macho Amur corktree	3–7
Phellodendron Amurense "Shademaster"	35–45	Shademaster amur corktree	3–7
Prunus Sargenti "Columnaris"	12–18	Columnar Sargent cherry	4–8
Prunus serrulata "Kwanzan"	20–25	Kwanzan cherry	5–6
Prunus subhirtella "Autumnalis"	15–30	Autumn flowering cherry	4–8
Prunus X "Accolade"	25–35	Accolade flowering cherry	4–7
Prunus X yedoensis	25–35	Yoshino cherry	5–8
Pyrus calleryana "Aristocrat"	25–35	Aristocrat Callery pear	5–8
Pyrus calleryana "Capital"	12–16	Capital Callery pear	5–8
Pyrus calleryana "Cleveland Select" (also known as "Chanticleer")	15–20	Cleveland Select Callery pear	5–8
Pyrus calleryana "Redspire"	20–25	Redspire Callery pear	5–8
Pyrus calleryana "Whitehouse"	15–20	Whitehouse Callery pear	5–8
Sorbus alnifolia	15–20	Korean mountain-ash	5–7
Sorbus thuringiaca "Fastigiata"	10–15	Columnar oakleaf mountain-ash	2–6
Ulmus parvifolia "Princeton Upright"	15–20	Princeton Upright Chinese elm	4–9
Short: Usually 25 feet or less			
Acer buergeranum	20–30	Trident maple	5–8
Acer ginnala "Flame"	15–20	Flame Amur maple	2–6
Acer tataricum	15–25	Tatarian maple	3–6(8)
Amelanchier sp. "Cumulus"	12–18	Cumulus serviceberry	4–8
Amelanchier sp. "Tradition"	15–20	Tradition serviceberry	4–8
Amelanchier X grandiflora "Robin Hill Pink"	12–15	Robin Hill Pink serviceberry	4–8
Amelanchier X grandiflora "Autumn Brilliance"	15–18	Autumn Brilliance serviceberry	4–8
Chionanthus retusus	15–20	Chinese fringetree	5–8
Cornus kousa	20–30	Kousa dogwood	5–8
Cornus X "Celestial"	20–30	Celestial dogwood	5–8
Cornus X "Constellation"	20–30	Constellation dogwood	5–8
Crataegus X lavallei	10–20	Lavalle hawthorn	4–7
Crataegus punctata "Ohio Pioneer"	20–25	Ohio Pioneer hawthorn	5–7
Crataegus viridis "Winter King"	25–30	Winter King hawthorn	3–7

(*continued*)

Table 2. (*Continued*)

Species/cultivar	Width	Common name	Zones
Halesia carolina "Rosea"	25–30	Pink Carolina silverbell	4–8
Halesia diptera "Magniflora"	25–30	Magniflora two-winged silverbell	4–8
Maackia amurensis	15–25	Manchurian maackia	3–7
Magnolia "Wada's Memory"	10–15	Wada's Memory magnolia	5–8
Malus baccata "Columnaris"	15–20	Columnar Siberian crabapple	2–8
Malus "Centurion"	15–20	Centurian crabapple	2–8
Malus "Donald Wyman"	20–25	Donald Wyman crabapple	2–8
Malus "Harvest Gold"	15–20	Harvest Gold crabapple	2–8
Malus "Prairifire"	15–22	Prairifire crabapple	2–8
Malus "Red Baron"	15–20	Red Baron crabapple	2–8
Malus "Red Jewel"	10–14	Red Jewel crabapple	2–8
Malus "Sentinel"	12–15	Sentinel crabapple	2–8
Malus "Snowdrift"	18–22	Snowdrift crabapple	2–8
Malus "Sugar Tyme"	14–16	Sugar Tyme crabapple	2–8
Malus "Zumi Calocarpa"	25–30	Redbud crabapple	2–8
Prunus "Newport"	15–20	Newport purple leaf plum	5–8
Prunus serrulata "Amanogawa"	5–7	Amanogawa cherry	6–8
Syringa reticulata "Ivory Silk"	15–20	Ivory Silk tree lilac	3–7
Syringa reticulata "Regent"	15–20	Regent tree lilac	3–7
Syringa reticulata "Summer Snow"	20–25	Summer Snow tree lilac	3–7
B. Trees for parks, windbreaks, and landscapes			
Evergreen, tall: 45 feet and up			
Abies concolor	20–30	Concolor or white fir	3–7
Abies holophylla	15–30	Needle or Manchurian fir	4–6
Abies nordmanniana	20–30	Nordmann fir	5–6
Cedrus atlantica glauca	35–60	Blue atlas cedar	6–9
Cedrus libani	35–60	Cedar of Lebanon	5–7
Chamaecyparis lawsoniana	10–15	Lawson cypress, Port Oxford cedar	5–7
Chamarcyparis obtusa	20–30	Hinoki cypress	4–8
Chamarcyparis pisifera	20–40	Sawara cypress	3–8
Cupressocyparis X leylandii	10–15	Leyland cypress	6–10
Picea abies	30–50	Norway spruce	2–7
Picea glauca	20–30	White spruce	2–5
Picea orientalis	20–25	Oriental spruce	4–7
Picea pungens	15–25	Colorado or blue spruce	2–7
Pinus resinosa	25–35	Red pine	2–5
Pinus strobus	30–50	Eastern white pine	3–8
Pinus taeda	30–40	Loblolly pine	7–9
Pinus wallichiana	30–50	Himalayan pine	5–7
Pseudotsuga menziesii glauca	20–35	Rocky Mountain Douglas-fir	4–6
Thuja plicata	15–25	Western red cedar, giant arborvitae	5–7
Evergreen, short or medium: 25 to 45 feet			
Ilex opaca[c]	15–25	American holly	5–9
Taxus baccata (tree cultivars)	15–25	English yew	6–7
Thuja occidentalis	5–10	Arborvitae, northern white cedar	2–7
Thuja occidentalis "Nigra"	5–10	Wintergreen arborvitae	2–7

Table 2. (*Continued*)

Species/cultivar	Width	Common name	Zones
Deciduous, tall: 45 feet and up			
Cercidiphyllum japonicum	25–60	Katsura tree	5–8
Ginkgo biloba	40–80	Ginkgo	3–8
Larix kaempferi	15–25	Japanese larch	4–7
Liriodendron tulipifera	35–50	Tuliptree, yellow poplar	4–9
Pseudolarix kaempferi	30–40	Golden larch	4–7
Deciduous, short or medium: 20 to 45 feet			
Acer palmatum	30–40	Japanese maple	(5)6–8
Aesculus pavia "Splendens"	15–25	Red buckeye	4–9
Cercis canadensis	25–35	Eastern redbud	4–9
Cornus florida	20–30	Flowering dogwood	(5)6–8
Cornus mas	15–25	Cornelian cherry	4–8
Lagerstroemia indica	10–15	Crape myrtle	7–9
Oxydendrum arboreum	15–20	Sourwood, sorrel tree	4–9
Stewartia koreana	10–20	Korean stewartia	5–7
C. Wet site trees			
Evergreen, medium: 25 to 45 feet			
Chamaecyparis thyoides	10–15	Atlantic white cedar	3–9
Thuja occidentalis "Nigra"	5–10	Wintergreen arborvitae	2–7
Deciduous, tall: 45 feet and up			
Acer rubrum and cultivars	20–60	Red maple and cultivars	3–9
Acer X freemani cultivars	20–50	Freeman maple (*A. rubrum x saccharinum*)	4–7
Acer saccharinum	40–60	Silver maple	3–9
Betula nigra and cultivars	30–50	River birch	4–9
Fraxinus pennsylvanica and cultivars	35–55	Green ash	3–9
Liquidambar styraciflua and cultivars	40–60	Sweetgum	5–9
Nyssa sylvatica	30–45	Tupelo, black gum	3–9
Platanus X acerifolia	60–80	London planetree	5–9
Platanus occidentalis	60–80	Sycamore	4–9
Quercus bicolor	50–80	Swamp white oak	3–8
Quercus nigra	40–60	Water oak	7–9
Quercus palustris	40–55	Pin oak	5–8
Salix spp.	20–60	Willow	Various
Taxodium distichum	25–45	Baldcypress	5–9
Ulmus americana cultivars	60–90	American elm	2–9
Deciduous, short or medium: 20 to 45 feet			
Amelanchier spp.	15–20	Serviceberry (shadbush)	4–8
Carpinus caroliniana	15–20	American hornbeam, ironwood	3–9
Magnolia virginiana	15–20	Sweetbay, swamp magnolia	5–9
D. Dry site trees			
Evergreen, tall: 45 feet and up			
Pinus spp.	20–45	Pine species	Various
Evergreen, short or medium: 20 to 45 feet			
Juniperus chinensis, tree cultivars	10–20	Chinese juniper	3–9
Juniperus virginiana	10–20	Eastern redcedar	2–9

(*continued*)

Table 2. (*Continued*)

Species/cultivar	Width	Common name	Zones
Deciduous, tall: 45 feet and up			
Fraxinus spp. and cultivars	20–55	Ash species and cultivars	3–9
Gleditsia triacanthos and cultivars	40–60	Honeylocust and cultivars	3–9
Gymnocladus dioicus	40–50	Kentucky coffeetree	3–8
Quercus coccinea	40–55	Scarlet oak	4–9
Quercus macrocarpa	50–70	Bur oak, mossycup oak	2–8
Quercus prinus	35–50	Chestnut oak, basket oak	4–8
Quercus velutina	50–70	Black oak	3–9
Robinia pseudoacacia	20–30	Black locust	3–8
Sassafras albidum	30–40	Sassafras	4–9
Ulmus X hollandica	30–50	Dutch elm	4–6
E. Salt-spray tolerant trees for seashore sites			
Evergreen			
Cryptomeria japonica	10–15[d]	Japanese cedar	5–6
Ilex opaca	15–25	American holly	6–9
Juniperus virginiana	10–15	Eastern redcedar	2–9
Magnolia grandiflora	15–20	Southern magnolia	6–9
Picea glauca	10–20	White spruce	2–6
Picea pungens glauca	10–15	Blue spruce	2–7
Pinus rigida	20–30	Pitch pine	5–7
Pinus sylvestris	20–30	Scotch pine	5–7
Pinus thunbergii	15–30	Japanese black pine	5–8
Pinus virginiana	15–20	Virginia pine	4–8
Deciduous			
Acer platanoides	25–35	Norway maple	3–7
Acer pseudoplatanus	25–35	Sycamore maple	4–8
Aesculus hippocastanum	20–30	Horsechestnut	3–7
Ailanthus altissima	20–30	Ailanthus, tree of heaven	4–8
Amelanchier arborea (wild type)	10–15	Serviceberry (shadbush)	4–8
Betula papulifolia	15–20	Gray birch	3–6
Broussonetia papyrifera	20–25	Paper mulberry	6–9
Catalpa speciosa	25–30	Catalpa	4–8
Celtis occidentalis	20–25	Hackberry	2–9
Gleditsia triacanthos	20–25	Honeylocust	4–9
Morus spp.	20–30	Mulberries	4–8
Nyssa sylvatica	20–25	Tupelo, black gum	3–9
Platanus X acerifolia	25–35	London plane	5–9
Populas alba	20–30	White poplar	3–8
Prunus serotina	15–25	Black cherry	3–9
Robinia pseudoacacia	20–30	Black locust	3–8
Styphnolobium japonicum	20–30	Scholartree	6–8
Tilia cordata	20–30	Littleleaf linden	3–7
Tilia euchlora	20–30	Crimean linden	3–7
Tilia tomentosa	20–30	Silver linden	4–7
Ulmus parvifolia	20–30	Chinese elm	4–9
Ulmus pumila	20–30	Siberian elm	4–9
Zelkova serrata	20–30	Zelkova	5–8

Table 2. (*Continued*)

Species/cultivar	Width	Common name	Zones
F. Tolerant to road salt			
Betula allegheniensis	30–40	Yellow birch	
Betula lenta	30–40	Black birch	
Betula papyrifera	30–40	Paper birch	
Betula populifolia	15–20	Gray birch	
Elaeagnus angustifolia	10–15	Russian olive	
Fraxinus americana	45–55	White ash	
Malus sylvestris	20–30	Apple	
Morus spp.	20–30	Mulberry	
Picea pungens	15–20	Blue spruce	
Populus grandidentata	15–20	Bigtooth aspen	
Populus tremuloides	15–20	Quaking aspen	
Quercus alba	60–70	White oak	
Quercus macrocarpa	60–70	Bur oak	
Quercus rubra	60–70	Northern red oak	
Quercus velutina	60–70	Black oak	
Salix spp.	20–30	Willow species	
Thuja occidentalis	5–10	Arborvitae	

[a]Mature sizes in urban settings vary, depending on site conditions. For full descriptions, see Gerhold *et al.* (1993), Dirr (1990), New Jersey Shade Tree Federation (1990), or nursery catalogs.
[b]Most D.E.D.-resistant (Townsend *et al.*, 1995).
[c]Occasionally used as a street tree along the shore because of its salt tolerance.
[d]Seashore widths.

30 feet if trees are to be planted under utility wires. Then if ascending branching is needed to provide clearance for pedestrians, the next table would sort out those trees that fail to meet this criterion. Additional tables would be needed for every important criterion, to conduct a thorough evaluation of candidate species. This would be cumbersome in situations where many criteria need to be considered. However, some experts have all this information stored in their minds and can use the method without actually consulting tables.

A faster method is to use a table containing a large matrix of information about all important selection criteria pertaining to many species. For example, Gerhold *et al.* (2001) contains such a table for selecting street trees, and the Landscape Institute (1989) has compiled a more extensive table of landscape trees and shrubs. To make such a table more compact, symbols are displayed that correspond to categories within selection criteria. The user can scan important characteristics rapidly and identify promising candidate species and cultivars.

In both these tabular methods, several candidate species are identified, hopefully not too many, which then must be reduced to the one which fits best. One or two alternates also should be selected in case the best choice turns out to be unavailable. At this time, it is advisable to consult references that provide thorough, definitive descriptions of the species and cultivars under consideration such as books by Dirr (1990), Gerhold *et al.* (2001), Hightshoe (1978), Jacobson (1996), Wandell (1994), Yiesla and

Giles (1992), and Zion (1995). Catalogs from large production nurseries also contain much useful information, including availability. Photographs and drawings can provide some types of information that are difficult to convey by words alone.

7. Computer Selection Methods

"Landscape Tree Factsheets CD" (Penn State School of Forest Resources, University Park PA 16802) contains all the information and photos in Gerhold *et al.* (2001). Several kinds of computer software have been designed to take the user through a selection process. "Plant Right" (Environmental Consultants, Inc., 207 Lakeside Drive, Southampton, PA 18966-4527) is one that was developed for northeastern United States. The user simply specifies the geographic location and chooses desired tree attributes. The software then generates a list of trees or shrubs that match the specifications and can print out the selected criteria and list of plants.

"Southern Trees—An Expert System for Selecting Trees" is more comprehensive. This CD-ROM software is interactive, guiding the user through a series of questions about the selection process. Responses set the selection criteria and search for matches, based on the wisdom of experts incorporated in the software, hence the expert system designation. The system includes factsheets of 700 species, drawings, and more than 2500 color photos that may be consulted at any time. Plans are under discussion for a northern version. Until it becomes available, "Southern Trees" can be used as a teaching tool about the selection process; it does contain some species that grow in the Northeast.

Computerized tree selection programs facilitate the process, making it more orderly and thorough, and help to avoid mistakes. Their effectiveness nevertheless depends on an experienced user who has carefully evaluated the purpose of a proposed tree, its site conditions, and care it will receive.

References

Dirr, M. A., 1990, *Manual of Woody Landscape Plants: Their Identification, Ornamental Characteristics, Culture, Propagation and Uses,* 4th ed., Stipes, Champaign, IL.

Gerhold, H. D., Lacasse, N. L., and Wandell, W. N., 2001, *Land scape Tree Factsheets,* Penn State College of Agriculture, University Park, PA.

Hightshoe, G. L., 1978, *Native Trees for Urban and Rural America,* Iowa State University Research Foundation, Ames, IA.

Jacobson, A. L., 1996, *North American Landscape Trees,* Ten Speed Press, Berkeley, CA.

The Landscape Institute, 1989, *Trees and Shrubs for Landscape Planting,* RIBA Publications, London.

New Jersey Shade Tree Federation, 1990, *Trees for New Jersey Streets,* 3rd ed., New Jersey Shade Tree Federation, Cook College, New Brunswick, NJ.

Schein, R. D., 1993, *Street Trees: A Manual for Municipalities,* Tree Works, State College, PA.

Wandell, W. N., 1994, *Handbook of Landscape Tree Cultivars,* 2nd ed., East Prairie Publishing, Gladstone, IL.

Yiesla, S. A., and Giles, F. A., 1992, *Shade Trees for the Central and Northern United States and Canada,* Stipes Publishing, Champaign, IL.

Zion, R. L., 1995, *Trees for Architecture and Landscape,* 2nd ed., Van Nostrand Reinhold, New York.

Chapter 12

Guide to Selecting and Specifying Nursery Stock

Jim Sellmer and Larry Kuhns

1. Introduction

Successfully planting of trees in urban areas is a difficult task. There are many things that must be done accurately, and a mistake in only one area can result in the failure of the plantation. Sites must be selected and evaluated thoroughly to determine if trees are appropriate for them, and if there is sufficient suitable soil to support their growth. Species that are adaptable to the site conditions must be selected and then located and purchased. The trees must be planted properly and maintained until established in the site. It is the objective of this chapter to define the factors that make a high-quality tree, and help urban tree managers locate and purchase high-quality trees for their communities.

2. Selecting Quality Trees

2.1. Plant Nomenclature

A lot of effort should go into selection of the best trees for particular sites. The best tree may be identified by its species, but in many situations it will be a named variety. When ordering the trees selected it is important to order by both the common and Latin names, giving the genus and species, and the variety name if appropriate. Many mistakes have been made over the years by both the tree managers in ordering the trees and by nursery owners supplying the trees. Tree managers must be extremely careful to order exactly what they want, and must check every tree that they receive to assure the species and variety delivered match what is ordered.

Jim Sellmer Department of Horticulture, The Pennsylvania State University, 102 Tyson Building, University Park, Pennsylvania 16802-4203.

Urban and Community Forestry in the Northeast, 2nd ed., edited by, J. E. Kuser.

Varieties are selected for certain characteristics that may be extremely important to particular sites. These characteristics include hardiness, size, form, tolerance to urban conditions, appropriateness to urban conditions, and the color of their foliage, flowers, and fruit.

There are several very commonly used tree species that are native to wide geographic regions. Red maple, for example, grows from New England to the Gulf of Mexico with broad hardiness between Zones 3b to 9. However, trees grown from seeds collected in the South will not survive in the North. If red maples grown from seeds are to be planted in the North, there must be some assurance that they were grown from seeds collected in the North. The converse is also true. For example, Northfire™ ("Olson") introduced by Bailey Nursery is rated as hardy in Zones 3 to 5. It would not be appropriate in southern sites in Zone 6b and upward. (Dirr, 1998) Conversely, "October Brilliance" rated for Zones 5 to 7 would be a poor choice for locations in Zone 3 to 4. There are a number of hardy named selections such as, Red Sunset® and October Glory®; however, care in choosing between the two remains important. While October Glory can be effectively used in the milder climates of Zones 6 to 8, it winter burns in Zones 3b to 5 due to slow acclimation (Dirr, 1998; Flint, 1997). Red Sunset is a better choice for planting in Zones 4 to 6. Newer cultivars ("Brandywine," "Somerset," and "Sun Valley") are rated hardy for Zones 4 to 8 and may offer greater flexibility for east coast planting situations.

There are also many examples of the significance of selecting and obtaining the correct variety. A tree planted in a narrow sidewalk planting bed between the street and a closely placed building is an example of placing a tree in a tight location. This situation leaves little room for the crown of the tree to develop and will likely result in the tree growing into the building and creating a maintenance problem. In this situation a tree with a narrow crown, such as "Armstrong" red maple should be used. Planting an October Glory red maple would be a mistake. Sugar maple varieties that originate in the cool, moist northeastern end of their range have beautiful fall color, but do not tolerate urban conditions, as well as the varieties from the hot, dry western part of their range. Varieties that originate in the western part of the range, such as "Commemoration," "Green Mountain," and "Legacy," have thicker leaves that tolerate the hot, dry, and windy conditions found in cities.

There are hawthorns that have few, if any, thorns. These can be used in cities, but should never be substituted with thorned varieties that could badly scratch pedestrians.

The points to remember are: (1) select the appropriate variety for a particular site and (2) substitutions for that variety by a nursery should only be allowed if the substitution is not substantially different from the variety ordered.

2.2. Nursery Sources

It is generally agreed that a business can offer only two out of the following three items: price, quality, service. If a nursery consistently offers the lowest price, it cannot consistently offer the best quality plants and excellent service. If they do in fact offer the best quality and service, they cannot offer the best price and remain in business for long. Since most trees are purchased through contracts that are awarded to low bidders, the specifications that are sent with the bids do much to determine the quality of the trees received. It is the responsibility of the urban tree manager to write bid specifications detailed enough to guarantee getting the quality and service needed. They must then find nurseries with good reputations for providing high-quality and service to include on their bid lists.

There are a number of electronic stock locators accessible through the worldwide web. These can be used to locate nurseries that list the trees needed. Web sites currently in operation can be found in Appendix A. To locate additional nurseries contact the state nursery and landscape associations in yours and surrounding states, and if possible the most prominent trade show in your region of the country. Collect catalogs that provide information on cultivars, hardiness zone, form, height, and other unique characteristics. Use them to learn about new trees you may want to use in future plantings in your community.

Nurseries, like all other businesses, develop reputations. These reputations can be determined by requesting a list of references from the nursery and then interviewing the referees. The best way to learn about a nursery is to visit it. Check to see if it is neat and clean, reasonably free of weeds, and free of insect and disease problems. Make sure the trees are labeled and spaced far enough apart to grow without serious interference from adjacent plants. Check the pruning practices to ensure that trees to be sold as street trees are limbed up to the appropriate height, and that corrective pruning has eliminated crossing and parallel branches.

From the references, determine how good a job the nursery does with the harvest, handling, storage, and shipping of their trees. The soil balls should be well shaped, and the burlap and wire baskets should be tied tightly around the balls. They should not be broken or damaged in any way from rough handling. An irrigated storage area should be available to hold the trees until shipped. The nursery's reputation for delivering trees on time should also be checked.

As important as any one of the above, is the reputation of the nursery for consistency. Obtaining some poor trees even every 2 or 3 years is not acceptable. Obtaining consistently high-quality trees is a necessity for any urban tree program.

If possible, establishing an ongoing relationship with a nearby nursery has many advantages. The shorter the length of time between harvest and planting, the better are the chances the tree will thrive in its new site. If the tree has been growing in the area for several years, it is very likely to be fully hardy in that area. The nursery can be visited on a regular basis, and trees to be purchased can be tagged. In addition, if there are problems with the trees in any particular year, they can be inspected by the nurseryman, and confirmed for replacement or credit much more easily. Finally, the nursery can be encouraged to grow species or varieties that the community wants but has trouble in locating.

2.3. Above Ground Quality

The first measure of quality of a tree is its size. Except at smaller sizes, trees are sold based on their caliper. Trees less than 4 inches in diameter are measured 6 inches above ground line; whereas, trees 4 inches and larger are measured 12 inches above the ground. The term diameter at breast height (dbh) is used to measure forest trees. This method should not be used for standard size ornamentals.

In addition to caliper, the ratio of height to caliper is also a measure of quality. Though it will vary somewhat with species, trees that are excessively tall for their caliper should not be accepted. This occurs when trees are either collected from the wild or are grown too close together in the nursery. Both situations produce relatively weak plants that do not survive transplanting as well as properly grown trees and may be susceptible to blowdown.

The form of the trees should also be closely inspected. It is important to select trees that have well-spaced structural branches, oriented uniformly around the trunk. Branches that are growing close together when young will grow into each other with age, and they will not be able to develop their full structural strength. Some trees, such as Bradford pear, have gotten a bad reputation for clusters of branches growing at acute angles that end up splitting in snowstorms. However, if the trees were pruned more carefully in the nursery, many of the problems could be avoided. Inspect the branch structure of the trees in the nurseries that you visit, and point out defects to the nursery operator that you feel should be corrected.

Many nursery operators trim trees like hedges to form dense canopies. This makes the trees look better temporarily, keeps them from growing into each other, and makes them easier to harvest, ship, and handle. However, it often leads to the development of many crossing and parallel branches. Again, you should encourage the operator to remove crossing and parallel branches during the production cycle. Otherwise, they will have to be removed at the planting site.

If the tree is a type that should have a strong central leader, avoid trees that have clustered growth at the top. These branches may develop narrow angles that will become weak sometime in the future, or they may grow together and form weaknesses in the tree.

For street use, always specify single-stem trees that are limbed up as much as possible in the nursery prior to purchase. It is not good to limb up too high, too fast, because it leads to weakness in the trunk. Trees must be limbed up enough to ensure that they do not interfere with pedestrian or vehicular traffic.

The general health of the tree must also be carefully inspected. First, inspect the trunk to ensure the bark has not been damaged during production or handling. Torn bark reduces the movement of water and nutrients in the tree and opens it up to infection by decay organisms. Bark can be damaged by equipment, frost cracks, or nonselective postemergence herbicides. Reward™ (diquat) or Finale™ (glufosinate) are contact herbicides that may kill bark they contact. This will result in sunken areas in the trunk where living wood and bark grow around dead bark. Bark damaged by Roundup Pro™ (glyphosate) splits open. The bark split can range from barely noticeable to several inches wide, reaching from ground level up into the canopy of the tree. The size of the split is dependent on the tree species, tree age, time of application, and the rate of herbicide applied to the tree.

Canker-causing fungi that enter the trunk through pruning wounds may also cause sunken areas or splits in the bark. Cankers caused by fungi can be diagnosed by the fact that they are centered on pruning wounds.

At the time of delivery, carefully inspect all trees for any signs of serious insect or disease problems. Look for cankers, scale, or borers on the trunk. Look for scale, caterpillars, severe aphid infestations, or diseases in the foliage. Plants that are found to have an insect or disease problem should be isolated and quarantined, so the problem is not spread to new or existing plantings. If the problem can be corrected, treat the plants accordingly. If not, they should be returned to the nursery.

The general health of the tree can be determined by examining the foliage color and density, and the length of shoot extension for the previous 2 or 3 years. The foliage should be lush green. Thin, light green or yellow foliage is a sign that the plants were grown under poor fertility conditions at the nursery. This results in trees that have poor nutrient and energy reserves at the time of transplanting. When trees are dug, 65% to 75% of their roots are left in the nursery, and their survival and growth in the landscape depend on the rapid regeneration of roots. The more nutrient and energy reserves in the plant, the faster it can regenerate a root system large enough to allow it to become well established in its new site.

With a little experience it is easy to identify the annual growth in length of the branches of trees. The growth should be stunted the year a tree is planted in the nursery, but each year after that the growth should increase in length. If the growth appears to be stunted the year before harvest you should try to determine the cause. It could be from drought, poor fertility practices in the nursery, or it could mean the tree was dug sometime during the previous year and was held above ground for many months. Again, a tree held this long will have poor nutrient and energy reserves compared to a freshly dug tree.

2.4. Below Ground Quality

There are many questions asked about the different types of burlap used to wrap soil balls. Natural, untreated burlap is the best. If buying locally it can be specified. However, it only lasts about 6 weeks when heeled-in in a mulch bed. Many growers use treated, or rot-resistant burlap. This burlap is treated with a copper compound that retards the decomposition of the burlap that is caused by fungi in the soil. It lasts 3 to 6 months in mulch beds or the soil. Both untreated and treated burlaps bind well with soil after planting and allow the free movement of water in and out of the soil ball. Both are acceptable.

Plastic burlap used to be green and shiny, and anyone with even a limited knowledge of plants knew it should be removed while planting. However, sometime around the mid-1980s a brown, woven plastic burlap with the same texture as natural burlap was introduced into the nursery trade. This introduction was a disaster for the industry. Many experienced landscapers did not recognize this new woven burlap as plastic and planted it with the root ball. Severe losses of trees and shrubs followed. When ordering trees, state specifically that you will not accept trees with root balls wrapped in plastic burlap.

The quality of the root system of a tree is the most important factor in determining how well the tree will survive transplanting, and how fast it will become fully adapted to its new site. As mentioned in Section 2.3 on height to caliper ratio, trees that are collected from the wild should not be accepted. The only roots close to the base of collected trees are large support roots. The smaller feeder roots are farther away and few are included in the soil ball when dug. Nursery grown plants are root-pruned several times during propagation and transplanting operations, prior to being dug for landscape use. This results in the development of many more fine roots close to the base of the tree, which are dug with the soil ball for movement to the landscape.

Trees are available in a variety of forms for planting in urban areas. They may be bareroot, balled in burlap (B&B), or grown in gro-bags or containers. Bareroot plants are not commonly planted in cities because they are highly susceptible to vandalism prior to establishment of a dense root system. However, bareroot plants are a very economical form when installing a large number of trees for planting in lightly traveled areas. Bareroot trees must be kept cool (35 to 40 °F) and moist prior to planting. These trees should be planted immediately after their arrival to ensure their survival. If the chance of long delays between arrival of the trees and planting exists, bareroot trees should not be ordered.

Balled in burlap is currently the standard marketing form for trees in the industry. There are several standards of quality by which the soil balls of B&B trees are evaluated. The size of soil balls should meet the minimum standards established by the American Landscape and Nursery Association (ANLA) in the publication, the American Standards for Nursery Stock ANSI Z60.1 (ANLA, 2004). The diameter and depth of the soil balls should both be measured and compared to the standards. It should be recognized that the standards published are minimum, and there is nothing wrong with receiving soil balls that are somewhat larger than the specified sizes.

In addition to size, there are several other characteristics of the soil balls that should be examined. The burlap and wire baskets used to wrap the soil balls should be very tight. Wire baskets should be crimped as tight as possible. The looser the wrapping, the greater the risk the ball will be damaged during handling. The trunks of the trees should be in the center of the soil balls. Sometimes trees dug with a tree spade are badly off-center in the ball. This reduces the tree's chance of survival, and may make it unsteady in the planting site. Ensure that the burlap and twine used to wrap the balls are not plastic or nylon. If they are, they must be removed at the time of planting.

The twine around the trunk of some of the trees should be removed to determine the location of the crown of the plant in relation to the top of the soil ball. The crown can be identified by the thickening of the trunk and the development of buttress roots. They should be right at the soil line as specified in the nursery stock standards (ANLA, 2004). If they are buried more than 2 or 3 inches, the tree will have an undersized root system. The widest part of a tree-spade dug soil ball is the top. If there are no roots in the top 6 inches of the ball, the root system may be reduced by 30% to 40% below the standards. These trees should not be accepted.

Some nurseries are now growing trees in gro-bags. These are bags made of geotextile (woven plastic) that allow some root penetration, but do not allow the expansion of large roots outside the bag. The theory behind this growing container is that the prevention of large roots from penetrating the bag will result in more fine roots forming within the bag assuring that a larger percentage of roots will be transplanted with the tree. Smaller balls with more roots mean lighter, easier to handle trees that establish faster than trees moved by the traditional B&B method. The minimum size bags required for different size trees are also specified in the nursery stock standards (ANLA, 2004).

Pot-in-pot (PIP) tree production is also popular. Trees are grown in containers, which are held below ground level in separate socket containers. This system allows for ready removal from the ground for planting in the landscape with all roots intact. As with other container-grown nursery stock, PIP plants are grown in soilless medium consisting of a mixture of peat moss and composted bark. Soilless media is designed to drain well, provide adequate aeration, and promote root growth. The trees grow well in the containers in the nursery; however, when transplanted to the landscape the container-grown plants initially require more frequent watering due to the well-drained soilless media and the vigorous yet less hardened roots than trees moved with a soil ball (B&B or gro-bags). Specifying and selecting container and PIP grown plants also requires inspection of the root system to assure the roots are healthy, non-circling, and intact. Trees grown in the PIP system should not be grown in the final container for more than 24 months. This timeline should assure a well-established root system without girdling roots, cut roots due to escape through container's openings, or reduced growth momentum due to limited container resources after being in the container for too long. Finally, container-grown plants are similar to bareroot plants in that they are easy to rip out of the ground soon after planting and should not be planted in areas in which vandalism is major concern.

2.5. A Summary of Nursery Production Practices Which Affect Postplanting Survival and Growth

2.5.1. Propagation or Planting Practices That Lead to Deformed Root Systems

Tree seedlings grown in containers for too long may develop circling root systems that are permanently hardened into that shape. As the tree grows in future years the roots can expand and girdle each other, resulting in a tree that is unstable and may die.

Twisting trees at the time of planting in the nursery to force long roots into a hole that is not large enough for them can also result in a circling root system in future years.

2.5.2. Hilling Trees or Planting Too Deep in the Nursery

There are two practices that can occur at the nursery, which can lead to a soil ball that has no roots in the upper part of the ball; hilling up" soil around the base of the tree, and planting too deep in the nursery. The practice of hilling up soil around the base of trees in the nursery was begun many years ago as a method of weed control. When trees were dug by hand, experienced diggers knew they had to take the time to move the "hilled up" soil away from the base of the plant before digging it.

Hilling up soil at the base of the trees no longer needs to be done because there are herbicides available that provide safe and effective weed control. However, some nurseries still do it. This results in the original root system of the plant being buried within 4 to 10 inches of soil. Another nursery practice that results in the original root system of the plant being buried is planting too deep. Some plants that have performed poorly in the landscape have been found to be planted up to 12 inches deep in the nursery.

Most trees today are dug with mechanical tree spades. Soil balls dug with tree spades are wider at the top than the bottom. If a tree has been planted too deep in the nursery, or if soil that has been hilled up around a tree is not removed prior to digging, the result is a soil ball that has no roots in the upper part of the ball. Since this is the widest part of the ball, it contains the highest proportion of the volume of the ball. Although the soil ball may be the correct size for the tree being moved, the actual root mass may be one-half or less of what it should be. The other problem that occurs is at the planting site. It is difficult for those planting the tree to discover the problem, so if they plant the tree properly, with the top of the soil ball level with the existing grade, they are actually planting the root system 4 to 12 inches deep. Trees with undersized root systems that are planted too deep rarely survive.

2.5.3. Postemergence Herbicide Damage—Roundup, Finale, Reward

Postemergence herbicides are an essential part of a weed control program in the production of trees. However, if used improperly they can damage the trunks of trees around which they are applied. Young trees with thin bark that show pigment are most sensitive to misapplications.

Roundup Pro is the most commonly used herbicide around trees. If too much is sprayed directly onto the trunks of sensitive trees it can be absorbed through the bark and move upward. Injury can be difficult to diagnose because it may not show up for a year or two after application. When it does, it appears as a splitting of the bark as it separates from the wood below it. Depending on the time and rate of application, and the sensitivity of the tree, the split may be small and barely noticeable, or it may develop to an inch or more wide and progress up into the canopy of the tree.

Damage to the bark by Finale and Reward results in sunken cankers that develop as the living tissue continues to grow and develop around the herbicide-killed tissue. Injury from these herbicides is much less common than injury from Roundup Pro.

Inspect plants for signs of bark splitting or sunken cankers. However, even if no injury is found, it is possible for latent injury from nursery applications to appear following planting in the landscape.

2.5.4. Poor Fertility

Plants that are grown properly develop a growth momentum. When a plant is grown in deep, well-drained soil with limited competition from weeds and is provided adequate amount of water and fertilizer; it develops a dense, dark green canopy of foliage that is highly photosynthetically active. It is in the process of photosynthesis that carbohydrates are produced. These carbohydrates provide the energy for additional plant growth. The longer the good growing conditions for a plant are maintained, the more energy is produced and stored, and the longer the growth momentum continues.

A tree that is grown under continuously favorable conditions will contain high-energy reserves that will be available to support the production of new roots, branches, and leaves following transplanting. This tree will survive and develop much better than a tree that was grown under conditions that would limit growth and photosynthesis. Growth momentum does move from the nursery to the planting site if trees are dug, shipped, and handled properly.

Visit nurseries to determine their soil types, the level of their weed management, and their fertilization and irrigation practices. Evaluate the growth and foliage of the trees to determine if they are growing vigorously. Eliminate a nursery as a potential supplier if poor production practices are encountered.

2.5.5. Growing on Poor Soils

Plants grown on poor soils may survive in the nursery, but they do not thrive. Their root systems will be limited, and they will not accumulate the high-energy reserves needed to develop the growth momentum that will help them become established in their new sites following transplanting. When dug with a tree spade the lower part of the soil ball may not have many, if any, roots in it. If grown in wet soils, the root system may be infested with root-rotting organisms that will further restrict the reestablishment of the tree in its new site.

2.5.6. Growing too Close in the Nursery

Plants grown too close together in the nursery have two major problems. They shade each other and reduce their level of photosynthesis, which results in lower energy reserves in the plants. A plant with low-energy reserves is not capable of producing as many roots as quickly following transplanting as a plant that has higher energy reserves. Trees grown too close together develop too much height in relation to their caliper. They are weaker trees that may need considerable support following planting.

2.5.7. Insect or Disease Problems

Some insect or disease problems have little effect on the survival and growth of trees in the landscape. A slight aphid or leaf spot infestation is of no concern. However, any plants that show signs of root-rotting diseases, trunk cankers, scale insects, spider mites, or borer or weevil infestations should be rejected. These problems become progressively worse on plants that are undergoing the severe stress that is associated with transplanting.

2.5.8. Digging an Undersized Ball

As mentioned in Section 2.4, the size of soil balls should meet the minimum standards presented in the nursery stock standards (ANLA, 2004). Undersized soil balls have less root mass and energy to grow the new roots that are essential for the survival and growth of newly planted trees.

2.5.9. Plastic Burlap

Untreated natural burlap rots within 6 weeks following planting; treated natural burlap rots within 6 months. Plastic burlap does not decay in strength after it is buried and no longer exposed to sunlight. It restricts water movement into the soil ball and root growth out of the ball. Roots that do penetrate the spaces between fibers are eventually girdled. Plants planted in plastic burlap establish slowly and generally die within 5 years.

When ordering B&B plants, specify that only untreated or treated natural burlap should be used to wrap the soil balls. Plants with soil balls wrapped in plastic burlap will be rejected. Removing the burlap entirely from a root ball prior to planting may result in damage to the soil ball. This is especially true if the burlap is covered with a wire basket and both the basket and burlap must be removed.

2.5.10. Nylon Twine

Though everyone in the industry should know that nylon twine must be removed at the time of planting, there are still far too many plants killed each year through girdling by nylon twine that was not removed from around the trunks of trees. Specifications should include a statement that soil balls must be tied with natural sisal twine; plants tied with nylon twine should be rejected. Natural twine can girdle trees if it is wrapped around the trunk of a tree too many times. Be sure to instruct planting crews to cut all twine that is wrapped around tree trunks.

3. Writing Bid Specifications

Preparing the bid specifications for a street tree planting, municipal park or for a new housing development is the backbone to assuring a successful and manageable planting program. Writing bid specifications requires two levels of understanding. The first includes experience in writing specifications; knowledge of the project and site; current drafts of the project; and a strong knowledge of plant materials and construction detailing. The second is an understanding that the specifications are legal documents that should express the specific requirements of the project.

In writing the bid specifications anticipate problem areas in the design (i.e., deicing salt usage in the area, traffic patterns, and overhead constraints) and the construction process (i.e., available planting depth along roadways and soil characteristics). Clearly define your requirements for all aspects of the project, especially in the anticipated problem sites and situations. In other words, write clearly, concisely, and precisely the requirements of the project to ensure proper, accurate, and responsive bids from contractors and to avoid misunderstandings and future problems on-site.

3.1. Realities

The bidding process includes two categories of projects; *private* or *public*. For most community projects, the bid is open which means the "call for bids" is publicly announced in the newspaper. Bids may also be sought from familiar and high-profile contractors (Carpenter and Walker, 1990). In addition to being an open bid, community projects are often constrained by law, community charter, or financial limits to accept the lowest acceptable bid tendered by a contractor. For this reason, a strong understanding of the plant materials and site are important. Coupling your intimate understanding of the project with a clear, precise, reasonable, and stringent set of bid specifications is the best way to assure proper plant selection, handling, site preparation, planting, and postplanting maintenance.

If you are not familiar with the plant materials or the site, then contact a reliable resource. Local horticulturists, landscape architects, landscape designers, consultants, city foresters, and your county cooperative extension educator can help in interpreting and defining the plant selection, site constraints, and planting steps which may affect the bid specification.

Finally, when possible, clearly state that bid acceptance will be based on factors other than low bid. Applying quality and competency standards to the bidding process allows you to compare the quality, workmanship, and capabilities of the bidding firms. By basing bid awards on two standards (low bid and quality), you maintain greater control over the project. In addition, you can assure that the competing firms are of the highest quality and that you are receiving the greatest quality for your budget.

3.2. Resources

Experience is probably the best resource in writing bid specifications followed by sample specifications used in other projects. In starting out, these two resources are often unavailable. Whether you are a novice or experienced in writing bids, it is important to maintain the most recent information on local ordinances, recommendations, industry standards, and plant materials. This will help to avoid a mistake when writing your specifications.

Many municipalities are instituting landscape ordinances to ensure uniform, environmentally sound, and attractive communities for their residents and visitors (Wolff and Womack, 1991). Similarly, urban forestry organizations are promoting the formation of tree commissions and the establishment of tree ordinances for municipalities to help in planning and managing our urban forests (Grey, 1993). Additional information for developing tree selection and planting specifications is available from state, county and municipal government agencies. Prior to planning a project contact your local government(s) to learn about the ordinances, recommendations, and specifications that may affect your planting program.

Historically, many growers' associations and trade organizations have developed standards for creating, growing, operating, packaging, and shipping their products to market. As mentioned earlier, the ASNS (ANLA, 2004) represents the consensus minimum guidelines for nursery stock. These guidelines include size measurements; above ground structural character (i.e., height of branching for street trees); and limits on root size and spread based on the nursery production method employed (i.e., bare root, B&B, and container specifications). These standards are revised and expanded from time to time to meet the needs of the industry (ANLA, 2004).

In the case of the landscape industry, there are several standards commonly employed and recognized for specifying, planting, and maintaining trees including the following:

1. *Landscape Specification Guidelines 5th Edition*, Landscape Contractors Association MDIDCIVA, Rockville, MD. These guidelines have been set forth as a guide for specifying, installing, and maintaining quality landscape projects.
2. *Principles and Practice of Planting Trees and Shrubs*, 1997. G.W. Watson and E.B. Himelick, International Society of Arboriculture, Champaign, Ill. This manual provides information on the various types of nursery stock production systems, planting, handling, and post-plant care information and specifications.
3. *American National Standard for Trees, Shrubs, and Other Woody Plant Maintenance—Standard Practice (Pruning)*, ANSI A300, 2001. American National Standards Institute (ANSI), New York, NY. The pruning standards describe the proper cutting techniques and a discussion of shade tree structure, health, and maintenance.

In writing bid specifications and planting plans, one of the most miswritten sections is the plant list. Misspelling and misidentification of the specified trees, shrubs, herbaceous perennials, and annuals can lead to the planting of inappropriate species and varieties at the planting site. The following are a listing of suggested guides to general plant nomenclature:

Published Resources:

1. *International Code of Nomenclature for Cultivated Plants*, 7th ed., 2004, (C. D. Brickell, B. R. Baum, W. L. A. Hettersheid, A. C. Leslie, J. McNeill, P. Trehane, F. Vrugtman, and J. H. Wiersema, eds.), International Commission for the Nomenclature of Cultivated Plants of the I.U.B.S. Acta Horticulturae, No. 647 (International Society of Horticultural Science), Leuven, Belgium.
2. *Datascape Guide to Commercial Nomenclature*, 2003, American Nurserymen Publishing Co., Chicago, IL.
3. Griffiths, M., 1994, *Index of Garden Plants: The New Royal Horticultural Society Dictionary*, Timber Press, Inc. Portland, OR.
4. Krussman, G., 1985, *Manual of Cultivated Conifers*, 2nd revised ed., (H. D. Warda ed.), Timber Press, Portland, OR.
5. Hillier, H. G., *Hillier's Manual of Trees and Shrubs*, 1973, 4th ed., (A. S. Barnes, ed.), South Brunswick, U.K.
6. Kelsey, H. P., and Dayton, W.A., 1942, *Standardized Plant Names*, 2nd ed., For and by the American Joint Committee on Horticultural Nomenclature, J. Horace McFarland Company, Harrisburg.

Internet Resources:

United States Department of Agriculture PLANTS program: http://plants.usda.gov/plants

International Plant Names Index (IPNI): http://www.ipni.org

As noted earlier, the success of the project begins with knowledge of the limits, requirements, character, and growth habits of the plants being used. Along with nursery catalogs, there are many helpful plant identification and cultural manuals available from libraries and bookstores. The following is a partial list of plant references that are available and helpful in choosing the proper trees for the site.

1. Dirr, M. A., 1998, *Manual of Woody Landscape Plants: Their Identification, Ornamental Characteristics, Culture, Propagation, and Uses*, 5th ed. Varsity Press, Charlotte, NC.
2. Flint, H. L., 1997, *Landscape Plants for Eastern North America*, 2nd ed., Wiley, New York, NY.
3. Staff of the L. H. Bailey Hortorium, 1976, *Hortus Third: A Concise Dictionary of Plants Cultivated in the United States and Canada*, Cornell University, Macmillan, New York, NY.
4. Gerhold, H. D., Wandell, W. N., and Lacasse, N. L. (eds.), 2002, *Landscape Tree Factsheets*, AGRS-56 Publication, Pennsylvania State University, University Park, PA.
5. Dirr, M. A., 2003, *The Interactive Manual and Photo-Library of Woody Landscape Plants* (DVD Version), Timber Press, Portland, OR.
6. Horticopia Professionals Database (CD), 2005, Horticopia, Purcellville, VA.

4. Specifying Quality Results

By clearly developing your plant and planting specifications for bid, you are maintaining control of the bidding process. The first attempt at this process may seem overwhelming with extensive detail required. Keep in mind that "sweating the details" prior to the bid process will reduce the chance for misunderstandings and errors in the bidding process. This type of preparation includes outlining solid specifications and contingency plans. Indecision and lack of preparation of the specifications can result in changes after the project is contracted. This can readily strain the relationship between client and contractor and can sequentially change other agreements, contracts, and stipulations of the bid. This can lead to delays in time and financial penalties once the contracts have been signed. Remember, the set of specifications is a legal document.

4.1. The General Form

A standard form for bid specifications does not exist. Bid specifications are not commonly separated into plant selection and planting specifications. Instead, a contractor will subcontract with a nursery to provide the trees for the project. Because of this relationship, a clear and detailed bid specification is a necessity to ensure that your requirements are met.

To start writing a bid specification the common recommendation is to look at other bid specifications. Sample bid specifications may be available from colleagues, contractors, published records, and references (Carpenter and Walker, 1990). The level of detail that you choose for your bid specifications rests with your decision on the level of expectation, requirements, time, and energy that the project requires to assure successful and satisfactory completion.

The following Sections 4.2 and 4.3 describe the characteristics and information that can be provided in a detailed bid specification through the point of receiving the plant material (Rideout, 2005). Chapter 13 details the planting process and should be taken into consideration in writing the planting specifications.

The bid specification is separated into two sections: the business section and the project section. The bid specification may also include detailed diagrams that are not part of the construction or planting plans.

4.2. Business Section Details

The business section describes the legal requirements, procedures, deadline dates, and limitations that can be expected from the contractor. The business section can include the following subsections:

1. Bid prequalification requirements
2. The proposal form
3. The details on awarding the contract
4. Procedures for examining project plans, specifications, and the site
5. Insurance requirements
6. Project inspection requirements
7. Payment plans and criteria for payment schedule
8. Completion dates and contingencies
9. Expected guarantees on work and labor expectations
10. The names of the contact person(s) for all specification questions

4.2.1. Prequalifying the Bidders

By requesting that the bidding contractors prequalify their bid with evidence of experience, you can maintain a greater level of control over the bidding process. In addition, you can review the types of work they have done; their references; and their ability to complete your project (i.e., labor, equipment, and experience) based on the documentation that they provide. Bidders that either are not interested or cannot provide documentation of their abilities and experience will be less likely to bid on the job.

Prequalifying contractors can reduce the likelihood that contractors with poor reputations in competence, quality, timeliness, and ethics will bid on your project. This can be accomplished with a simple statement, such as: "*All bidders, prior to award of the contract must present satisfactory evidence that they have been regularly engaged in this type of work and they are prepared with the necessary labor, materials and equipment to satisfactorily execute the work*" (Rideout, 2005). This can be enough to inform the bidding contractors of your requirements. Keep in mind, passively requesting evidence is not enough. It is your responsibility to interview the references and visit the facilities or example work sites of the bidding contractor.

4.2.2. Bid Proposal Format

The format and language used for bidding on the project is defined here. By standardizing the form and language of the bid, the proposals can be rapidly, clearly, and easily reviewed, and this reduces the chance for misunderstanding or confusion.

Standardization can include: providing specifically designed forms to be used by bidding contractors; establishing a hierarchy of importance in cases of discrepancy; and requesting a breakdown of prices per activity (i.e., tree price, planting price, and total price). The request can be as simple as, "*Bidders must submit prices for the bid in both words and figures. In case of discrepancy, written words shall prevail. Prices shall be given by tree price, planting price, and total price*" (Rideout, 2005).

4.2.3. Awarding of the Contract

This subsection defines how the contract will be awarded, whether more than one contract will be awarded, and further clarification as to the services, which contractors should be proposing to provide. Finally, this section can be used to reserve your right to reject or accept proposals based on which is most advantageous to your project.

4.2.4. Expectations, Limitations, and Insurance Requirements

This subsection accomplishes two important goals: (1) inviting contractors to bid on the project and (2) placing the burden of clearly reviewing the project on the contractor. By emphasizing the prebid review, the landscape contractor can also assist you in uncovering overlooked problems in the plan or on the planting site.

In this subsection, limits and allowable procedures for deviating from the bid specification should be clearly defined. Proof of insurance coverage and minimum limits for Worker's Compensation, public liability, and property damage insurance should also be requested.

4.2.5. Inspection, Completion, Guarantees, and Contacts

This subsection outlines a timetable for inspection and identifies the inspector. In all cases, final payment for the project should be contingent on a final, complete, and satisfactory inspection. Often, periodic and progressive inspections of the site have been limited due to the time and cost of having a dedicated inspector. By building a regular inspection schedule into your bid specification, you continue to maintain control of the project and can prevent miscommunication with the contractor. In contrast, inspecting the site when you are not comfortable or knowledgeable enough about the project can lead to slowed progress and poor relations with the contractor. If you are not comfortable acting as the inspector, you can hire a knowledgeable consultant.

In this subsection, project deadlines, late penalties, and the procedures for negotiating changes in deadlines and unexpected delays should be outlined. The contractor will require a solid deadline in order to calculate their cost and ability to complete the job. In addition, outlining the penalties for incomplete work and noncompliance of the specifications ensures that the contractor understands prior to bidding the expectations of the project and defends your position when a penalty is levied.

This subsection should contain your expectations on guarantees and warranties to be observed and fulfilled by the bidder. In other words, the guarantee period for plant materials purchased and the environmental and physical conditions that must be documented for the guarantee to be redeemed.

Finally, a list of contact people including names, positions, and phone numbers should be provided so that the bidders have ready access in asking questions about the project and the specifications. If there is a primary contact on specifications, then that person should be clearly identified to eliminate confusion and assure good communication.

4.3. The Project Section Details

The project section describes in detail the project, applicable standards, materials and procedures to be employed, and the parameters for acceptance of the project at completion. The following subsections can be included in this section:

1. The scope of the project
2. A list of applicable standards for the contractor to consult
3. A timeline for planting
4. A detailed outline of materials
5. Compliance and certification requirements
6. Procedures for selection, tagging, digging, and handling of plant materials
7. Delivery deadlines and procedures
8. Methods and timeline for inspection and acceptance of the planted materials
9. Guarantee and replacement parameters
10. The final inspection and acceptance process

4.3.1. Scope of the Job

The scope of the job includes the necessary elements for execution and completion of the planting based on the specifications, drawings, and plant materials list. The expectation should be that all required equipment, labor supervision, materials and supplies necessary for the job will be provided by the contractor at no extra charge. In addition, if existing plants, infrastructure, and physical features may be impacted by the project, it is important for you to define the areas to be protected and establish protection guidelines and accepted methods that the contractors must address in their bid submission.

4.3.2. Listing of Applicable Standards

This subsection provides the contractor with a specific source for the standards that are expected to be employed. Clearly stating what standards are to be followed assures that you and the contractor are following the same standards and talking the same language. By having this section in your specifications, you can reduce the likelihood of miscommunication. Commonly used standards are listed in Section 3.2 of this chapter.

4.3.3. Project Materials

This subsection should include details on the materials to be used on the project on which the contractors base their bids. You should provide to the contractors the following; a complete list of quantities and sizes of required plants, planting schedules, and materials required for the project to be satisfactorily completed. In addition, state clearly which takes precedence in case there is a discrepancy in size, quality, or quantity between the drawings and plant lists.

A list of proposed nursery sources should be provided by the bidders. You should clearly state your requirements for the adaptability, freshness, architecture, training, production method, and overall above and below ground health that the nursery stock should meet. By clearly outlining the parameters for acceptable plant materials you prevent misinterpretation and miscommunication between bidder and client. The following specification statement accomplishes that goal:

> All plant material shall conform to the American Standard for Nursery Stock. Plants shall be true to species and variety specified and nursery grown in accordance with good horticultural practices under climatic conditions similar to those in the locality of the project for at least 2 years. They shall have been freshly dug during the most recent favorable harvest season. Plants shall be so trained in development and appearance as to be unquestionably superior in form, compactness, and symmetry. They shall be sound, healthy, vigorous, well branched and densely foliated when in leaf, and free of disease and insects (egg, larvae or pupae). They shall have healthy, well-developed root systems and shall be free from physical damage or other conditions that would prevent thriving growth (Rideout, 2005).

Above and below ground acceptable and unacceptable plant characteristics as discussed in Sections 2.3 and 2.4 should be outlined clearly for the contractors. Expected size measurements should also be defined. The following specification is a good example.

> Containerized plants shall be well established in the container with a root system sufficiently developed to retain its shape and hold together when removed from the container. Plants shall not be pot bound, nor have kinked, circling or bent roots. Caliper measurements shall be taken on the trunk 6 inches above the root flare for trees up to 4 inches in caliper, and 12 inches above the root flare for trees over 4 inches in caliper. Height and spread dimensions specified refer to the main body of the plant and not from branch tip to branch tip. Plants shall be measured when branches are in their normal position. If a range of size is given, no plant shall be less than the minimum size, and no less than 50 percent of the plants shall be as large as the maximum size specified. Plants that meet measurements but do not possess a normal balance between height and spread shall be rejected." (Rideout, 2005).

The procedures for substituting plant materials should be outlined for the contractor to understand both the limitations and the procedures for negotiating a compromise if the client agrees. An example statement could consist of the following form:

> Substitutions of plant materials will not be permitted unless authorized in writing by the Purchaser. If proof is submitted, substantiated in writing, that a plant specified is not obtainable, consideration will be given to the nearest available size or similar variety, with a corresponding adjustment of the contract price (Rideout, 2005).

During the actual project there is no time to learn about organization. One of the most confusing times in a project is the period between selecting and preparing plants for digging and organizing the planting area on site. Often plants are misplaced, mislabeled, and misplanted because of poor organization. By requiring the bidding nursery or contractor to properly label plants, bundles, and containers much of the initial confusion can be reduced. Your requirements can include the type of label, form, and printing process.

If the contracting company is expected to carry out the planting process, this section can also include planting specifications. See Chapter 13 for a more detailed discussion on planting and maintaining trees in the urban landscape.

4.3.4. Phytopathological Certification

Legally, all nursery stock shipped into each state must have a valid nursery, dealer, or agent certificate attached. Both the shipper and the purchaser should be listed on the certificate as well as the contents of the shipment. The Phytopathological certificate verifies that the shipment is pest-free, based on inspection by the Plant Industry Inspection division of the Dept of Agriculture of the state of origin.

To protect your project from the introduction of quarantined pests on your newly purchased plant materials, a statement requiring certification should be distinctly included in the specifications. Assurances by the nursery or contractor should be required to the extent that all plant materials delivered comply with both state and federal laws governing the inspection, shipping, selling, and handling of nursery stock. If you have questions about the plant inspection process and plant-pest quarantines which exist in your state, contact the Plant Industry Division of your state Dept of Agriculture and talk with a nursery inspector for more details.

4.3.5. Requirements for Selection, Tagging, and Inspection of the Plant Materials

Although rarely done in practice, if you are concerned that the quality and health of the plants for the project meet your standards this can be assured by placing a statement in the specifications requiring your presence or the presence of your representative during the selection and tagging of the plant materials. If you are not familiar enough with the plant materials, consider hiring an inspector to do the job for you. Plan this in advance of soliciting bids to ensure the cost of the inspector is included in the bid.

4.3.6. Procedures for Handling Plant Material

Clearly state your expectations on the handling of the nursery stock. The ASNS (ANLA, 2004) includes information on the proper handling of container, B&B, bareroot, balled and potted, in-ground fabric grown bags, and processed balled plant materials.

The handling of plant materials also includes how they are transported and stored prior to planting. Nursery stock should be transported in a covered truck or covered securely with tarps to minimize leaf and stem desiccation. In addition, consider how the trees and shrubs are being handled from nursery to truck and truck to planting site. All precautions should be taken to prevent damage to the ball, trunk, roots, and branches of the plants. Transit guards should be used to protect the trunks of trees, and the branches of trees should be tied up to prevent branch breakage during transportation. Nursery stock should be protected from drying out while on the site. Specific and detailed instructions on the protection practices to be implemented, including wind and sun protection, irrigation, and mulching of root systems should be provided in the specification guidelines. In addition, a timeline for allowable storage on site prior to planting can be required or a specified limit for storage prior to planting can be stated in the specification guidelines.

4.3.7. Deadlines for Delivery of the Plant Materials

In writing your specifications, protect yourself by outlining the delivery requirements and expectations on cost for delivery. One approach is to write in the bid that the bid prices must include delivery to your receiving site. Provide clear directions for the delivery of the plants. Directions should include the complete address and the name of a contact person in case of delivery problems. In addition, clearly outline the deadlines for delivery, acceptable methods of delivery, and the penalties for not following these instructions.

4.3.8. Guarantee and Replacement Requirements and Limitations

For success of this project, be specific and realistic. Clearly specifying your expectations on guarantees should be common sense and strictly related to the materials and handling, as outlined in your specifications. Specifications should state the length of the guarantee, plant condition during the guarantee period which is acceptable or would trigger warranty enforcement, identification of responsibility for maintenance of plants in the guarantee period, definition of acceptable replacements, types of damage or plants not covered under warranty, and whether improper care by the client (you) would void the warranty (LCA, 2000).

Plant guarantees or warranties are commonly based on a specified time period after planting (e.g., 1 year or one growing season). The warranty may begin upon the date of inspection and acceptance of the project by you with the understanding that all plants shall be alive and show satisfactory growth at the end of the guarantee period. Alternatively, the warranty may be written more specifically to state that any plant material in which a certain percentage of the crown is dead (e.g., 25%) shall be considered dead and must be replaced at no charge.

Guarantees can easily become sources of bad feelings between contractor and client. This is especially true where responsibility for the maintenance of the plant materials during the warranty period is not clearly stated. Bids may be cheaper in overall price if the client accepts responsibility for the plant maintenance; however, clients are often not familiar with, or are unwilling to provide, proper maintenance.

To avoid problems, provide the necessary information and requirements in the specifications to allow the contractor to make an informed choice whether or not to bid on the project. Remember that extreme requirements and expectations may result in no interested reputable bidders. Conversely, writing loose specifications on the procedures to be followed, guarantee requirements, and acceptable replacement options can lead to poor-quality or wrong plant materials being used for replacements.

5. Conclusion

The success of a tree-planting project is directly tied to your ability to select the best of species and varieties for the areas to be planted, and to clearly specify the quality of the nursery stock that will be considered to be acceptable for the project. The process begins by conducting an analysis of the planting areas and selecting trees that are adapted to those conditions. Space available for the crowns; pH; exposure to high amounts of salt, direct and reflected sun, or high winds; and volume of soil that can be explored by the roots must all be considered when selecting a specific tree for a site. After the trees are selected, the proper nomenclature must be used to assure the best trees for the site are obtained.

To meet the overall project goals, after you have identified suitable tree varieties for the site it is important to understand the nursery production systems used to produce trees. Trees can be bought as balled and burlapped, bareroot, or produced in containers or gro-bags. Understanding the production process, and the installation and maintenance requirements of nursery stock produced in the different ways, will help ensure the success of the planting project. Combined with a clear perspective on what constitutes above and below ground quality of nursery stock, and an understanding of the different production systems will allow you to write clear specifications to assure that the plant materials ordered and installed meet the expectations of the job.

After the tree types have been selected, and specifications for the quality of the nursery stock have been written, you must clearly and precisely specify the requirements of the installation so that bidding contractors can accurately and knowledgeably bid on your project. Your specifications should include defining the deadlines, limitations, exceptions, materials and methods to be employed, methods of communicating, levels of responsibility, and your terms for payment and acceptance of the work carried out by the contractor. The specifications must be concise, precise, and strict in communicating your financial and legal expectations to the bidding contractors.

Selecting the best plants for a site, and writing the specifications that stipulate their acceptable quality, planting procedures, and the guarantees that ensure their survival and growth requires a significant amount of education and experience. When that education and experience is not available with the staff of the municipality, it can be acquired by consulting a landscape architect, horticulturist, urban forester, or reputable landscape contractor. In many cases a lawyer may be needed to prepare a template of specifications that cover the plants and planting practices.

Appendix A. Plant Locator Resources on the Internet

1. Find Plants.Com (membership required): http://www.findplants.com
2. Plant Source Journal Nurserystock eBooks (membership required): http://www.theplantsource.com
3. HortiPlex: http://hortiplex.gardenweb.com/plants
4. Associated Landscape Contractors of America: http://www.alca.org Recently ALCA merged with the Professional Lawn Care Association to create the

Professional Landcare Network: http://www.landcarenetwork.org/, Both sites remain active.

5. Plant & Supply Locator (membership required): http://www.plantlocator.net
6. Pennsylvania Department of Agriculture: Nursery Dealers and Stock Locator: http://www.agriculture.state.pa.us/agriculture/nurseryapp/index.asp
7. Pennsylvania Landscape & Nursery Association: Buyers Guide and Membership Directory: http://www.plna.com/content.asp?/resources/567.html

References

American Nursery and Landscape Association, 2004, *American Standard for Nursery Stock, ANSI Z60.1*, American Nursery and Landscape Association (ANLA), Washington, DC.

American National Standards Institute, 2001, *American National Standard for Trees, Shrubs, and Other Woody Plant Maintenance-Standard Practice (Pruning), ANSI A300*, American National Standards Institute (ANSI), New York, NY.

Bailey, L. H., and Hortorium Staff (eds.), 1976, *Hortus Third: A Concise Dictionary of Plants Cultivated in the United States and Canada,* Macmillan, New York.

Brickell, C. D., Baum, B. R., Hettersheid, W. L. A., Leslie, A. C., McNeill, J., Trehane, P., Vrugtman, F., and Wiersema, J. H. (eds.), 2004, *International Code of Nomenclature for Cultivated Plants,* 7th ed., International Commission for the Nomenclature of Cultivated Plants of the I.U.B.S., Acta Horticulturae No. 647 (International Society of Horticultural Science), Leuven, Belgium.

Carpenter, P. L., and Walker, T. D., 1990, *Plants in the Landscape*, 2nd ed., W.H. Freeman, New York.

Datascape Guide to Commercial Nomenclature, 2003, American Nurserymen, Chicago, IL.

Dirr, M. A., 1998, *Manual of Woody Landscape Plants: Their Identification, Ornamental Characteristics, Culture, Propagation, and Uses,* 5th ed., Stipes, Champaign, IL.

Dirr, M. A., 2003. *The Interactive Manual and Photo-Library of Woody Landscape Plants* (DVD Version), Timber Press, Portland, OR.

Flint, H. L., 1997, *Landscape Plants for Eastern North America*, 2nd ed., Wiley, New York.

Gerhold, H. D., Wandell, W. N., and Lacasse, N. L. (eds.), 2002, *Landscape Tree Factsheets*, AGRS-56, Pennsylvania State University, University Park, PA.

Grey, G. W., 1993, *A Handbook for Tree Board Members*, The National Arbor Day Foundation, Nebraska City, NE.

Griffiths, M, 1994, *Index of Garden Plants: The New Royal Horticultural Society Dictionary,* Timber Press, Portland, OR.

Hillier, H. G., 1973, *Hillier's Manual of Trees and Shrubs* (A. S. Barnes, ed.), 4th ed., South Brunswick, UK.

Horticopia, 2005, *Horticopia Professionals Database (CD)*, Horticopia, Purcellville, VA.

Kelsey, H. P., and Dayton, W. A. (eds.), 1942, *Standardized Plant Names*, 2nd ed., American Joint Committee on Horticultural Nomenclature, J. Horace McFarland Company, Harrisburg, PA.

Krussman, G., 1985, *Manual of Cultivated Conifers* (H. D. Warda, ed.), 2nd revised ed., Timber Press, Portland, OR.

Landscape Contractors Association of MDIDCIVA, 2004, *Landscape Specification Guidelines*, 5th ed., Rockville, MD.

Rideout, R., 2005, *Developing Tree Purchase and Planting Specifications for Bid*, Wisconsin Department of Natural Resources, Urban Forestry Program, Madison, WI, http://www.dnr.state.wi.us/org/land/forestry/uf/resources/Developing%20Planting%20Specs%209911.pdf

Watson, G. W., and Himelick, E. B., 1997, *Principles and Practice of Planting Trees and Shrubs*, International Society of Arboriculture, Champaigne, IL.

Wolff, T., and Womack, J., 1991, *Guide to the Chicago Landscape Ordinance: Regulations and Guidelines Relating to Title 10, Chapter 32 and Title 17, Chapter 194A of the Chicago Municipal Code*, City of Chicago.

Chapter 13

Planting and Maintenance

James W. Consolloy

1. Introduction

Trees planted in the urban street setting are constantly fighting for their survival. Trees need sunlight, oxygen, water, and soil volume for nutrients and stability. As our urban population centers grow, technology and planning are needed to help establish trees in this environment. As new planting materials are developed and transplanting techniques are modified, it has become necessary to change our method of planting to meet the urban forests' challenge. Using the Princeton area as a model for planting and maintenance specifications in the Northeast will show both the seasonal considerations and problems involved in keeping a healthy and sustainable tree population.

Starting out with a management plan that fits into the community's budget (Chapter 8) is the first step needed to ensure a healthy tree community. The budget should cover costs for labor, materials, equipment, and maintenance needs. Today, newly developing communities that are still growing can most easily adapt to the latest specifications suggested for tree spacing, as well as introducing new mixes of trees. Older communities often do not have the luxury of providing an adequate growing space for shade trees and therefore are forced to crowd street trees between the curbs and sidewalks along the public right-of-way. This chapter also will consider both extremes and unusual situations faced by planners and municipal shade tree organizations in plant selection, shipping and unloading, planting techniques, transplanting, and early maintenance practices.

James W. Consolloy Department of Grounds and Building Maintenance, Princeton University, Princeton, New Jersey 08544-5264.
Urban and Community Forestry in the Northeast, 2nd ed., edited by, J. E. Kuser.

2. Site Selection

A study showing the planting locations is the first step in planning the plant material list. Considerations should be given to various environmental factors that will impact the first 2 to 3 years of plant growth and establishment. Exposure to various environmental factors, such as the temperature, light, water, soil conditions, air quality, and biological and physical influences, should be assessed at each location. Urban areas are faced with many man-made influences as well. The following details some of the natural and man-made factors that in many cases have a direct correlation with a plant's sustainability:

1. Natural factors
    - Temperature (hardiness zone)
    - Water (flooding, annual rainfall)
    - Light (shade, partial, and full sun)
    - Air (percent 02, wind, and salt spray)
    - Soil (pH, percent sand, silt, clay, and organics)
    - Biologicals (native pests, diseases)
    - Physical (wind, lightning, and snow/ice)
2. Urban factors
    - Building vents, utilities
    - Drainage systems, irrigation
    - Reflection, streetlights, signs, and overhead wires
    - Exhaust emissions, blowers
    - Gas leaks, road salt, and oil spills
    - Heating, ventilation, and air conditioning
    - Imported pests on plants
    - Accidental scars, smoke/fire

2.1. Temperature and Hardiness

The hardiness rating according to zones (Chapter 11, this volume) has been generally accepted as one standard measure for a plant's ability to adapt to its environment. The temperatures apply mainly to the hardiness of the portion of the plant aboveground. The root systems are not as hardy as the aboveground portion and therefore should be protected during the winter seasons (Perry, 1989). Trees with fibrous root structures are especially susceptible to freeze-drying during the late fall and winter. When handling trees in the bare-rooted condition, it is very important to insulate the roots while transporting, and storing, during the planting process to ensure that the roots do not desiccate or the tissues freeze-dry.

Excessive heat from building ventilation systems will create artificial microenvironments having both negative and positive effects on plant growth. Aboveground blowers will create adverse drying conditions and eventually kill the plant, as will underground heat pipes; however, under some conditions plants that normally grow in more southern climates will survive close to these heat sources in the northern temperature zones.

2.2. Water as a Factor

Annual rainfall is certainly a consideration in plant selection. Drought tolerance, which is related to the plant's transpiration rate and the plant's ability to survive low-soil oxygen levels due to wet conditions, plays a key role during the first few years of establishment. See Chapter 11, this volume, for lists of drought-tolerant species and plants that will tolerate wet soil conditions (Flint, 1983).

In the urban environment, the rapid collection of rainfall through storm water drainage systems should be a prime consideration for the long-term survival of city trees. Likewise, the new installation of lawn basins to collect rainwater runoff may adversely affect existing trees dependent on that water. There also should be a careful design of impervious surfaces to allow large quantities of water to flow away from the tree pit (Whitcomb, 1987).

2.3. Light Considerations

Trees will dramatically respond to the presence or absence of light. This is a primary factor in deciding the species of tree to be planted. The successful growth of the tree will depend on the amount of natural light available. For example, American beech and sugar maple are climax forest trees that can live in dense shade under other trees and can be used in underplant aging street trees that are destined for replacement within the next 10 years. Less shade tolerant species, such as oak, hickory, and sweet gum, will not grow in a straight, upright fashion, but will lean toward the available sunlight, thus creating future problems for arborists to correct.

Night lighting, creating long days, can delay dormancy and result in continued elongation of vegetative shoots. *Betula, Catalpa, Platanus*, and *Tilia* were found to be the most sensitive to all sources of light. It was found in *Rhus, Koelreuteria*, and *Zelkova* that of the artificial light sources from most to least effective are incandescent, high-pressure sodium, metal halide and cool-white fluorescent and clear mercury vapor lighting (Harris, 1983). Reflected light from mirrored surfaces will scorch leaf tissue and result in distorted growth if the planting is too close to the source.

2.4. Air Quality

The natural effects of wind, salt spray, and wind-driven hail have a stunting and damaging effect on the growth habit of most tree species. Tree species with thicker leaves and bark and compact growth habits will survive under some of these conditions. Low-oxygen levels either as a result of higher elevations or the presence of higher levels of other gases will produce stunted growth or reduced chlorophyll production. The loss of the chlorophyll pigment will be one of the first signs of low-oxygen levels. Natural gas pipeline leaks and trapped sewer gas from unvented manholes will result in soil oxygen displacement.

Carbon monoxide exhaust from motor vehicles has chronic effects on vegetative growth. Avoid planting street trees too close to busy intersections. Exhaust from trucks that are forced to stop for traffic lights scorches and bums leaves and stems, with the inevitable loss of trees in that planting zone. Exhaust fans and large ventilation equipment, as well as steam and chilled water exhaust will have negative effects, the first signs of which are infestations of plant pests and diseases (Johnson and Lyon, 1979; Sinclair *et al.*, 1989).

2.5. Soil Conditions

Preliminary soil tests prior to planting will eliminate future developmental problems with new plantings. Soil compaction, pH, percentage of sand, silt, clay, drainage, and the overall depth of the soil are important. Areas disturbed by construction are prime examples of overcompacted soils that need to be excavated and amended for increased porosity (see Chapter 10, this volume).

Topsoil for new plantings is frequently purchased from developers who are converting previously farmed land to new homes. These untested soils could contain root-feeding nematodes and residual herbicides.

When replacing street trees, if the cause of death is unknown, soil testing for gas leaks or road salt buildup and other contaminants could eliminate having to replace the tree for a second time. Soil volume, or the lack of, is a leading cause of street tree decline. There are design techniques for increasing soil volume under sidewalks and plazas using air-entrained soil mixes (Arnold, 1993).

2.6. Biological Factors

Planting trees in a new environment can present potential conflicts in terms of survival. For example, introducing European little-leaf linden into an area with high concentrations of Japanese beetles will add additional maintenance on those trees. Trees purchased from certified nurseries and planting resistant trees for the location will eliminate many future maintenance problems. Also, knowing what diseases are spreading into the area will help decide the plant selection.

2.7. Physiological and Morphological Factors

Tree location often determines future physical and structural problems. Will bicycles be locked to the trunks leaving scars? As the tree develops, will large trucks interfere with the branching? Will overhead wires result in severe pruning? Will snow and ice slide from rooftops onto the newly planted trees? Will low-branched trees be sandblasted during the winter deicing of roads and snow piled and pushed against the plantings? A preliminary study of the area, its usage, and the future impact on maintenance is very helpful (see Chapter 11, this volume).

3. Site-Specific Factors

The size of the planting space ultimately determines the tree's size and its ability to grow to maturity. Research has shown that there is a direct correlation between soil volume and crown size and sustainability (see Chapter 10, this volume). The depth of soil is not as critical as the total volume of the root zone (Bassuk and Trowbridge, 1989).

The crown size and growth habit must be considered when planting under utilities, close to building structures adjacent to roads and walks, and in competition with other vegetation. For example, planting trees with a multileader growth habit will allow for little future interference by overhead wires and minimal top pruning. Special purpose trees, widths of trees, and lists of tree species can be found in Chapter 11, this volume.

Longevity of a species is a frequently asked question of arborists and nurserymen. There are documented street tree plantings well over 200 years old (i.e., Stamp Act Sycamores, ca. 1768, planted along Nassau Street, Princeton, New Jersey), while others, planted in New York City 7 years ago, cease to live. The problem of surviving in harsh city conditions, regardless of a tree's genetic makeup, has been the subject of ongoing investigation. We know that certain popular, short-lived street trees, like the Callery pear and Norway maple, are still planted in large numbers despite the vast menu of other species. There also is a strong correlation between longevity and planting methods (Arnold, 1993).

The starting size for a street tree depends largely on the municipality's budget, and the total number of trees that can be planted with the resources available. Other factors to consider are survivability of certain tree species and planting times. The cost to plant a 2-inch caliper bare-root street tree can be as little as 40% of the cost to plant the same size balled and burlapped. This is partly due to the added labor costs for digging, handling, and shipping.

There are customized trees available in containers that work quite well if the planting job is delayed. From a maintenance point of view, containerized plant material is easier to handle and keep watered and can be planted at any time of the year. Bare-root trees must be installed without delay in order to ensure their survival.

4. Transportation, Care, and Handling

Care and handling of trees in shipment should follow the same guidelines as that of preserving perishable goods. Live plants need to be protected from desiccation and wind damage while being transported on open bed trucks. Branches of balled and burlapped trees should be carefully tied together before loading and bundles of bare-root trees wrapped prior to storage and shipment. Balled and burlapped trees are often stacked on flatbed trucks and trailers and then tied down and tarped. This practice has many advantages, including that of preventing desiccation of plant tissue and wind damage to buds, leaves, and other structures. Tree trunks should be padded when in direct contact with other hard surfaces, such as tree balls, headboards, and other trunks. Even the tie-down ropes must be padded to prevent any girdling effect on bark tissue.

Trees shipped in closed trailers are normally refrigerated and kept cool, dark, and moist. They will stay dormant under these conditions. However, if left under these conditions for too long, as with cross-country shipments, there may be signs of fungus growth on the plants, particularly with branches and evergreen shrubs that are tightly packed together.

4.1. Inspection

Upon arrival, plants should be inspected for broken and bruised roots, trunks and limbs. Balled and burlapped plants especially should be inspected for signs of rough handling, such as exposed sapwood from bruising, rope burns on the bark, damaged soil balls, exposed roots, and soil loss from torn burlap. All plants must be unloaded immediately. The plants will dry out quickly and the tied up portions of the trees will start to heat up if not untied. At this time, any broken branches can be removed. Reports of damaged plants must be filed with the shipping company and the seller as soon as the plants are unloaded. Pictures of the damage will be very helpful in recovering your losses.

If the trees are not ready to be planted, they should be spaced apart as they were at the nursery and kept in the upright position and watered frequently. When they are left lying on their sides, phototropism will take over and the terminal shoots will start bending toward the light. The terminal shoots of evergreens will start growing in this curved mode as soon as they are left in a prone position. When trees are planted with the terminal shoots crooked, it may take several growing seasons to correct this deformity. The other solution would be to prune the tops back to a lower terminally growing bud. The risk is that these buds may not show apical dominance. This is another advantage of planting trees that are dormant and not actively growing.

Nurserymen and landscapers in the past would "puddle" the roots of bare-rooted plants by dipping them in a slurry of mud and "heeling" the roots into sawdust or mulch hay while they were waiting to be planted. Today, nurseries have cold storage facilities and holding areas to protect the plants from extreme temperatures. This process helps to ensure the freshness of the trees being shipped.

Before planting bare-rooted trees, public works and landscape crews can use water gels that are available in most horticultural supply houses to protect the roots from desiccation. These water gel granules or powders, when added to water, swell to many times their size. They will adhere to the root fibers when the tree roots are dipped into a tub or trough of the solution. The gel will remain with the roots when planted allowing newly developing roots to absorb water from this source. The water gel granules can also be added to the backfill mix when planting or transplanting balled and burlapped and containerized trees.

5. When and How to Plant

Tree planting is most efficient when soils are "workable," neither too wet nor too frozen. In early spring when the frost disappears and the soil temperatures start rising above 40xF, the soil will begin to dry. Conditions are optimum for planting when the soil falls off the shovel or spade as it is backfilled into the tree hole. Likewise, in the late summer and early fall before the soil temperatures begin to decline, planting will be optimum. Trees will establish new roots more quickly when the soil temperatures are higher. This is an important aspect of fall planting, especially when planting evergreens. The sooner a tree sets roots in the fall, the less likely that it will be blown over by the winter winds.

When planting during the dry fall months, watering is critical so as to assure that the roots do not freeze-dry during the winter. Soils should be "settled in" around the rootball of a balled and burlapped tree and more thoroughly around the roots of a bare-root tree. This eliminates any voids and air pockets that may form. These voids often harbor insects and rodents that will feed on root tissue and nest during the winter months.

5.1. Soils: Native versus Amended

The probability of finding undisturbed native soils in urban planting locations is low, and therefore excavated materials should be examined, tested if necessary, and removed. Ideally, any native topsoils should be saved and stockpiled for use in contaminated sites for just such occasions.

Native soils contain important soil micronutrients and organisms necessary for good root development. Structurally, these soils should support newly planted trees and evergreens with compaction rates sufficient to stabilize the plant and keep it from slipping or leaning during periods of high winds. Higher density soils, that is, with more silt and clay particles, will "firm up" and keep plants from moving.

Native soils also contain detrimental organisms such as *Phytophthora* and root-feeding nematodes. Soil samples should be bioassayed prior to use. The enzyme-linked immunosorbent assay is one such test for the presence of harmful soil organisms (Smiley, 1994).

Artificial soils, if amended with the necessary nutrients and microorganisms, are quite effective in establishing new root systems. Because of their low density; however, they generally have a lower compaction rate and cannot give the same long-term structure as native soils that have higher proportions of silt and clay.

Trees that were planted in artificial soil mixes will settle in the tree pit as the organic matter in the mix starts to decompose. This will leave the basal root flare below grade, and in some cases the trees will be sitting in a depression. If this depression is filled back up to grade, it is very important to keep the fill material, in some cases mulch, away from the trunks where wood decaying organisms can destroy the root collar tissue (Fig. 1).

5.2. Planting Hole

One of the most important steps in tree planting is preparing the tree hole or tree pit. The hole size is based on the size of the root ball (balled and burlapped) or the root mass (bare root). The diameter of the hole should be two times (2×) the size of the rootball for a small to medium sized street tree and three times (3×) for larger and taller trees such as ash, elm, oak, locust, maple, and sycamore.

The depth of the hole is determined by the height of the soil ball (balled and burlapped) or the height of the root mass (bare root). Check the soil ball before planting and locate the top of the root mass by excavating some of the soil around the trunk. Follow the root flare from the trunk to determine whether the tree was transplanted with enough of the root system. Nursery-grown trees that are mechanically dug should be examined closely (Fig. 2). Keeping the top of the root mass slightly

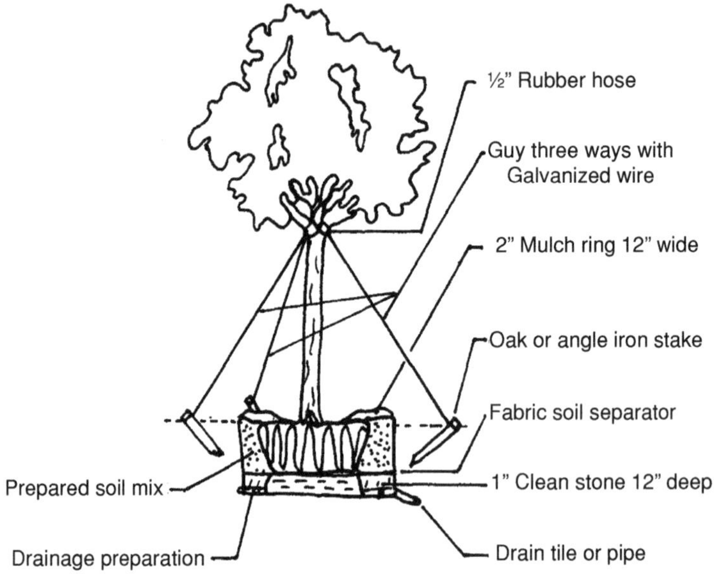

FIGURE 1. Guying and mulching detail.

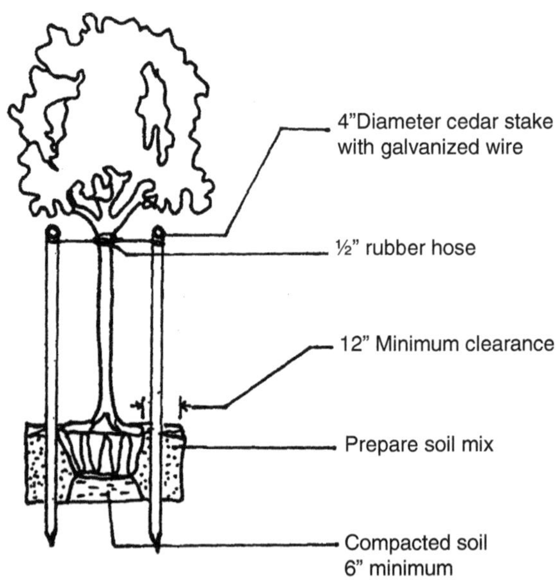

FIGURE 2. Planting and Staking detail.

above grade will ensure that the roots will have the right oxygen levels needed and that the bark tissue and the base of the trunk will not decay.

5.3. Pruning

There are two widely practiced methods of pruning young plant material following transplanting. The first method removes only dead or broken branches and, in the case of bare-rooted trees, torn roots and any root decay. The theory behind this method is to retain as much root and branch structure as possible and therefore the stored energy reserves.

The second method requires more precision pruning. Not only are the broken, decayed, and dead portions of the tree removed but the branching is also cut back, with several buds and some inner branching removed. The theory behind this method is to reduce the amount of vegetative growth that cannot be supported by the loss of root mass following transplanting. This is also the principle behind the cultural practices of the bonsai in containers and pollarding in some European cities to offset low-soil volume in confined urban settings.

It is an especially important practice to make clean, sharp cuts when pruning branches and roots. This will ensure proper tissue callusing, thus preventing disease formation at the wound entry sites. New roots form quickly from sharp cuts versus jagged, torn root structures. Arborists also recommend sanitation of the pruning equipment, including chain saws and power scythes to prevent the spread of bacterial, viral, and fungal diseases.

5.4. Fertilizing

Fertilization of newly transplanted trees is best carried out during the planting process. The incorporation of decomposed organic matter in the backfill mix still should be a part of the planting process. The addition of mycorrhizal spores in the soil mix adjacent to the root ball will "jump start" the rooting process to ensure increased root regeneration at the start. The spores can be applied either in a solution sprayed or poured on the outside of the root ball or using biodegradable packets placed in the backfill mix.

Granular fertilizers, such as 10–6–4 (50% organic nitrogen) or 20–12–8 (50% organic nitrogen), have been dependable formulations. The latter, higher formulation is used at half the rate, which amounts to 1/2 pound per 1-inch diameter of trunk size. The decreased volume of fertilizer can be labor saving when you have hundreds of trees to fertilize.

Water-absorbent gels incorporated in the backfill mix are an added assurance during drought periods when trees cannot be watered on a regular basis. As the gel crystals absorb water and expand, the increased water-holding capacity provides sites for newer roots to feed. These sites continue to trap water as it passes through the soil.

Granular wetting agents also ensure water penetration to the root zone and will decrease the initial surface runoff. If these wetting agents are not present in the upper layers of soil, adding a soluble wetting agent to water supplied from tank trucks will accomplish the same result.

Soluble tree fertilizers can be applied at any time provided that the nitrogen source is nearly 100% organic. These slow-release nitrogen formulations can be applied to newly planted trees in the fall without having phytotoxic effects or dramatically increasing the vegetative growth, which could delay the onset of dormancy. They are essential for continued root development that should continue throughout the dormant period.

The negative effects of increased vegetative growth during the late fall are manifest in continued stem growth, leaf retention, cambial growth, and consequently water movement in the young trunks. This can result in frost cracking of the basal trunks and branch breakage from early wet snowfall before leaf drop. Frost cracking will occur when the daytime temperature drops 40x at the bark surface within a few hours, before the water within the vessels can flow back to the root zone. These cracks will appear on trees with thin bark such as sycamores, maples, and fruit trees.

Recent evidence suggests that some soil amendments have had little effect on additional growth to the crowns of newly planted trees when compared to non-amended back-fill mixes (Gilman, 2004).

5.5. Bare-root Planting

The care and inspection of bare-root trees up until the time they are planted is critical to their survival. One distinct advantage in specifying bare-rooted trees is seeing the complete root structure before the tree is planted. Another advantage in using these plants is the ability to make sharp cuts at the ends of the major roots, increasing root regeneration at these sites. This should be the case with balled and burlapped trees if the digging crews used a sharp spade or loppers during the balling process.

Once the planting hole is prepared (Fig. 2), the tree can be positioned properly in the hole. Another advantage is being able to see the root flare and adjust the height of the root collar making it slightly visible above the soil line.

The condition of the backfill mix is best when "friable," meaning that it is dry enough to flow in and around the roots, displacing all voids and potential air pockets. Using water afterward will help to accomplish this important step in the planting process. Inspections of unsuccessful bare-root plantings reveal damaged roots when these air pockets are not eliminated. The finer roots and root hairs often are dried up or rodents have found shelter and food among the roots during the winter months. The bark of fruitwoods, maples, and some needled and broad-leafed evergreens is very attractive to rodents and rabbits as the winter food. Again, roots left to freeze-dry over the winter will result in crown dieback, which will appear in the late spring and the early summer.

5.6. Balled and Burlapped Planting

Trees dug with a soil ball, either by hand or mechanically, will have a higher survival rate than bare-root trees. The added soil will protect the root structure during the transplanting process. Balled and burlapped trees can be successfully maintained for longer periods aboveground before planting. The soil ball adds stability to the tree and retains water to ensure against root tissue desiccation.

Either hand- or machine-dug trees with calipers of 2 to 32 inches are the easily handled sizes for transplanting and have shown satisfactory survival rates (Kressel and Burck, 1995). They also are much easier to handle and maintain at those sizes than larger ones. Balled and burlapped trees in the 2 to 3 1-inch caliper range will establish new root systems, resume vigorous growth faster, and can reach an environmentally beneficial size faster than larger trees (4 to 12-inch caliper) planted at the same time. The time required for trees to regain pretransplanting growth is approximately 1 year for each 1-inch caliper (Watson, 1996). Young trees will need less post-transplant support (i.e., straightening, staking, and guying).

Natural untreated burlap, sisal, and manila rope on hand-dug soil balls are decomposed within weeks by soil microbes, allowing easy root penetration into the surrounding soils. Machine-dug trees are placed in a burlap liner inside a wire metal basket. At the time of planting, the wire basket should be removed or cut away once the tree is in position. In all cases, the burlap and rope should be removed or folded down away from the top of the soil ball. If left in place, the burlap will interfere with proper water uptake and can retard basal root development. Leaving the top lacing of rope around the trunk will result in girdling and the eventual death of the tree (McNeil *et al.*, 1982).

As the burlap is removed from the top of the soil ball, inspect the root crown or top of the root mass to make sure it is visible. If the root crown is not visible, the tree was planted too deep in the nursery and therefore not dug to a proper depth. Most likely, many of the bottom feeder roots were left at the nursery and not recovered with the soil ball. This is a leading cause of crown dieback and goes unnoticed by many plantsmen.

Balled and burlapped trees are best handled by the root ball and not the trunks. Trunk bruising is a common cause of wounding and bark dieback, leaving entry points for disease and insect attack. The use of a hydraulic ball grabber or a set of steel forks will safely handle the smaller-size trees, as well as help position the tree within the planting hole.

Trees that are skinned or bruised through rough handling will show signs of bark dieback within a few months after planting. At that point, a typical maintenance procedure by arborists called *tracing* will allow for faster wound closure and callusing of the cambial tissue. This is normally done with a sharp utility knife, cutting or tracing around the wound just into the live tissue, and then removing the dead bark from the sapwood layer (Conover, 1977).

5.7. Container Planting

A recent method of growing trees in containers is referred to in the nursery industry as the "pot-in-pot" method. The only difference from the original method of container growing is the placement of the containerized plants into other containers that are set in the ground in rows and connected to the proper feeding and drainage systems. The plants are easily removed from the stationary pot and fitted with a new one when sold. The advantages of this method of production are controlled growth of roots and stems, stable upright growth, and more importantly ready availability of the plant at any time of the year. Not all tree species grow well in a container environment, so there are some drawbacks to this method of tree production. Transplant shock is lessened when using containerized plants because the root and shoot growths are relatively balanced (Flemer, 1972). However, they are irrigated on a regular basis and the new shoot growth can be very soft and subject to wilting when moved from the containers. In such cases, it is necessary to remove some of the soft tissue growth and maintain a regular watering schedule until the root system adapts to the new soil conditions.

Container plants have been grown in artificial soils which are less dense and will drain readily. Therefore, it becomes necessary to dig the planting hole twice as wide and at least 6 inches deeper than the container soil ball. Break up the sides of the hole to allow the roots to penetrate the native soils more easily. The soil under the plant should be tamped to prevent settling or shifting over time.

When the plant is removed from the container, examine the roots carefully. If they formed a solid mass around the inside of the container, trim approximately 2 inches off the roots and break apart the root mass evenly. If the root mass is minimal, make four or more even slices 1 inch deep around the root ball before planting. Another very important procedure is to break down the "shoulder" of the root ball at the top to allow water penetration into the top of the ball.

Use a light backfill mix of about 25% peat moss or leaf compost to 75% soil. Never plant containerized plants in waterlogged soils because the container soil mix will act as a sink and wick the surrounding water. Look for a better-drained site.

5.8. Large Tree Transplanting

Moving larger trees (4- to 12-inch caliper) is accomplished efficiently using tractor- or truck-mounted mechanical tree spades. The larger spades typically can move up to an 8-foot soil ball with little effort. One of the benefits of this method is minimal root, trunk, and site disturbance, since the tree is only handled one time. This can be a large savings in terms of time, labor, and equipment. Two people can complete the entire operation with only one piece of equipment to dig, transport, and plant the large trees.

Some of the disadvantages are digging in shale- and rock-littered soils, crushing and tearing of large framework roots when caught between spades, and planting the spaded root ball into a predug hole where a "plug" of soil was previously removed. In heavier clay soils, these predug holes have glazed and compacted sides that can form barriers to future root penetration and proper development.

To ensure a better planting environment in poor soils, use a backhoe to predig a hole larger than the maximum size of the tree spade and backfill it with an improved soil mix. Compact the new mix in 2-foot intervals as it is added using the back of the hoe bucket until the hole is filled 6 to 12 inches above existing grade. Allow the site to settle and add water if needed. After settling has occurred, remove the soil "plug" to match the size of the root ball being transplanted.

At the time the new tree is being set into the hole, the addition of root-building material applied to the walls of the hole is very beneficial. There are many new root-building supplements available from horticultural suppliers: water gels, wetting agents, ectomycorrhizal fungal spore solutions, liquid chelated micronutrient compounds, and humic acids. These products can be used alone or mixed with a soluble root fertilizer and sprayed or injected into the root zone area. All these products will help reduce transplant shock and promote faster root growth and establishment. Most landscape contractors and tree-moving companies include these products in their transplant costs.

6. Staking and Guying

All newly transplanted trees should be staked or guyed for a minimum of 1 year or until the tree has developed its own support roots. Maintenance of the stakes and guy wires is minimal if properly installed. Shade trees with calipers up to 3 inches planted either bare root or balled and burlapped will require two stakes. In sandy soils and windy sites such as shorelines the trees will require three stakes. An 8-foot stake is minimum length for shade trees and is driven into the soil 2 feet. For smaller flowering trees and small bare-root shade trees less than 2 inches in caliper, a 6-foot stake driven into the soil 2 feet will suffice.

Trees larger than 3 inches in caliper need a minimum of three guy wires. Lighter, 16- to 14-gauge galvanized steel wire will support trees in the 3- to 4-inch caliper range. Use of 12-gauge galvanized steel wire for trees with excess wind resistance or height is recommended for support and is cost-effective. Larger-caliper trees with large crowns will require more elaborate guying techniques. Additional guying wires may be required and use of 1/8- to 9/16-inch aluminum aircraft cable with 3/8- to 9/16-inch galvanized steel turn-buckles for tensioning, again depending on the site conditions, will support trees up to 12 inches in caliper. The use of "deadmen" instead of stakes will be needed to support these heavier structures (Fig. 2).

Maintenance of the guys over several years is necessary because of the slower reestablishment rate of larger trees. To assure the trees' stability, use the turnbuckles to determine whether the tree can stand on its own. Lessening the tension on all guy wires allowing for minimal slack on each one and watching for any shift in the tree's position will determine how much longer the guying is needed. If any of the wires or cables start to tighten on their own, then it is too soon to remove them.

Proper trunk protection is also needed at points along the trunks where wires and cables are attached. Materials range from soft rubber hose, polypropylene chain, and sling material made from plastics to burlap-wrapped blocks of wood tied around large trunks for cable protection. Regular inspection during the growing season will prevent any girdling from taking place. Retensioning of the wires is critical at this time. Girdling occurs more readily in soft wooded trees and conifers than it does in oaks, ash, and maples.

Trunk protection using paper or burlap tree wrap will help prevent insect damage and vandalism. The wrap is applied in a spiral overlapping pattern and tied with plastic garden tie. There are plastic spiral basal trunk wraps available that are useful in areas of high rodent or rabbit populations. All wraps should be regularly checked for any signs of girdling; when they are removed, any insect eggs or nests under the wrapping should be identified and eliminated at that time.

7. Mulching

The single-most important benefit of mulching is water retention; along with weed retardation it is a valuable step in tree planting. A 2-inch layer of mulch will save on additional irrigation and will trap needed rainfall before runoff can occur.

Overuse of mulch has negative effects on plant growth and establishment. Large quantities of mulch piled up around the base of the trunk will harbor wood-decaying organisms that damage bark tissue and upper root collar tissue and can eventually cause plant dieback. Mulches should be properly aged before using in the landscape. New mulch will render nutrients (specifically nitrogen) unavailable for the plant. Excess mulch also will invite unwanted rodents during the winter periods. They will feed on the bark and cambial tissue at the base of the trunk, totally girdling the tree below the mulch line.

Therefore, it is recommended that the mulch does not come in contact with the trunk of the tree. As the tree matures, it is a healthy sign when the upper root flare of the tree is exposed to the air. Mulch rings can be extended but should not cover the buttress roots (Sinclair *et al.*, 1989).

8. Watering

Adding water to the backfill mix as it is added into the planting hole is extremely important and will ensure sufficient contact between the root ball and the adjacent soils. It will settle the soil around the root ball, helping to eliminate any voids or crevices created during the backfilling process. Normal rainfall in the Northeast will not suffice during the summer months, and supplemental watering is critical during the first year of establishment.

Using a wetting agent in the tank mixture and adding granular water gels to the backfill mix will add to the efficiency of the watering process. The wetting agents help reduce surface tension on water droplets so that they can penetrate hydrophobic conditions on the soil surface and the burlap. The granular water gels will swell up when they come in contact with water in the soil and act as water-holding sites for the new feeding roots. Other products on the market not only will enhance soil penetration, but also add to the efficacy of biological chemicals that will penetrate the root tissue and aid in absorption of systemic pesticides.

Water bags and plastic rings that hold and release water slowly have proved to be a very effective means of watering newly planted trees. These structures can be used during drought periods and then stored for future use.

Watering is most effective if done either early in the morning or late in the day to prevent wilting. The evaporation rate is minimal during these cooler periods. Many municipalities in urban areas have water restrictions during the summer months but allow some watering at night. Most will allow tree watering by tank if bags and mulch rings are in place.

Adding nutrients and fertilizers to the water tank mix in the late summer or early fall will save an additional maintenance step. All plants should be thoroughly watered before the ground freezes. Many young trees die in their first year because of the winter root losses from desiccation.

Determining how much to water depends entirely on the available water-holding capacity of the soil within the root zone and the tree's water needs. Soils with good tilth that are friable (crumbly) and well structured have the best water-holding capacities and infiltration (Kelsey, 1996). Strategically placed rain gauges and soil sampling probes will help assess how much additional watering is needed during periods of drought.

9. Tree Care and Maintenance

Trees require four-season maintenance and inspection to ensure survivability. A typical tree-preventive maintenance schedule will involve a program dedicated to total plant health care. The size and nature of the municipality will determine how to structure the plant health care program. For small towns, volunteer groups can adopt individual streets, assist municipal works departments with the summer watering, and watch for vandalism. As the areas increase in size, the municipality can subcontract some of the pruning and watering, especially if the trees were planted by subcontractors (see Chapter 19, this volume). Large cities will need to staff full-time arborists and landscape crews to look after their larger investment of shade trees. And all municipalities need to share common problems with neighboring municipalities at annual regional shade tree meetings.

The first 3 years are critical for the establishment of newly planted trees. A plant health care program should include all newly planted trees each year as they are recorded in the inventory. The plant health care program should include integrated pest and plant management monitoring (see Chapter 18, this volume) by a trained individual who will be responsible for the reporting and treatment of all tree and environmental problems. Most tree maintenance companies offer a full plant health care program and the services of a monitor or field scout who can report to the municipality's public works director or department superintendent. The monitoring period should extend from early April through late November. During this period reports should be filed weekly and recommendations for treatments discussed on a regular basis.

The landscape crew should take full responsibility for the maintenance of all trees newly planted over the past 5 years. After that, they should be included in the scope of work for the arborist crew. During the initial 5-year period the landscape crew's scope of work should include: tree straightening; checking stakes, guys, and wraps; weed control; pruning; fertilizing; remulching; watering and insect and disease control; and if necessary tree replacement. Insect and disease control is normally coordinated with the integrated pest management monitor and the arborist crew to minimize the use of pesticides.

References

Arnold, H. F., 1993, Sustainable trees for sustainable cities, *Arnoldia* **53:**3.

Bassuk, N., and Trowbridge, P., 1989, Urban islands, *Landscape Architec.* **79:**130–131.

Clark, and Rapauno Associates, 1965, Planting Specification for Princeton University, not published.

Conover, H. S., 1977, *The Grounds Maintenance Handbook*, New York: McGraw Hill, p. 108.

Flemer, W. III, 1972, Growing street trees in containers, *Am. Nurseryman,* 136:75–83.

Flint, H. L., 1983, *Landscape Plants for North America*, Wiley-Interscience, Purdue University, IN.

Harris, R. W., 1983, *Arboriculture: Care of Trees, Shrubs, and Vines in the Landscape*, Prentice-Hall, Englewood Cliffs, NJ.

Johnson, W. T., and Lyon, H. H., 1979, *Insects That Feed on Trees and Shrubs*, Cornell University Press, Ithaca, NY.

Kelsey, P., 1996, Diagnosing nutrient deficiencies in hardwood trees, *Grounds Maintenance* **31**(6):28–33.

Kressel, S., and Burck, S. (eds.), 1995, From the Proceedings from *ASIA Urban Street Tree Planting Symposium,* Boston Society Landscape Architects, Boston, pp. 17–31.

MacDonald, J. D., Costello, L. R., Lichter, J. M., and Quickert, D., 2004, Fill soil effects on soil aeration and tree growth, *J. Arboric.* **30**(1):19–27.

McNeil, R. E., Fountain, W. M., and Witt, N. L., 1982, Tree decline in the landscape when the trees were dug BNB, in *Proceedings of the Urban Trees and Forest: Pest Management Problems, Needs and Prospects*, (unpublished).

Perry, T. O., 1989, Tree roots: Facts and fallacies, *Arnoldia* **49**(4):3–29.

Sinclair, W. A., Lyon, H. H., and Johnson, W. T., 1989, *Diseases of Trees and Shrubs*, Cornell University Press, Ithaca, NY.

Smiley, E. T., 1994, *Plant Diagnostic Report–Princeton University*, Bartlett Tree Research Laboratories, 13768 Hamilton Road, Charlotte, NC 28278.

Watson, G. W., 1996, Tree transplanting and establishment, *Arnoldia* **56**(4):11–16.

Whitcomb, C. E., 1987, *Establishment and Maintenance of Landscape Plants*, Lacebark, Stillwater, OK, pp. 251–262.

Other Suggested Reading

Birchell, R. S., 1996, Saving trees—Conservative construction, *Grounds Maintenance* **31**(5):48–54.

Cranshaw, W., 1997, The pros and cons of biological control, *Grounds Maintenance* **32**(3):44–48.

Cranshaw, W., 1997, The pros and cons of biological insecticides, Part II: Microbials, extracts and IGRs, *Grounds Maintenance* **32**(4):45–56.

Fraedrich, B. R., 1995, *New Concepts in Tree Planting*, Bartlett Tree Research Laboratories Bulletin, Charlotte, NC.

Fraedrich, B. R., 1995, Solutions to tree—sidewalk conflicts, *Tree Sci.* (Winter) **1995:**3–6.

Gilman, E. F., 2004, Effect of amendments, soil additives, and irrigation on tree survival and growth, *J. Arboric.* **30**(5):301–310.

Grabosky, J., and Gilman, E., 2004, Measurement and prediction of tree growth reduction from tree planting space design in established parking lots, *J. Arboric.* **30**(3):154–164.

Herms, D. A., 1996, Growing growing gone, *Am. Nurseryman* **184**(7):52–57.

Hummel, Jr., N. W., 1995, Assess soil conditions on site, *Grounds Maintenance* **30**(6):22–28.

Percival, G. C., 2004, Evaluation of physiological tests as preditors of your tree establishment and growth, *J. Arboric.* **30**(2):80–91.

Pill, W. G., 1997, Filling the voids, *Am. Nurseryman* **185**(12):30–37.

Roger, M., 1994, Trees need fertilizer, too, *Grounds Maintenance* **29**(3):74–78, 95.

Ruter, J. M., 1996, Regulating growth of woody ornamentals, *Grounds Maintenance* **31**(5):42–46.

Shigo, A., 1988, *A New Tree Biology: Facts, Photos, and Philosophies on Trees and their Problems and Proper Care*, Shigo and Trees, Associates, Durham, NH.

Shigo, A., 1991, *Modern Arboriculture: A System Approach to the Care of Trees and their Associates*, Shigo and Trees, Associates, Durham, NH.

Shoup, S., Reavis, R., & Whitcomb, C. E., 1981, Effects of pruning and fertilizers on establishment of bare-root deciduous trees, J. *Arboric.* **7**(6):155–157.

Smiley, T. E., and Shirazi, A. M., 2003, Fall fertilization and cold hardiness in landscape trees, *J. Arboric.* **29**(6):342–346.

Smiley, T. E., and Shirazi, A. M., 2003, Fall fertilization and cold hardiness in landscape trees, *J. Arboric.* **29**(6):342–346.

Tubesing, C., 1997, Another look at first rate plants, *Am. Nurseryman* **186**(1):62.

Urban, J., 1989, New techniques in urban tree planting, J. *Arboric.* **15**(11):281–284.

Weddle, A. E. (ed.), 1983, *Landscape Techniques*, Van Nostrand Reinhold Company, New York, pp. 147–157, 183–201.

Weiner, M. A., 1992, *Plant a Tree*, Wiley, New York, pp. 53–74.

Chapter 14

Pruning

David L. Johnson

1. Introduction

Pruning is the removal of plant parts to produce a desired effect. Pruning is both an art and a science, allowing the practitioner to add their individual touch while following basic rules of tree physiology. If done correctly, pruning can provide many benefits to the tree including increased flower and fruit production, stronger branch structure, and reduced disease potential, as well as create a more aesthetic appeal to people. If done incorrectly, pruning can set in motion a series of responses that will affect the health or structural stability of the tree over its lifetime. The importance of understanding how the tree will react to pruning is critical before removing even a single branch.

2. Trees in an Urban or Suburban Environment

Trees in an urban or suburban environment tend to develop a canopy that has a larger, more open form than their counterparts in the forest. This comes from less competition for sunlight. Trees in a built environment also have to withstand the force of the wind without any buffering from neighboring trees, so in many instances they develop a different structure, and need to be cared for differently. Because of the proximity of people and property to trees in a built environment, public safety warrants that these trees are inspected periodically to determine when pruning or removals should be done.

David L. Johnson New Jersey Forest Service, Community Forestry Program, Trenton, NJ 08625
Urban and Community Forestry in the Northeast, 2nd ed., edited by, J. E. Kuser.

3. Reasons for Pruning

There are several reasons for pruning trees. Many of them are to correct problems or defects affecting the structure of the tree due to genetic characteristics, natural or man-made stresses, or conflicts with the built environment. Before any pruning is started, there should be a stated reason and an expected outcome.

3.1. Pruning for Form

Pruning alters tree form by changing the height, width, crown density, and branch direction of the tree. Unless the trees have been growing close together, wider, more rounded crowns are found on trees growing in a built environment. This form has become the template for how trees should look in a streetscape or landscape setting. It is important to know the species of the tree and its normal shape before beginning to prune. What might be an unacceptable form in one tree may be a normal growth shape in another. Pruning to enhance the natural shape of the species is easier and more beneficial to the tree than trying to establish a form that is at odds with the tree's genetic blueprint.

3.2. Pruning for Strength

A tree's strength can be increased through proper pruning of branches predisposed to breaking or contributing to weak areas in the tree. The angle that a branch attaches to a parent stem can indicate a potentially weak attachment and can also indicate the presence of included bark. Included bark occurs when bark is forced between two limbs of a tree crotch instead of out and above it. Trees having narrow branch angles that mimic a "V" will often grow against each other and cause bark to be included in the crotch. Removal of these weakly attached V-crotches in favor of crotches that grow in a "U" shape will strengthen the tree. Areas where double or triple leaders occur are prone to failure. This develops when two or more branches of similar diameter grow from the same area of the trunk. The branches should be pruned so that a balanced set of scaffold branches can develop. These scaffold or main branches will form the shape of the tree, and they should be chosen so that they are well spaced along the trunk from both a horizontal and a radial perspective.

3.3. Pruning for Tree Health

Over the course of a tree's life many branches can develop which may negatively impact the tree's health if they are not removed. Dead and dying branches should be removed so that the tree can seal off the area and minimize potential entranceways for insects and disease. Branch stubs left due to incorrect pruning should be removed for the same reason. Any crossing or rubbing branches should be pruned out of the canopy in order to eliminate a recurring wound and to help develop better scaffold branch development.

Branches that appear diseased or infested by insects should be removed if feasible to negate the onset of a more serious problem. Pruning of a branch infested with insects may be more prudent than spraying the tree with a pesticide.

Water sprouts, which are weakly attached branches that develop on the trunk, branches, and at the base of the tree should be removed. They can develop because of incorrect pruning cuts, problems at the tree's graft union, injury to the root flare, or when the canopy of the tree is suddenly opened up. Water sprouts are faster growing than normal branches and will use resources that are better allocated elsewhere in the tree.

Trees in public places may be the targets of vandalism from time to time. It is oftentimes advisable to remove the lower branches within the reach of children or young adults before they are broken and the bark is torn. This preventative pruning will minimize injury to the tree by uncaring or malicious people. Properly timed removal of lower branches (crown raising) to provide adequate clearance for pedestrians and vehicles will also prevent the breaking or tearing off of branches by the public that interfere with their daily life.

Trees growing on public property can often grow into private property and initiate action by a private homeowner. Because the public is not trained in proper pruning techniques, the result is often deleterious to the health and structure of the tree. A well-run municipal tree care program can minimize impacts to the tree's health and maximize the tree's usable life.

3.4. Pruning for Safety

The safety of people and property should be the major concern for anyone in charge of a town's tree resource. It is important to be as proactive as possible to minimize the risk of injury or damage caused by trees on public property. Pruning for safety removes the large dead, dying, or diseased limbs from the canopy. This will lessen the chance that a falling limb will damage property or injure someone. Limbs that interfere with important signage, pedestrian clearance, or vehicular clearance should also be removed. Branches that block streetlights or lights positioned to enhance the visibility of walkways or building entrances may also need to be pruned back if they do not allow the targeted areas to be properly illuminated. Limbs that could damage the facades of buildings, windows, or other structures should be removed before they grow large enough to interfere.

Occasionally, trees may be planted in an area where the fruits or flowers of a tree become a hazard. Pruning that discourages fruit or flower production may be utilized so that the tree does not have to be removed. Elevating lower branches to discourage climbing may be necessary to ensure the safety of neighborhood children. Also the removal of lower branches containing thorns or flowers that attract stinging insects may be a wise precaution in certain locations.

3.5. Pruning for Aesthetics

Since many trees are planted to enhance the aesthetic appeal of an area, it is not surprising that pruning is done for aesthetic reasons. Pruning can improve the shape and appearance of a tree. It can also alter the size or amount of flowers and fruit. Pruning may be employed to open up a section of the crown so that a tree's unique bark is visible or to accent other features in the landscape. Pruning may provide a means to reveal interesting views or enhance landscape vistas. The density of a tree's canopy can often be a detriment to growing turf or other plants underneath it. Thinning the canopy will allow more light penetration and may promote better growth to understory plants or denser turf. Shearing conifers or broad-leafed evergreens will have an opposite effect by increasing the density of the foliage. This can allow the trees to provide a backdrop for other plants or landscape features or create a privacy screen.

3.6. Specialty Pruning

There are several different types of specialty pruning that are done to enhance the aesthetic appeal of plants. Pollarding is an ornamental pruning technique that consists of cutting back branches periodically to the same location on a selected limb. This creates a swollen, ball-like area at the end of the limb. Pollarded trees can be seen lining many of Europe's major boulevards.

Espalier pruning begins with the attachment of a tree's branches to a flat surface. It is then trained to develop into a two-dimensional shape through the use of pruning. This technique can shade the sunny wall of a building as well as accent it.

Topiary pruning is used to create animals and other shapes using the foliage of plants. It is done by shearing young branches or foliage continuously in order to achieve and maintain the desired shape.

Another specialty pruning technique is bonsai training. Bonsai training is pruning the branches and roots of a tree in order to maintain it in a dwarf size.

4. Seasonal and Developmental Pruning

Timing of pruning can be critical in tree care because trees react differently at different times of the year. The type of pruning and the time of year it is done can influence growth habit and energy cycles, as well as flower and fruit production. Generally, tree growth will be greatest if pruning is done just before a period of rapid growth (spring in the Northeast) and least if done during or immediately after a rapid growth period (late spring or summer). However, decisions to remove dead or damaged branches should not be influenced by the time of year. They should be removed as soon as possible to prevent additional damage to the tree and to alleviate hazardous conditions.

4.1. Pruning Evergreens

Evergreen trees are often pruned to create a more dense plant and retain the pyramidal shape commonly associated with them. Many evergreen trees in the landscape are also pruned to limit their height and width, as in the case with evergreen hedges. This type of pruning is known as shearing and should be done immediately following the new growth's (candle) full expansion. Shearing the candle growth will cause the tree to become denser. Larger scaffold branches or intermediate branches should be removed during the dormant season (winter).

4.2. Pruning Deciduous Trees

It is preferable to prune deciduous trees during the dormant season when insects and diseases are not active. The dormant season allows the practitioner to get a more complete view of the tree's branch structure because of the absence of foliage and to better identify problem limbs. You will probably see a significant difference in price if you contract out your tree pruning in the off-season, so there is an economic benefit as well. The dormant season varies according to where you are located in the Northeast, but it is initiated by leaf drop and colder weather in the fall and continues up to bud enlargement in the early spring. Some species of trees exude sap or "bleed" when pruned too late in the dormant season. These include maples, birches, and elms. Although bleeding is mostly an aesthetic problem, it may attract unwanted insects or animals. Pruning of these species in the early winter will prevent this problem.

The dormant season may or may not be the most appropriate time to prune flowering trees depending upon the species and where the flower buds develop. Early spring flowering trees like dogwood, redbud, and callery pear have their flower buds on the previous season's growth, so pruning during the dormant season will decrease springtime flowering. Other species, such as crepe myrtle, which flowers later in summer on new growth, will not be as adversely affected by dormant season pruning (Gilman, 2002).

It is best not to prune trees during times of low energy reserves or when soluble food is being converted into energy reserves. These times coincide with spring leaf expansion and autumn senescence (leaf drop). Pruning during these times may not allow the tree to create chemical boundaries or mount a strong defense against disease organisms (Shigo, 1999).

4.3. Pruning Newly Planted Trees

Newly planted trees should only receive a minimum amount of pruning because of the need to direct resources toward developing the root system. In general, only dead branches, broken branches, or basal suckers should be removed. While many planting details and specifications call for the tree's crown to be reduced by one-third, it is better to retain as much of the live crown as possible to maximize photosynthesis and provide resources for root growth and plant establishment.

4.4. Pruning Established Young Trees

Young trees that have become established in the landscape (1 to 3 years after planting) are ready for pruning to develop the crown structure. After young trees have become established, it is important to begin to identify the scaffold or main branches that will form the shape of the tree. Branches that have strong attachments and are well spaced along both the horizontal and the radial axes throughout the tree should be selected. These branches may be directionally pruned to ensure that they grow in the chosen direction. Make the directional pruning cut on small branches at a slight angle about 1/4 inch past a bud that will grow in the desired direction. Temporary branches should be allowed to remain but may need to be reduced in size so that they do not outgrow or conflict with the scaffold branches. Temporary branches provide resources to the tree and help in the development of branch and trunk taper. Crown cleaning of dead, dying, diseased, or crossing branches should also be done. As the tree becomes more established in the landscape (4 to 7 years after planting) the temporary limbs should systematically be removed. Crown cleaning should be continued during these years to eliminate problem branches.

4.5. Pruning Mature Trees

Pruning any live branches from a mature tree should be done judiciously. Mature trees need as much foliage and live tissue as possible in order to generate and store food. The removal of live tissue in the tree forces it to expend energy to compartmentalize the wound and can cause a number of problems (Gilman, 2002). Crown cleaning of dead branches is the most frequent type of pruning needed on mature trees, but occasionally there may be a need to reduce the length of a branch (crown reduction) that poses a threat. Thinning branches in the outer portions of the canopy may also be desirable to reduce wind resistance in the crown (see Section 7.2).

5. Proper Pruning Cuts

All pruning cuts should be made outside the branch bark ridge and the branch collar without leaving a stub (Fig. 1). The branch bark ridge is a raised area of bark tissue at the union of a branch and a trunk or the union of two branches. The branch collar is the swollen area at the base of a branch. It is the area where the branch wood and the trunk or larger branch wood come together in an overlapping pattern (Shigo, 1999). Protection zones form in the bases of branches and resist the spread of decay pathogens from the branch into the trunk. Cutting into the branch collar can remove the branch protection zone and injure the trunk tissue. This usually results in decay or cracks developing in the trunk wood, leading to hazardous trees in the future.

Dead branches that remain a long time on the tree often develop pronounced branch collars. Only the dead portion outside of the branch collar should be removed when pruning, even if the collar is aesthetically unpleasing.

When pruning larger branches, the three-cut method should be used (Fig. 1) to prevent bark tear from the weight of an improperly supported limb. The first cut is an undercut, the second cut is further out on the branch and removes the limb, and the third cut removes the remaining stub.

5.1. Types of Pruning Cuts

A heading cut reduces the length of a currently growing branch back to a stub or small twig. Heading cuts are properly used when shearing evergreens, pinching

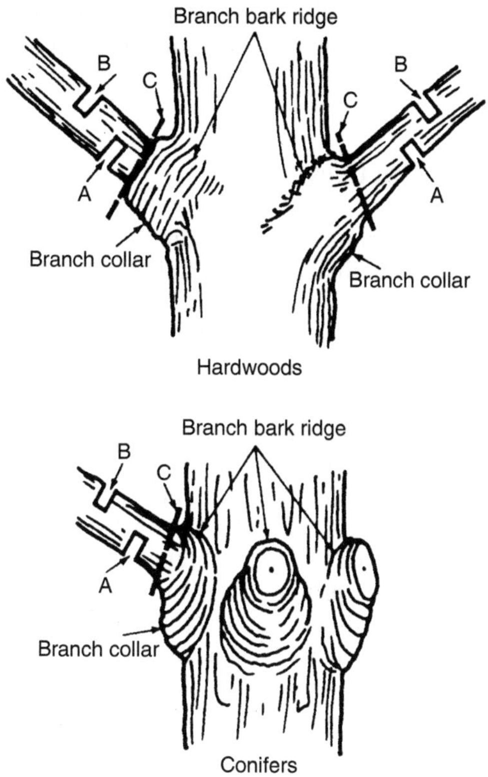

FIGURE 1. Pruningcuts. To remove a large branch cut A undecuts the branch, cut B removes the branch, and cut C removes the stub just outside the branch bark collar.

back, or tip pruning young trees and pollarding deciduous trees. This type of pruning cut encourages denser canopies, trains young trees to grow in a certain manner, and maintains a desired tree size, respectively. When heading cuts are used on mature trees, however, a variety of negative effects usually cause the tree to begin a downward spiral of health. The use of heading cuts on mature trees is also referred to as topping, stubbing, hat racking, and rounding-over and is not considered a proper method of pruning.

A reduction cut reduces the length of a branch back to a lateral branch that is at least one-third to one-half the diameter of the one cut. It is the proper method to use when reducing the size of the tree's crown. This type of cut directs the growth of trees by allowing the smaller branch to assume apical dominance and maintain the general shape of the tree. Reduction cuts, however, injure the branch or stem and can cause decay to extend down into the stem in trees that are poor compartmentalizers. Crown reduction pruning using reduction cuts is also known as drop-crotch pruning.

A removal cut, sometimes called a collar cut, removes a branch from the trunk or parent limb without injuring the tree protection zone in the branch collar. Since different tree species have different collar shapes and locations, there is no standard angle at which to make a correct pruning cut. Most maintenance pruning should be done using removal cuts.

6. General Pruning Practices

Although specific circumstances may require unique pruning methods, there are some general rules for pruning that should be followed. Since removal of foliage and live woody tissue impacts the tree's capacity to produce and store food, care should be taken not to remove too much at a time. Although the general rule is that no more than 25% of the canopy should be removed at one time, even this amount may be too much. Tree species can tolerate differing amounts of pruning. The age of the tree also plays a part in pruning tolerance. Mature trees do not tolerate the same amount of pruning as young or medium aged trees. If possible, removal of live branches should be done over a period of years. Removal of too much live tissue at one time can result in formation of compartmentalization wall 4s, cracks, decay, sprouting, and stress (Gilman, 2002). Remember, once you take a branch off, you can not put it back on.

When doing crown-raising pruning, it is important to try and maintain a live crown ratio of at least two-thirds. Live crown ratio is the height of the tree's canopy compared to the tree's overall height. This means that even after the crown has been raised, the remaining live canopy should occupy more than 66% of the total height.

Tree responses to different types of pruning are predictable in most cases. Pruning-off terminal buds promotes lateral branches to develop. Similarly, when lateral branches are pruned, the leader emerging from the terminal bud will elongate. These predictable responses should guide the tree care professional in determining which branches should be removed to create the desired effect.

7. Maintenance Pruning

Maintenance pruning is recommended when the primary objective is to maintain or improve tree health and structure. Maintenance pruning should consist of one or more of the following pruning types:

7.1. Crown Cleaning

Crown cleaning is the removal of dead, dying, diseased, decayed, or other objectionable branches from a tree's crown such as crossing or rubbing branches. The size of the branches and degree of work should be explained in the pruning specifications to assure the correct amount of work is being done.

7.2. Crown Thinning

Crown thinning is the selective removal of branches in order to increase light penetration and air movement or to reduce branch weight. At least one-half of the foliage should remain on branches that are in the lower two-thirds of the tree. Thinning consists primarily of removing lateral and parallel branches growing toward the ends of the main branches. Thinning of only the interior branches is known as "lions-tailing" and is an improper pruning method that damages the tree.

7.3. Crown Raising

Crown raising is the removal of the lower branches of a tree in order to provide clearance underneath, prevent vandalism, or access to the crown of the tree by children. This type of pruning is preferably done over a number of years to minimize the impact to the tree. The removal of too many lower limbs at one time can cause water sprout formation, cracks or sunburn on the lower, exposed trunk.

7.4. Crown Reduction

Crown reduction is the reduction of the top, sides, or individual limbs by removing the leader or the longest part of a limb back to a lateral branch of sufficient size. Crown reduction is necessary when a tree has outgrown the space in which it was planted and/or interferes with human activity in some way. It is also done to reduce shade and lessen the chance of windthrow. Crown reduction should only be done after other pruning methods have proven inadequate.

7.5. Vista Pruning

Vista pruning is thinning, crown raising, crown reduction, or a combination of any of these pruning types for the purpose of enhancing a view.

7.6. Crown Restoration

Crown restoration is pruning done to improve the structure and form of a tree that has been severely damaged by either natural disasters or man-made causes. This type of pruning should be done over a number of years and may not completely restore the natural form of the tree.

8. Pruning Specifications

In order to assure that pruning is done correctly when it is contracted out to a tree care company, it is important to specify exactly what is to be done and to reference a set of standards. The American National Standards Institute (ANSI) A300 standards are the national standards for tree care operations (American National Standards Institute, 1995). These standards are compiled by representatives from the tree care industry and represent the most up-to-date arboricultural practices. In addition to the A300 standards, the ANSI Z133 standards should also be referenced to ensure that the tree care company is performing the work in a safe manner.

9. Utility Pruning

Utility or line clearance pruning is done to prevent or minimize service outages and prevent damage to equipment. Many utilities contract out their line clearance work to companies specializing in tree care work around electrical lines. Different techniques have been used as arboricultural knowledge has advanced over the years but the inherent conflict remains (see Chapter 15).

10. Pruning Tools

Three general types of pruning tools are used in tree care. They are pruners, saws, and shears. Pruners are designed to cut smaller branches in a single action using a movable single blade against a fixed stop. The bypass type of pruner is preferred when cutting woody material. It is designed so that the cutting blade goes past the fixed stop when the cut is made and does not crush the stem like an anvil type of pruner would. Hand pruners are commonly used when training young trees and doing detailed tree work involving branches 1/2 inch and smaller. Pruners can also be attached to fiberglass telescoping poles that allow for cutting of small branches higher up in the crown of the tree. Less expensive models employ a gear driven mechanism that is operated manually using a rope. More expensive pole pruners can be pneumatically or hydraulically powered to increase worker production and lessen fatigue.

Pruning saws come in several different sizes and shapes and are used to remove larger branches. Saw blades are designed to cut branches by removing woody tissue incrementally as the teeth gouge out a path. Handsaws have a variety of cutting teeth configurations depending upon the manufacturer and the size of the branch that is going to be removed. Saw-blade lengths may range from about 8 inches to over 24 inches. Saw blades are also commonly attached to telescoping poles and used to remove larger branches that cannot be reached by hand.

Chain saws can dramatically decrease the amount of time and effort needed to remove tree limbs. They basically operate in the same way that a fixed saw blade does but utilize a series of saw teeth connected to a movable chain. Chain-saw models vary greatly in size and power. Chain-saw technology has been adapted so that a chain-saw bar and chain can now be mounted and used on telescoping poles.

Pruning shears are tools that use two cutting blades to remove small branches. Shears work like traditional scissors and are mostly used for pruning soft, new growth off of evergreens. They are the appropriate tool to use when shaping Christmas trees and doing topiary work.

11. Treatment of Pruning Wounds

The practice of painting pruning cuts was once believed to seal out disease organisms and help in the closure of wounds. Research has proven this practice to be ineffective in helping the tree and can actually be detrimental if the paint contains substances that are toxic to the living tissue of the tree.

References

American National Standards Institute, 1995, *Standard Practices for Tree, Shrub and Other Woody Plants Maintenance, ANSI A300-1995*, American National Standards Institute, Washington, DC.

Argent, R. A., 2000, *Handbook of Urban and Community Forestry in the Northeast*, Kluwer Academic/Plenum Publishers, New York, NY, pp. 205–214.

Gilman, E. F., 2002, *An Illustrated Guide to Pruning*, Delmar Division, Thompson Learning, Albany, NY, pp. 86–87.

Harris, R. W., Clark, J. R., and Matheny, N. P., 1999, *Arboriculture-Integrated Management of Landscape Trees, Shrubs And Vines*, Prentice-Hall, Upper Saddle River, NJ, pp. 400–460.

New Jersey Board of Tree Experts, 1997, *Pruning Standards for Shade Trees*, New Jersey Board of Tree Experts, Jackson, NJ.

Pirone, P. P., Hartman, J. R., Sall, M. A., and Pirone, T. P., 1988, *Tree Maintenance*, Oxford University Press, New York, NY, pp. 115–136.

Shigo, A. L., 1999, *A New Tree Biology: Facts, Photos, and Philosophies on Trees and their Problems and Proper Care*, Shigo and Trees, Associates, Durham, NH, pp. 183–84.

Shigo, A. L., 1989, *Tree Pruning: A Worldwide Photo Guide*, Shigo and Trees, Associates, Durham, NH.

Chapter 15

Trees, Utilities, and Municipalities

Robert F. Lee and Richard S. Wolowicz

1. Introduction

The purpose of this chapter is to investigate the problem-prone relationship between utility companies and municipalities. The nature of the controversy stems from the placement, planting, or location of trees and the presence of utility conductors. The dispute can be mitigated by the utility and municipality working together to promote a robust environment. By working together, both parties will contribute to the survival of the urban forest, to healthier trees, and will have citizens and consumers having their needs met with an uninterrupted power supply.

Trees, whether overgrown or falling down, are one of the major causes of power interruptions. This chapter will explore different cooperative ventures that promote the coexistence of each party's programs.

2. Background

People live in a complex world of fast-paced advancement in technology. Life has progressed to a degree that has converted nonessential conveniences to mandatory standards. Look at some of the changes over the past 20 years in manufacturing: Processes have improved, efficiency has increased, and new advancements in automation have revolutionized the industry. Office environments also have changed. They are filled with multiple computers with the latest technological advantages. Homes as well have been influenced. While not all homes have

Robert F. Lee Jersey Central Power and Light Co., Morristown, NJ 07960
Richard S. Wolowicz Public Service Electric & Gas Co., Newark, NJ 07102
Urban and Community Forestry in the Northeast, 2nd ed., edited by, J. E. Kuser.

computers, most have appliances with digital circuitry and computer chips to provide value-added features. All these progresses have put new emphasis on the need for reliable electrical power.

The demand for electricity has increased greatly with technological advances in industry and home. This increased demand along with resistance to build new transmission lines has resulted in taxing the existing facilities built many years ago. In response to this, utility foresters have a greater responsibility to manage the right-of-ways to insure reliability of the electrical system.

Today's utility customers demand nothing less than an uninterrupted power supply that is both safe and inexpensive. Any slight variation in that supply can result in a loss of data, can shut down synchronized manufacturing processes, and can require digital equipment to be reset.

The local utility plays a vital role in delivering this reliable electrical energy. Electricity is transmitted and distributed along underground cables and/or overhead wires. In the deregulated environment of the utility companies, electrical conductors will continue to be maintained but not necessarily by the same utility that generates the power.

3. Distribution and Transmission

Distribution circuits and transmission lines are the basic components of the electrical delivery system. Classification of each is determined by the line voltage. The transmission system delivers high voltage power over long distances from one geographic area to another. The Distribution system uses lower voltages to deliver power to businesses and residences. Construction also varies between the two systems, with overhead distribution circuits normally supported by wooden poles.

Transmission right-of-way agreements (easements) have documented language granting the utility the right to remove trees. Utility Companies have the right to remove trees to provide safe and reliable electrical service. Unlike the distribution system, contacts or close proximity to vegetation can have a catastrophic effect on the electrical system. The last three major cascading outage events in North America, including two major western US power failures and the August 2003 Northeast Blackout, involved trees. The latter was the largest power failure in the country's history (Guggenmoos, 2004).

4. Tree and Wire Conflict

Tree and wire conflict with each other. This conflict raises serious safety and reliability concerns that need to be addressed. There are customers whose lives depend on the uninterrupted flow of energy. There are commercial customers who cannot tolerate a single momentary blip or interruption without suffering enormous damage to their products. Consequently, the utility must maintain a consistent clearance around energized conductors through vegetation management. This minimizes damage to trees and conductors, and helps to ensure safe, reliable electrical energy.

Closer examination of this tree and wire conflict shows that electricity takes the path of least resistance along conductive materials to reach the ground. Nothing magical; it simply wants to reach the earth. When a tree makes contact with energized conductors, electrical energy will use that tree as pathway to the ground. This may cause the fuse to operate, which will disrupt the electrical circuit. Trees can also rub against the wires, causing wires to burn down. When this occurs, there is physical damage to the conductors and the tree. Along with the interrupted circuit, a safety hazard is created to people and property, since the wire may have fallen on the ground or may be hanging off an insulator.

In delivering electrical energy, the utility must perform various maintenance operations. Maintenance of lines includes the inspection of poles and an assortment of electrical equipment. Probably the most visible maintenance program performed by utility companies is tree trimming around electrical conductors. Because of the hazards involved, specialized crews must be used. The line clearance tree trimmer has the basic knowledge of electrical safety and how to prune the trees safely without endangering himself or herself. Federal regulations stipulate the need for a strict regimen of training and techniques involved in these operations.

5. Regulations

If a tree is within 10 feet of an energized wire, a qualified tree trimming company that has the proper expertise and training to work around energized wires must be used (Allard and Johnson, 1996). However, even these qualified line clearance tree trimmers must follow strict safety rules in personal proximity. Table 1 gives the minimum

Table 1. Line Clearance Tree Trimmer Minimum Approach Distances*

	OSHA 1910.269		
Nominal voltage phase-to-phase (kV)	Elevation factor sea level to 5,000 feet (feet/inch)	Elevation factor, 5,001 to 10,000 feet (feet/inch)	Elevation factor, 10,001 to 14,000 feet (feet/inch)
0.05–1.0	Avoid direct or indirect contact	Avoid direct or indirect contact	Avoid direct or indirect contact
1.1–15.0	2′4″	2′8″	2′10″
15.1–36.0	2′9″	3′2″	3′5″
36.1–46.0	3′0″	3′5″	3′9″
46.1–72.5	3′9″	4′3″	4′7″
72.6–121.0	4′6″	5′2″	5′7″
138.0–145.0	5′2″	5′11″	6′5″
161.0–169.0	6′0″	6′10″	7′5″
230.0–242.0	7′11″	9′0″	9′9″
345.0–362.0	13′2″	15′0″	16′3″
500.0–550.0	19′0″	21′9″	23′6″
765.0–800.0	27′4″	31′3″	33′10″

*Minimum separation in various voltage classes (Abbott, 1994).

Table 2. Persons Other Than Qualified Line Clearance Tree Trimmers Minimum Approach Distances[**]

Nominal voltage phase-to-phase (kV)	OSHA 1910.3333 and ANSI Z133—1994 (feet/inch)
0.0–1.0	10′00″
1.1–15.0	10′00″
15.1–36.0	10′00″
36.1–50.0	10′00″
50.1–72.5	10′09″
72.6–121.0	12′04″
138.0–145.0	13′02″
161.0–169.0	14′00″
230.0–242.0	16′05″
345.0–362.0	20′05″
500.0–550.0	26′08″
765.0–800.0	35′00″

[**]Indicates the proximity of nonline clearance tree trimmers to energized conductors (Abbott, 1994).

separation distances stipulated by the Occupational Safety and Health Act (OSHA) regulations for these personnel when working on different voltage classifications. American National Standards Institute (ANSI) and OSHA regulations also stipulate that at 12,000 V, nonqualified personnel must maintain at least 10 feet of distance from any energized conductor. Table 2 shows how this distance changes as voltages increase.

Should a private property owner, municipality, or contractor be required to work on a tree within 10 feet of an energized conductor? The local utility can be contacted for assistance. The electric company has three options in this case: deenergizing the line, covering up the conductors, or providing the trimming clearance to the requesting party. Deenergizing the line is a last resort and undesirable solution. Covering up the liens provides a false sense of security. It is impossible for a utility to guarantee that the area is safe from electrical hazards. Therefore, the utilities most often provide the necessary clearance to the customer. More utilities will probably be charged for this type of service in the future.

6. Myths About Utility Line Clearance Programs

Lack of trust is often a problem between municipalities and utility contractors. Utilities have been blamed for doing unprofessional work. To be perfectly truthful, there has been much unprofessional work done by all groups, including municipalities, private tree companies, and homeowners. It is not only the utility that should be blamed for poor arboricultural techniques. Major emphasis and conscientious efforts are being made by most, if not all, utilities to manage vegetation in a professional and scientific manner. There are benefits from this method: it is the correct way to do things; generates good results in public relations, and is financially prudent. It must be realized that pruning trees in the vicinity of power lines does not have the same objectives as standardized pruning categories (see Chapter 14, this volume). The intent of utility line clearance is to direct growth away from the wire zone and remove limbs that may pose a safety hazard.

A sense of professionalism is increasing among the line clearance contractors. Through the Utility Arborist Association, which is a chapter of the International Society of Arboriculture, a utility tree trimmer certification is being developed (McNamara, 1995). This plan involves a thorough testing process, resulting in a certification for those trimmers who pass the test. This ensures that the trimmer is aware of the latest and best methods to trim for wire clearance and tree health. Figures 1 and 2 are examples of modern methods utilized in line clearance.

A line clearance training manual has been developed by an arboricultural consulting and training company, ACRT, Inc. (Abbott, 1994). The Ohio firm developed the manual with the cooperation of a Pennsylvania utility (Abbott, 1994). These two certification programs, which do not compete with one another in any way, are examples of how professional advancement is gathering momentum. This will promote quality workmanship in the industry.

Another example of industry advancement can be seen within the utility companies themselves. Many companies have hired foresters to manage tree growth. This adds credibility to the utility's reputation in vegetation management activities. Customers, citizens, and the public observe that the utility is serious about environmental issues.

Trimming work done by the utilities also serves as a testimony to improvements in arboricultural industry standards (Kempter, 1996). Sincere efforts are made to trim

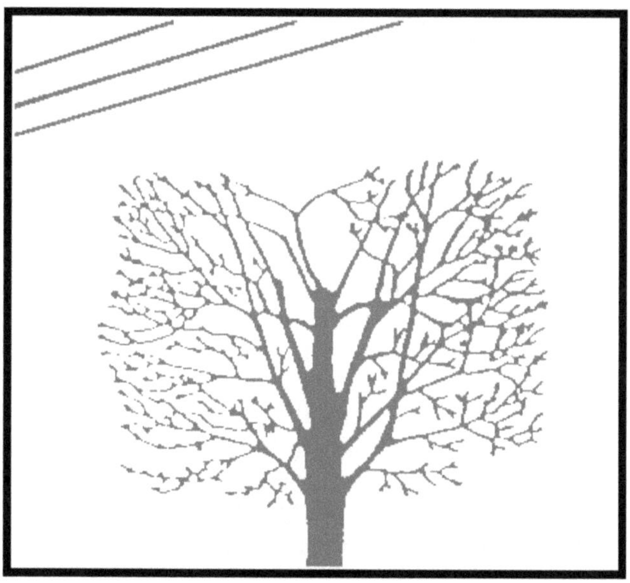

FIGURE 1. Natural trimming—promotes growth of existing branches rather than rapid growth of many new suckers. (Courtesy of Environmental Consultants, Inc.)

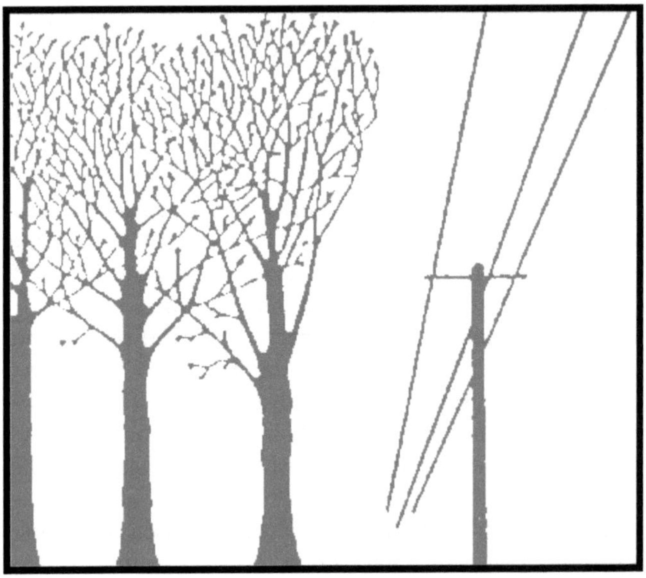

FIGURE 2. Natural pruning. This technique also lends itself to side trimming. This pruning will promote the growth of existing branches rather than rapid growth of new suckers. (Courtesy of Environmental Consultants, Inc.)

trees to provide wire clearance, yet not hinder the tree from healing. Tree branches are cut to lateral branches rather than stubbed. Trees are directionally pruned to direct growth away from the lines. Trimmers are using drop crotch pruning methods to avoid topping. As in private tree care, final cuts are not being made flush with the stem, but are in accordance with the latest techniques. Trees also are not being rounded-over, a practice that produces superfluous shoot growth and weak branch attachments. Figures 3 to 5 show examples of improper line clearance techniques.

7. Utilities and Municipalities Working Together

When you think about it, municipal and utility arborists have a lot in common. Although they trim trees for very different reasons, both are in the business of managing tree resources in the best interest of the communities they serve (Todd, 1992).

Utility and municipal arborists maintain trees in an effort to maximize their benefit to the community. Just as a street tree that is full of deadwood and decay may be of little value; so too is the tree that has grown into the electrical wires to the extent that it causes power outages. It is the responsibility of the respective arborist to schedule and perform the appropriate corrective maintenance. In either case, such maintenance, whether pruning deadwood or performing line clearance trimming, will increase the value of that tree to the community.

It is also the responsibility of the utility and municipal arborist to minimize the liability that trees may present through maintenance programs. Certainly, a tree that

FIGURE 3. Side trimming stubbing. This undesirable work is done by stubbing off portions of limbs along the side of the tree to obtain clearance. Cutting off portions of limbs (leaving stubs) creates many fast-growing suckers that become a serious line clearance problem. Additionally, stubbing off branches is not conducive to wound healing after the pruning operation. (Courtesy of Environmental Consultants, Inc.)

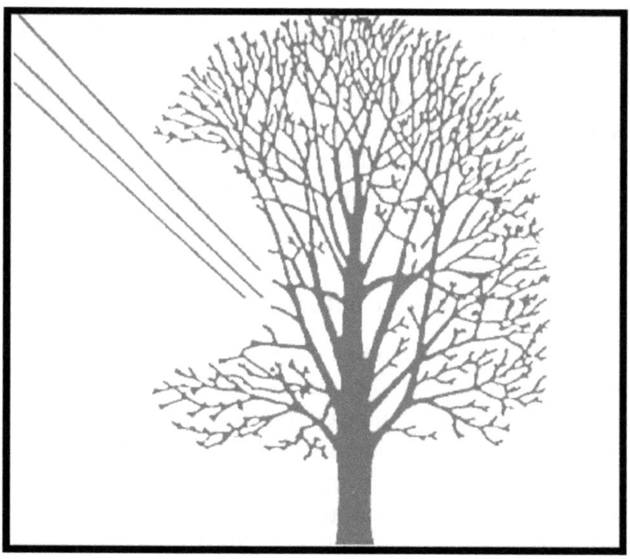

FIGURE 4. "Shaping" around lines. This is accomplished by trimming limbs in an arc to obtain clearance. This pruning method leaves branches above the conductors that could bend or break, causing outages. Shaping also creates many fast-growing suckers. (Courtesy of Environmental Consultants, Inc.)

FIGURE 5. Side trim "shelf." Leaving a "shelf" below the conductors when side trimming allows the branches to grow back toward the lines. While not always possible, these limbs should be removed to minimize future trimming. (Courtesy of Environmental Consultants, Inc.)

blocks a traffic sign, or causes a sight obstruction at a traffic intersection has the potential of becoming a public liability. By the same token, trees that are in direct contact with energized lines may cause arcing and sparking with the very real possibility of catching fire. Again, in both cases, the utility and municipal arborists have the responsibility of minimizing these types of tree liabilities through adequate maintenance programs.

Keep in mind that utilities and municipalities share the same client base. The tax-paying citizens of a municipality, who fund the shade tree program, can be the same rate-paying customers who are served by the local utility that provides line clearance. Both the municipality and utility are working in the customers' and citizens' interests. By working together, we can serve them better.

7.1. Avoiding Municipal–Utility Conflicts

In many cases, utilities perform line clearance tree trimming on planned maintenance cycles. Most often, this work includes entire-circuit trimming that covers a large geographic area. Planned maintenance cycles are usually scheduled well in advance, and the scope of the cycle is easily identified through reviewing the circuit maps and project specifications. Circuit or grid maps are a kind of electrical road map that details the location of the electrical circuit (Shriver, 1979).

Often, conflicts between municipality and utility line clearance operations can be avoided by conducting joint meetings prior to the start of any major planned line clearance program. At the meeting, the utility arborist is given the opportunity to present the project to the concerned municipal representatives. He or she should address such key points of information as: where the maintenance work will be done, proposed start date, and what type of work will be needed (trimming, removals, chemical treatments, etc.). The municipality also should be provided with the names and numbers of the contractor and/or utility representatives. The municipality is given the opportunity to go over any parameters that may exist, such as morning start-up time restrictions, road flagging and signage requirements, or issues related to the municipal shade tree ordinance.

The practice of holding a preproject start-up meeting has many benefits. It allows all involved parties to develop personal relationships that create open lines of communications. The most significant benefit of these prework meetings is that it eliminates the element of surprise. When the call comes into the municipal official from a resident questioning why tree trimmers are working in the neighborhood, it is always nice to have the answers.

Utility arborists have learned that conflicts often arise where there is a lack of knowledge about utility line clearance practices. It is incumbent on the utility to educate its customers, municipal representatives, and public work employees to communicate the how and why of utility line clearance. This may be accomplished through model electrical hazard demonstrations performed at schools, firehouses, public works garages, or wherever needed. Utility arborists must be available to shade tree commissions to go over line clearance policies and procedures and answer any questions or concerns. It is also important that utility arborists attend and speak at various shade tree conferences and conventions. These are ideal settings to meet with industry professionals, municipal representatives, and concerned citizens. Utility arborists are always searching for a soapbox from which to disseminate line clearance information to the public.

7.2. Joint Ventures

Speaking and educating the public on utility programs are means to promote positive municipal and utility relationships (Barnes, 1988). Action steps in which the two can work together are equally important to promote the urban forest. Quite often, utilities and municipalities work jointly on projects that require mutual cooperation, such as road or housing projects, or during emergency conditions brought on by storms or catastrophic events. Mutual benefits also can be derived through working jointly on tree maintenance programs. A joint street tree program may involve a municipal beautification project or Arbor Day program that would call for interest from the local utility. Or it could be a part of a capital improvement project where curbs and sidewalks are to be installed, and mature trees located in the wire zone need to be removed and replaced. If a municipal project involves electrical wires, contacting the local utility can prove to be a win-win situation.

It is in the interest of everyone to ensure that the "right tree in the right location" rules are followed. To this end, many utilities offer small tree replacement programs (Rossman & Harrington, 1986). The programs will vary in detail, but the objectives usually involve removal of large, overpruned trees that are located directly under primary electrical wires and replacing them with more suitable low-growing trees (Dirr, 1990) (Table 3 lists tree species compatible with overhead power lines that have proven successful in areas with climatic conditions similar to that of New Jersey). Tree replacement programs enable the utility to remove a costly nuisance tree from its

Table 3. List of "Low-Growing" Species, Which are Compatible with Overhead Conductor Lines

Low-growing trees	Species
Amur Maple	*Acer ginnala* 'Flame'
Hedge Maple	*Acer campestre*
Serviceberry Hybrids	*Amelanchier x grandiflora* (Tree Form)
American Hornbeam	*Carpinus caroliniana*
Hawthorne	*Crataegus* (spp.) Tree Forms
Goldenrain Tree	*Koelreuteria paniculata*
Flowering Crabapple	*Malus* (spp.)
Amanogawa Oriental Cherry	*Prunus serrulata* 'Amanogawa'
Kwanzan Oriental Cherry	*Prunus serrulata* 'Kwanzan' (High Branched)
Shubert Chokecherry	*Prunus virginiana* 'Shubert'
Cumulus Shadblow	*Amelanchier cumulus*
Pink Shadblow	*Amelanchier canadensis* 'Robin Hill Pink'
Ivory Silk Tree Lilac	*Syringa amurensis japonica* 'Ivory Silk'
Dogwood	*Cornus* (spp.)
Ruby Red Horsechestnut	*Aesculus carnea briotti*
Redbud varieties	*Cercis* spp.
Carolina Silverbell	*Halesia carolina*

Other low-growing trees may be available in different geographic areas.

maintenance cycle. The municipality enjoys the aesthetic improvement of having a street lined with trees in their natural form (New Jersey Shade Tree Federation, 1990). Replacement programs are financially flexible and may include cost sharing, which may be direct dollar contribution or in-kind costs, such as providing workers and equipment to assist in the removal and planting of trees.

Another example where utilities and municipalities can work together is in the coordination of the pruning cycle for street trees. This can provide safer working conditions and many cost benefits. Municipalities that have regular planned maintenance cycles should contact their local utility prior to beginning work to find out whether there are any plans to do line clearance in the near future within the same area. If the project can be coordinated, it would be best to have the local utility provide the line clearance prior to normal street tree maintenance pruning. Although there are functional differences between utility and municipal tree pruning practices, having the utility remove dangerous trees and timber and clearing limbs from the wire zone will provide not only a much safer work environment but also substantial cost saving to the municipality.

There are certain principles to follow in any joint venture to enhance the chances of success:

1. Keep the project simple. With current downsizing of staff levels in both the utility industry and the municipalities, arborists can ill afford to get involved in a major, time-consuming project not directly related to their daily duties and responsibilities. Keep the project cost as low as possible. The more funding that is required, the higher up the management structure you need to go for approval. This involves more time and getting more people to sign on.

2. The joint project proposal should be presented to the community by a member of the community. The utility is often viewed as an outsider. A utility representative proposing a project to the community would most likely be met with hesitancy and suspicion. The community would be much more open to one of its own.
3. Finally, there must be open lines of communication. One of the major benefits of a joint program is to improve communication between the utility and municipality. Working together allows us to get to know one another and builds relations that benefit the entire community.

8. Summary

Trees and wires will continue to be a source of complaints if municipalities and utilities hinder one another from resolving mutual conflicts. The only sure way to prevent a conflict between municipalities and utilities is to avoid it in the first place. Specifically, for the utility this means do not place wires in or around trees, and for the municipality it means do not plant tall-growing trees under or near the wires. Certainly, this is a very simple solution with a difficult degree of complexity to carry out. Be realistic, and recognize that neither the municipality nor the utility will go away. Make a commitment to work with one another and resolve the problem. There will never be a resolution if two agents use their muscles to hinder true solutions.

Positive changes are taking place in the utility industry and gaining momentum. More electrical facilities and conductors are being placed underground, primarily in new construction areas. More people are understanding the need to plant small trees under overhead wires. Planting trees beside the conductors and not directly underneath the wires is another avenue that can be explored. Trees planted alongside the conductors will only require their side branches to be trimmed. This is a better, albeit no perfect solution. This also gives added future benefits of minimizing sidewalk and curb upheavals caused by tree roots (see Chapter 16, this volume). Planting the right tree in the right location makes a lot of sense, and is a practical solution to avoid future conflicts and frustrations with overhead conductors. See Fig. 6 for practical planting locations.

Fast-growing and high-maintenance trees are a problem to both the municipality and the utility if the conductors are in close proximity. For a municipality, these trees require more frequent trimming and cleanup. For the utility, these trees have a tendency to be weak natured and pose threats to the abutting power line. A selective tree replacement program can prove a benefit to the utility and municipality. Such a program will avoid drawing the battle line every time the utility trimmers appear on the scene. Creative thinking can achieve positive results.

It is important for the town to have an active shade tree program. This promotes a link to the utility to seek out and hopefully work together to address each group's mutual problems. It is neither the responsibility nor it is possible for the utility or the municipality to work by itself and arrive at a solution without the other's participation.

It is doubtful whether the tree and wire conflict will go away in the near future. If constructive energies are put in place, opportunities will exist to make the best of a

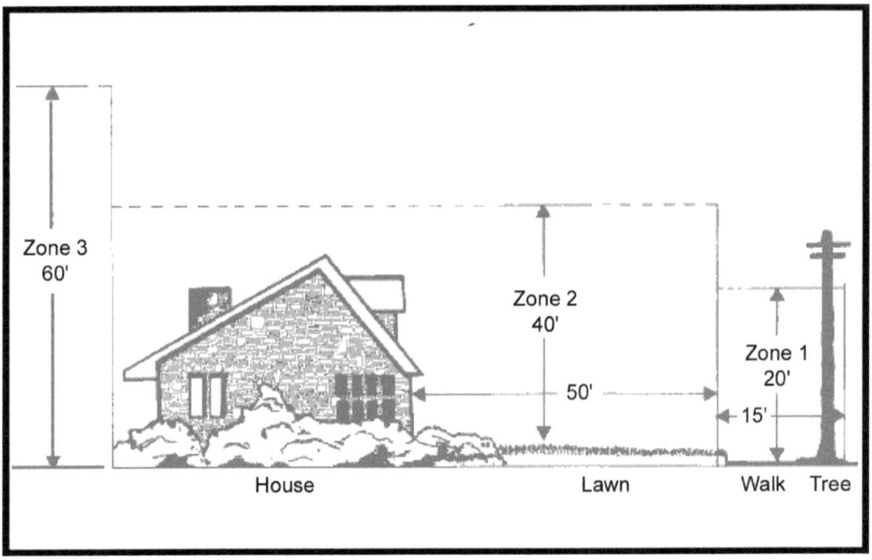

FIGURE 6. Showing the relationship between three planting zones. Zone 1 will accommodate ornamental trees to avoid a conflict between trees and power lines. In Zone 2, medium sized trees are suggested as the best for improving the way the house and the property look. Zone 3 is the land behind the house, which is suitable for a wide selection of trees, including larger kinds not recommended for the other zones. (Diagram courtesy of GPU Energy Co., 1990.)

tough situation. Some battles will be won and some will be lost. But many will hopefully produce a compromise responsive to customers and citizens.

References

Abbot, R., 1994, *Line Clearance Tree Trimmer Certification Manual*, ACRT, Inc., PO Box 401, Cuyahoga Falls, OH.

Allard, J., and Johnson, E., 1996, Regulatory compliance: Working near wires, *Tree Care Ind.* **7**(10):64–66.

Barnes, B., 1988, Community involvements/public outreach in line clearance, *J. Arboric.* **14**(12):298–301.

Dirr, M. A., 1990, *Manual of Woody Landscape Plants*, Stipes Publishing, Champaign, IL.

Environmental Consultants Incorporated, 1989, *Distribution Tree Trimming*, 301 Lakeside Drive, Southampton, PA.

Gerhold, H., Lacasse, N., and Wandell, W., 1995, *Compatible Tree Factsheets for Electric Lines and Restricted Spaces*, Penn State University, University Park, PA.

GPU energy Co., 1990, *Trees: Practical Planning*, 300 Madison Avenue, Morristown, NJ 07040.

Guggenmoos, S., 2004, Managing trees to improve the bottom line, *Energy Pulse*. Transmission & Distribution World, 4.26.04 issue.

Kempter, G., 1996, ANSI A300 Meets the Real World, *Arboric. Age* **16**(12):28–31.

McNamara, M., 1995, Pennsylvania Electric Company's line clearance tree trimmer certification program, *Utility Arborist Assoc. Q.* **3**(2):1–5.

New Jersey Shade Tree Federation, 1990, *Trees for New Jersey Streets*, New Jersey Shade Tree Federation, New Brunswick, NJ.

Rossman, W., and Harrington, C., 1986, An attractive alternative to tree trimming for line clearance, *J. Arboric.* **12**(1):20–23.

Shriver, J., 1979, Distribution line tree-caused disturbances, *J. Arboric.* **5**(8):191–192.

Todd, A. P., 1992, Municipal/utility synergy: Opportunities for partnerships, *J. Arboric.* **18**(2):79–80.

Chapter 16

Tree Roots Versus Sidewalks and Sewers

William R. Comery

1. Introduction

Everyone blames trees for damaging sidewalks and sewers. This chapter will explain how trees grow and how and why they conflict with concrete. It is best to start at the beginning. Citing the fundamental structures of a tree and its general physiology will show different growth patterns and what happens when they conflict with artificial conditions. We will take into account the most common problems and list the conditions that can cause those problems to occur.

Property owners, government officials, arborists, and even municipal foresters disagree about how the problems occur, but these disagreements are small when compared to how they should be resolved. This chapter will provide analyses of many concerns regarding conflicts with tree roots, sidewalks, and sewers. **It will also provide solutions**. These solutions will not fit all of the varied situations that do occur, but with some degree of modification a solution can be found for specific problems. This does not necessarily mean that they can solve all problems to the satisfaction of all parties.

Legal concerns regarding safety and liability have caused sharp disagreements on how to correct and prevent sidewalk and sewer problems. These legal stumbling blocks have stalemated active pursuit of meaningful resolutions to these very difficult situations. These have caused abandonment of many worthwhile and productive tree maintenance programs. This "do-nothing" stance leads to larger problems with greater liabilities some of which they cannot resolve without enormous expense—if at all.

William R. Comery, C.L.A. Shade Tree and Parks Commission, Borough of Paramus, Paramus, New Jersey 08625.

Urban and Community Forestry in the Northeast, 2nd ed., edited by, J. E. Kuser.

2. Root Systems

2.1. Identification of Root Systems

Contrary to what you may think, a tree's root system generally occupies only the top 12–18 inches below the surface. Many of the surface roots are found within the top 6 inches of soil. There are two principal root types, surface and tap. So named because of their shape and depth into the soil. In the Northeast, the most widely planted street trees with a tap root system are the Oaks (Quercus). A tap root is the primary root and anchors the tree. A tap-root tree has transport and support roots near the surface for additional stability and provides life-giving water and nutrients to the tree.

The surface root system fundamentally differs in that it does not possess a single major anchoring root. In most other respects the root systems are the same except that transport and support roots alone provide the anchoring and stability of the tree. In the Northeast, the most widely planted street trees with a surface root system are the Maples (Acer), Ash (Fraxinus), Locust (Gleditsia), and Elm (Ulmus).

In both root systems the transport and support roots grow larger as the tree matures. These woody roots attach to the trunk of the tree and divide becoming smaller until they lose their woody character and become fine rootlets whose function is absorption. These are the roots that are incompatible with artificial structures such as sidewalks and sewers.

2.2. How and Why Root Systems Grow?

Roots, especially healthy roots, are critical for the survival of trees. They provide stability and transport water and nutrients from the soil to the tree. How and why this occurs is one of the least understood functions in the life cycle of trees. To perform these vital functions, tree roots must reach out into the surrounding environment, namely the soil. During this natural growth process, the tree roots not only find needed nutrients but also develop stability for the tree. Their movement accomplishes this in a downward and outward direction. The extent of this root growth determines the future development of the tree. Under optimal conditions tree roots will spread as needed to provide what is necessary for continued growth. As growth continues, they require more nutrients thus expanding the root system. "Roots typically spread up to twice the height of a tree—or farther! However, the essential mass of roots is usually considered to be within the dripline, the area underneath a tree's branches."

Ideal soil and root interactions are seldom found in suburban ecosystems and almost never in urban areas. Therefore, less than ideal conditions impede the natural development process. Different tree species have limits to how large they can grow depending on their genetic makeup. This will limit the growth of the tree roots as well. For example, the Norway Maple (Acer platinoides) has an average growing height at maturity of 60 feet with a corresponding root system spread of 120 feet. The species Amur Maple (Acer ginnala) has an average growing height at maturity of 25 feet with a corresponding root system spread of 50 feet.

FIGURE 1. An uprooted tree indicates a surface root system. Because of poor drainage, this tree uprooted in a wind storm. Photo, William R. Comery.

2.3. What Problems Impede Root System Growth?

A healthy tree must have a healthy root system. The health of the root system is affected by environmental factors such as soil, weather, and conditions created by man. As tree roots spread through the soil searching for the necessary water and nutrients, the ease with which these supplies are found is dependent upon the above-mentioned factors. Ideal soils are those that allow for regular root penetration and contain sufficient amounts of organic materials to provide for proper growth of the tree.

Nutrients required for proper tree growth are the following basic elements: (1) in larger quantities nitrogen, phosphorus, potassium, calcium, sulfur, and magnesium; and (2) in lesser quantities manganese, zinc, boron, copper, iron, molybdenum, and chlorine. These are found in the soil. Hydrogen, oxygen, and carbon are obtained

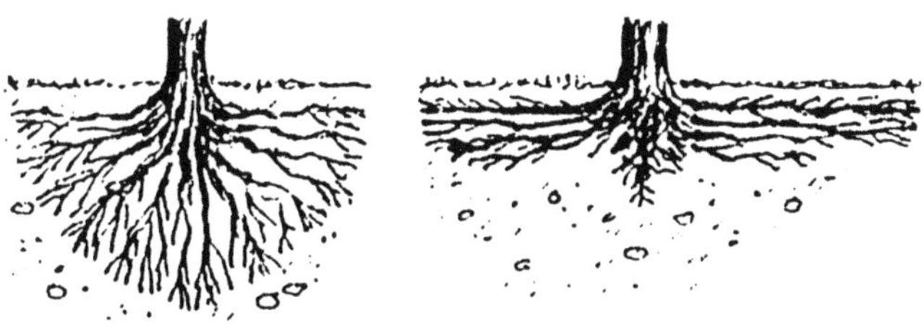

FIGURE 2. Typical root system of a tree

from air and water. In appropriate proportions they are essential for proper tree growth. Alkaline and acidic soils are dependent upon the proportions of nitrogen, hydrogen, and oxygen in the soil moisture known as pH. Different trees require varied soil pH depending upon their genetic makeup. Almost no tree can grow in severe alkaline or acidic soils.

Soil texture varies from small claylike particles (dense) to larger sandlike particles (loose). Clay particles prevent water and nutrients from penetrating the soil. Sandy soil allows too much diffusion through the soil. In both cases the necessary materials are not available to the tree roots, and they must search elsewhere.

Soil structure is a combination of soil particles, organic matter, air, and water. In proper proportions they form a medium that is conducive to superior root growth.

Adequate soil depth enables the tree roots to comfortably extract the needed elements for continued growth. Specific species require different soil depths that depend upon its ultimate size at maturity. Compacted subsurface soils or rocks do not allow for root penetration nor would they provide any nutrients.

Weather conditions, for the most part, are dependent upon geographical location, atmospheric conditions, and nature's fancy. The elements of weather, such as rain, wind, and temperature, affect tree growth both positively and negatively. Certain trees do well in a warm and dry climate and others in cool and moist areas. Weather conditions in the Northeast vary from season to season and year to year. The combination of weather and soil conditions, along with tree species, affects the normal health and growth of the tree.

Artificial and man-made conditions, when in proximity to trees, adversely affect their root systems. These man-made obstructions interfere with the tree's ability to obtain what it needs from the soil and cause the roots to deviate from a normal growth pattern.

2.4. Common Sidewalk Problems

When tree roots come in contact with concrete only bad things can happen. The obstruction to the tree inhibits the tree's ability to continue with a normal growth cycle. When tree roots are restricted, they are unable to obtain the means needed for proper growth. In this quest for nutrients, the tree will use any means to survive including misaligning, cracking, and raising sidewalks. When tree roots are restricted and confined, the natural force exerted cannot be understated.

A raised sidewalk occurs when a root or roots grow under a sidewalk and exert pressure causing the upward movement of the sidewalk slab. A misaligned sidewalk develops when pressure from a root or roots shifts the slab. This pressure can be lateral and/or from below. Cracked sidewalks occur when the roots infiltrate under the slab and the concrete breaks from its inability to withstand the pressure. All sidewalk problems are not caused by tree roots. Poor construction, weather, soil, and drainage, along with abuse, create similar conditions. Remember, it is not always the tree's fault.

Safety is a major concern when sidewalks and trees are in conflict. When sidewalks are deemed unsafe and the courts are asked to decide, the degree of safety and the desire to correct the problem becomes secondary to liability claims. Accidents generally result in liability claims. Accident prevention results in safe conditions. In today's society, liability and finding fault do not always provide safe conditions. Many worthwhile and productive tree maintenance programs have been abandoned because well-meaning legal advisors have cautioned against undertaking problem-solving measures. These include sidewalk repair, root cutting, and even acknowledging that a problem exists. Tree and sidewalk conflicts are addressed differently from community to community. Not all communities have the resources available to provide a comprehensive tree/sidewalk program.

The State of New Jersey has recently adopted the "New Jersey Shade Tree and Community Forestry Assistance Act" (N.J.R.S. 40:64-14 and P.L. 1958, c.41) that greatly reduces the liability to municipalities and counties and allows for a defined tree maintenance program throughout the state. This far-reaching legislation provides for continued tree planting and maintenance programs with increased safety and reduced liability.

There is a direct correlation between property values and aesthetics. Conservative estimates put the increased value between 7% and 10% for most residential properties when properly landscaped. Additionally, neighborhoods can have a higher real estate value of an added 10% when the streets are lined with trees. A canopy of street trees is a proven asset and often the deciding factor in the purchase of a home. The benefits of a well-planned and well maintained street tree planting are self-evident to most concerned sellers and buyers. Raised, misaligned, and cracked sidewalks can detract from property values and are the result of poor planning and poor maintenance. If for no reason other than aesthetics, well cared for sidewalks should be given a high priority.

2.5 Common Sewer Problems

Infiltration of tree roots into sanitary sewer lines is a major problem in many older towns. In these towns, which were generally developed before World War II, the construction materials and techniques did not allow for the force of tree roots. When tree roots are close to sanitary sewers there is a struggle for space. Early sanitary sewer systems in the Northeast were usually made of clay. These pipes facilitated the movement of waste, but were not always completely water tight. Their strength was from their thickness not from the durability of the material from which they were made.

As technology improved, more durable materials were utilized to minimize the effects of outside sources. Tree roots enter sanitary sewers in their search for water. How they enter the sewer system has always been a mystery. The answer is simple. A tree root cannot penetrate a properly constructed durable and impervious pipe. In all cases, where defects appear in the pipe the tree root wins.

Until *recently*, storm and sanitary sewers used a common pipe conduit. Increased environmental awareness has eliminated this undesirable practice in most areas of the Northeast. Storm sewers are generally larger in size than sanitary sewers because they convey huge volumes of water. The pressure within a storm sewer, under surge, can reach levels significantly higher than most sanitary sewer lines found along tree-lined residential streets. Because most storm sewers are constructed closer to the surface than sanitary sewers, they are placed in direct conflict with tree roots. Storm sewers in low, wet, and poorly drained areas have pipes that can lay within 18 to 24 inches from the surface. This is the same area where the majority of a tree's roots are growing. In time, the battle for those 2 feet will begin.

Sewer pipes can be cracked and damaged in many ways. Briefly stated, the most common are as follows:

2.5.1. Old systems

Old systems, both storm and sanitary, are subject to the same external elements. The forces that cause pipe deterioration in older systems are, for the most part, the pipes themselves. Early technology, design, and knowledge, along with the availability of local raw materials, limited the choice of pipe (clay, wood, concrete, and so on). For these reasons, pipes were unable to withstand the forces because these were inflexible.

2.5.2. Construction techniques

Construction techniques were very simple in the early days; pipes were just butted end to end. Later, as leaks were found, a pipe system utilizing spigot and bell ends was developed. This system was still not fool proof, so a system utilizing gaskets was designed. Today, in some cases, this gasket system has been refined to include solid joint or welded connections. The quality of workmanship throughout the years has resulted in both reliable and substandard installations. Modern technology, design, and knowledge, along with new materials such as PVC, make pipe installation more uniform, easier, and more reliable, with fewer chances of leaks.

2.5.3. Unforseen perils

Unforseen perils, such as earthquakes, floods, soil movement, and accidents, caused by the presence of man, crack and damage sewer lines. With proper design and material selection, the stability of the system can be enhanced. With increased urban sprawl in the Northeast, chances of accidents are more likely to occur. These accidents can be the result of neglect, poor planning, and/or poor judgment.

2.6. What is the Government's Responsibility When Trees Conflict with Sidewalks and Sewers?

The onset of sidewalks and sewers as part of the suburban and urban landscape in the Northeast began about 75 years ago. Before that, basic needs of the community did not require sophisticated improvements. As communities grew, the need for such improvements became necessary. With improved technology, sidewalks and sewers became commonplace. When city dwellers moved to suburbia, they came to expect the presence of sidewalks and sewers. Little thought was given to the potential for conflict between trees and these new structures. In fact, the conflict does not occur with a newly planted tree until the next generation moves in. In the case of existing trees, more consideration was given to the easiest, quickest, and cheapest installation of the slabs and pipes, giving minimal, if any concern for the trees. As a result, our first problems occurred with existing trees because they had already established their root systems. Governments were slow to act when these conflicts first appeared. The reasons for their inaction were that it was something new, the problem was not widespread, and officials were reluctant to admit that it was a governmental responsibility.

There are three positions that municipalities have taken with respect to the tree/sidewalk problems. We will state the following basic examples and show how they evolved. In many ways, the more complex positions grew from the first, which is the "do-nothing" approach.

2.6.1. "Do-Nothing": No Tree Requirements

Local governing bodies, planning boards, and zoning boards make no provisions for trees under this system. They allow development to occur without consideration for environmental concerns, good planning, aesthetics, and increased property values. The post-World War II building boom provided sidewalks but left tree planting and landscaping to the purchasers of the new homes. The developer was given no direction from the local government and the local government's primary purpose was to increase its tax base and provide needed housing. **The result was treeless streets**.

2.6.2. "Do-Something": Requires Trees but no Tender Local Care (TLC)

Municipalities, in an attempt to provide a suburban/rural setting, required developers to plant trees within the borough right-of-way. They left to the developer the decision to plant between sidewalk and curb or to a lesser degree, behind the sidewalk. The town can only dictate plantings in the municipal right-of-way (ROW). No thought was given to aftercare or maintenance of the trees, or what could happen when the meeting of the trees and sidewalks resulted in a stormy conflict. This approach is a beginning effort to provide a more desirable look to the community. However, it does not yet establish the criteria for a well-planned suburban community. The missing components are maintenance and responsibility. This incomplete attempt to regulate development fails to provide what is needed. **The result was no uniform street tree planting**.

2.6.3. "Do-It-All": Requires trees and provides Tender Local Care

Communities that provide TLC for their trees within the municipal ROW are those that have recognized their responsibility for how trees affect their surroundings. There are significant benefits from a street tree planting where trees are planted behind the sidewalk. A comprehensive street tree program provides for placing trees, whenever possible, in this more desirable place allowing for more room and less conflict with sidewalks. It also makes for safer conditions concerning overhead wires, visibility, and the need for more frequent trimming. The establishment of a tree canopy over the street by trees planted this way is more aesthetically pleasing and requires less maintenance. The evolution from "Do-Nothing" to "Do-Something" to "Do-It-All" approaches are each giant leaps in the field of Urban Forestry. In the thousands of communities in the Northeast, relatively few have taken the third leap. The visionary communities that have done so are on the cutting edge of Urban Forestry. Commitment to this approach and the giant step it takes is the promise of future success. The returns of this will affect our children's children.

Determination of fault between trees and sewers is much more difficult to assess than trees versus sidewalks. The principal reason for this is that the sewer lines are all buried in the ground and the point of conflict is not visible. The age of the sewers, as previously described, is an indicator of the frequency of problem incidents. Two distinct kinds of problems are those that occur in the street ROW and those that occur on private property.

Not Beyond This Point approach. Local governing bodies make no provisions for relief beyond the street ROW. They provide inspection, maintenance, and repair of the sewers in the street ROW. Financial considerations and legal restrictions generally prevent most sewer agencies from doing more extensive maintenance and repairs. In their search for water, tree roots will enter a defective pipe. Since the municipality installs the main line and hookup spur, the responsibility, clearly, is theirs. Because the line from the spur hookup to the building is installed and usually paid for by the property owner, responsibility is his. Generally this is the most common occurrence. Exceptions do exist. Sometimes, the sewer authority is responsible for the entire installation and chooses to maintain all of the sewer pipes.

Storm sewers installed within the street ROW and on private property, where easements have been obtained, must be maintained by the municipality. These easements, when required, serve to provide needed drainage for the area. Because of this, repair and maintenance is not an individual responsibility. When storm sewers are obstructed by tree roots the municipality must provide relief. **Repair of the faulty pipe will frustrate unwanted intrusion**.

3. Choosing the Right Tree for the Right Spot

The evolution of street tree planting locations began with the choice of planting between sidewalk and curb. In most urban communities, residential districts have limited space for a street tree planting. Even this space must be shared with utilities. With the coming of suburbia, old standards still prevailed.

Planting trees between sidewalk and curb continued to be the norm until someone said, "Look at all the room we have!" This revelation inspired people to plant trees farther back whether sidewalks existed or not. When problems arose, planners began to question the wisdom of planting between sidewalks and curbs. Trees were then placed behind the sidewalk within the street ROW. The standard street ROW usually varied from 40 to 50 feet in width. The width of the pavement varied—the 10-foot easement on either side of the road remained constant. Today's standard is generally a 50-foot ROW with 30 feet of pavement measured from the center of the road.

As suburban communities mature, sidewalk problems develop. Utilizing the same old standards cannot find the response to these problems. Innovative solutions must be sought. Tree roots and sidewalks conflict when roots mature and grow closer to the sidewalk. This is an inevitable result. The solution to the problem is to provide more growing room. Because the ROW provides no more than 10 feet of planting area on either side of the paved roadway, and sidewalks take almost half of this, added space for the tree is needed. This new additional space is for tree planting and maintenance only. This does not limit the property owner from using this space for permitted residential uses (shrubs, fences, and lawn and garden areas). Voluntary acceptance of this space limitation can be effective if adequate information is provided to the property owners. Because no binding restrictions are in place, this unforced agreement has no legal memory and cannot be enforced. The result can be no uniform street tree planting in the future.

The "Planting Easement" ideally should be an additional 10 feet on each side of the street ROW. A restrictive legal description of this area must be a part of the subdivision of the property in all new developments. Trees placed within the planting easement are to be treated the same as all other street trees under the community's care. Reluctance to adopt the "Planting Easement" concept is overcome by the irresistible logic of the solution. This new easement is the answer. It allows for unimpeded growth. **Everyone wins!**

3.1. Tree Species Selection

The trees planted along the streets in the early part of the 20th century were chosen because of their availability and hardiness. Little attention was given to their growing characteristics because street tree plantings were as new as the science of Urban Forestry. The trees found in the Northeast were primarily Maple, Oak, Chestnut, Hickory, Linden, Elm, and Sycamore. They were the few choices available. Little was known about the impact of their vigorous root systems because they had not previously made sidewalks, sewers, and trees to coexist in a limited area. Because of insufficient knowledge and technology in the field of arboriculture, at that time, little attention was given to root systems. Raised and misaligned sidewalks were not perceived as problems and were accepted as the norm with large trees. Since there was no evident problem, there was no clamor to find a solution. Lawsuits against municipalities and property owners for injuries as the result of accidents became more frequent when people began to blame someone else for these occurrences. Legal advisors, seeking compensation for their clients, would name anyone even remotely connected with the incident in their lawsuit. This bombardment of lawsuits put everyone on the defensive. The result was diminished tree care and more lawsuits.

Increased awareness and the ability to produce new species and improved cultivars created an explosion of varieties better suited for street tree planting. Most of these new trees were extensions of the old species. These new trees were less prone to insect infestations and disease. Further scientific refinement in technology resulted in improved appearance and allowed arborists to plant trees in areas otherwise considered unacceptable. As awareness for aesthetics heightened, people began to expect more and more, including problem-free trees. New species (sub species), such as Emerald Queen Norway Maple, Green Mountain Sugar Maple, Greenspire Linden, Armstrong Red Maple, and even new hybrid Elms, do not address the expansive root system problems. Very few, if any, tree species in the Northeast possess narrow columnar root systems that will grow close to sidewalks or sewers without problems. Future research and development must consider all aspects of tree growth including the root system and its effect on the environment.

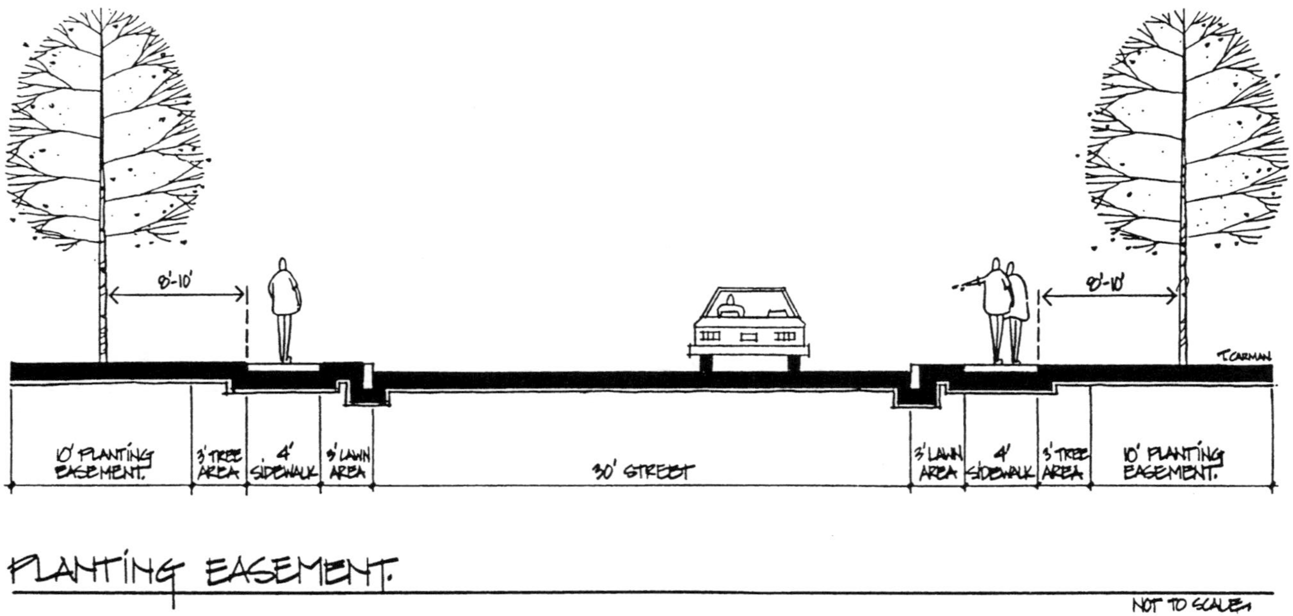

FIGURE 3. Planting easement sketch. Sketch, Tom Carman.

3.2. Planning for the Future

Adequate space is a major consideration in determining proper species selection. The root system of large street trees is almost impossible to restrict in confined spaces, without jeopardizing the health of the tree. Many ornamental and flowering trees are not suited for street tree plantings. A **street tree** needs to provide shade and be of a size at maturity that is appropriate for the characteristics of the street where it is planted. In harsh urban areas where space and soil are compact, extraordinary effort is needed for the root system to fight the hostile environment in the search for water and nutrients. This causes erratic growth of the roots that results in a worsened condition than is characteristic. No tree is going to do well in all conditions when there are extremes. A balance of soil structure, moisture, and sufficient room is needed to avoid the clash of tree versus concrete. Experience has shown certain trees do well in specific areas.

3.2.1. Confined Urban Space

- Princeton Sentry Ginkgo—Ginko biloba Princeton Sentry PNI 2720 P.P.2726
- Fastigiate European Hornbeam—Carpinus betulus "Fastigiata" Tree Form
- Columnar Norway Maple—Acer platanoides—"Columnare"
- Armstrong Freeman Maple—Acer x freemanii "Armstrong"
- Fastigate English Oak—Quercus robur "Fastigata"
- Scholartree—Sophora japonica
- Hedge Maple—Acer campestre

3.2.2. Restricted Suburban Space (Within ROW)

- Greenspire Littleleaf Linden—Tilla cordata "Greenspire" PNI 6025 P.P.2086
- Green Vase Japanese Zelkova—Zelkova serrata "Green Vase" P.P.5080 TM
- American Hornbeam—Carpinus caroliniana Tree Form
- Katsura Tree—Cercidphyllum japonica Tree Form
- Skyline Honey Locust—Gleditsia triacanthos 'Skycole' P.P.1619 TM
- Blackgum or Black Tupelo—Nyssa sylvatica
- Sawtooth Oak—Quercus acutissima

3.2.3. Unrestricted Suburban Space (Planting Easement)

- October Glory Red Maple—Acer rubrum "October Glory" PNI 0268
- Red Sunset Red Maple—Acer rubrum "Franksred"
- Green Mountain Sugar Maple—Acer saccharum "Green Mountain" PNI 0285 P.P.2339
- Summit Green Ash—Fraxinus pennsylvanica "Summit"
- Dawn Redwood—Metasequoia glyptostroboides
- Pin Oak—Quercus palustris
- Northern Red Oak—Quercus rubra
- Hybrid American Elm—Ulmus americana New Hybrids

Remember all trees have limitations. The above list is a guide to suggest trees for different locations. There may be other species and varieties available with similar characteristics that you may want to consider. Local conditions vary and may limit the suitability of some of these trees. Consult with your State Department of Agriculture, State University or State Foresters Office for their assistance.

3.3. Proper Planting Location

The "Lawn Area" or green space between the sidewalk and curb is not recommeneded for the planting of street trees. In most communities this area is too narrow and restricted for the establishment of a street tree planting. Whenever possible avoid planting in the lawn area which is a "Confined Urban Space." When the lawn area is a minimum of 4 feet in width, a tree may be selected from the "Confined Urban Space" list, if and only if no other space is available.

The "Tree Area" is the space behind the sidewalk within the ROW. The unpaved portion of the ROW is usually 10 feet (3 feet between curb and sidewalk, 4 feet sidewalk, and 3 feet tree area). The tree area differs from the lawn area in that growth is usually unrestricted on the far side of the sidewalk. The root system will seek the area of least resistence to satisfy its needs. The trees that can successfully establish themselves in this "Restricted Suburban Space" are limited. Experience has shown that many trees planted in this space will create problems in the future.

The "Planting Easement" is a new space beyond the ROW for the establishment of a uniform street tree planting under the care and maintenance of the municipality or local authority. This concept, as described earlier, allows for unrestricted and relatively trouble-free tree plantings. It also provides for expanded street tree selection formerly denied in the old "Lawn Area" and "Tree Area." This "Unrestricted Suburban Space" permits the planting of many trees that will grow into large specimens at maturity.

3.4. Proper Construction Practices

Properly constructed sidewalks and sewers resist the intrusion of tree roots. Modern practices, techniques, and materials in sewer construction have significantly reduced the incidence of tree root infiltration. All new sewer construction should be done following the latest methods. Systems needing reconstruction must be done using the most modern methods to prevent problems.

Sidewalk construction, where trees exist or could be planted, require specific design and installation. A gravel base of 4 inches in depth, or 6 inches in depth in wet areas, is needed for proper stability, and also to act as a cushion for tree root growth. Remember, tree roots need moisture and organic material to grow. This cushion of gravel will not support root growth. The root will extend under the gravel to find food, and as it grows it will displace the gravel. If the cushion is not deep enough, the concrete will be adversely affected. Class B 3500 lb/psi air entrained 6% concrete must be used for sidewalk construction. Expansion joints should be placed at a maximum of every 5 feet. This closer than usual spacing adds a buffer to the sidewalk. This smaller sidewalk slab is also a benefit when sidewalk repairs are needed in the future.

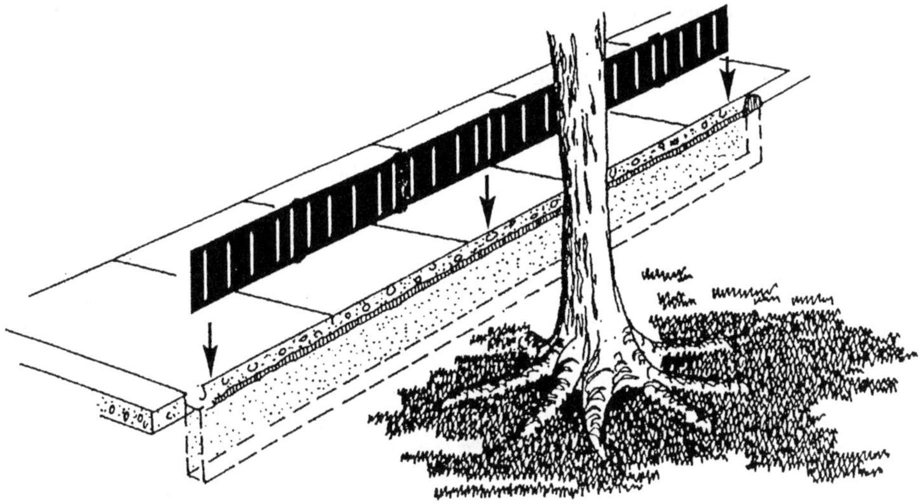

FIGURE 4. Atypical root barrier installation. Source, deep root tree barrier.

When dealing with new sidewalks that are being constructed or repaired near trees within the street ROW, the installation of a tree root barrier is suggested. A typical tree root barrier is a control device that limits root growth and/or directs the roots down. The installation of root barriers at the time of construction can reduce, delay, and possibly eliminate root and sidewalk conflict. Positive results are being observed with root barrier products and installations in the Northeast.

3.5. Options to Fix Sidewalks

3.5.1. Tree Removals

Why sidewalks became broken has been discussed previously. Now the suggested solutions to fixing them will be reviewed. The first solution that comes to mind is to remove the tree or trees. Sidewalk slabs are repaired, and tree replacement is an often overlooked option. The negative impacts of tree removal are, *no shade, no uniform street tree planting, and the lessening of property values*. Alternative methods should be explored before tree removal is decided.

This solution is the last resort.

3.5.2. Root Cutting and Tree Trimming

This is another alternative to solve the tree root/sidewalk conflict. You cannot do one without the other. If this option is selected, the tree must be inspected by a qualified arborist for health, stability, soil conditions, safety, and alternative suggestions. Root cutting halts future growth by limiting water and nutrient absorption and additionally diminishes tree support and energy storage. The possibility of windthrow is also increased when exposed to adverse weather conditions. Tree trimming and crown reduction are essential elements of this solution. Without trimming, trees are susceptible to greater stress and can succumb to the changing weather. Trees under stress are not good candidates for root cutting because poor initial health make them more prone to the attack of insects and disease. Root cutting is done best with a commercial "root cutting machine", and caution must be taken when cutting near the root collar. Other equipment, such as a backhoe, is not suited for root cutting. Clean, precise cuts, not ripped wood, heal faster. Root barriers can be placed in a trench to prevent regrowth.

FIGURE 5. Example of contoured sidewalk installation. Who said must be straight and level? Sidewalks must be safe, functional, and appropriate fir their location. Photo, William R. Comery.

3.5.3. Contoured Sidewalks

This solution disrupts the trees less than any other. This method acts on the concrete and not the tree and should be the primary option. Who said, "Sidewalks must be straight and level?". Sidewalks must be safe, functional, and appropriate for their location. This option creates "breathing room" for the tree and eliminates, for a considerable time, more drastic action. Contoured sidewalks can be constructed in such

FIGURE 6. Contoured sidewalk creates "breathing room" for the tree. Photo, William R. Comery. Note: For most trees 10 to 12 feet of DeepRoot Barrier will be adequate to protect the sidewalk from future damage.

a novel way that they can accomodate the tree and comply with new regulations such as the Americans with Disabilities Act.

During contoured sidewalk construction, root barriers can be placed in a trench to prevent regrowth. Depending on the site conditions, different pavement materials can be used in contoured sidewalk construction. Dramatic developments in paving materials and construction, such as concrete pavers, brick pavers, and asphalt, have fostered the design and installation of contoured sidewalks.

3.5.4. Replacement Tree Planting Policy

Whenever a tree is removed, for any reason, it must be replaced with a new tree. This ensures the integrity of the community. A uniform street tree planting is an essential part of planning for the future. Once the problem has been identified, no excuse for less trees is acceptable. Choice of species, quantity, and location are the only determining factors in replacing street trees.

"NO TREES" is not an option!

Chapter 17

Hazard Tree Inspection, Evaluation, and Management

E. Thomas Smiley, Bruce R. Fraedrich, and Peter Fengler

1. Introduction

Wherever trees and people cross paths, the potential for personal injury or property damage resulting from some type of tree failure is of public concern. When the potential for damage is great and likely to occur relatively soon, immediate and drastic measures may be required to mitigate the problem. If the potential for failure is slight and the extent of probable damage minimal, no immediate action may be necessary. Determining the appropriate course of action in these situations is often complex and problematic.

With the urban forest aging at an increasing rate and society becoming more litigious, factors such as budgetary considerations, questions of liability, and conflicting community goals often play as important a role in the decision-making process as the arborist's expertise in tree care. As a result, the detection, evaluation, and management of hazardous trees have become a major concern for urban foresters and park managers throughout the country.

Advancements in our understanding of tree structure and function, along with the development of new technology to examine the effect of natural and man-made stress on the tree's ability to function have made it possible to more readily identify potential tree problems. As we continue to gain more knowledge and experience with trees, it becomes increasingly important for urban tree managers to include tree structure evaluation as part of their overall management programs. In so doing, they provide their constituents with an essential community service: reducing the potential for personal injury and property damage resulting from tree failure.

E. Thomas Smiley, Bruce R. Fraedrich and Peter Fengler (Works For Asplundh Tree Expert Co.)
Bartlett Tree Research Laboratories, 13768 Hamilton Road, Charlotte, NC 28278, 704-588-1150.
Urban and Community Forestry in the Northeast, 2nd ed., edited by, J. E. Kuser.

2. Methods of Tree Evaluation and Hazard Assessment

According to Clark and Matheny (1991), the primary goal of tree structure evaluation is to identify potentially hazardous trees so that they can be treated before failure occurs. By evaluating trees and rating the hazards associated with them, the arborist can set priorities and schedule abatement procedures in order to significantly reduce the risk involved with living in an urban forest. To help resource managers with this task, various formulas and methods of evaluating trees for hazards have been developed.

Researchers from the US Forest Service such as Paine (1971), Sharon, and Robbins, among others, have been involved in tree hazard identification and assessment in parks and recreation areas throughout the country. Clark and Matheny (1991) have developed a photographic guide to the evaluation of trees in urban areas. Other researchers, such as Smiley and Fraedrich (1992) and Mattheck and Breloer (1993), have concentrated their efforts on determining the strength characteristics of trees under normal and abnormal conditions of stress. In each case, the objective is to develop methods that will help to mitigate the potential damage trees can cause.

Costello and Berry (1991), on the other hand, have taken a somewhat different approach. They have developed a program aimed at establishing species "failure profiles" in order to more accurately assess the failure potential for different taxa. By accumulating data on failed trees, such as identifying and documenting the frequency and type of failure for a particular species, trends for that species may become evident.

3. Identification of Hazardous Conditions

What makes a tree a hazard?

For a tree to be considered unsafe to the public, it must meet either of the following criteria

1. It must possess some type of structural defect (Fig. 1) that predisposes it to failure, and it must be associated with a target such as a building, road, walkway, or recreational area where there are people and property present. For example, a tree with a large basal cavity would not be hazardous if growing in a forested area, away from people or property. However, it could be considered a hazard on a residential homesite, a campground, or if growing along a street or in a park where it would come in contact with the public.
2. On the other hand, a tree that is structurally sound may also be considered hazardous if it interferes with the routine activities of people. Examples of such situations include the following:
   - Tree stems or branches obstructing motorists' vision,
   - Tree roots raising sidewalks, and
   - The interruption of utility services resulting from branches contacting the wires. Deterioration or weakness in any of the structural components (stems, branches, and roots) of a tree may result in a hazardous condition (Fig. 1).

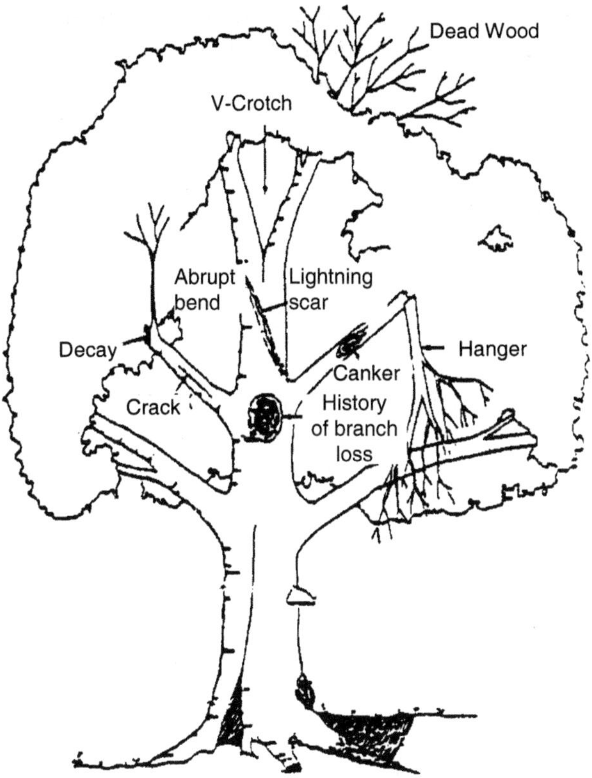

FIGURE 1. Schematic diagram of defects associated with hazardous trees.

Since both a tree with defects and a healthy tree in an inopportune location may be problematic, a systematic and thorough inspection of the entire tree and its surrounding area is necessary to detect a potentially hazardous condition. The following is a description of conditions to evaluate.

3.1. Tree Location

Although the location of a tree does not directly render it hazardous, hazardous trees are more likely found in certain locations or circumstances, including:

1. At the edge of wooded areas (edge trees): Native trees that were present on site prior to development frequently have a high incidence of defect and disorder after the site has been developed. Trees bordering the edge of natural (wooded) areas, or roads, drives, and parking areas, generally are exposed and subject to stress during storms. These trees may also have experienced root damage during land clearing. It is a good practice to inspect trees in wooded areas that are 1.5 times their height from a potential target. For example, a 60-foot tall tree is inspected if it is within 90 feet from the edge. Even if the trunk appears undamaged, the roots may have severe problems.

2. Lone trees: *Especially* when very tall, these trees are more prone to lightning strikes. If they are the only remnants of a previously existing wood lot, lone trees may have damaged or undersized root systems and be prone to windthrow. Lone trees with less than one-third of their height with live branches (low live crown ratio) are considered severe risk trees.
3. High traffic areas: Trees in areas of high vehicular or pedestrian traffic are prone to soil compaction, root wounding, and subsequent root decay. Root disease and similar disorders are common in high traffic areas.
4. Wet sites and shallow soils: Trees growing on wet sites generally have a shallow root system that predisposes them to windthrow from ground or root failure. Sites that have been altered suddenly by grade changes resulting in poor drainage are predisposed to root mortality from suffocation. Trees in areas that receive excess irrigation are also likely to have root rot.

3.2. Problem Conditions and Locations

Structurally sound trees may create hazardous conditions if they are in inappropriate locations. Some of the problems caused by trees due to their locations rather than conditions include the following.

1. View obstructions: Obstructed view of motorists, resulting in automobile crashes, is one of the leading causes of large lawsuits against municipal forestry departments. Areas to be examined include locations in front of stop signs, signal lights, and other essential traffic signs. Clear vision must also be maintained at all intersections and curves in the roadway.
2. Road edge hazards: The distance between the edge of the roadway and a tree is regulated in some states. In general, the higher the legal speed limit, the greater the required space between tree and road should be.
3. Sidewalk interference: Roots from many species of trees grow under sidewalks. As the roots growing under sidewalks increase in diameter, they may crack and/or lift sections of pavement, creating tripping hazards for pedestrians. This problem can be reduced by proper species selection, not planting in areas that restrict root development, appropriate sidewalk design and construction, or careful root growth management. Great care must be taken in any root-cutting operation used to correct sidewalk problems so that the tree retains its structural support system.
4. Power line conflict: The primary reasons for maintaining adequate clearance between electric wires and trees are to ensure public safety and reliable electric service to the community. When trees contact wires, an electric current may flow down the tree to ground. Contacting an electrically charged tree can result in serious injury.
5. Susceptibility to lightning strikes: Trees are susceptible to lightning strikes mainly because of their height. While there are species differences (Table 1), any tree can provide a path for lightning. Trees of greatest concern are those under which people may take refuge when on a golf course, campground, or other recreational site. Shelters under large trees are also susceptible since lightning will sometimes jump or side flash from tree to structure.

Table 1. Lightning Strike susceptibility of non-protected, temperate zone trees. (from Smiley et. al. 2002)

Tree species	Susceptibility to Lightning Strikes
Acer (maple)	moderate
Aesculus (horsechestnut)	low
Betula (birch)	moderate to low
Catalpa (catalpa)	moderate
Fagus (beech)	low
Fraxinus (ash)	high
Ilex (holly)	low
Liriodendron (tulip popular)	very high
Picea (spurce)	moderate
Pinus (pine)	moderate
Platanus (sycamore)	moderate
Populus (poplar)	moderate
Quercus (oak)	high
Robinia (black locust)	very high
Tsuga (hemlock)	moderate
Ulmus (elm)	moderate

3.3. Tree Defects

A thorough inspection of the tree's branches, stem, and root collar is essential in detecting hazardous conditions. The following is a list of commonly encountered defects that require periodic examination in order to determine what type of hazard, if any, they pose to the public

3.3.1. Crown

- Broken branches and hangers: Broken branches that remain attached to the tree as well as 'hangers' (non attached broken branches) can easily be detected by their brown wilted leaves or their out-of-place look within the crown of the tree. Look for broken ends and unusual angles.
- Dead and weakly attached branches: Look for dead leaves, defoliated branches, dead or missing buds, blunt branch tips, fungal structures, and loose bark on branches over 1 inch in diameter. Pay special attention to crown areas over roadways, drives, walkways and other man-made structures. Inappropriate pruning practices, such as topping, often cause the affected branch to dieback to its parent stem producing new sprouts in an effort to supply the energy needed to support the branch. However, these new sprouts, as well as those branches with a narrow angle of attachment and included bark are more prone to breakage of their weak attachment to their parent stem. This is due to decay moving down the branch below improperly made cuts often undermines the structural integrity of the wood and inheretantly weaker wood (Greg Dable, personal communication).
- Unbalanced crown and low live crown ratio: Trees with unbalanced asymmetrical crowns have weight distributed poorly over the stem. These trees are more prone to failure when combined with other stem and root defects. Crowns can be

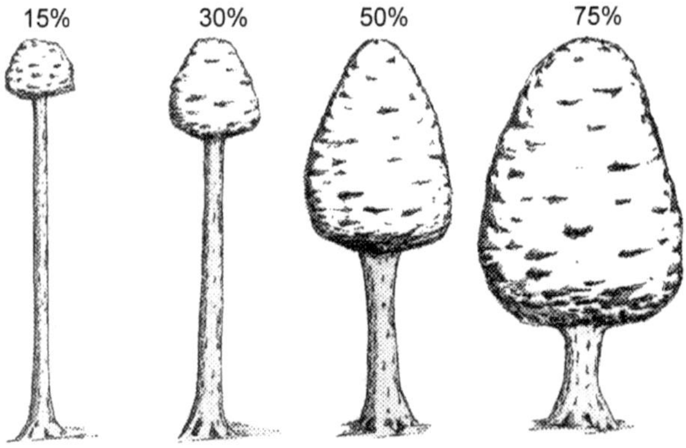

FIGURE 2. Live crown ratios < 30% are at a higher risk of failure.

unbalanced due to leaning stems, crown development on one "side" of the stem, or low live crown ratios (Fig. 2). Excessive end weight or low live crown ration are often caused by poor pruning practices ("lion's tailing") and can lead to branch and stem failures. Stems and branches with low live crown ratios and heavy branch ends often have poor taper. Trees with live crown ratios 30% or that have been lion's tailed pose at least a moderate risk of stem and branch failure.

- Abrupt bend and weakly attached branches/sprouts: Branches with a narrow angle of attachment with included bark are more susceptible to breakage. Branches that have developed from topping cuts are poorly attached. Abrupt bends indicate old wounds that are prone to decay and drying of wood tissues. Typically, these defects represent a moderate failure potential. If cracks in the wood are present, they should be considered a high to critical risk.

3.3.2. Trunk and Scaffold Branches

- Lightning injury: Vertical bark injuries extending in a spiral pattern from the top to the base of the tree usually indicate lightning injury. Wood affected by lightning decays rapidly. Contrary to popular belief, trees that have been struck once by lightning are more likely to be struck again.
- Weak junctions ("V"-crotches): Leaders and major branches that form "V"-crotches are very prone to breakage. The weakest attachments are those between two or more nearly equal leaders or branches (Gilman, 2003; Rob Ferrell, personal communication). If there is bark included between these branches the strength is reduced an additional 10% (Smiley, 2003). As opposing leaders grow in size, stress, and the likelihood of breakage increase.
- Leans: Trees that have grown in a leaning position are not hazardous in and of themselves. However, when associated with decay or other defect, a lean adds considerable stress on the stem and increases the likelihood of failure. Lean can be monitored over time using a digital level (Smiley, 2005).
- Multiple stems from the same root system: Multiple stemmed trees originating from stump sprouts (coppice growth) often have extensive internal decay in the

area of the root collar. Sprouts formed low on healthy stumps are less likely to be decayed.
- Abnormal flare root collar: Abnormal proliferation of the lower trunk tissues often is associated with lower stem and root decay. If the buttress roots are not clearly visible, there may be a suppression of the root system that could make it incapable of supporting the tree.

3.3.3. Roots/Root Collar

- Fungus Fruiting Structures: Any fungus fruiting structure including mushrooms or conks growing on the root collar may indicate extensive root deterioration. Mushrooms growing from the soil adjacent to trees may be those of a root pathogen or of beneficial mycorrhizal fungi. Positive identification of the mushroom away from the root collar is necessary to determine its possible role in decay.
- Loose/dead bark: When the cambium is killed by a pathogen or by mechanical means, the bark outside of it will become less firmly attached to the tree. Root decay is often associated with loose or dead bark on the root collar.
- Lack of root flare: Trees that lack a distinct basal flare may be "buried" by fill soil. This will predispose them to deterioration or decay of wood on the lower trunk and root collar area. A post hurricane study found that one-third of the root- or soil-failure trees had soil above the buttress roots. This was compared to 8% of the surviving trees (Smiley *et al.*, 1998).
- "Mounded" trees: Trees growing on raised mounds may have been affected by a grade cut. This may result in loss of support roots, and thus predispose the tree to windthrow.
- Trunk "bleeding": Bleeding from the trunk or low pruning wounds is occasionally associated with root decay.
- Abnormal Flare: There are many causes for abnormal trunk flare, including wounds associated with decay.

3.3.4. Stumps

Stumps left in areas where pedestrian, mower, or vehicle traffic occurs can cause tripping, mechanical damage, or vehicle rollover. The most dangerous stumps are those between 2 and 8 inches high, which may be difficult to see in tall grass. To eliminate stump hazards, they should be ground to below the level of the soil surface whenever possible. If this cannot be done, then the stump should be cut flush to the soil surface or it should be left high enough to be visually obvious.

3.4. Decay

Decay in the trunk or a major branch is the most common hazardous defect encountered in urban trees. Symptoms of decay, such as open cavities, may be quite obvious or they may be inconspicuous and require careful examination. When there is sufficient decay present inside a tree, it will be more prone to failure. Small amounts of decay cause no structural problem for most hardwood trees. Symptoms of decay include as following.

- Open cavity: Cavities represent deterioration of bark, sapwood, and heartwood; they are usually located where the initial injury, which started the decay process, occurred (Fig. 3).
- Fungus fruiting structures: Most types of fungus fruiting bodies, including shelf like conks, mushrooms, or bracket structures growing on live trees, indicate extensive wood decay.
- Loose dead bark or wounds: Loose or dead bark typically indicates that the cambium and often the wood beneath are dead. This condition may occur with decay or prior to it.
- Cracks: Cracks are radial separations in wood and bark. They may be associated with extensive wood decay, minor wounds, rapid growth rate, or sudden temperature changes. Cracks allow the wood beneath to dry and become more brittle. When fresh cracks are associated with decay, the tree may be in the process of falling apart and should be assessed immediately.
- Seams: Seams are radial separations of the bark tissues that have been closed by callus growth. Like cracks, seams may result from rapid growth rate or sudden temperature changes or may often be associated with wood decay. Two seams on opposite "sides" of the tree usually indicate extensive internal decay.

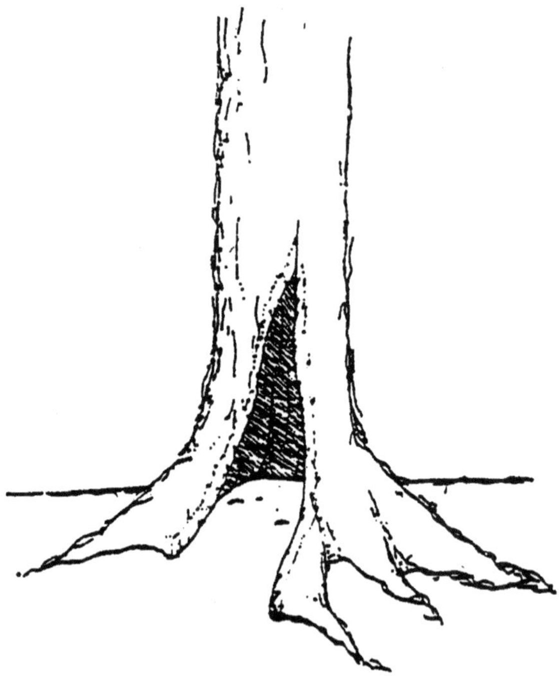

FIGURE 3. Severe decay and cavity opening.

- Burl: An abnormal proliferation of wood and bark tissue on localized areas of trunks and branches may be associated with decay. They also may be caused by pathogens, including bacteria, viruses, and fungi that do not cause decay or reduce structural integrity.
- Cankers: Cankers are localized dead areas on the bark of trunks and branches that are caused by fungal and bacterial pathogens. Wood beneath cankers may be decayed or the structural strength may be reduced if the canker inhibited wood growth for a number of years.

3.5. Root Defect Evaluation

Typically, one-third of all tree failures are related to the root system. However, up to 75% of failures are root related in some areas. The majority of tree failures occurs when winds exceed 50 mph as during hurricanes and thunderstorms. However, failure may occur under any wind conditions if the roots are sufficiently weakened.

Two types of failure have been identified: root failure and soil failure. *Soil failure* is extremely difficult to predict. Failure occurs when the soil does not have enough strength to keep the roots down. This type of failure usually occurs in saturated soils. Failure is more common on very shallow (<2-feet deep) soils. Soil failure also occurs when trees are surrounded by pavement that does not allow the root system to develop sufficiently to support the tree.

Root failure occurs when roots break or are cut and therefore unable to provide the necessary support for the tree. Root failure occurs more readily on trees that have root decay, where roots have been cut for construction, or when other root problems exist. This is the most common cause of failure when trees fall over on "calm" days. The two factors are involved with root decay are the number of roots that have decay and the degree of decay in each root.

Groups of trees, recently opened stands, and newly created edge trees are more susceptible to windthrow due to lack of root spread and increased susceptibility to root disease. Although root disease can be detected, this is a relatively difficult procedure to perform.

3.6. Symptoms of Root Problems

Trees with extensive root decay often show little or no symptoms of decline. Tree health should never be confused with tree stability. External indicators of root decay include:

— Dead (loose) bark on the roots, root flare, or lower trunk
— Fungal fruiting structures around the root flare. These include mushrooms, conks, and bracts on or immediately adjacent to the tree
— Oozing from the root flare, lower trunk, or wounds on the lower trunk
— Cuts or fill soil moved beneath the tree
— Cracks in the soil above or beside major roots

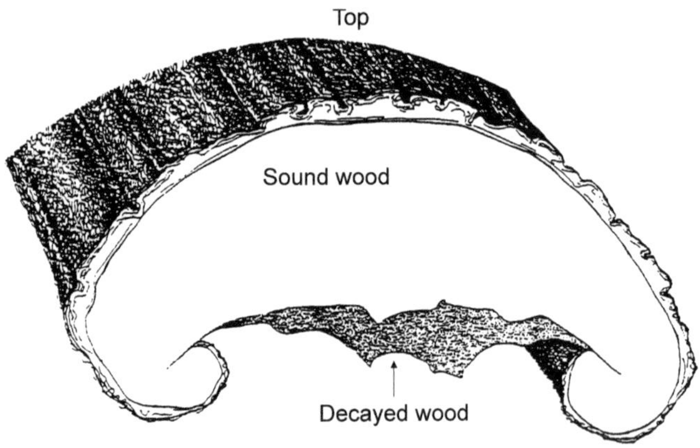

FIGURE 4. Typical pattern of root decay, starting from the lower side working upward.

Although many defects including dead branches, hangers, and "V"-crotches can be detected and fully assessed based solely on visual inspection, a more thorough examination is required to evaluate the impact of wood decay and root disorders. For wood decay, evaluation involves probing the stem or branch to determine the extent of decay and then calculating the strength loss resulting from deterioration of the woody tissue. For root decay, the process is more difficult, since decay starts at the bottom and tips of the root and progresses upward and toward the trunk (Fig. 4). The root collar area may require excavation and examination of the buttress roots for discoloration and decay (Fig. 5). For further information regarding the detailed procedures for evaluating wood decay and root disorders in living trees, see Smiley and Fraedrich (1991) or Fraedrich and Smiley (2002).

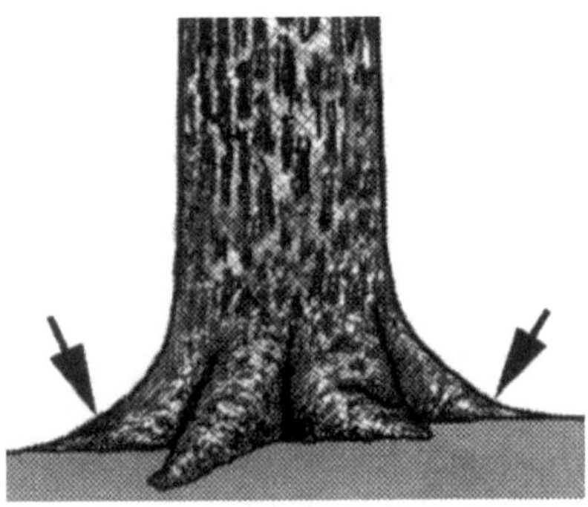

FIGURE 5. Sampling for decay in the roots is done slightly below the midpoint in the transition from horizontal to vertical (see arrows).

4. Management of Hazard Trees

At what point does the threat of personal injury or property damage posed by a tree growing along a city street or near a playground become severe enough to remove that tree? The answer to this question requires experience, good judgment, a thorough knowledge of tree biology, and a basic understanding of the legal responsibility associated with the management of publicly maintained trees. However, it also requires a management program that is being implemented. The most sophisticated plan and inventory database will not mean a thing if it is not being utilized.

By implementing a systematic approach to the hazard evaluation process, a number of important management objectives can be achieved. First, prioritization is possible, allowing for timely hazard abatement. Second, the data collected can become an important component in the development of the administrator's tree care budget. Third, this information becomes essential in supporting existing policies and procedures, particularly those focusing on issues of liability. When based on accepted industry standards and work practices, tree hazard evaluation provides resource managers with the necessary foundation to balance the risks and benefits associated with the management of trees in their communities.

But why should a community be asked to "balance" the obvious risks of personal injury or property damage resulting from tree failure against its perceived benefits? Should that large red oak on Main Street, whose roots have been lifting the sidewalk or invading the sewer line, and whose branches are interfering with the electrical conductors, be removed before it causes major property damage or serious personal injury?

This is the type of question each city forester, resource manager, and shade tree commission must wrestle with on a daily basis. Urban land managers have a duty to inspect trees periodically in order to spot the dangerously defective ones. This obligation is of special importance in developing suburbs, where more trees are left during construction, often in poor condition (Anderson and Eaton, 1986). Even trees located in more sparsely populated rural areas of the country are now coming under increasing scrutiny in order to avoid potential liability and costly litigation resulting from various forms of tree failure.

In a landmark case involving the State of Connecticut in 1987 (see Chapter 3, this volume), the court ruled that even rural public lands—for which a lesser standard of care had typically been allowed—were subject to inspection for hazardous trees under certain circumstances. As a result, public agencies responsible for tree maintenance have been put on notice to pay closer attention to the management of their trees, even in rural areas, or suffer possible legal and financial consequences.

4.1. Decision-Making

After a tree has been inspected and hazards have been identified, a decision must be made on how best to remedy the problem. Choices may range from taking no immediate action to removal of the entire tree, depending upon the nature and severity of the problem. The decision-making process often becomes more difficult when tree maintenance budgets are limited. Insufficient funds to perform all of the required work in one budget year may force resource managers to make tough choices. As a result, priorities must be developed and thresholds for action implemented based on each community's special circumstances.

The following are some general guidelines for dealing with hazards associated with trees.

1. Dead trees: If a tree is dead and has a target, or more than 50% of its major branches are dead, it should be removed.
2. Dead branches/hangers: As a general rule, all hangers and dead branches 2 inches in diameter or greater should be pruned or removed from the crown of a tree. For a comprehensive description of pruning practices and procedures, refer to the current revision of the American National Standards Institute (ANSI, 2001) A300 pruning guidelines.
3. Decayed branch/overextended branch: When decay has significantly weakened a branch, remove the branch. Otherwise, prune to reduce weight and/or install cable to support weight. An overextended branch can be defined as one that is long for its species and does not have branches or leaves growing along the inner/lower two-thirds of its length. This type of branch should be correctively pruned or removed.
4. Weak junctions (V-crotches): On young trees, remove or subordinate one of the leaders. On mature trees, bolt and cable branches if overall health of the tree warrants it. If severe decay is present, remove affected branch(es).
5. Trunk decay: Trees with significant internal decay should be removed. Removal of trees with lesser amounts of decay should be considered if other structural defects are present including leans, one-sided crown, cankers, and high-value targets.
6. Lightning protection: Large trees where people may congregate during storms and trees within 10 feet of a house or other building are of major concern. Lightning may sideflash to the house, often starting a fire, or branches broken by a strike may fall on the structure. Properly installed lightning protection systems are extremely effective at reducing the damage to trees and adjacent property. Lightning protection systems must be inspected visually every year. At least every 5 years the tree should be climbed, and all conductors should be inspected. Repairs should be made as soon as faults are discovered. They should be made using the same materials as the original system to avoid electrolysis-induced corrosion of the system. Standards for lightning protection installation are presented in the ANSI A300 Part 4.

4.2. Strength Loss Thresholds From Stem Decay

Wagener (1963) states that a conifer can tolerate up to a one-third loss in strength without affecting the safety of a tree if the weakening agent is heart rot uncomplicated by other defects. He emphasizes that the one-third strength loss limit is not absolute but only a general guideline. He also states that a specific strength loss standard is less applicable to hardwoods because of variations in basic form between conifers and hardwoods, the variant influence of leverage on the breakage potential, and the high mechanical strength of the wood of many hardwood species.

Table 2. Trees with Structurally Weak Wood and Decay Susceptibility*

Botanical name	Common name
Acer negundo	Box elder
Acer platanoides	Norway Maple
Acer pseudoplatanus	Sycamore Maple
Acer rubrum	Red Maple
Acer saccharinum	Silver Maple
Aesculus spp.	Horsechestnut/Buckeye
Betula spp.	Birch
Liriodendron tulipifera	Tulip tree/Yellow Poplar
Populus deltoides	Cottonwood
Salix spp.	Willow
Tilia spp.	Linden

*Conifers including pine, spruce, fir, Douglas fir, and arborvitae have relatively weak wood but are not particularly prone to decay.

Smiley and Fraedrich (1992) surveyed hardwood trees that were broken during 1989's Hurricane Hugo in Charlotte, North Carolina. Sustained winds were 69 mph with gusts to 90 mph during the storm. He found that 52 of the 54 broken trees had internal decay. Using formulas proposed by Wagener (1963) and modified by the Bartlett Tree Research Laboratory, strength loss of broken trees averaged 33%. These results have been verified, although a different expression of the formula is presented by Mattheck and Breloer (1993). The Bartlett thresholds are presented in Table 3.

A strength loss index of 33% is equivalent to a shell of sound wood that is 15% of the stem/branch diameter at the point of defect, assuming no cavity opening. These values are the maximum to be tolerated (high-risk category). Strength loss threshold often must be reduced after considering the following factors:

- Wood strength: Species such as oak are much stronger than red maple or tulip tree (Table 2).
- Severity of stress and exposure: Lone or widely spaced trees are more prone to failure than protected trees.

Table 3. Minimum Thickness of Sound Wood Surrounding Decay Columns on Stems with and Without Cavity Openings

Cavity opening % of circumference	Minimum thickness of sound wood surrounding decay (wood diameter X)	
	High risk	Critical risk
0	0.15	0.10
5	0.17	0.11
10	0.18	0.12
15	0.20	0.14
20	0.23	0.15
25	0.26	0.17
30	0.33	0.18

- Proximity of decay to main branch crotches: Where decay is associated with a branch crotch, especially "V" crotches, trees will be more prone to failure.
- Lean: Although a sound tree that leans is not inherently more hazardous than a non-leaning tree of the same type, when decay occurs in leaning trees, stem failure is more likely. Decay of reaction wood can increase the risk of failure.
- Size and density of crown: Large, dense, or uneven crowns increase the "sail effect" of the crown, which increases the risk of failure in decayed stems.
- Target value: Decayed trees become more of a concern in locations that are heavily used by people. Examples include playgrounds, schools, parks, and roadways.
- Tree value: A declining tree, or one of low aesthetic or functional value, is often removed more readily than a healthy tree that contributes greatly to the landscape.

4.2.1. Sapwood Decay and Mechanical Damage to the Trunk

Occasionally, the sapwood of a tree is removed from the outside toward the center of the tree. This type of damage is done by beavers chewing on trees, cars crashing into trees, or by certain decay fungi. If one-third or more of the trunk diameter has been decayed or mechanically removed, the tree should be taken down (Fig. 6). As with internal decay, if there are other factors present that may predispose the tree to failure, this threshold value should be reduced.

4.2.2. Roots

Assessing root decay is complicated by the fact that the decay is frequently more severe than detection procedures will indicate. Subsequently, whenever any root or basal decay is encountered, the tree care specialist should be aware that root disease may be more severe than anticipated. There is always a danger of soil failure (windthrow) when root decay is encountered.

What is a significant amount of decay in roots? A study of tree failure after Hurricane Fran in the Raleigh, Durham, and Chapel Hill area of North Carolina in 1996 by Smiley attempted to determine how much root decay was too much. These

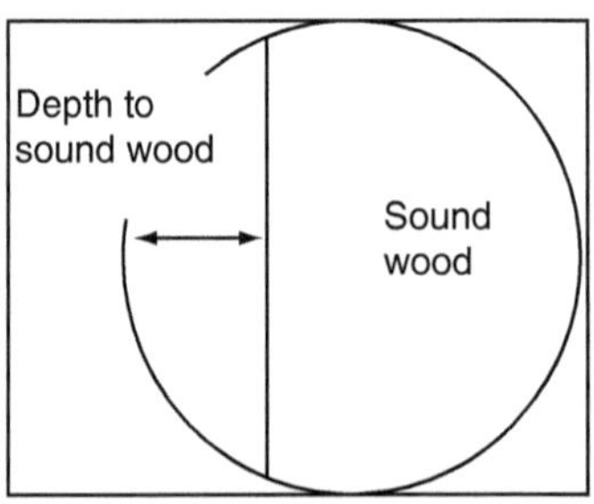

FIGURE 6. If more than one-third of the cross-sectional area has been decayed or was removed, the tree is prone to failure and should be removed.

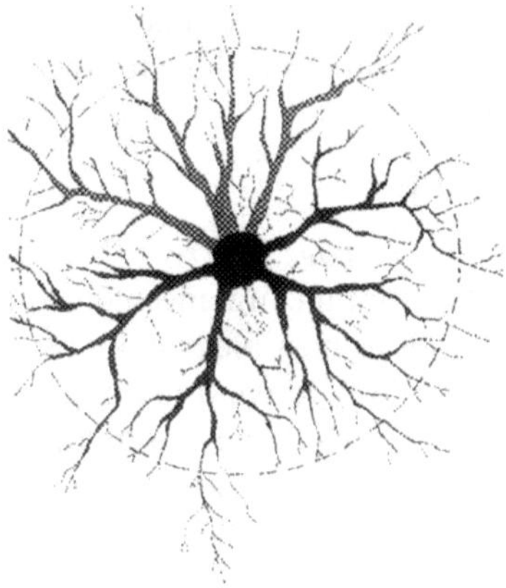

FIGURE 7. If one-third or more of the root system is missing or significantly decayed, the tree should be removed.

findings indicate that if the amount of sound wood present on major buttress roots was less than diameter at breast height (DBH)X 0.15, then the tree was more prone to failure. Therefore, the threshold for root decay is if there is less than DBH X 0.15 inches of wood on one-third or more of the root system, the tree should be removed (Fig. 7).

High-risk trees may tolerate a lower percentage of root decay. High-risk trees include the following:

1. Leaning trees
2. Tree with limited root space
3. Trees at the edge of recently cleared areas where severe windstorms frequently occur
4. Trees with large and/or dense crowns
5. Trees which have soil fractures associated with one or more major roots

4.3. Utility Line Clearance and Tree Hazards

In order to maintain a reliable network of transmission and distribution lines throughout the country, specially trained line clearance crews prune trees in order to prevent them from becoming a hazard to the electrical system. The type of pruning required for maximum electrical reliability sometimes leaves a tree looking deformed and unattractive. Trying to balance the interests of electrical system integrity with tree health and beauty is often difficult (see Chapter 15, this volume). Deciding whether the hazard a tree poses to a transformer or electrical line warrants its removal (rather than some other form of remedial action) requires the professional judgment and experience of the tree care manager and the tree care practitioner. The action taken should be based on a maintenance plan that takes into consideration factors such as local geography, climate, weather, tree species, tree maintenance funding, and community preferences. As a result, line clearance practices may vary regionally and throughout the country.

4.4. Storms and Tree Hazards

In their book, *The Body Language of Trees*, Mattheck and Breloer (1993) eloquently describe trees as "self-optimizing mechanical structures, which make as economic a use of their materials as possible and are as strong as necessary." It seems that trees have the capacity to adjust to wind loading and release, meaning they can biologically adapt their structures by adding woody material, where necessary, to the stem, roots, and branches, in order to adjust to ever-changing environmental conditions. As a result, they become directionally "wind firm," able to withstand what would be considered "normal" wind and weather conditions (Coder, 1989). However, severe storms, such as blizzards, ice storms, tornadoes, and hurricanes, regularly produce abnormal conditions that stress trees beyond their ability to protect themselves against structural damage.

What can urban tree managers and the public do to minimize the effect of storms on trees? Potentially, quite a bit. Since various factors such as a tree's genetic makeup, physical condition, and maintenance history play a part in its ability to survive severe weather, it is important to use proper cultural practices in the establishment of an urban forest. First, select the appropriate tree species and location for each new planting. Second, install only healthy nursery stock. Third, follow the correct planting techniques (ANSI A 300 part 6) and provide appropriate aftercare to ensure optimum tree health and vigor. Finally, prune trees as required for proper growth and structural development (ANSI A300 part 1).

Once the storm has passed, the following steps should be taken, to mitigate the resulting damage:

1. Inspect and assess each tree's roots, trunk, and crown condition. Follow the ANSI A300 standard part 1 for pruning methods to remove dead, damaged or diseased branches, stems, or roots on those trees worth keeping; remove trees that pose a significant risk to the community.
2. Evaluate all planting sites before replacements are installed. Do not replant an inappropriate species for the site; don't plant at all if the site is inappropriate.
3. Inform and educate the public about your tree maintenance program objectives.

The sooner these steps are taken, the more likely surviving trees will benefit from your efforts; perhaps even more importantly, newly planted trees will have a better chance to survive the next major storm.

4.5. Learning More About Predicting Tree Failures

In 2005 the United States Department of Agriculture (USDA) Forest Service in conjunction with private industry and the ISA have established the International Tree Failure Database (ITFD). This project is intended to collect information about failed trees so that future tree failures can be more accurately predicted. Regional training and meetings will provide information on this project. The website is http://ftcweb.fs.fed.us/natfdb/. Once the database is populated, information will be available on many aspects of tree failure including species profiles, wind speeds and failures and other factors associated with tree failure. Tree failure information in welcomed from all municipalities and should become part of routine pest storm activities.

5. Summary

In order to take advantage of the physical, environmental, and psychological advantages trees provide, we must be willing to accept a certain amount of risk. That risk be calculated, not ignored negligence or apathy. A community has a responsibility to maintain its trees, as it does the other parts of its urban infrastructure. Roads are continually being repaired or resurfaced, bridges torn down and rebuilt. Trees, however, are often left to take care of themselves, sometimes in extremely harsh urban environments. This benign neglect of a key component of our urban infrastructure has produced an urban forest that for many communities has become a liability to its residents rather than its most important asset.

The threat of personal bodily injury or property damage resulting from a hazardous tree is all too real and often a consequence of poor or nonexistent maintenance practices. As a result, the potential for municipal liability becomes greater. Therefore, each community needs to provide those individuals charged with the care and maintenance of its tree resource with sufficient financial support and operational leeway to make informed decisions, using the latest information and technology available to them. There are numerous outstanding publications throughout the literature that focus primarily on the identification of defective trees, some of which have been referenced in this chapter. There are fewer educational and reference materials that deal with how and why to set up a tree risk management program. One that addresses these issues is available from the USDA Forest Service and called "Urban Tree Risk Management: A Community Guide to Program Design and Implementation." This publication stands out because of its emphasis on establishing a program that will provide communities with a step-by-step process for the detection, assessment, prevention, and correction of hazardous tree defects.

Will we ever be able to prevent trees from causing further damage to people and property? Certainly not. However, we can reduce the number and severity of tree-related damage claims and also extend the useful life of many of our urban trees if we take advantage of the research and experience available to us.

References

American National Standards Institute (ANSI), 1995, *Standard Practices for Tree, Shrub, and Other Woody Plant Maintenance*, ANSI A300-1995, ANSI, Washington, DC.

Anderson, L. M., and Eaton, T. A., 1986, Liability from hazardous trees, *J. Arboric.* **12**(8):189–195.

Clark, J., and Matheny, N., 1991, *A Photographic Guide to the Evaluation of Hazard Trees in Urban Areas*, International Society of Arboriculture, Champaign, IL.

Coder, K. D., 1989, Should you or shouldn't you fill tree hollows, *Grounds Maintenance* **24:**9:68–70, 98–100.

Costello, L. R., and Berry, A. M., 1991, The California tree failure report programs: An overview, *J. Arboric.* **17**(9):250–255.

Fraedrich, B. R. and E. T. Smiley. 2002. Assessing the failure potential of tree roots. *In* Smiley, E. T. and K. D. Coder (ed) 2002. Tree Strucutre and Mechanics Conference Proceedings. Intern. Soc. Arboric. Champaign IL pp. 159–163.

Gilman, E. F. 2003. Branch to stem diameter ration affects strength of attachment. J. Arboric. 29(5) 291.

Lightning Protection Institution, 1987, *Standards of Practice*, LPI 175, Lightning Protection Institute, 3335 N. Arlington Heights Road, Suite E, Arlington Heights, IL 60004.

Mattheck, C., and Breloer, H., 1993, *The Body Language of Trees*, HMSO Publishing, London.

National Arborist Association, 1987, *Lightning Protection Installation Standard*, National Arborist Association, Amherst, NH.

Paine, L., 1971, Accident hazard: Evaluation and control decisions on forested recreation sites, in *USDA Forest Service Pacific Southwest Forest and Range Experiment Station Research Paper*, PSW-68.

Pokorny, J. D., 2003, *Urban Tree Risk Management: A Community Guide to Program Design and Implementation*, NA-TP-03-03, USDA Forest Service Northeastern Area *State and Private Forestry*

Smiley, E. T., and Fraedrich, B. R., 1991, *Hazardous Tree Evaluation and Management*, F. A. Bartlett Tree Expert Company, Charlotte, NC.

Smiley, E. T., and Fraedrich, B. R., 1992, Determining strength loss from decay, *J. Arboric.* **18**(4):201–204.

Smiley, E. T., Martin, T. R., and Fraedrich, B. R., 1998, Tree root failures, in *the Landscape Below Ground II: Proceedings of a Second International Workshop on Tree Root Development in Urban Soils* (G. W. Watson and D. Neely, eds.), International Society of Arboriculture, Champaign, IL, pp. 131–135.

Smiley, E. T., W. Graham, S. Cullen. 2002. Best Management Practice for Tree Lightning Protection. International Soc. Arboric. 29pp

Smiley, E. T. 2003. Does included bark reduce the strength of codominat stem? J. Arboric. 29(2) 104–106.

Smiley, E. T. 2005. Determining change of lean. Arborist News 14(1) 22–23.

Wagener, W. W., 1963, *Judging Hazards from Native Trees in California Recreational Areas: A Guide for Professional Foresters*, US Forest Service Research Paper PSW-P1.

Chapter 18

Integrated Pest Management

Deborah Smith-Fiola

1. Introduction

Insect, disease, and weed problems are familiar to those maintaining urban trees. Since the 1950s, the use of pesticides has been the primary means to control these pests. Pesticides are easy to use; they are very effective, have a quick kill rate, and are simple to use compared to the more complex and labor intensive nonchemical options. These Pesticides are also readily available and relatively inexpensive. Pesticides have simplified pest control, and in fact have become the primary means of insect, disease, and weed control in tree maintenance programs.

Traditional tree maintenance to manage insects and disease in the past has centered upon involves reliance on broadcast pesticide cover sprays. With this method, all plants in the landscape are covered with synthetic pesticides in order to kill existing pest populations. Cover sprays are typically applied according to the calendar or to a spray guide. For example, 5 to 10 all-purpose (insecticide/fungicide) cover sprays a season may be scheduled, to all trees on the property, on a monthly or bimonthly basis. Cover spray maintenance programs can achieve high levels of pest control and are an easy way to give the customer a pest-free environment.

It is important to realize that no one method of pest control, including pesticides, can keep a landscape totally free of insect, disease, and weed problems for extended periods of time. Intense pesticide reliance by the green industry (lawn, landscape, greenhouse, nursery, and arborist) created rising societal concerns by the late 1960s about potential human health and environmental hazards.

Deborah Smith-Fiola M.S.; President, Landscape IPM Enterprises, Keedysville, MD 21756.
Urban and Community Forestry in the Northeast, 2nd ed., edited by, J. E. Kuser.

1.1. Pest Resistance

Over 450 insect species are now resistant to one or more pesticides. If a pesticide is sprayed again and again onto a pest population, eventually (through mutations) a resistant strain will develop. For example, it took only 1 year before the first insect became resistant to DDT (Dichloro diphenyl trichloroethane).

Pesticide resistance can develop within just a few insect generations. For example, if a specific pesticide is sprayed onto an aphid population, the vast majority will be killed. A few, however, may survive if they have a gene for a substance that can metabolize the pesticide. The offspring of these individuals will be more highly represented in the next generation. If sprayed again, this further increases the selective pressures, and the third generation has an even higher representation of this special gene. Hence, this special aphid population is well on its way toward resistance to this pesticide. There are at least 36 known examples in the landscape of insects and mite pests resistant to one or more pesticides (Raupp *et al.*, 1994), including black vine weevil, bag-worm, spider mites, some aphids, several scale insects, Japanese beetle, and others.

1.2. Target Pest Resurgence

Natural enemies (also called beneficial insects or beneficials) are important for keeping pest populations in check. When pesticides are used as cover sprays, or if the control timing is wrong for the pest, the beneficials will be killed. Meanwhile, the pests rebound from lack of competition. A common example is the sequence of events when pesticides are applied for control of armored scale insects outside of their immature crawler emergence period: In this scenario, the naturally occurring beneficial insects are killed, allowing for population explosions of the scale. It has been shown that beneficial insects recover more slowly from pesticide treatments—up to 1 year before full recovery—than do the pests (Smith, 1970).

1.3. Secondary Pest Outbreaks

Pesticides often kill not only the target pest but also any beneficial insects present. Sometimes minor or secondary pests were present on the plan and that had been out-competed by the target pest and kept under control by natural enemies. These secondary pests may not be killed by the spray, and subsequently their population explodes. A good example is the use of the insecticide carbaryl (Sevin) for control of aphids. Aphids and spider mites often coexist on the same plant. Carbaryl is a weak toxin against aphids; it may kill some aphids, yet will kill all beneficial insects present, while not affecting mites at all. The lack of competition often results in a damaging outbreak of spider mites. In fact, the carbaryl label encourages the user to add a miticide to the spray tank to prevent spider mite flare-ups.

In order to discourage the resurgence of target pests and outbreaks of secondary pests, applicators often increase pesticide dosages to higher, and/or levels, increase the frequency of application. This "pesticide treadmill" has had an effect on human health and environmental pollution.

The effects of long-term, low-level dosages of pesticides to human health is the subject of ongoing research, since pesticides have been implicated in cancer

(Wassermann *et al.*, 1976; Eckholm, 1977). Nonfatal human poisoning cases from pesticide exposure are estimated to be more than 100,000 per year (Pimentel *et al.* 1978). Persistent pesticides (including DDT) accumulate in the food chain, where they become more concentrated and may significantly affect wildlife and honeybees. Run off of rain and irrigation water can further contribute to pesticide spread to adjacent land areas, into bodies of water, and into groundwater. Public concern about pesticides has led to more stringent regulation and registration. These drawbacks have not been demonstrated for all pesticides, yet underscore the importance of proper usage, in order to minimize their side effects.

Reliance on pesticides to eradicate all pests from the landscape thus is not justified from an ecological, cost, material, or labor perspective. The Federal Government has reviewed and removed many pesticides from use via FIFRA (the Fed Insecticide, Fungicide & Rodenticide Act.) The pesticide industry has responded to these issues by offering insecticides with narrower spectrums and low persistence, as well as implant and injection methods of application (Raupp *et al.*, 1994). Many commercial horticulture professionals have also turned to an alternative pest control strategy called integrated pest management (IPM). The IPM stresses the balanced *management* of quality landscape plants, as opposed to chemical annihilation of pests. If methods are used correctly, IPM is *the* most efficient and economical method of pest control in the lawn and landscape.

1.4. IPM-Multiple Strategies

The word "integrated" means that numerous techniques or approaches are used to manage plants in the landscape as a balanced ecosystem. Pesticides are often used in IPM, but the emphasis is on keeping pests at low levels through the judicious use of other alternative pest control strategies so that pesticides are used less often and the least toxic product is used. Rather than trying to eliminate all pests, a variety of control strategies are used to reduce pests to low, tolerable levels in order to maintain populations of natural enemies (beneficial organisms). The IPM differs from traditional pest control in that it employs a variety of control strategies, rather than relying on a single pesticide strategy. These strategies include the use of biological control, resistant plant varieties, the use of biorational, low-toxicity pest control products, cultural control, proper plant selection for the site, mechanical, and chemical controls, usually with two or more techniques employed at a time.

Integrated Pest Management IPM focuses on the long-term suppression or prevention of pest problems below a damaging level, with minimal impact on the environment, nontarget organisms, and human health. The IPM programs first stress the cultural considerations of growing healthy plants (Plant Health Care) to naturally reduce pest problems. A healthy plant is much less susceptible to severe pest problems, and the presence of pests usually is an indication that management efficiency has broken down somewhere. Pests attack to take advantage of a change within the plant: perhaps it is under environmental stress (drought stress, improperly sited, and so on.), or conversely the plant is extremely vigorous (overfertilized, excess new growth, and so forth.). The IPM manager therefore can alter the landscape to the benefit or detriment of pests by manipulating the plant, the site, the pests, and so forth.

2. Monitoring

For IPM to be successful, a monitoring program is required. Monitoring is the close inspection of plants for insects, disease, and abiotic problems on a regular basis throughout the growing season. The IPM managers use this compiled plant and pest information to detect, appraise, and predict pest outbreaks.

Traditional cover spray programs are based on the calendar. Sprays are applied at certain times of the year, regardless of whether the pest is actually present. An industry shift has occurred within the last decade toward a "see it and spray it" approach to pest control, in which a pest is sprayed when it is noticed. Unfortunately, in this method damage often precedes obvious observation of the pest. Additionally the pest may be at population levels where the only management option is the use of pesticides. What makes an IPM program superior to See and Spray Programs is the utilization of monitoring, or scouting, of plants in the landscape.

Monitoring is close inspection of the plant from top to bottom, looking for both obvious and not so obvious signs and symptoms of pests and poor health. This requires more knowledge on the part of the monitor: knowledge of plant identification, damage symptoms, insect and disease life cycles, population thresholds, and the advantages of various control measures. Monitors look for indicators of insect activity, such as frass from caterpillars or honeydew from aphids, to pinpoint a problem or potential problem. Regular plant inspections (every 2 to 4 weeks) are an essential way to keep track of changes in pest problems and plant health.

Pest control decisions are made according to the actual presence of the pest and on their most susceptible life stage, not on their hypothesized presence. For example, instead of spraying every arborvitae for bagworms in July, (when perhaps they were a problem in the past), the IPM monitor inspects plants months beforehand for the telltale overwintering bag. The monitor also handpicks any bags found (which may contain up to 800 eggs each), and re-inspects the plants closely in early-mid June, when the eggs would hatch. If young bagworms are spotted at potentially damaging levels, a spray could be made in June, while they are immature, before noticeable damage occurs, and while some alternative pesticide products (such as the entomopathogenic bacteria *Bacillus thuringiensis*), could be very successfully used.

Note that not every plant is sprayed; only plants which have a damaging level of the pest present are treated. This is called *spot treatment*. Spot treatment can significantly reduce pesticide usage merely by not spraying every single susceptible plant and focusing on those with moderate levels of the pest. Spot treatment maintains naturally occuring populations of beneficials.

Regular monitoring prevents pest problems since potential infestations are discovered when pests are small and damage is low; before they become serious. Damage is thus prevented. Alternative controls to traditional pesticides are also best used on smaller/immature pests. Monitoring also tends to alert the landscape manager to previously undetected pest problems, so plant quality improves. In addition, when plants are inspected on a regular basis it is possible to keep track of beneficial organisms (predators and parasitoids) that may be naturally reducing the pest population.

Record keeping (Fig. 1) is an essential part of monitoring. Good records are kept of all monitoring actions: early signs and symptoms of pest activity, pest levels related

SITE ________ DATE ________ SCOUT ________ TASKS ________________________

GDD TOTAL ____ CLOUD COVER ____ TEMP ____ RAINFALL ____ ________________________

SITE SPECIFIC
INDICATOR PLANTS/STAGE

INDICATOR PLANTS/STAGE	B[1]	F[1]	E[1]	KEY PEST$/BENEFICIALS	STAGE ($)

EVALUATION OF PREVIOUS TREATMENTS:

PLANT TYPE/REF #	VITALITY[2]	PROBLEM	STAGE/ TYPE[4]	LEVEL[3]	COMMENTS	EXTENT OF DAMAGE	PROBLEM INTENSITY[6]	RECOMMENDATIONS

[1] Bloom: B - Beginning bloom <50% anthers showing. F - > 50%, E - Ending bloom > 50% petal fall
[2] Vitality: Healthy, stress, dieback, defol.
[3] Level: Slight, moderate, heavy
[4] Stages: O - Overwintering, EG - Egg. El - early stage, ML - mid stage, LL - late stage, P - pupa, A - Adult
[6] Intensity: Low, moderate, high
[4] Type: A - abiotic, D - disease, I - insect, M - mite, N - nematode, W - weed

FIGURE 1. Sample IPM monitoring form (from the Rutgers Cooperative Extension Landscape IPM program).

to the condition of the plant, beneficial organisms present and their levels of control, predictions of when and where the pest may attack, temperature and rainfall, life stage(s) of the pest, plant vitality, and results of control tactics. A formal record-keeping system that is consistently used while monitoring is essential to a successful IPM program.

Some IPM practitioners use rough landscape maps that are copied and notes written on the copy at each monitoring visit. Others use simple checkoff charts, which highlight the presence, location, and level of the pest. Handheld computer data loggers are popular, as are computer recordkeeping systems for the office. Monitoring records should certainly be a part of an updated street tree inventory. Records are critical in order to evaluate the effectiveness of the program and predict future pest activity.

In addition to plant inspections, insects are monitored through the use of insect *traps* (blacklight traps, pheromone traps, pitfall traps). These traps can indicate when an insect is first present in an area and how its population is changing. Traps are often used to monitor pest activity levels to better time controls. Pheromone traps, for example, can determine when adults of gypsy moth, Japanese beetle, clearwing borers, and tip moths are active and would begin egg laying. Sticky traps are useful to monitor emergence or population levels of adult insects (such as the birch leafminer, whiteflies) or scale insects (immature/crawler stages). Burlap bands are refugia that can monitor levels of insect pests in trees, such as elm leaf beetle or gypsy moth.

Augmenting plant pest monitoring observations is *environmental monitoring*. Environmental monitoring is keeping track of environmental conditions that favor a particular disease or insect. For example, sycamore anthracnose foliar outbreaks are common under conditions of wet weather and temperatures between 16° and 20°C. On the other hand, sycamore anthracnose twig blight/canker outbreaks are associated with cool temperatures (12° and 13°C) during spring budbreak, which slows foliar growth while allowing the pathogen to gain a foothold. When temperatures are warm in the spring, the quick growth of twigs reduces their susceptibility to this disease (see Case Study A). Forecasting systems and models to keep track of rainfall and temperature levels have been developed and are valuable for predicting disease. Knowing whether the conditions are conducive for a disease outbreak may influence the need to apply a fungicide, not apply a fungicide, or wait to see how both the weather and the pathogen progress. The use of accumulated heat units (growing degree days) uses temperature information to predict insect emergence. Plant phenological or developmental models to predict pest emergence according to plant flowering are also used. For example, the bagworm is expected to hatch once 600 growing degree days have accumulated, about the time the mountain laurel, catalpa, and mockorange bloom. Weather conditions condusive for pest at breaks can be placed in computer databases.

Based on monitoring, pest control decisions can be made according to what pests are known to be present, as opposed to when they are thought to be present. This becomes a decision-making process, based on the knowledge of the IPM manager weighing the success of all the management options. Regular monitoring can therefore prevent pest damage, since potentially serious infestations are discovered and managed while they are still minor. In addition, when plants are inspected on a regular basis it is possible to keep track of population levels of beneficial organisms, such as ladybird beetles or mite parasites, which may be slowly controlling the pest. Many

IPM monitors make timely plant control treatments on the spot in order to save time returning to a site.

An IPM monitor might find it useful to carry the following monitoring equipment: 10x hand lens, pocket knife, white clipboard, pruners, plastic zip-lock sample bags, vials (some with alcohol), flagging tape, trowel, flashlight, and pest control guidebooks. For record keeping, essentials are monitoring forms (Fig. 1), a monitoring notebook or pocket tape recorder, or a modern handheld computer data logger. Optional equipment includes camera, aspirator, pH meter, burlap bands, pitfall trap and other traps, drop cloth, thermometer, sunscreen, pocketsized textbooks/references. Monitors also have a backpack or handheld sprayer back in their vehicle, available if needed for spot treatment applications.

Monitoring also alerts the landscaper to previously undetected pest problems, so ultimately plant quality improves. Monitoring records are very useful in evaluating IPM programs at the end of the year for predicting future problems and ordering pest control materials. Marketing IPM begins with marketing monitoring services.

2.1. Thresholds

The most difficult task facing the landscape manager is identifying the pest in the landscape and then determining whether it is abundant enough to warrant control. Because an ornamental plant is valued for its aesthetic appeal, the degree of damage that is tolerated is called the aesthetic injury level. This is the threshold at which damage is perceived and a remedy is required. Research shows that most people, including customers and professionals, perceive that a plant is damaged when exhibiting only 10% damage symptoms; on young plants this drops to 5% damage (Sadof & Raupp, 198-). Plants are thus treated for pests once damage exceeds 10%. Customers, however, may have different expectations (or pest tolerance) of the degree of maintenance for a high-valued plant, which must be incorporated into the IPM program. Thus, an IPM manager may monitor priority sites/plants versus low-priority sites/plants at different aesthetic thresholds.

3. Key Plants and Key Pests

Any one landscape site contains many numbers and species of trees. A survey of municipal foresters found that over 200 species or cultivars comprised the urban forest in Western cities (Kielbaso and Kennedy, 1983).

Of these numerous trees & shrubs, some are more prone to pests than others. The reason(s) for this could be innate (genetic; native species versus introduced species) or acquired (sub par siting/planting, monoculture situation, and so on).

A key plant is defined as a plant that has an increased level of pest susceptibility. It is likely to incur pest problems year after year. Key plants are more pest prone and require greater, repeated annual pest control measures. They must be monitored more closely, and long-term maintenance is more costly. Key plants should be highlighted during the landscape design process or the initial walk-through because of the maintenance consequences of planting and maintaining them. They also should alert the

IPM practitioner to concentrate monitoring attention on these plants; to confront problems before they become oppressive.

Many of these key plants are the ones that have been repeatedly sprayed over time. In many cases, environmental stress predisposes the plant to be more susceptible to insects and disease. Correcting the site conditions and their associations with the pests can reduce the severity of the problem and the likelihood of repeat occurrences.

Key plants may also be plants which are favored by the customer, or significant to the client for some reason. An example is a mildew-ridden lilac that was planted in honor of the deceased/a loved one. Because of the emotional attachment to the plant, it must be kept healthy and pest free at any cost. Another example is a disease-prone flowering cherry located at the front entry of the house. As a focal point plant, it is significant to the client & must be kept healthy.

Many key plants are common species that are widely planted. They may be inexpensive & available at mass market retailers.

1. **Key plants in the Northeast:**
   Austrian pine
   Azalea
   Crab apple
   Dwarf Alberta spruce
   Eastern flowering dogwood
   Euonymus
   Hawthorn
   Hemlock
   Rhododendron
   Sycamore
   White birch

The pests that commonly attack these key plants ate called key pests. Key pests are the most common or most frequently encountered insects and diseases causing damage in the landscape. The list of key pests may vary from region to region, depending on what plants are common:

2. **Key pests in the Northeast:**
   Anthracnose
   Aphids
   Bagworm
   Birch leafminer
   Black vine weevil
   Bronze birch borer
   Crabapple scab
   Elongate hemlock scale
   Euonymus scale
   Hemlock woolly adelgid
   Lacebug
   Sphaeropsis (*Diplodia*) tip blight
   Spider mites

Research shows that a limited number of pests create the majority of landscape problems that are damaging enough to warrant control. In Maryland, data from IPM programs in suburban residential areas found that 10 pest species accounted for more than 83% of the total insect and diseases found in the landscapes (Raupp *et al.*, 1985). A national survey of arborists found that 10 insect species accounted for 63% of the total insect problems encountered (Wu *et al.*, 1991). Plants that are host to these key pests are often key plants; they must be monitored more closely and may be more costly to maintain.

4. IPM Pest Control Strategies

A well-trained IPM manager can answer these questions *before* a pest control tactic is used:

- IS: *the pest present.*
- WHERE: is *the pest located, on how many plants.*
- WHEN: is *the most susceptible lifestage of the pest*?
- WHICH: *control product is most effective and least toxic to use*?

4.1. Cultural Control/"Plant Health Care"

The IPM principles are based on holistic approaches to bolster plant health via manipulation of the environment/site. Plants are managed to maintain balanced, healthy growth, which invigorates them, making them less susceptible to insect and disease problems. Thus, the IPM practitioner is first and foremost a good horticulturist.

Cultural management tactics are those practices that are performed to change the site to favor the plant, as opposed to the pest (see Case Study B). For example, excessive fertilizer may cause new growth flushes of highly nutritious foliage, which is readily fed upon by many sucking and chewing insects, as well as more prone to attack by disease organisms. It has been demonstrated, for example, that the performance of the sucking insects hemlock woolly adelgid and elongate hemlock scale on hemlock improved under fertilization (McClure, 1991). Underfertilization, on the other hand, may weaken a plant and make it more susceptible to insect borers or canker diseases. Therefore, a balance of fertilizer, based on soil test results, is needed to favor the plant versus the pest. Proper mulching, fertilization, watering, soil preparation, staking, and so forth, are all important to plant longevity.

Since proper siting is crucial for optimal plant growth, advance planning for landscape designs may eliminate the need for future pest treatments by choosing a species that can tolerate the stresses in the specific environment. Plants that are tolerant of drought conditions, such as Ginkgo or Japanese Zelkova, will not be as predisposed to pest attack in a dry site as would a nonadapted, more sensitive tree such as Sugar Maple. Additionally, selection of a plant inherently resistant to specific pest problems is wise.

The landscape design process is often artistry at the expense of pest management. Monoculture planting might be pleasing to the eye, yet repetitive placing of the same plant species or cultivar makes it easy for a pest to attack first one, plant and then the entire planting. It is thus important to diversify the planting to include many different types of plants; this diversity will minimize potential losses to any one host-specific insect or pathogen.

Using good pruning techniques to prune out deadwood is another cultural control tactic. No stubs are left that could act as an entry port to pests. Thinning of dense, overcrowded branches can increase air movement for drying of foliage, which is important to minimize diseases spread by water. Removing sucker growth can reduce the number of aphids feeding on the succulent foliage. On the other hand, pruning an elm prior to spring activity of the European elm bark beetle (the vector of Dutch elm disease) may increase its susceptibility to this pest, as it is attracted to fresh wounds (Barger and Cannon, 1987).

Reducing drought stress can also enhance tolerance to pests. For example, watering and mulching a drought-stressed white birch increases its vigor, helps it defend itself from the bronze birch borer that is attracted to stressed trees. Sites may be modified to reduce overwintering sites for insects (such as removing a nearby weedy hedgerow) or to encourage beneficial insects (by planting flowers under a tree to provide a food source for adult beneficials).

Existing problems often can be minimized by improving the growing conditions. Proper mulching to conserve moisture, proper fertilizer, and proper watering are all basics of good horticulture. A healthy plant also can compensate on a short-term basis for stressful changes in the environment, such as a period of drought stress.

Some of the pest problems seen in the landscape are actually secondary problems, the direct result of plant stress from poor soil conditions, poor plant quality, improper plant site (wrong amount of sun, soil moisture, and so on), poor planting, improper pruning, fertilization, and watering, lack of weed control, and so on. Regular soil tests can provide a great deal of useful information about plant growing conditions that the landscaper can use to prevent pest problems. Mulching to retain moisture, aeration to control compaction, and pruning out deadwood/suckers will improve the health of most plants. Arborists must be aware of the role that temperature, humidity, irrigation, and fertilization can play in the development of plant diseases. While the first two cannot be controlled, the latter two can be manipulated to help reduce the incidence of diseases.

4.1.1. Resistant Plant Varieties

Many landscape plants, including some native plants, have evolved to be resistant to many pest problems. Other plant cultivars and varieties have been bred to be resistant to insects or diseases. Use of these plants can have a positive impact in the reduction of pest problems. For example, while crab apple scab is a key pest of crab apple, there are many crab apple varieties and cultivars that are resistant to scab. Likewise, there are varieties and cultivars that are also resistant to rust and fireblight (Smith-Fiola, 2004).

These pest resistant plants can be used to "relandscape" a site, particularly to replace key plants (those plants that incur the most problems year after year). For example, if a white birch is planted in full sun in a droughty site, it is quite prone to attack from the bronze birch borer. Why not substitute the 'Heritage' river birch or the newer cultivar 'DURA HEAT' in its place? River birch (*Betula nigra*) is adapted to floodplains and can withstand both flood and drought conditions. While the river birch species has brown bark, the cultivar 'Heritage' has been selected for white bark, which peels off as does that of the white birch. Thus, 'Heritage' can be used as a substitute in the landscape design for a white, multistemmed tree, without the high maintenance of controlling the bronze birch borer.

The use of a plant resistant to one pest may not make it resistant to another pest (see Case Study C). Limited breeding programs, reluctance of the nursery industry to increase production of these varieties, and the failure of landscape architects to recommend these plants all impede the more widespread use of resistant plant material (Raupp *et al.*, 1989).

4.2. Mechanical Control

The IPM methods performed manually, by hand specifically, to reduce the potential of attack, from a specific pest are called mechanical controls. Sanitation, or removal of undesirable plant parts, is one mechanical control tactic. For example, numerous pathogens overwinter in infested plant material. Many leaf spot diseases overwinter in fallen leaves. These can be raked up and removed from the site, in order to reduce the inoculum required for the pathogen to reinfest the following year. Likewise, many cankers and twig blights can be pruned out when noticed, to remove the origin of the problem and minimize its future spread.

Sometimes the pest itself can be controlled by hand. Bagworms, as previously mentioned, can be handpicked off the plant. Considering that hundreds of eggs of the next generation are inside the bag 10 months of the year, handpicking can be feasible on small plants. Egg masses of other insects, such as gypsy moth and Eastern tent caterpillar, can be also handpicked/scraped and destroyed. Some web-making caterpillars, such as mimosa webworm, Eastern tent caterpillar, or fall webworm, can be pruned out by hand or by using pole pruners to efficiently remove early infestations before major damage has occurred.

Washing a plant with a strong stream of water will destroy a certain number of aphids and mites and knock others off onto the ground where they are attacked by spiders and other predators. Barriers, such as burlap or sticky tree bands, also can be somewhat effective to reduce the size of an insect population. Insects, such as gypsy moth caterpillars, black vine weevils, and elm leaf beetles, will hide under the bands during the day, where they can be destroyed, or get stuck in the sticky trunk barrier as they migrate up and down the tree.

Some diseases require two alternate plant hosts. Cedar apple rust, for example, requires an apple/crab apple host as well as a juniper host for the pathogen to complete its life cycle. The disease has noticeable foliar symptoms on the ornamental crab apple, while native cedars (*Juniperus virginiana*) may be in the vicinity harboring the sporulating stage of the pathogen. In this example, the alternate host (the cedars) could be removed from the surrounding area to break the life cycle of the disease. (On the other hand, beware of planting susceptible groundcover junipers beneath susceptible crab apple trees!)

Many pest problems can be prevented with physical means. Sticky traps are used not only as a population monitoring tool, but are sometimes used as a means for direct control of insects such as potato leafhoppers, aphids, and other small flying insects. A dropcloth placed early in the morning under plants infested by Japanese beetles/black vine weevils and shaken can dislodge pests present to drop onto the cloth, from which they can be removed and disposed of. Success of utilizing mechanical controls relies on early recognition of the problem, and early intervention.

4.3. Biological Control

Biological control is the use of one organism to control another organism. Within the balance of nature, there are abundant, naturally occurring predators and parasitoids that consume insect pests. These natural enemies have evolved with the pest as a natural control mechanism. Thus, if a pest has been introduced into this country (e.g., Japanese beetle, gypsy moth), no natural enemies typically exist in the new habitat; and with no natural checks the pest population often explodes, resulting in significant pest outbreaks. In some cases, scientists have returned to the country of origin to collect natural enemies, import them, mass rear them in a laboratory, and release them in this country. Many natural enemies of the gypsy moth, for example, have been imported and mass reared out through cooperative US Department of Agriculture (USDA) programs (see Case Study D).

A predator is an insect that usually consumes its insect prey completely. Common predators include lady beetles, green lacewings, spiders, ground beetles, and predaceous mites. An example of an introduced predator is *Chilocorus kuwanae*, a lady beetle that feeds on euonymus scale.

A parasitoid is an insect that lays an egg(s) in or on an insect prey; the egg hatches and feeds as an immature on the host insect, slowly killing it. Parasitoids are usually tiny wasps or flies. Many parasitoids are so small that they are seldom seen, such as the *Trichogramma* wasp, which is the size of a speck of dust. Monitoring for parasitoid activity commonly detects evidence of activity, such as swollen, parasitized aphid mummies, or scale insects with holes in their waxy cover (exit holes of new adult parasitic wasps).

Entomopathogenic nematodes, which are microscopic worms that carry lethal insect killing bacteria in their gut, are effective on many larval insects, including Japanese beetle grubs, clearwing borers, and black vine weevil. These are purchased commercially in an infective juvenile stage, mixed with water and sprayed through conventional spray equipment. Nematode species must be carefully chosen according to the target host and applied under specific environmental conditions to achieve levels of control similar to that of pesticides. Entomopathogenic nematodes kill their prey within 48 h and are extremely safe to the applicator, the environment, and nontarget organisms. Research using these nematodes on clearwing borers has shown from 20% control to 84% control, for example, depending on the nematode species chosen and the time of year (see Case Study E).

When the use of pesticides is limited, conserving naturally occurring predators and parasitoids is fostered. For example, a large pine needle scale population was naturally reduced by the twice-stabbed lady beetle, *Chilocorus stigma*, during a 3-year study in Maryland at a Christmas tree plantation (DeBoo and Wiedhaas, 1976). Natural enemies also may be purchased from insectuaries and introduced/augmented into the site. Note that success of using this latter method is complex, weighing the right beneficial(s), the correct timing, ideal weather conditions for release, and the appropriate prey present. Landscape successes with biological control have recently involved the release of predaceous mites against pest mites, green lacewing larvae against lacebugs (Shrewsbury and Smith-Fiola, 1998), *Pseu-doscymnus* (an Asian lady beetle) against hemlock woolly adelgid (see Case Study F), and *Cybocephalus* (a lady beetle) against the euonymus scale (Drea and Carlson, 1988).

Biological control may be the ideal pest control method, since it is environmentally safe and ecologically sound. However, biological control usually does not completely eliminate a pest population but maintains low levels of the pest. There remains a lack of practical research on using biological control organisms in landscape settings (see Case Study G).

4.4. Microbial and Viral Control

The use of certain microbes, for example, bacteria, which attacks specific stages of specific pests, is a very safe alternative to the use of pesticides in IPM. For example, a naturally occurring virus (nuclear polyhedrosis virus) attacks gypsy moth larvae when populations get too large or crowded. A pathogenic, antagonistic bacterium is now commercially available to control crown gall disease.

The most popular microbial is *Bacillus thuringiensi* (BT), a bacterium that is ingested by young caterpillars and slowly kills them by rupturing their stomach lining. The BT works well on populations of *young* caterpillars (1st to 3rd instars) under dry conditions. Caterpillars ingest the bacteria and die slowly over 3 to 10 days. The BT kills only immature caterpillars, although new strains also kill some immature beetles (elm leaf beetle, Japanese beetle) and mosquito larvae. The BT as been a major weapon in the battle against the gypsy moth nationally.

New research has shown some promising new microbial pesticides, many of which are derivatives of bacterial fermentation. Examples include Abamectin, a miticide, and Spinosad, an insecticide.

4.5. Pesticides and Biorational Pesticides

As mentioned earlier, pesticides are part of an IPM program, but they are used differently than in a conventional pest control program. Pesticides are not applied preventatively but in accordance with monitoring results. By monitoring specific plants/areas of the landscape, where a problem is serious enough to warrant a control, can be pinpointed. Treatments are timed for application when the most susceptible life stage of the pest is present. Spot sprays of individual infested plants only can represent a great savings in the amount of pesticide applied to a property when compared with a conventional program in which all plants would be sprayed. When pesticides are necessary, an IPM program makes their use more efficient.

Integrated pest management programs also stress the use of the least toxic pesticides that have minimal negative environmental impact because of short residual times and negligible effect on beneficials. Biorational pesticides are naturally occurring products or synthetic mimics of naturally occurring products that exercise pesticidal action. Examples of these least toxic alternatives include horticultural oil (petroleum oils similar to baby oil, that suffocate insects), insecticidal soap (fatty acids of potassium salts, which dessicate insects), neem oil (activity includes insect growth regulator, repellent, and antifeedent), pyrethrin (synthetic mimic of pyrethrum, a contact insecticide), silica gel (dessicant), and others. Research shows results similar to those of pesticides with proper use of these products.

5. Summary

By the 21st century, it may become illegal to spray pesticides in the same manner as in the past. Thus, landscape pests will be managed differently, with an IPM approach representing the most feasible option. The IPM programs across the country have proven to be economically and environmentally viable. Pesticide use is often significantly reduced, without a loss of plant quality. In Maryland, landscape IPM programs have reduced pesticide use by 40% to 83% in residential communities. In California, an IPM program on municipal street trees reduced pesticide use by over 90%, resulting in a $22,500 savings. In New York greenhouses, pesticide use has been reduced by 45% under IPM. The National Park Service has implemented IPM at all park sites across the country, reducing pesticide use by 70%.

Reduced pesticide use will save money, and such savings can be rolled over to offset any additional labor costs from monitoring. The IPM is a concept that is growing in acceptance and is here to stay.

References

Barger, J. H., and Cannon, Jr., W. N., 1987, Response of smaller European elm bark beetles to pruning wounds on American elm, *J. Arboric.* **13:**102–104.

Cloyd, R. A., Nixon, P. L., and Pataky, N. R., 2004, *IPM for Gardeners: A Guide to Integrated Pest Management*. Timber Press, Portland, OR. 204 pp.

DeBoo, R. F., and Weidhaas, J. A., 1976, Plantation research: XIV. Studies on the predation of pine needle scale, *Phenacaspis pinifolia*, by the coccinellid, *Chilochorus stigma*, Canadian Forest Service Report CCX-119.

Drea, J. J., and Carlson, R. W., 1988, Establishment of *Cybocephalus sp.* (Coleoptera: Nitidulidae) from Korea on *Unaspis euonymi* (Homoptora: Diaspididae) in the Eastern United States, *Proc. Entomol. Soc. Wash.* **90:**307–309.

Eckholm, E. P., 1977, *The Picture of Health*, Norton, New York.

Jones, R. K., and Benson, D. M., 2001, *Diseases of Woody Ornamentals & Trees in Nurseries*, APS Press, St Paul, MN.

Kielbaso, J. J., and Kennedy, M. K., 1983, *Urban Entomology: Interdisciplinary Perspectives*, Praeger, New York.

McClure, M. S., 1991, Nitrogen fertilization increases susceptibility to hemlock woolly adelgid, *J. Arboric.* **17:**227–230.

Pimentel, D., Androw, D., Dyson-Hudson, R., *et al.*, 1978, *Environmental and Social Costs of Pesticide Use*, unpublished manuscript, Department of Entomology, Cornell University.

Raupp, M. J., Koehler, C. S., and Davidson, J. A., 1994, Advances in Implementing IPM for woody landscape plants, in *Handbook of IPM for Turf and Ornamentals* (A. R. Leslie, ed.), CRC Press, Boca Raton, FL., pp. 125–142.

Raupp, M. J., Davidson, J. A., Holmes, J. J., and Helman, J. L., 1985, The concept of key plants in integrated pest management for landscapes, *J. Arboric* **11:**317–322.

Sadof, C. and M. Raupp

Shrewsbury, P., and Smith-Fiola, D., 1998, *Biological Control of the Azalea Lacebug*, Annual Meeting of the Entomology Society of America, Nashville, TN, 1997.

Smith, R. F., 1970, Pesticides: Their use and limitations in pest management, in *Concepts of Pest Management*, (R. L. Labb, and F. E. Guthrie, eds.), North Carolina State University, Raleigh, pp. 103–118.

Smith-Fiola, D. C., 2004. *Pest Resistant Key Trust Shrubs*, landscape IPM Enterprises, Keedysville, MD. 55 pp.

Wasserman, M., Nogueria, D. P., Tomatis, L., *et al.*, 1976, Organochlorine compounds in neoplastic and adjacent apparently normal breast tissue, *Bull. Env. Contam. Toxicol.* **15:**478–483.

Wu, Z., Jamison, S., and Kielbaso, J., 1991, Urban forest pest management, *J. Arboric.* **17:**150–158.

Case Study A

Sycamore Anthracnose, *Apiognomonia veneta*

by John E. Kuser

Approach: Resistance Breeding and Selection

The most conspicuous and probably best known of leaf anthracnoses in the Northeast is sycamore anthracnose, *Apiognomonia veneta*, which attacks native sycamores (*Platanus occidentalis*) severely and London planes (*P. X acerifolia*) to a lesser extent. With the coming of warm weather in the spring, sycamore buds swell, and if the warm weather holds on for a few days, the buds burst and new growth begins. Then (usually) cool, wet weather comes back. At lower temperatures (below 12° to 13°C) shoot growth and leaf expansion slow down, and conditions are ideal for *Apiognomia* to attack and kill the tender new growth. This often causes sycamores to be bare as late as June, long after other trees have leafed out. It may kill several cycles of the trees' attempts to resume new growth, if warm spells are followed by cool spells. It also causes formation of witches' broom at the ends of branches, a characteristic that can be used to differentiate sycamores from London planes. Later in the spring, as warmer temperatures (above 15° to 16°C) prevail, growth of new shoots and leaves resumes, twig killing ceases, and the trees leaf out normally (Sinclair *et al.*, 1987). Because this warm weather recovery can reliably be expected, fungicides are not recommended to control anthracnose.

Sycamore grows to be one of the largest, most massive trees in the Northeast in spite of yearly defoliation. Sanitation (raking up leaves when they drop) can be used to lessen the amount of inoculum.

Because annual defoliation of sycamore makes it so unsightly, most municipalities prefer to plant the London plane tree, a hybrid of sycamore crossed with anthracnose-resistant Oriental plane (*P. orientalis*). Caution is in order when doing this, however, because much nursery stock of London plane is seed-grown and seedling clones vary widely in resistance. Only vegetatively propagated trees of the resistant cultivars Columbia and Liberty and the original anthracnose-resistant clone of Bloodgood (which also resists powdery mildew) should be used (Smith-Fiola, 2004).

Many other trees, including various species of ash, oak, maple, walnut, hickory, elm, hazel, redbud, birch, and so forth, suffer from anthracnose or leaf-spot diseases. Outbreaks are usually associated with wet spring weather. There is probably clonal and species variation in susceptibility; this is mentioned as to oaks by Sinclair *et al.* (1987), and exists among specimens of white ash on the Rutgers University campus in New Jersey. Because no systematic observations or selection for resistance have been reported (Sinclair *et al.*, 1987), the best recommendation is to follow cultural practices that maximize tree health. These include selection of a site appropriate for the species, provision of sufficient soil, good drainage, fertilization, irrigation (if necessary), and protection from harmful insects.

References

Sinclair, W. A., Lyon, H. H., and Johnson, W. T., 1987, *Disease of Trees and Shrubs*, Cornell University Press, Ithaca, NY.

Smith-Fiola, D., 2004, *Pest Resistant Key Trees and Shrubs* Landscape IPM Enterprises, Keedysville, MD. 55 pp.

Case Study B

Eastern Tent Caterpillar, *Malacosoma americanum*

by Steven Rettke

Approach 1: Cultural & Mechanical Controls
Approach 2: Biological Controls (Disease, Natural Enemies, & Starvation)

The eastern tent caterpillar, *Malacosoma* spp., is native to North America, found throughout the United States east of the Great Plains, and in the southern part of eastern Canada. This pest is considered to be one of the most significant defoliators of deciduous shade trees (Penn State Cooperative Extension, 1991). Their preferred hosts include isolated, open-grown trees, especially wild cherry, crab apple, and apple. During outbreak years, which frequently occur at 8- to 10-year intervals, this pest will also occasionally attack pecan, hawthorn, beech, willow, and other shade trees (USDA Forest Service, 1990).

Eastern tent caterpillars construct conspicuous silk nests in the forks of trees, which are easily recognized during the spring months. The noticeable tents/nests cause an exaggeration of their impact as pests. Defoliation by this caterpillar will rarely cause tree mortality, as trees will refoliate. The reduced aesthetic value of trees in urban and suburban areas is the primary harm created by the activities of this pest (Coulson and Witter, 1984). Outbreak years also arouse much concern among area residents when the caterpillars migrate en mass across landscapes in search of new food or a place to complete their development. Nevertheless, unlike the gypsy moth, the eastern tent caterpillar has never been a major threat to the vitality of our forests and rarely reach large populations in ornamental trees (Shetlar and Chatfield, 1995).

Tent caterpillars spend the winter in brown masses of 150–350 eggs that the adult female attaches around small twigs. These shiny brown bands are readily recognized and removed by hand. During the late 1800s and early 1900s, whole communities mobilized to combat the perceived threat of infestation by the eastern tent caterpillar. In an early eradication campaign in Connecticut in 1913, for example, more than 10 million egg clusters were destroyed when the extension service offered a $25 prize to the school child who collected the most (Fitzgerald, 1995). Such mechanical control efforts still overlooked as many as 20% of the eggs. Therefore, after larvae hatched in the spring and the silk tents formed, prize money was again offered to school children in numerous towns for the number of tents they collected. In one such town, the children collected nearly 17,000 tents, weighing over half a ton. Then, usually with great vengeance and satisfaction, these tents were torched by flames or viciously stomped upon. Many of these same towns also offered a reward of 10 cents per quart filled with pupal cocoons of this insect. For instance, in 1899, at Glens Falls, New York, 1350 quarts containing the cocoons were turned in (Fitzgerald, 1995).

The simple mechanical methods of removing and destroying egg masses, tents, and pupal cases of tent caterpillars were and continue to be environmentally friendly ways of effectively suppressing their numbers in small, localized areas. However, even though these cultural practices should continue to be encouraged where they are practical, their limitations will always be evident.

Disease, natural enemies, and starvation are the primary environmental factors involved in the suppression of tent caterpillar species, which act to curb cyclical outbreaks

(*continued*)

Case Study B (*continued*)

of tent caterpillars. Numerous predators (i.e., spiders, ants, yellow jackets, birds) and parasitoids (i.e., braconid wasps, ichneumonid wasps, tachinid flies) attack tent caterpillars, but in some years these beneficials do not arrive in time or in sufficient numbers to adequately control them every season (Coulson and Witter, 1984).

During tent caterpillar outbreaks, the higher competition invariably results in lower food quality and greater vulnerability to infection by viral (i.e., NPV), bacterial (*Clostridium*), and fungal (i.e., *Entomophaga*) diseases. Pupal parasitism of the caterpillar also increases with outbreaks of long duration (Coulson and Witter, 1984). Finally, weather factors often play a key role toward the collapse of tent caterpillar outbreaks. For optimal survival, egg hatch in the spring is synchronized with the development of host tree foliage. Observations have indicated over 99% mortality of tent caterpillars from starvation in regions where unusual weather patterns caused the forestalling of the development of the leaves on the trees (Fitzgerald, 1995).

Occasionally cultural and biological control strategies will fail to adequately keep their populations in check in specific areas. Environmentally friendly, biorational products such as horticultural oil and insecticidal soap can suppress young larvae upon direct contact. The biological insecticide material of choice against *young* tent caterpillars is the bacterium (*Bacillus thuringiensis*). Costly protection efforts to prevent damage to trees are rarely justified. Pesticide treatments may then be used as the option of last resort and then only on trees of high value or in areas of important recreational uses.

References

Coulson, R. N., and Witter, J. A., 1984, *Forest Entomology—Ecology and Management*, Wiley-Interscience Publications, New York.

Fitzgerald, T. D., 1995, *The Tent Caterpillars*, Comstock Publishing Associates and Cornell University Press, Ithaca, NY.

Penn State Cooperative Extension, 1991, *The Eastern Tent Caterpillar. The Pest Sheet* (G. A. Hoover and P. R. Haller, eds.), Department of Entomology, Penn State College of Agriculture, University Park, PA.

Shetlar, D. J., and Chatfield, J. A., 1995, *Woody Ornamental Insect and Disease Slide Set Script*, The Ohio Nursery and Landscape Association, Westerville, OH.

USDA Forest Service, 1990. The Eastern Tent Caterpillar. Maryland Department of Agriculture and USDA Forest Service, US Government Printing Office, Washington, DC.

Case Study C

Dutch Elm Disease, *Ophiostoma ulmi*

by John E. Kuser

Approach: Selection for Resistance

Until the advent of Dutch elm disease (DED), American elm, *Ulmus americana*, was one of the most widely used shade trees in the United States (Harlow *et al.*, 1996). Cities and towns in the Northeast and Midwest lined all or nearly all of their streets with elms, because they were tall, stately, fast-growing, easy to transplant, and tolerant of a wide range of soils and sites (Del Tredici, 1998). In addition, elm was one of the few species known in the early 1900s to grow well in the upper Midwest. Consequently, when Dutch elm disease arrived in 1930 on veneer logs imported from Europe (Schreiber and Peacock, 1980), disaster ensued.

Symptoms of DED include wilted terminals and flagging of infected branches with leaves all turning yellow beyond where the vascular system is blocked. Occasionally the whole tree may die at once. Laboratory examination is necessary to determine whether the cause of flagging is drought, senescence, phloem necrosis, or DED (several branches from affected parts of the tree should be examined, because some may not show the characteristic streaking).

Up to this point, the history of DED runs parallel with that of chestnut blight: a pathogen coevolved with its host in Eurasia, came to the United States and devastated a related host that had never been exposed. But from here on the story is different: DED did not rapidly or completely destroy its host. *Ophiostoma ulmi* has two means of spreading: (1) its spores are carried by two species of bark beetles: *Hylurgopinus rufipes* and *Scolytus multistriatus*, (USDA, 1971; Johnson and Lyon, 1976), and (2) by root grafts (Schreiber and Peacock, 1980). Where elm trees were not plentiful, the bark beetles did not always find them; also, the beetles' preference for larger, older trees meant that many elms could reach seed-bearing age and perpetuate their species before being attacked.

Some temporary success in controlling DED has been achieved by rigorous sanitation programs (Schreiber and Peacock, 1980) and by annual fungicide injections. For example, Princeton University has maintained 120 elms on campus for several decades by promptly removing diseased limbs and dead trees to eliminate the inoculum. Beetle populations are monitored with pheromone traps to determine population levels and treatment timing; trees were formerly (1970s) sprayed with methoxychlor for beetles when necessary but are not sprayed now. They may be sprayed again if a suitable pyrethrin becomes available. Campus elms also are injected with fungicide. The cost of the combined treatments is estimated at $300/tree per year (Consolloy *et al.*, 1998; J. W. Consolloy, personal communication). This approach is not feasible for elms growing in the wild, nor is it affordable for many communities. What else can be done?

Because elms were not all wiped out at once, many were thought to be resistant. The USDA's research station at Delaware, Ohio, under the leadership of Drs. Joseph Kamalay and Alden Townsend, tested hundreds of these and found that most were merely escapes. But eventually the continued screening turned up two resistant clones, now introduced as 'Valley Forge' and 'New Harmony.' Their resistance has been compared with that of several widely planted elm cultivars, and is far better (Townsend *et al.*, 1995).

(*continued*)

Case Study C (*continued*)

Several other elm species and relatives are resistant. In general, Asian species of *Ulmus* are resistant, so it is thought that the disease originated in Asia. Siberian elm, *U. pumila*, is highly resistant, but weak-wooded, highly susceptible to elm leaf beetle, and not an acceptable substitute; Chinese elm, *U. parvifolia*, Urban elm (a *U. glabra* hybrid), *Zelkova*, and hackberries (*Celtis* spp.) are resistant and are useful trees in their own right, but do not look quite the same as American elm.

The best hope for restoring American elm lies in planting resistant clones of *U. americana* and perhaps some of its *U. wilsoniana* look-alikes. Rutgers University is undertaking to field test several of these for their combination of disease resistance, growth rate, appearance, and geographic adaptation (Kuser, 1998).

References

Consolloy, J. W., O'Kelley, P. W., and Hockreither, L. E., 1998, *Princeton University's Campus American Elm Survey*, Princeton University, Princeton, NJ.

Del Tredici, P., 1998, The ecology and economics of elm replacement in Harvard Yard, *Arnoldia* 1998 Spring, pp. 27–33.

Harlow, W. M., Harrar, E. S., Hardin, J. W., and White, F. M., 1996, *Textbook of Dendrology*, 8th ed., McGraw-Hill, New York.

Johnson, W. T., and Lyon, H. H., 1976, *Insects that Feed on Trees and Shrubs*, Cornell University Press, Ithaca, NY.

Kuser, J. E., 1998, *Elm Comparison Test*, (unpublished manuscript), Rutgers University, Department of Ecology, Evolution and Natural Resources, New Brunswick, NJ.

Schreiber, L. R., and Peacock, J. W., 1980, *Dutch Elm Disease and Its Control*, US Government Printing Office, Washington, DC, 13 pp.

Townsend, A. M., Bentz, S. E., and Johnson, G. R., 1995, Variation in response of selected American elm clones to *Ophiostoma ulmi*, *J. Environ. Hort.* **13:**126–128.

USDA, 1971, *Diseases of Forest and Shade Trees of the United States*, USDA Handbook No. 386, US Government Printing Office, Washington, DC.

Case Study D

Gypsy Moth, *Porthetria dispar*

by John E. Kuser

Approach: Biological Controls (Parasitoids, Predators, Virus, Fungus)

In 1869, gypsy moth was brought from Europe to Medford, Massachusetts, by a Professor Trouvelot, an astronomer who tried to interbreed it with silkworms. A few moths escaped, and 20 years later the first serious outbreak occurred in Medford. Gypsy months are defoliators, with some of their preferred hosts being oak, apple, and birch. Conifers are less preferred, but hungry late-instar caterpillars will strip a pine or larch after other foods are gone, and one defoliation kills the conifer. Tulip tree, *Liriodendron tulipifera*, is repellant to gypsy moth caterpillars; if one drives through a defoliated oak forest in June and a few tulip poplars are present, they stand out in striking contrast.

In 1890, the Massachusetts legislature appropriated $25,000 and formed a three-person commission to exterminate the moth (USDA Forest Service, 1981). Within the next 100 years, *Porthetria* had spread over the entire Northeast and infested new areas in the South and West. In 1981, a peak year, it defoliated over 5 million acres of forestland. Possibly no other forest insect has been studied as thoroughly or has been the target of such intense containment, control, or eradication strategies (USDA, 1985), and none has cost more to control (Johnson and Lyon, 1976).

Many strategies were used in the battle against gypsy moth, and most failed because of the insects' high reproductive capacity. DDT was sprayed on forests in northeastern Pennsylvania but quickly abandoned when it killed fish in lakes. Egg masses were painted with creosote. After DDT came Sevin, which provided acceptable control but also killed bees and hymenopterous beneficial insects. *Bacillus thuringiensis* and nuclear polyhedrosis virus (NPV) were sometimes effective and other times less so, depending on strain, spray formulation, timing, and weather.

When the gypsy moth was introduced to the United States, it left its natural array of parasitoids and predators behind in Europe. In 1905, the USDA, in cooperation with several affected states, started a program of importing natural enemies of the moth. Forty-five species have been tested, and this has resulted in successful establishment of 10 species of parasites, two predators (USDA, 1985), and one highly effective fungus (USDA Forest Service, 1993; Reardon and Hajek, 1993). There also is a naturally occurring pathogen, NPV—that attacks high-density gypsy moth populations and causes them to crash; NPV has been observed and studied as long as gypsy moth has been a problem, and much work has been done to improve its potency. Other control strategies, such as the sterile insect technique, have been developed (Reardon and Mastro, 1993) with success.

Recently the fungus *Entomophaga maimaiga* has been the cause of a dramatic drop in gypsy moth populations in the Northeast. Originally brought from Nishigahara, Japan, in 1909, the fungus failed to establish then, perhaps because of unfavorable weather and an NPV outbreak. In 1984, it was reintroduced by Soper and Shimazu; isolates were evaluated in the laboratory and the most virulent one was released in New York in 1985 and in Virginia in 1986. In 1989, it was found causing extensive epizootics in seven northeastern states; and by 1990, in three other states and Ontario. It is prevalent in both low- and high-density gypsy moth populations, particularly during wet springs, causing up to 100% mortality of late stage larvae (Reardon and Hajek, 1993).

(*continued*)

Case Study D *(continued)*

In one northeastern state (New Jersey), ten predators and parasitoids had become established, but never had enough impact to prevent peak defoliations from recurring. In that state, acres defoliated totaled 258,425 in 1973; 798, 790 in 1981; 431,235 in 1990; but only 28,000 in 1993 after *Entomophaga* became epizootic and never more than that since (J. Kegg, personal communication)

The unanswered question is, what long-term effect will the fungus have? Is the current gypsy moth collapse due to a lucky combination of weather, fungus, and virus? Nobody is yet sure (Twardus, 1996), but one lesson is that perseverance with biological controls does pay.

References

Johnson, W. T., and Lyon, H. H., 1976, *Insects That Feed on Trees and Shrubs*, Cornell University Press, Ithaca, NY.

Reardon, R. C., and Hajek, A., 1993, *Entomophaga maimaiga* in North America: A Review, USDA Forest Service, Washington, DC.

Reardon, R. C., and Mastro, V. C., 1993, *Development and Status of the Sterile Insect Technique for Managing Gypsy Moth*, USDA Forest Service, Northeastern Area, Morgantown, WV.

Twardus, D. B., 1996, *Gypsy Moth Populations Plummet in 1996 While the Fungus Skyrockets, Gypsy Moth News* No. 42, USDA Northeastern Area, Morgantown, WV.

USDA, 1985, *Insects of Eastern Forests*, USDA Publ. No. 1426, US Government Printing Office, Washington, DC.

USDA Forest Service, 1981, *The Gypsy Moth. Research toward Integrated Pest Management* (C. C. Doane and M. L. McManus, eds.), Forest Service Tech. Bull. No. 1584. USDA, Washington, DC.

USDA Forest Service, 1993, *Maimagia* Mania, *Gypsy Moth News* **31**:2–5, USDA Northeastern Area, Morgantown, WV.

Case Study E

Banded Ash Clearwing Borer, *Podosesia aureocincta*

by John E. Kuser

Approaches: 1) Chemical
2) Biological Control

Ash trees grown for use as street and shade trees are a valuable landscape and nursery crop. A key pest attacking green ash, *Fraxinus pennsylvanica*, is the banded ash clearwing borer (BACB). A 1971 survey showed that approximately 50% of the green ashes in the cities of the Canadian Prairies were attacked by clearwing borers, as were 33% of the boulevard trees in Grand Forks, North Dakota (Dix *et al.*, 1978). Economic losses approached $5000/acre per cropping cycle in Ohio nurseries (Purrington and Nielsen, 1977). Traditional synthetic, long-residual insecticide controls have relatively narrow treatment windows. A viable biological control option for controlling the BACB is the use of entomopathogenic nematodes (Smith-Fiola *et al.*, 1996).

There is one generation per year of BACB, with adults emerging in late August and September in Ohio, Virginia, and Maryland (Gill *et al.*, 1994). Females deposit eggs on tree branches and trunks, and larvae tunnel through the bark into the cambium before feeding and excavating upward and inward into the sapwood, where they overwinter. Mining causes branch dieback, disfiguration, structural weakening, and death of trees (Solomon, 1975).

Chemical control options exist during a relatively narrow window of time, just before egg deposition. Residual insecticides, such as Lindane or Dursban (chorpyrifos), are commonly applied as a protectant bark spray 10–14 days after first adult male capture in pheromone traps. A single, properly timed insecticide application provides effective control (Bone and Koehler, 1991); however, once larvae are under the bark, pesticide treatments are ineffective.

Entomopathogenic nematodes in the family *Steinernematidae* have been shown to be effective in controlling the dogwood borer, peach tree borer, western poplar clearwing borer, and clearwing borers in alder and sycamore (Davidson *et al.*, 1992; Gill *et al.*, 1992; Kaya and Lindegren, 1983; Kaya and Brown, 1986). The humid larval galleries are ideal for nematode searching and survival (Kaya and Brown, 1986). Nematodes are applied directly on the woody portions of trees with conventional spray equipment. The nematodes enter the borers' feeding galleries, search, and attack borer larvae. Larvae are typically killed within 48 hours of attack by the *Xenorhabdas* bacteria symbiotically sustained by the nematodes (Kaya and Gaugler, 1995).

In trials on borer-infested green ash in Maryland and New Jersey, two species of nematodes reduced the number of pupal skins per tree by 54% (*Steinernema glaseri*) and 46% (*Steinernema feltiae*), compared with 74% reduction in skins by Dursban treatment. These results are consistent with those from the California alder clearwing borer study (Kaya and Brown, 1986), where *Steinernema* trunk sprays in late September provided 77–84% control. Nematode treatments applied to dry bark did not provide acceptable control. In New Jersey, October nematode treatments (targeting newly hatched larvae) tended to give better control than summer treatments (to mature larvae). Fall treatments may be preferable to summer treatments, because trees have yet to sustain major damage and cooler temperatures are not as hostile to nematodes. Additionally, nurserymen and landscape managers are not as busy at this time of year and should welcome the opportunity to widen the spray window.

(*continued*)

Case Study E (*continued*)

References

Bone, P. S., and Koehler, C. S., 1991, Study describes ash borer infestations, tests management method, *Cal. Agric.* **45**(5):32–33.

Davidson, J. A., Gill, S. A., and Raupp, M. J., 1992, Controlling clearwing moths with entomopathogenic nematodes, *J. Arboric.* **18:**81–84.

Dix, M. E., Tagestad, A. D., Nielsen, D. G., and Purrington, F. F., 1978, Biology and control of the lilac borer in the northern Great Plains, USDA Forest Ser. Res. Note RM-358.

Gill, S., Davidson, J., MacLachlan, W., and Potts, W., 1994, Controlling Banded Ash Clearwing Borer using Entompathogenic nematodes, *J. Arboric.* **20**(3):146–149.

Gill, S. A., Davidson, J. A., and Raupp, M. J., 1992, Control of peach tree borer using entomopathogenic nematodes, *J. Arboric.* **18:**184–187.

Kaya, H. K., and Gaugler, R., 1995, Entomopathogenic nematodes, *Ann. Rev. Entomol.* **38:**181–206.

Kaya, H. K., and Brown, L. R., 1986, Field application of entomogenous nematodes for biological control of clearwing moth borer in alder and sycamore trees, *J. Arboric.* **12:**150–154.

Kaya, H. K., and Lindegren, J. E., 1983, Parasitic nematode controls western poplar clearwing moth, *Cal. Agric.* **43:**31–32.

Purrington, F. F., and Nielsen, D. G., 1977, Biology of *Podosesia syringae* (Lepidoptera: Sesiidae) with description of a new species from North America, *Ann. Entomol. Soc. Am.* **70:**906–910.

Smith-Fiola, D. C., Gill, S. A., and Way, R. G., 1996, Evaluation of entomopathogenic nematodes as biological control against the banded ash clearing borer, *J. Environ. Hort.* **14**(2):67–71.

Solomon, J. D., 1975, Biology of an ash borer, *Podosesia syringae*, in green ash in Mississippi, *Ann. Entom. Soc. Am.* **68:**325–328.

Case Study F

Hemlock Woolly Adelgid, *Adelges tsugae*

by John E. Kuser

Approaches: 1) Systemic Insecticides
2) Biological Control

The hemlock woolly adelgid is an aphidlike insect that is a serious pest of eastern and Carolina hemlocks. Native to Japan and China, it appeared in the Pacific Northwest in 1924, on western hemlock, where it is of little consequence in the forest but sometimes weakens and kills ornamental trees (Annand, 1924; Furniss and Carolin, 1977). It was discovered in Virginia in the 1950s, and has since spread throughout Pennsylvania (1960s), Connecticut, and Massachusetts (1980s), killing eastern and Carolina hemlocks in forests and landscapes from North Carolina to New England (Smith-Fiola, 1995). Eastern hemlock is important both as a forest tree and as an ornamental. It reaches impressive size in the wild, with one giant in western North Carolina measuring 169 feet tall (Turnage, 1996). It often grows in ravines, where its deep shade keeps trout stream waters cool in midsummer. As an ornamental, it is relatively easy to grow, makes a dense screen, and can be trimmed into a hedge. Infested hemlocks look grayish-green from a distance, and close inspection shows small, cottony, white adelgids at the base of each needle.

What can be done about this pest that has caused moderate to severe defoliation in 44% of a 1267-km^2 study area in New Jersey between 1984 and 1994, killing 5% of the trees (Royle and Lathrop, 1997)? It has caused one nursery owner a financial loss of \$20,000 to \$30,000 due to destruction by the insect (Smith-Fiola, 1998). It has been found that insecticidal soaps and horticultural oil give as good control as diazinon, malathion, and other insecticides if applied during June and July when the insect is in its settled nymph stage before it becomes woolly. Coverage must be complete, however; because that is difficult to achieve, the adelgid population rebuilds quickly. Therefore, soil drenching with systemic insecticides has been found to be more effective (Smith-Fiola, 1994). The cost of thus protecting a 60-foot hemlock is estimated at \$30 per year, using the insecticide imidichloprid (MERIT) and assuming that application would be made every other year (W. Porter, personal communication).

Because of the impracticality of soil treatment under forest hemlocks, Dr. Mark McClure of the Connecticut Agricultural Experiment Station began a search for a biological control in 1992. The search led him to hemlock forests in the remote mountains of Japan, where he found mites, flies, and beetles feeding on woolly adelgids. One of these beneficial insects, a lady beetle called *Pseudoscymnus tsugae*, showed the greatest promise. The New Jersey Department of Agriculture's beneficial insect laboratory is now (July 1998) producing as many beetles as it can, and has released them at ten public and private sites that will be monitored for 3 years to determine success. In earlier studies in Virginia and Connecticut, *Pseudoscymnus tsugae* reduced adelgid populations by 47% and 100%, and it also feeds on other adelgid species such as pine bark adelgid. There may be hope for hemlock.

(*continued*)

Case Study F (*continued*)

References

Annand, P. N., 1924, A new species of Adelges (Hemiptera, Physlloxeridae), *Pan-Pac. Entomol.* **1**:79–82.

Furniss, R. L., and Carolin, V. M., 1977, *Western Forest Insects*, Misc. Publ. No 1339, USDA Forest Service, Washington, DC.

Royle, D. D., and Lathrop, Jr., R. G., 1997, Monitoring hemlock forest health in New Jersey using landsat TM data and change detection techniques, *Forest Sci.* **43**:327–335.

Smith-Fiola, D., 1995, *The Hemlock Woolly Adelgid: Life Cycle, Monitoring, and Pest Management in New Jersey,* Rutgers Cooperative Extension Bull. FS 751, New Jersey Agricultural Experiment Station, Rutgers, New Brunswick, NJ.

Smith-Fiola, D., 1998, Hemlocks suffer from woolly adelgid, *The New Jersey Farmer,* July 15, 1998, pp. 1, 16, 19.

Turnage, S., 1996, Trees with a record, *Our State,* Sept. 1996, pp. 16–18.

Case Study G

Chestnut Blight, *Cryphonectria parasitica*

by John E. Kuser

Approaches: 1) Resistance Breeding
2) Biological Control of Pathogen

A hundred years ago, American chestnut, *Castanea dentata,* was the most important tree in the eastern hardwood forest. Its growth was rapid and its wood was as durable as redwood, slightly lighter than oak but easier to work. Unlike many other nut trees, it was a reliable annual producer of heavy nut crops that were a mainstay for deer, turkeys, squirrels, and humans. It was the most dominant hardwood species throughout its Appalachian range, from Maine to Georgia, often making up 25% of the forest. Chestnuts lived as long as 600 years, and mature trees were sometimes as large as 4–5 feet in diameter (Harlow *et al.,* 1996; American Chestnut Foundation, 1988, 1996a).

In 1904, a foliar blight was noticed on trees lining the avenues at the New York Zoological Garden. Pruning and a spray program using Bordeaux mixture failed to control or contain the new disease (Murrill, 1906). Within 50 years this had spread over the whole range of chestnut, eliminating the species as a tree. The blight did not kill the roots, however, because of the presence of soil organisms antagonistic to *Cryphonectria;* so sprouts arose repeatedly and grew rapidly until they in turn were killed. Soon after the blight was introduced, the US Department of Agriculture and the Connecticut Agricultural Experiment Station launched vigorous programs of resistance breeding, using the Chinese chestnut, *C. mollissima,* as a source of resistance genes. The latter had apparently coevolved with chestnut blight, so that *Cryphonectria* caused only cosmetic damage to it but killed American chestnut, which had never been exposed to blight. The government programs did not succeed in producing chestnuts combining the forest tree form of American chestnut with the blight resistance of Chinese, and were discontinued by 1970 (American Chestnut Foundation, 1996a).

In 1983, the ACF was formed; one of the original organizers was Dr. Charles Burnham, the eminent Minnesota plant geneticist familiar with methods of resistance breeding by crossing disease-sensitive crop plants with their disease-resistant wild relatives (American Chestnut Foundation, 1996a). In this strategy, the two plants are crossed to yield an F_1 hybrid intermediate in resistance, which is then backcrossed to the desirable parent and these B_1 progeny are selected for resistance. The most resistant are again backcrossed to form a B_2 generation, and the cycle is repeated to form a B_3. Two of the most resistant B_3s are then crossed, and some of their progeny inherit two sets of resistance genes and should be as resistant as their wild ancestors, while having the other characteristics of the desirable parent (American Chestnut Foundation, 1996b). The ACF is following this backcross breeding plan, and their Dr. Fred Hebard (personal communication) states that they expect to release B_3 nuts by 2005 or 2006.

Another approach to overcoming chestnut blight is by the use of hypovirulence, a condition caused by a virus that infects *Cryphonectria* and causes it to weaken. Hypovirulence was first noticed around 1950 in Europe, when the Italian plant pathologist Antonio Biraghi noticed that blighted chestnut trees were healing themselves and recovering (Biraghi, 1951). In 1964, French mycologist Jean Grente visited Italy, took

(*continued*)

Case Study G *(continued)*

samples from healing cankers to his laboratory, and from them he isolated strains of *Cryphonectria* that looked different. He called these hypovirulent (Grente, 1965). These hypovirulent forms cured existing cankers when they were inoculated into them. This suggested that in the host hyphae of the hypovirulent strain anastomosed with those of the virulent strain and passed some genetic determinant to them, which turned out to be double-stranded RNA (dsRNA) (Day *et al.*, 1977; Anagnostakis and Day, 1979).

In France and Italy, hypovirulence has spread naturally. It was found that tree climbing slugs, *Lehmannia marginata,* effectively transport hypovirulent inoculum when feeding on the fungal stromata in cankers (Turchetti and Chelazzi, 1984). In the United States, however, hypovirulence has not spread. This may be because there are many vegetative compatibility groups of *Cryphonectria* here, and many are incompatible; they do not form anastomoses with others and transmit the dsRNA. The recent introduction of biotechnology into hypovirulence research promises novel methods for overcoming problems such as this (Nuss, 1996). It is clear that we are only beginning to understand the complex ecology of this system and much work remains to be done (Anagnostakis, 1987).

References

American Chestnut Foundation, 1988, *The American Chestnut Story,* American Chestnut Foundation, P.O. Box 4044, Bennington, VT 05201-4044.

American Chestnut Foundation, 1996a, *History, Methodology, Successes, and Organization,* American Chestnut Foundation, P.O. Box 4044, Bennington, VT 05201-4044.

American Chestnut Foundation, 1996b, *The ACF's Backcross Breeding Program,* American Chestnut Foundation, P.O. Box 4044, Bennington, VT 05201-4044.

Anagnostakis, S. L., and Day, P. R., 1979, Hypovirulence conversion in *Endothia parasitica, Phytopathology* **69:**1226-1229.

Anagnostakis, S. L., 1987, Chestnut blight: the classical problem of an introduced pathogen, *Mycologia* **79**(1):23–27.

Biraghi, A., 1951, Caratteri di resistenza in Castanea sativa nei confronti dei *Endothia parasitica, Boll. Staz. Patol. Veget.* anno 8, Serie Terza. 5 p.

Day, P. R., Dodds, J. A., Ellison, J. E., Jaynes, *et al.,* 1977, Double-stranded RNA in *Endothia parasitica, Phytopathology* **67:**1393–1396.

Grente, J., 1965. Les Formes Hypovirulentes d'*Endothia parasitica* et les espoirs delutte contre le chancre du châtaignier, *C. R. Hebd. Seances Acad. Agr. France* **51:**1033–1037.

Harlow, W. M., Harrar, E. S., Hardin, J. W., and White, F. M., 1996, *Textbook of Dendrology,* 8th ed. McGraw-Hill, New York.

Murrill, W. A., 1906, Further remarks on a serious chestnut disease, *J. New York Bot. Gard.* **7:**203–211.

Nuss, D. L., 1996, Using hypoviruses to probe and perturb signal transduction processes underlaying fungal pathogenesis, *The Plant Cell* **8:**1845–1853.

Turchetti, T., and Chelazzi, G., 1984, Possible role of slugs as vectors of the chestnut blight fungus, *Eur. J. For. Pathol.* **14:**125–127.

Chapter 19

Tree Appraisal: The Goal Is Equity

James B. Ingram

1. Introduction

What do you think of when you hear the words "tree appraisal?" The replies of most professional arborists, consulting foresters, or any horticultural appraiser would probably include words such as "storms," "damage," "destruction," and "failure," and with good reason, these are the times when horticultural experts are most often contacted for their appraisal services. Unfortunately, these are the worst times to try to establish an accurate estimate of a tree's worth. The tree may be so heavily damaged or totally destroyed, so that it is difficult to assess its former positive and negative qualities and how it appeared prior to the damaging event. If an entire site has been damaged or destroyed, the problem becomes many times more difficult.

Given these drawbacks, it is more equitable to evaluate and appraise a landscape and its supporting plants before any damage occurs (Ingram, 1993). This is especially critical for such urban forest properties as arboreta, botanical gardens, public parks, and other sites whose total value is heavily or exclusively dependent on such natural features. An existing appraisal can be important to have not only in the event of a natural disaster; man-made events, such as construction, may heavily damage or destroy plants and landscapes as well. Knowing the work of such features in advance can be a powerful incentive to protect and preserve them from such predictable harm.

There are other, equally valid reasons that argue for the use of appraisals as a proactive measure, performed on a routine basis. More and more municipalities are conducting tree inventories in order to quantify the contribution that trees make to the quality of life in a community and to help them make provisions for protection of this resource. One component of such inventories is to place a value on each tree and

James B. Ingram Bartlett Tree Experts, Osterville, Massachusetts 02655-0177. *Handbook of Urban and Community Forestry in the Northeast*, edited by Kuser.
Urban and Community Forestry in the Northeast, 2nd ed., edited by, J. E. Kuser.

on the community forest as a whole. These values are then used to establish and justify budget requests for maintenance, removals, and new plantings, in order to ensure the continuance of a healthy community forest.

Care of the community forest depends on more than just municipalities, since the definition of such a forest is that it includes "all trees in the community, regardless of ownership," and it is estimated that "60%–90% of the trees in urban forests in the United States are found on privately owned land" (Clark *et al.*, 1997). Just as knowing tree values can aid cities and towns in budgeting for maintenance, knowing the appraised value of a landscape in the urban environment also can be a convincing argument to private owners to take care of their tree resources and to budget adequate funds for that purpose.

There are other reasons for performing appraisals of trees, shrubs, and the community forest. Insurance companies may require appraisals on either public or private properties, in order to both establish coverage levels before any damage occurs and assess the amount of damage in the event of a claim (Ingram, 1993). Appraisals may be needed for tax purposes in the event of a loss. Liability issues may become significant in the event of injury to persons or property caused by tree failure or loss of tree or vegetation on a site. Tort cases involving a wrongful act or a failure to act may require appraisals as evidence in courts of law.

In the plant appraisal process, values of individual plants and the landscape as a whole are reasonably and closely dependent on the value of the land they occupy (Bartlett, 1982). A major ingredient for the professional horticultural appraiser, that is, consulting arborists, consulting forester, or any other horticultural consultant to consider when making an appraisal is the need to ascertain some estimate of a property's total value, a process that may require assistance from a qualified real estate professional. A corollary to support this fact is the number of research studies that have proven trees and other landscape features contribute substantially to both assessed value and selling price of a property and have quantified the actual percentages of value added by attractive landscaping. The potential for horticultural appraisers to offer appraisal services for this purpose is substantial and as yet largely untapped [American Society of Consulting Arborists (ASCA), 1997]. Adopting a proactive approach to tree appraisal before damage occurs, rather than a reactive stance after the fact, offers yet another advantage; when horticultural appraisers systematically take the time to develop reasonable values to aid them, they are able to establish even more accurate and thoughtful appraisals.

While the proactive approach represents the ideal case, it is not always the realistic one. Sudden and unexpected disasters, both natural and man-made, do occur. After a catastrophic loss or any other unforeseen detrimental event, they require the professional appraisal expertise of an experienced horticultural appraiser on an emergency basis. However, the more the appraisals are performed ahead of time, the more the public will be educated to understand that plants can have substantial value and that it is best established before disaster strikes. There are many factors to consider in plant appraisal and there are various appraisal techniques we have to employ given a specific situation. We have designed a format to help the beginning consulting arborist, consulting forester, or any consulting horticultural appraiser learn about the accepted industry procedures and techniques to follow.

2. Appraisal Procedures

The process of tree and plant appraisal is a complicated one, involving many factors and many possible variables within each factor. Part of the process consists of procedures, a set of established methods that can be learned on a step-by-step basis (Council of Tree and Landscape Appraisers, 1992). The other part of the process, not so easily quantified, requires a mature ability to exercise sound judgment through forensic fact finding while maintaining ethical standards.

2.1. Qualifications of the Tree and Plant Appraiser

An appraiser must have a core of practical knowledge about plant structure and maintenance, as well as familiarity with appraisal procedures and record-keeping. A competent appraiser also needs to have confidence, gained from both education and experience, to be able to support and validate the conclusions and the final appraised values (ASCA, 1997). Personal integrity is of utmost importance, since clients and courts of law rely heavily on the honesty and professionalism of the consulting arborist or any other horticultural appraiser to help them establish values that are just, impartial, and fair for all parties concerned. The credibility of the horticultural consultant profession depends on the public perception that each appraiser will establish values that are both reasonable and equitable. If a client wants an appraiser to misrepresent a situation, to decrease or increase plant values for personal gain, or to adjust true value in any way, the appraiser should refuse to work with that client. The ability to weigh the ethical considerations and uphold the highest standards are critically important qualifications for a horticultural appraiser.

2.2. Diagnostic Instruments and Tools

One of the basic tasks in plant appraisal is to diagnose the condition of the plant. This is done by a thorough inspection of the leaves, twigs, branches, trunk, root collar, and roots. The appraiser will need a variety of instruments and tools to accomplish this task. Some tools, such as a trowel and a shovel, are simple to use; other more sophisticated instruments, such as increment borers and Shigometers (which measure resistance of tissue to direct current and help differentiate healthy from decayed wood), require some training and experience. Table 1 provides a list of scientific equipment often used in plant appraisal; they are widely available from horticultural and forestry supply companies.

Because of the complicated and sometimes subjective nature of the plant valuation process, it is essential that the horticultural appraiser follow a systematic procedure. Accurate field notes and photographs are important in recording a professional appraisal or diagnosis. The Council of Tree and Landscape Appraisers (CTLA) is composed of six trade and professional organizations whose members are often required to appraise plants and landscapes. These organizations include American Association of Nurserymen, American Society of Consulting Arborists, Associated Landscape Contractors of America, Association of Consulting Foresters of America, International Society of Arboriculture, and National Arborist Association. In addition

Table 1. Diagnostic Instruments, Tools, and Supplies[a]

Recording	Collecting and examining specimens
Clipboard*	Disinfectant
Compass*	Labels*
Distance meter*	Microscope
Engineer's pocket scale	Plastic, glassine bags*
Photography equipment*	Vials for insects*
Pocket calculator	Tree problem diagnosis
Report forms	Battery-operated drill
Tape recorder	Binoculars*
Size of plant	Chisel gouge*
Diameter tape measure*	Entrenching tool*
Height meter*	Hand lens*
Mallet*	Ice pick*
Pole pruner*	Increment borer
Pruning knife*	Diagnosing soil problems
Pruning shears*	Gas detector
Shigometer, PIRM	pH meter
Shovel, spade*	Soil auger, profile tube*
Small saw	Soil moisture meter
Trowel*	
Trunk calipers	

[a]From Council of Tree and Landscape Appraisers (1992).
* Denotes equipment most critically needed.

to the *Guide for Plant Appraisal* (CTLA, 1992), the CTLA has developed two reporting forms designed to aid professional appraisers with essential record-keeping and calculations as they make their observations. Printed in a manila file for convenience in filing and easy reference, both the "Field Report Guide for Trunk Formula" and the "Field Form Report for Cost of Cure" provide a format to record the steps followed in the appraisal process. The result is a clear and accurate representation.

3. Two Basic Methods of Plant Appraisal

There are several methods for establishing the monetary value of a plant (ASCA, 1997). One of the appraiser's tasks is to select the appropriate one given the particular situation; sometimes a second method may be used to confirm the findings from the first one. The two methods most often used by appraisers are replacement cost and trunk formula.

3.1. Replacement Cost

The replacement cost method is used when a tree or shrub or other vegetation being appraised is small enough that it can be replaced by one of similar size. This method is based on the cost of the largest transplantable, survivable plant that is commonly available from growers or suppliers in a particular region. The appraisal plant, and hence its replacement also must be of a size that can be legally transported on public roads without special permits (CTLA, 1992). The replacement cost may be adjusted mathematically after evaluation of the condition and location (see Sections 4.1 and 4.2) of the plant being replaced. However, each situation is different and must be carefully assessed to determine which, if either, of these adjustments are applicable.

3.2. Trunk Formula

The trunk formula is based on replacement cost and generally is used to appraise trees considered too large or unreasonable to replace with nursery or field stock. The trunk formula is composed of a number of specialized terms, all of the factors that affect the appraised tree value. Definitions of these terms are provided below, since many of them may be unfamiliar.

1. The formula for the trunk formula method is: *appraised value* = *basic value* × *condition* × *location*
2. The definition of basic value is also a formula: *basic value* = *replacement cost* + (*basic price* × [$TA_A - TA_R$]) × *species* (%)
3. Definitions
   - *Condition*: A percent rating based on structural integrity and plant health.
   - *Location*: A percent rating that is the average of the percent ratings assigned for each of three categories, *site, contribution*, and *placement*.
   - *Replacement cost*: The cost to buy and install the largest normally available, transplantable, survivable tree; varies according to region.
   - *Basic price*: The cost per unit trunk area (measured in square inches or square centimeters) of a replacement tree; figured from diameter measured at the height prescribed by the American Nursery Standards. The cost may be wholesale, retail, or installed, depending on how the tree is grown, marketed, and transplanted. These factors and hence basic price will vary according to region.
   - TA_A: Size of *trunk area* of an existing tree whose diameter is 30 inches (75 cm) or less, using diameter measured at 4.5 feet (1.4 m) above ground.
   - TA_R: Size of *trunk area* of a replacement tree (the largest normally available, transplantable, survivable tree) using diameter measured at either 6 or 12 inches (15 or 30 cm) above ground.
   - *Species*: Species rating of the appraised tree; a percent rating that varies according to region as well as species.

4. Factors That Affect Tree Value

4.1. Condition

As a factor in tree appraisal, condition can be quantified and expressed as a percentage that is multiplied into the trunk formula. Two principal attributes define the overall condition of a landscape tree: (1) the structural integrity of the tree and (2) its general health.

4.1.1. Structural Integrity

Assessing a tree's structure should include careful examination of the roots, the trunk, and the branches (Fraedrich, 1993). The root system should be evaluated for general soundness, stability, and signs of potential problems such as broken or dead roots, large areas of missing roots, raised soil around the tree, and leaning of the trunk. Indications of trunk problems include longitudinal seams, externally visible cavities, and outward signs of inner cavities, such as bracts of wood-decaying fungi. Obvious branch problems are codominant stems, a weak V-crotch with included bark, several branches arising close together on the trunk, and dead limbs. Lightning scars, hangers, cankers, frost cracks from wounds (Shigo, 1989), sharp bends, and deadwood also may be signs of branch problems and possible hazards. The presence of one or more of these problems will affect and lower a tree's condition rating.

It may happen that one or more of the above problems is so severe that the tree is in danger of failing and constitutes a hazard or an unreasonable risk to persons or property (See Chapter 16). If this is the case, it should be removed before it causes harm. Such a tree may actually have a negative value, if the cost of removal and cleanup exceeds any value of the wood gained from its sale for timber or firewood (CTLA, 1992).

4.1.2. Plant Health

An important skill required of the professional plant appraiser is the ability to diagnose plant health. Often such diagnosis involves detective work, to uncover past causes of current ills or even to determine the health and history of a plant that is no longer there.

The general vigor of an existing plant can be assessed by examining the comparative amounts of annual shoot growth for the 3 to 4 preceding years. Progressively less growth for each of the past several years may indicate a tree suffering from stress. In addition, signs of declining or poor health are often exhibited in the foliage: leaves may be smaller than is normal for the species or cultivar; they may be distorted, withered, or scorched-looking; they may be off-color or yellowed, or may change to fall colors earlier than normal; they may show signs of insect or disease attack; and the leaf buds may exhibit decreased size, texture change, discoloration, or distortion. Other visible symptoms of a plant's poor health include dieback, insect frass, fungi or conks, and disfigured stems or roots.

If any symptoms of problems are identified as part of the observation or an existing plant's general appearance, the appraiser's challenge is to try to find the cause, in order to rate the current condition and help establish the plant's dollar value. When a plant has been destroyed or damaged beyond recognition, the detective work intensifies. In addition to careful examination of any remaining parts of the plant in question and of the environment around it, the appraiser may want to ask the property owner about the previous appearance of the plant before the final damage occurred, whether the owner remembers seeing any of the above symptoms, and whether any pictures or tree inventories are available that include the plant in question. Neighbors and local officials may sometimes be sources of information. The appraiser also may want to investigate whether recently there was any construction in the area and whether any chemical treatments were applied on the property or on a neighboring property.

When seeking out causes of damage to plant health, an appraiser can look to three principal sources: diseases caused by biotic (living) agents, noninfectious disorders caused by abiotic agents, and direct mechanical injury.

4.1.2.1. Disease-Causing Biotic Agents. Fungi, viruses, bacteria, mycoplasmas, nematodes, and insects invade, infect, and affect plants, causing poor health (Harris, 1992). In addition to causing disease directly, insects may weaken a plant by defoliating it or otherwise physically damaging it, and thus may cause it to be more vulnerable to diseases from other sources.

Knowing the appraised worth of existing plants and being aware of the value they add to their real estate can be incentives to owners to care for those plants on a regular basis. Many of the problems caused by biotic agents can be limited or prevented with a plant health care (PHC) program. A PHC program will include general plant maintenance, such as fertilization and pruning, regular inspection and monitoring of insect and disease problems, and treatment of those problems before they become severe. A landscape insurance study (Ingram, 1993) of properties where maintenance included a regularly scheduled PHC program showed that when vegetative loss was monitored on a yearly basis, the percentage of plant losses relative to the overall value of landscape plants was extremely low.

A PHC program actually may increase plant and landscape value as it improves health and general landscape tree and shrub appearances; it may also diminish the risk of harm, since healthy plants tend to be more resistant to damage from natural causes. Establishing dollar value thus creates opportunity for the professional arborist or practicing landscaper to offer and provide the service of a regular PHC program as an aspect of general property maintenance, another argument for the proactive approach to plant appraisal before any possible damage may occur.

4.1.2.2. Noninfectious Disorders Caused by Abiotic Agents. Aside from biotic concerns, abiotic agents can inhibit growth and also may cause significant harm to plants. Abiotic agents include poor soil, high alkaline or high acid levels in soil, insufficient soil volume, natural or landfill gases leaking into the soil or air, polluted air or water, prolonged drought or excessive rainfall and flooding, spray or runoff from salt used to deice roads and sidewalks, girdling roots, and competition from other plants. Another type of abiotic damage occurs when trees whose root areas are covered by pavement or blocked in by buildings suffer from lack of access to moisture, oxygen, and nutrients. Soil compaction, a frequent problem in heavily trafficked areas of the urban forest, is also an abiotic cause of poor soil drainage, oxygen deprivation, and limited nutrient access. Finally one of the biggest culprits among the abiotic causes of damage to trees is planting too deeply. This all-too-common mistake buries the root collar, causing plants to suffocate (Fraedrich, 1993).

When plants are stressed or weakened by one or more of these abiotic agents, they become more susceptible to attack by secondary biotic agents. The regular monitoring provided by a PHC program can be useful to alleviate or correct problems that may be initiated by abiotic agents such as soil compaction and root collar suffocation.

4.1.2.3. Direct Mechanical Injury. While biotic and abiotic agents may affect plants over a period of time before any damage is actually incurred, the third source of plant damage—mechanical injury—occurs during a short time span (usually less than a few minutes). Although it may be possible to pinpoint the exact minute of injury, damage from the injury may not be visible for some time afterward. A common (and preventable) source of injury to trees is new construction projects and renovations, which often cause large-scale cutting of roots, or injury to the trunk's bark and cambium layer. Either type of injury will cause decline over time. Other causes of injury are weed whackers, lawn mowers, vehicles, vandals, lightning, and storms in general.

Chemicals, such as herbicides, pesticides, fertilizers, and air or water pollutants, also have the capability of injuring plants and causing visible harm and death, either immediately after an application, on a delayed basis, or with frequent exposure over time. When a chemical overshoots a target area, accidentally or intentionally, this infringement is termed `chemical trespass.' Such trespass may have legal ramifications if it causes injury to plants on another's property.

Injuries may occur to the roots, the trunk, or the crown of a tree. It is obviously most difficult for the appraiser to discover and assess root damage and easier to assign a percentage loss for extent of crown injury. The reduction in condition rating due to trunk injury is more complicated and depends on the extent of bark damage, the general health, the species, and whether it is a vertical or horizontal injury. If a tree is completely girdled and the cambium layer destroyed all the way around the circumference, the water and nutrient flow interruption will almost always kill the tree. Table 2 lists percentage guidelines for adjusting the condition rating based on the amount of bark injury and whether the tree is healthy or in a weakened state. It can be seen from this table that

> ... if a vigorous, healthy tree were to lose less than 20% of its bark around its trunk circumference, the main effect would be primarily visual .... A tree may be considered a total loss if bark and cambium on more than 50% of the trunk circumference were lost on a weak tree, or more than 80% on a healthy tree (CTLA, 1992).

Table 2. Influence of Trunk Bark and Cambium Injury/Loss on the Diminution in Tree Value[a]

	Diminution in tree value (%)	
Bark and cambium injury loss percent circumference	Weak tree	Healthy tree
Up to 20	At least 20	10–20
25	25	21–25
30	35	26–30
40	70	31–45
50	100	46–65
65	—	66–85
80–100	—	86–100

[a]From Council of Tree and Landscape Appraisers (1992).

4.1.3. Calculating the Condition Rating

The two condition attributes—structural integrity and plant health—can be further defined using five condition factors and then referring to a list of specific items created to help evaluate each factor. The CTLA has developed a systematic and workable procedure for assessing condition in this manner and for calculating its percentage factor in large tree valuation. The process is well explained in Table 3, in an easy-to-use worksheet (CTLA, 1995).

4.1.3.1. The Five Condition Factors. The condition rating of a plant is determined by the sum of the rating scores for each of five general categories, called factors: (1) roots, (2) trunk, (3) scaffold branches, (4) small branches and twigs, and (5) foliage and/or buds. The appraiser is expected to perform a root collar inspection when evaluating the roots factor and is expected to do a climbing inspection when evaluating the trunk and scaffold branches factors if a score of 100% or five of a total possible of five is attainable.

Each of the five general factors includes up to nine specific items to be considered in the appraiser's inspection of that factor. Each factor is to be given a rating on a scale of 0 to 5; a rating of 5 indicates no problems while a 0 indicates major problems. There are 25 total possible points. The percentage rating for condition is calculated by dividing the number of designated points by the number of possible points. For example, if a plant was assigned 19 points through the five aspects of the tree, its condition rating would be $19 \div 25 = 76\%$. The CTLA encourages the appraiser to take good notes on the present condition of the tree for future documentation.

Checklists help appraisers with both the assessment of plant values and the accurate documentation of the findings. Plant appraisers are often called on to testify in court cases as expert witnesses, and the CTLA field reports are designed so that, as noted at the bottom of each cover page, "Information entered on this form may be admissible as evidence." When the goal is equity, the standardization of plant appraisal practices and record keeping made possible by the CTLA (1995) "Field Report Guide for Trunk Formula" represents a large step forward in the continuing effort to assign fair and reasonable values to landscape plantings.

4.2. Location

Location is the second of three items for which a percentage rating is determined for use with the trunk formula of the plant appraisal process. Used together with the percentage ratings for species and condition, the location percentage allows the appraiser to adjust and modify tree value based on the unique situation of a particular plant.

Three factors are used to determine location values: the *site* of a property or landscape, the plant's functional and aesthetic *contribution* to the landscape, and the *placement* of that plant in the landscape. The location value is calculated by assigning a separate percentage to each of these three factors and then averaging them to calculate percentage for location overall.

Table 3. Guide for Judging the Condition of Landscape Trees[a]

Note: A separate hazard tree evaluation may be required for trees in poor condition.

Scoring System

No problem[c] 5
No apparent problem(s) 4
Minor problem(s) 3
Major problem(s) 2
Extreme problem(s) .... 0 or

TREE NUMBER

Factors[b]	1	2	3	4	5	6	7	8	9	10
ROOTS	POINTS									
Root anchorage S.										
Confined relative to top S.										
Collar soundness S.H.										
Mechanical injury S.H.										
Girdling & kinked roots S. H.										
Compaction or water-logged roots H.										
Toxic gases & chemical symptoms H.										
Presence of insects or disease H.										
TRUNK	POINTS									
Sound bark & wood, no cavities S.H.										
Upright trunk (well tapered) S.										
Mechanical or fire injury S.H.										
Cracks—frost, etc. S.H.										
Swollen or sunken areas S.H.										
Presence of insects or disease H.										
SCAFFOLD BRANCHES	POINTS									
Strong attachments S.										
Small diameter than trunk										
Vertical branch distribution										
Free of included bark										
Free of decay and cavities S.H.										
Well-pruned, no severe heading back S.H.										
Well-proportioned—tapered, laterals along branches S.										
Wound closure H.										
Amount of dead wood or fire injury S.H.										
Presence of decay, insects, or diseases H.										
SMALLER BRANCHES & TWIGS	POINTS									
Vigor or current shoots, compared to that of 3–5 previous years H.										
Well-distributed through canopy H.										
Normal appearance of buds—color, shape, & size for species										
Presence of weak or dead twigs H.										
Presence of insects or diseases H.										
FOLIAGE AND/OR BUDS	POINTS									
Normal appearance—size & color H.										
Nutrient deficiencies H.										
Herbicide, chemical or pollutant injury symptoms H.										
Wilted or dead leaves H.										
Presence of insect or diseases H.										
Total points										
Condition %										

[a]From Council of Tree and Landscape Appraisers (1995), with permission.

[b]Give one rating for each factor. The items listed under each factor are to be considered in arriving at a rating for that factor.

S. = item is primarily structural; H. = item is primarily health; S.H. = item may involve both structure and health

[c]A rating of "5" indicates no apparent problems found having done a root-collar inspection and/or climbing the tree to inspect the trunks an major limbs.

Condition % = total points divided by 25 possible points.

4.2.1. Site

Rating a site involves assessing it on the basis of several criteria, including the quality of development and the prosperity and general appearance of the area. Site rating also will be affected by the design and quality of the structures on the site and in the surroundings and by the landscape design and healthy appearance of the plants on the site and on other landscapes in the area. The level of maintenance on both structures and landscape and the intensity of site use are additional factors to be considered.

A site may be residential, commercial, industrial, or agricultural. The type of site does not influence the site rating; any site may earn a high or low rating as compared to other sites of the same type. For example, a strip mall whose paint is peeling and whose parking lot has no plantings except for being overgrown with weeds will merit a low site percentage rating. In contrast, a relatively high rating will be awarded to a similar mall property where the buildings are well maintained and there are trees, shrubs, and annual plantings in good condition that complement the parking areas.

4.2.2. Contribution

Contribution refers to how the attributes and characteristics of a particular plant enhance or detract from its position on a specific landscaped site. "Attributes" are generic functions that the plant fulfills in the landscape. An attribute indicates the purpose that the plant serves in the landscape. That purpose may be to help solve an environmental or engineering problem on the site, such as a planting of junipers to prevent erosion on a steep bank. It may be to play an architectural role, such as planting a hedge of evergreens to form a wall screening of an unsightly view. It may be to attract wildlife to conservation land or to welcome birds to a private backyard. A rare, unusual, or historically significant plant also may be an important attribute to a landscape or community.

"Characteristics" are factors that are unique to a particular species or cultivar but which only have implications when considered relative to the plant's position in the landscape. Some of these characteristics can be categorized into four groups: growth characteristics, resistance or tolerance, maintenance requirements, and aesthetic values. For example, a tree such as honey locust, with large, twisted seed pods, is a litter nuisance when it is placed next to a walkway or a driveway; the pods become a decorative feature and contribute seasonal interest when the tree is located away from traffic patterns. Similarly, a tree such as European beech, which typically has a low branching habit, is undesirable when it is planted where the lower branches interfere with pedestrian or vehicular traffic; but it becomes an asset when planted as a focal tree in an expansive lawn area. A white pine has a moderate tolerance for salt and will be stressed if planted near a roadside or near the salty air of the seashore; but it thrives on a large scale in the inland areas of the Northeast.

4.2.3. Placement

The placement of the plant being appraised will determine whether it serves its intended purpose. If a row of trees is planted specifically for a windbreak, it will only be effective if it is planted where it will intercept and protect the site from the prevailing winds. If a tree is needed for summer cooling of a house and patio, planting the tree on the north side of the house will not fulfill that need; it will never cast shade on that house. If a large-growing shade tree is planted under utility wires, frequent improper pruning employed to contain it will probably result in a misshapen tree with unreasonable risk rather than a specimen. If a plant fulfills the intended purpose, it is more valuable than one that is planted for a reason but not in the right place.

There are other aesthetic considerations in plant placement that will affect the value of a plant. A single tree, such as a specimen quality Colorado blue spruce, at the end of a walk is worth more than one blue spruce of similar size growing crowded among several in a woods. One white oak in a row of white oaks lining a driveway would probably have a greater value if it were lost than the entire row if it were lost, due to the gaping hole left and the difficulty of replacing the lost plant with one that matched the remaining ones. In these cases, the placement of a plant adds to its value. The placement of a plant also may detract from its value; a Norway maple tree intended for shade but planted too close to a driveway poses a potential risk to a car (Fig. 1).

4.2.4. Calculating the Location Rating

The location percentage is an average of the ratings determined by the appraiser for site, contribution, and placement. The following example adapted from the CTLA (1992) *Guide for Plant Appraisal* illustrates how this percentage is computed: the plant in question is a solitary, 14-inch diameter, 25-foot tall Norway maple tree (see Fig. 1). It is located, in the direction of the morning sun, 25 feet from the front corner of a five-bedroom Cape style home in a moderately affluent neighborhood. The tree shades the drive. It also screens a view, otherwise visible from the kitchen and from the living area inside the house, of industrial smokestacks located 1000 yards to the southeast. An extensive lawn is bordered by well-kept shrubs, bedding plants, and small trees.

1. Site rating
   Well-maintained suburban residence.
   General area is moderately high quality.
   Residences and landscapes are attractive.
   Site rating selected: 85%
2. Contribution rating
   Tree has interesting structure.
   Tree has a well-thinned crown that allows summer breezes and winter sunshine.
   Contribution rating selected: 90%
3. Placement rating
   Tree protects house from unsightly view.
   House receives some winter morning sun.
   Placement rating selected: 80%
4. Location rating
   Average of the tree ratings: 85 + 90 + 80 = 255 ÷ 3 = 85%

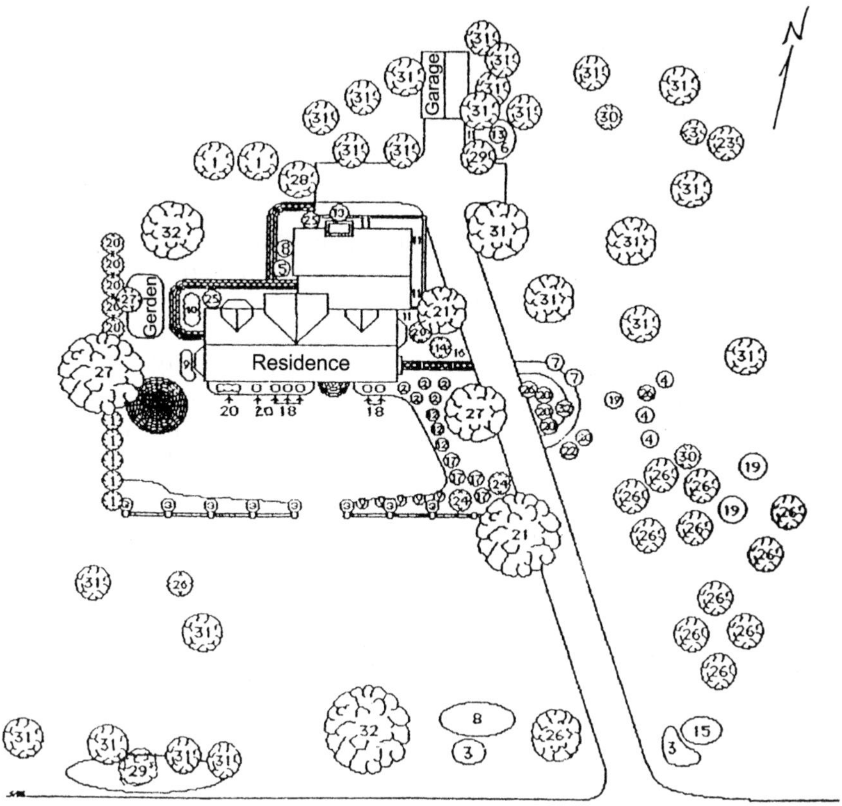

Shrubs:

1. Arborvitae
2. Azalea
3. Rosa Ragosa
4. Eastern Red Cedar
5. Clethra
6. Virginia Creeper
7. Hanoki Cypress
8. Honeysuckle
9. Hydrangea
10. Japanese Ilex
11. English Ivy
12. Leucothoe
13. Lilac
14. Cutleaf Japanese Red Maple]
15. Mulberry
16. Periwinkle
17. Rhododendron
18. PJM Rhododendron
19. Spirea
20. Enlish Yew

Trees:

21. Ash
22. Birch
23. Cherry
24. Dogwood
25. American Holly
26. Locust
27. Norway Maple
28. Red Maple
29. Black Oak
30. English Oak
31. Pitch Pine
32. White Spruce

FIGURE 1. #27 Norway Maple Near Drive.

Another example, taken from the `Field Report Guide for Trunk Formula' (CTLA, 1995), is helpful to contrast with the above case and to illustrate how the placement factor is distinguished from site and contribution. The plant in question is a tree that screens the view of a local landfill (dump) from an abutting neighborhood. The ratings would be represented the following way:

1. Site rating
   Quality, appearance, and use are unattractive and therefore merit low rating.
   Site rating selected: 20%
2. Contribution rating
   Tree screens undesirable views (moderate rating).
   Contribution rating selected: 70%
3. Placement rating
   Tree is planted on dump property line (high rating).
   Tree provides dust reduction for neighbors (high rating).
   Placement rating selected: 90%
4. Location rating
   Average of the three ratings: 20 + 70 + 90 = 180 ÷ 3 = 60%

In this example, although the tree rates a moderately high contribution percentage (moderately high because it is a single tree and therefore provides only partial screening), it merits a high placement percentage because it is planted in exactly the right place to accomplish its purpose. These two relatively high ratings balance out the extremely low site rating, resulting in a 60% overall location rating and graphically illustrating that a tree may be deemed important even to a landfill.

These two examples illustrate how the "Location Chart," which is page 3 of the CTLA's (1995) "Field Report Guide for Trunk Formula," is available to assist the appraiser in evaluating the location factor. It represents another step in completing this standardized form, resulting in a professional document that supports the credence and credibility of the valuations with property owners as well as courts of law (see Table 4).

4.3. Species

The species rating is one of three factors in large-tree appraisal that the CTLA has advised should be determined on a regional basis by a regional committee of tree professionals. The other two factors—replacement cost and basic price (the square inch value of a tree's cross-section)—are derived from species rating and therefore are dependent on it. A number of regional supplements are now available to help appraisers with the assignment of percentage ratings to each species and with the ensuing steps of determining replacement cost and basic price in their particular region. For the Northeast area of the United States (the scope of this volume), species rating guides are available for the regions including Pennsylvania and Delaware, mid-Atlantic states, New England, New York, and New Jersey.

The three factors are considered specialized according to region because the value of a particular species often varies geographically, depending on its relationship to a particular environment (CTLA, 1992). Differences in species value may be attributable to hardiness zone, soil type, and climatic tolerances. For example, a tree such as blue atlas cedar will thrive in the warmer areas of southern Connecticut and near the southeastern Massachusetts shoreline, but the same species will struggle and may not survive when planted farther north on the edge of or beyond its hardiness zone. Other considerations not necessarily regional in character but which may affect a species rating include longevity, structural integrity, and susceptibility to insects and diseases. There are other characteristics, such as branching habit, foliage color,

Table 4. Location Chart: Site/Contribution/Placement Factors to Consider in Determining Location Values[a]

Functional and aesthetic contribution factors suggested rating range 10% to 100%	
Functional attributes	Aesthetic attributes
Environmental and engineering	Architectural and plant characteristics
Sun radiation and reflection control	Attractive bark, flowers, foliage, fruit, fragrance
Wind control	Accents structures
Drifting snow	Screens undesirable views
Safety barrier	Frames view
Light and glare shield	Defines space
Privacy	Creates vistas
Erosion control	Attracts wildlife
Dirt and dust absorption	
Traffic control	Other considerations
Noise attenuation	Historic, rare, or unusual specimen
Air purification	Unusual site situation
Transpiration cooling	

[a] *Location:* The location rating considers the site of a property, the plant's functional and aesthetic contribution, and the placement of the plant in the landscape. *Site:* The rating of a site is determined primarily by: (1) The quality of development, the general appearance and the intensity of use of the area in which the site is situated; (2) The design and quality of structures and landscapes in the area; and the landscape design and quality of the planting and maintenance of the site; and (3) The type of area (residential, mall, etc.) is not particularly helpful in rating a site. *Contribution:* The functional and aesthetic contribution of a plant influences its value in a landscape. Tree characteristics largely determine contribution and value. *Placement:* The position of a tree in relation to how effectively it provides its functional and aesthetic attributes determines the placement rating of the tree. A single specimen tree has greater value than would the same tree as one of many. The placement of a tree can also have an unfavorable as well as a favorable effect on its contribution, such as proximity to overhead wires, street lights, and buildings.

production of fruit litter, and intolerance of salt residue, that once were considered species factors but now are used in appraisal to help determine the contribution adjustment for location (see Section 4.2.2). The positive or negative aspects of the latter characteristics are relative to the plant's location in the landscape; a plant that is a positive addition in one place may be a serious liability in another.

Like the ratings for condition and location, species rating is expressed as a percentage that is multiplied into the trunk formula. The appraiser may not elect to use a species rating if it is already included as part of the replacement cost from the grower. In this case, the species value is a reflection of current market value that determines basic price or the square inch value of a specific species.

4.4. Size

The size of a transplantable tree is normally expressed by its trunk diameter at a specified distance above the ground. This distance is either 6 inches (15 cm) above the ground for trunk diameters up to and including 4 inches (10 cm), or 12 inches (30 cm) above the ground for larger trees still considered to be transplantable. These heights are prescribed by the American Nursery Standards, and diameter at these heights above ground is generally referred to as "caliper."

To measure size for appraisal purposes of a tree considered too large to replace with a transplantable, survivable equivalent, accepted industry practice uses the area of a cross-section of the trunk as calculated from its diameter or circumference. In this case, diameter is measured at 4.5 feet (1.4 m) above the ground. This diameter often has been referred to as diameter at breast height (DBH) through the years because of the forestry profession's quest to determine merchantable tree wood fiber value. The horticultural appraiser may take measurements that may include aspects of a tree that a conventional forester would deem undesirable.

4.4.1. Trunk Area and Adjusted Trunk Area

The cross-section of the trunk is assumed to be a circle; therefore, the formula for calculating the area of a circle is used to determine cross-section area. This formula can be expressed using the radius (r): area = r^2.

It can also be expressed using the diameter of (d or $2r$):

$$\text{Since } \pi = 3.14 \text{ and } r = (d \div 2)$$
$$\text{Area also} = (d^2 \div 2^2) = 3.14(d^2) \div 2^2 = 3.14(d^2) \div 4 = 0.785(d^2)$$

In order to simplify the process, horticultural appraisers, including consulting foresters and arborists alike, generally use diameter to calculate cross-sectional area. The diameter is easily obtained with a special measuring tape that converts the circumference (the distance around the outside trunk) to diameter. A "diameter tape," as it is called, has standard English or metric units on one side and those units divided by 3.14 (called pi) on the other. When the above formula is solved, the result is trunk area (TA), the cross-sectional area expressed in square inches.

The TA for a tree with a trunk diameter over 30 inches (75 cm) can be but is not dictated to be calculated somewhat differently. This is called adjusted trunk area (ATA) and is based on the premise that a large mature tree will not increase in value as rapidly as its trunk area will increase, and on the general experience that such larger trees are more likely to fail (CTLA, 1992). Formulas for ATA have been developed based on increase in tree size, expected longevity, anticipated maintenance, and structural safety. Detailed information on calculating TA and ATA and tables with specific values may be found in the *Guide for Plant Appraisal* (CTLA, 1992).

4.4.2. Other Factors in Determining Tree Size

There are certain variables that may need to be considered in calculating TA. In some cases, the circle formula will not yield a precise enough measurement of cross-section area. Tree trunks are often elliptical (oval) rather than circular, a particularly common occurrence on leaning trees, closely planted trees in a windrow, and those subject to prevailing winds. In these cases the formula is adjusted by using both the largest diameter and the smallest diameter, measurements that can be obtained using a caliper.

Another consideration in TA calculation is bark thickness, which varies according to species and age and can account for a significant percentage of trunk diameter in a larger tree. An appraiser may choose to reduce the trunk measurement by one-third to one-half the bark thickness in order to obtain a more accurate true size for valuation purposes. Conversely, an appraiser may not choose to reduce bark thickness because they may want to give maximum credit to an unusual species with aesthetically pleasing bark. An example of such a tree would be a feature redwood found in someone's yard in Mill Valley, north of San Francisco.

4.4.3. Accurate Measurement of Diameter at 4.5 Feet (1.4 m)

Because the accuracy of TA and ATA (and the resultant tree value, as one can see later) depend on the accurate measurement of the diameter, it is important that this measurement be as precise as possible. If a tree has even a slight trunk taper, measuring the diameter slightly above or below the usual 4.5-feet (1.4-m) height will alter the trunk area calculation.

If a tree is leaning or is growing on a slope, the diameter measurement should be made 4.5 feet (1.4 m) along the center of the trunk axis, so that the distance is the average of the shortest and longest sides of the trunk. The measurement is made at right angles to the trunk. This method of measuring, generally used by consulting arborists, takes the whole tree into consideration, including the flare. Traditional foresters, on the other hand, whose concern is the amount of harvestable timber, usually measure DBH from the uphill side. The latter method reduces the amount of root flare area and therefore the overall tree size and value as well. Regional differences in forestry become readily apparent in the terminology; the wedge that is cut out below and opposite the backcut in a tree's root flare when felling a tree is variously referred to as a scarf, a wedge, a notch in parts of New England, and in Pennsylvania, New Jersey, and New York, an undercut.

If a tree has large, trunklike branches close to the ground, it may not be possible to measure diameter at 4.5 feet (1.4 m). In this case there are several solutions. One is to measure the smallest trunk diameter below the lowest branch (CTLA, 1992). Another is to determine the sum of the cross-sectional areas of the two branches measured about 12 inches (30 cm) above the crotch, then to average the sum of the two branch areas plus the smallest cross-sectional area of trunk below the crotch. A third possibility is to use the trunk measurements of nearby trees of the same species and similar crown volume, either in place of the measurement of the tree in question or as a check of its measurement.

There are other special problems that may be encountered in accurately determining tree size. Excessive trunk flare, trees cut off below 4.5 feet, and multistem trees all present challenges and require the appraiser's careful assessment and judgment. As these cases illustrate, different trees develop different growth peculiarities, and it may take many years of experience for a horticultural appraiser to become skilled at assessing size and all the other elements that contribute to tree value.

4.5. Replacement Cost and Basic Price

Replacement cost is the second of the three factors that are calculated using regional information. It is the median price for purchasing and installing the most commonly available, transplantable, survivable tree. Information detailing and documenting the most common-size trees sold to the layperson is easily understood and makes realistic and believable evidence to the laypersons who serve on juries in courts of law. Reasonable values are continuously stressed by judges in the courts and are perceived as being fair for all concerned.

In order to establish the median price (replacement cost) of a given species, the appraiser will contact professionals at three reputable nurseries in the region to ask what is the common-size tree available for that species and what is the cost of that size. After choosing the most commonly available size for the species, the appraiser will list the three prices in the region for that size and species, and from that will determine the median (middle) value. The appraiser will document the name of the specific plant supplier or nursery representing the median size and the price for that size, as well as the name of the contact person from whom the information was obtained. The appraiser now has one documented source and one expert witness (the contact person at the nursery) to call if a particular valuation is called for question in court. This value, because it is the middle or median price and not an average, is judged to be a reasonable and real price that one would actually pay to purchase a replacement tree. In a court of law, most judges appreciate only taking time with one expert instead of three or four experts.

Most of the regions represented in the International Society of Arboriculture (ISA) have developed these figures and published them for use by appraisers in the region. In New England, based on data provided by the ISA chapter, a 3-inch caliper has been established as the median size for the most commonly available, transplantable, survivable tree. Using the formula in Section 4.4.1 for cross-sectional trunk area results in a median square inch figure of 7 square inches for New England:

$$r = d \div 2 = 3 \div 2 = 1.5 \text{ and}$$
$$A = r^2 = 3.14(15^2) = 7.065$$

In New England, research has determined that the average most commonly sold 3-inch tree found in the nursery costs \$252.00—this leads to an average basic price per square inch of \$252.00 ÷ 7 square inches = \$36.00 per square inch.

The cost may be wholesale, retail, or installed; practices vary by region as to which cost is used. In New England, the median landscaper cost of installation is figured into the replacement cost, using a multiplying factor of 2.5 times the average retail price: \$252.00 × 2.5 = \$630.00.

Thus, in New England, the median price or replacement cost is \$630.00, the amount it costs to buy and install an average 3-inch caliper tree, using a retail purchase price derived from price information given by independent plant growers or suppliers and landscapers. Basic prices throughout the Northeast are established and reviewed in this manner on an ongoing basis.

5. Summary of Trunk Formula Method

Section 4 has provided a detailed discussion of the factors that affect tree value, all the individual elements needed to solve the equations of the trunk formula

Figure 2

European beech 28" diameter at 4.5' from grade. Largest commonly available transplantable tree that will survive or replacement tree = 3" caliper or 7 square inches. Cost of replacement tree = $252.00

Council Of Tree & Landscape Appraisers

TRUNK FORMULA METHOD FORM

Appraised Value = Basic Value x Condition % x Location %
Basic Value + Replacement Cost x (Basic Price x [TA_A-TA_R] Species %)

1. **Replacement Cost:** largest transplantable tree* $ 630
2. **Basic Price** of replacement tree* $ 36 /in²(cm²)
3. Difference in trunk areas of appraised & replacement trees
   A. Appraised tree trunk area (**TA_A** or **ATA_A**)** 615 in²
   B Replacement tree trunk area (**TA_R**)* — 7 in²
   C. Difference in trunk areas 608 in²
4. Multiply **Basic Price** difference in trunk areas
   $ 36 /in²(cm²) x 608 in²(cm²) = $ 21,888 .
   (Line 2) (Line 3C)
5. Adjust Line 4 by **Species** rating* 90 % = $ 19,699 .
6. **Basic Value** = $ 630 . + $ 19,699 . = $ 20,329 .
   (Line 1) (Line 5)
7. Adjust Line 6 by **Condition** 72 % = $ 14,637 .
8. Adjust Line 7 for **Location:**
   Location = (Site + Contribution + Placement) + 3
   =(95 % + 60 % + 40 %) + 3 = 65 % = $ 9,514 .
9. **Appraised Value** = Round Line 8 to nearest $100 = $ 9,500 .

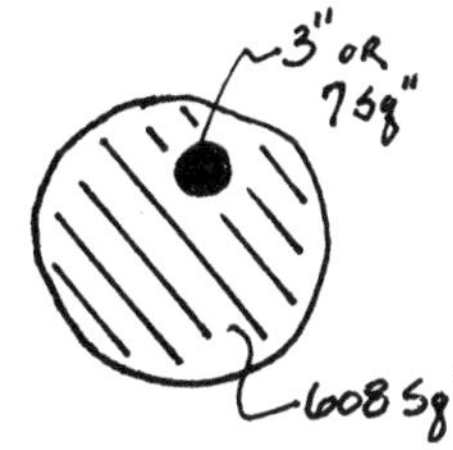

Replacement cost* = Cost to buy & install largest-commonly-available-replacement tree but with no adjustments.
Basic Price* = Cost per unit trunk area of replacement tree. Can be based on the wholesale, retail or installed cost or cost of tree + unit trunk area of replacement tree.
TA_A = Trunk Area of appraised tree (diam. 30" [75 cm] for less). See Table 4-1, 4-2, 4-3 or 4-4 or use the equation below: $TAA = 0.785d^2$ or $0.080c^2$
ATA_A = Adjusted Trunk Area of appraised tree more than 30" in diameter. See Table 4-1, 4-2, 4-3, or 4-4 or use the equation below:

English units for diameter > 30":	Metric units for diameter > 75cm:
$ATA = 0.335d^2 = 69.3d - 1087$	$ATA = 0.335d^2 = 69.3d - 1087$
for circumferences > 94":	for circumferences > 240cm:
$ATA + 0.0333C_2 + 22.1c\ 1087$	$ATA + 0.0333C_2 + 22.1c\ 1087$

TA_R* = Trunk Area of largest-commonly-available-replacement tree
Species* = Species rating of appraised tree
Conditions = Structural integrity and health rating
Location = Site, Contribution & Placement rating averaged

CTLA's Sponsoring Organizations:
American Association of Nurserymen • 1250 I Street, N.W. • Suite 500 • Washington, DC 20005 • (202) 789-2900
American Society of Consulting Arborists • 5130 West 101st Circle • Westminster, CO 80030-2314 • (303) 466-2722
Associated Landscape Contractors of America • 12200 Sunrise Valley Drive • Suite 150 • Reston, VA 22091 • (703) 620-6363
International Society of Arboriculture • PO Box GG • Savoy, IL 61874 • (217) 355-9411
National Arborist Association • P.O. Box 1094 • Amherst, NH 03031 • (603) 673-3311

FIGURE 2. Trunk formula method form.

method. After evaluating each of these elements, the appraiser is finally ready to calculate the appraised value (Fig. 2).

Once the concepts and the many details and exceptions are learned, all the information required to calculate value is found, exact figures can be entered into the formula. Figure 2 illustrates a completed form for a hypothetical situation using the New England area dollar figures for replacement cost and the basic price.

When the worksheet has been completed, the appraiser can transfer the relevant information on it to the back page of the *Field Report Guide for Trunk Formula* (CTLA, 1995), completing the document. This report then comprises a professional, standardized valuation within an evaluation that is available for use by clients, insurance companies, and so on, and that may be admissible as evidence in a court of law. The CTLA recommends that the appraiser should not use trunk formula methodology for IRS tax valuations; the cost of cure method is more suitable (see Section 7).

6. Compounded Replacement Cost

While replacement cost and trunk formula are the two most commonly used methods in plant appraisal, there are often more complicated situations that require a more extensive approach. In these cases, there are additional tools that the professional horticultural appraiser may use in order to formulate more detailed, precise, and accurate estimates. The first of these is compounded replacement cost.

When replacing large plants, a reasonable approach is to plant the most survivable-sized plant and determine the estimated years it will take for the replacement plant to reach an equivalent size (parity). The information needed to figure the compounded replacement cost Method include:

> (1) the replacement cost for the largest plant of the same or comparable species normally available from suppliers for the region; (2) the cost of installing the plant; (3) the number of years estimated for the plant to grow to the same size (or a smaller size if more desirable) as the plant being appraised; (4) the interest rate that would be expected to be earned for investing the replacement cost for the number of years to parity estimated in Step 3 above; and (5) determining the cost of maintaining the replacement plant in the landscape by adding the compounded maintenance cost for each year for the number of years remaining until the plant is expected to be the size of the plant being appraised (CTLA, 1992, Spicer, 1949).

The individual appraiser may be faced with the task of assessing in which areas it is appropriate to apply compounded interest factors in order to obtain a compounded appraised value. Compounded interest methods may be used not only to figure plant replacement costs but to calculate several other aspects of an appraisal situation as well. These include the cost of plant pruning and/or removal, the cost of installation, and the cost of annual maintenance.

7. Cost of Repair and Cost of Cure Methods

The cost of repair method is used when a plant is damaged in such a way that repairing it will restore it to a close approximation of its former condition within a reasonable amount of time. Examples of repair procedures include wound treatment, cabling, bracing, pruning, amending soil, stump sprout management, irrigation, insect and disease management, improving compacted soil, and follow-up care. Plant appraisal in this situation consists of determining the specifications and the cost estimates for the recommended repair treatments. There may be instances in which the cost of repair appraisal also recommends additional compensation for the property owner depending on the nature of the damage, the circumstances under which it occurred, and the anticipated time for recovery.

The cost of cure method is used when extensive damage has occurred, including loss of plants and destruction of ancillary features such as walks, roads, driveways, landscape structures, and shaped terrain. Although the cost of cure method is often used for postdamage assessment, it also may be used to estimate landscape value for the purpose of a routine inventory. Cost of cure determines the cost of replacing and/or repairing plants, as well as all the costs required to restore the property to its precasualty condition. Among the recommendations for site restoration that may be included in a cost of cure appraisal are: posting the area to stop any additional damage; regrading soil that has been displaced; reestablishing damaged plants worth saving; replanting trees, shrubs, flower beds, ground cover, and turf; perhaps replacing a large plant with one or more smaller plants; rebuilding roads, walks, and/or landscape structures; and doing whatever else is needed to restore the property to near its original condition without inflicting more damage.

Cost of cure recommendations always should diminish any future threat to a damaged site. Clearing debris with large heavy machinery such as skidders or bulldozers may disturb an already fragile site well beyond the original scope of damage. In this situation, the CTLA recommends clearing by hand; thus, any planting, restoration, or establishment task should be designed to maximize the chances for physical and ecological renewal.

As it has done for the trunk formula method, the CTLA has developed a user-friendly tool for appraisers for the cost of cure method as well. This form divides the costs of cure into three categories: (1) debris removal and hardscape restoration cost; (2) plant replacement cost; and (3) plant restoration and establishment cost (CTLA, 1997).

Because a landscape appraisal situation that merits using the cost of cure method is generally broad in scope, it requires a high degree of expertise to identify and integrate all the facts needed to determine value in such a complicated situation. The *Field Form Report for Cost of Cure* (CTLA, 1997) provides the knowledgeable, experienced appraiser with a systematic procedure to evaluate each factor needed to develop an accurate estimate. When completed, the field form report constitutes a professional document that is easily understood by clients, insurance representatives, especially tax personnel, and others. Like the *Field Form Report for Trunk Formula*, (CTLA, 1995), the *Field Form Report for Cost of Cure* also may be admissible as evidence in a court of law. The *Manual for Plant Appraisal* (CTLA, 1986) and the two field form reports all contain additional helpful information; the three resources are designed to be complementary tools to assist both beginning and experienced appraisers as they work through the process. Often, many experts collaborate together to determine cost of cure value.

8. Plant Appraisals and Property Worth

The use of techniques to value the real estate of the appraisal plants in question represents a relatively new approach that is increasingly recognized as an important one. It is not necessarily a separate method, but another factor that enters into the overall equation. Since there are guidelines but no hard and fast formulas, it is an area in which the research and judgment of the horticultural appraiser play significant roles with real estate professionals.

8.1. Estimating Total Property Value

There are two principal ways that an appraiser may obtain an estimate of the worth of the property. The first is real estate tax assessments. Annual property tax statements contain the value of the property as assessed for tax purposes; such tax records are available from local governments for public inspection. Although an assessed value does not constitute an appraisal, it is a rough indication of property value.

The second way for an appraiser to learn property value estimates is to contact local real estate brokers, real estate appraisers, and others in the local area with knowledge of property sales. Information on specific selling prices will provide an understanding of the market and will help establish the likely range of property value.

In most cases, it will be impractical for the plant appraiser to obtain a precise estimate of overall property value to include as part of the plant estimate. Property estimates from tax statements or market information from brokers and appraisers, however, will provide sufficient data to lend even more credence to a competent valuation report.

8.2. Contributory Value of Trees, Shrubs, and Other Vegetation

Studies have established that attractive landscaping adds value to a property; conversely, it has been determined that lack of landscaping or landscaping in poor condition may detract from value.

8.2.1. Market Evidence

A number of studies of tree cover as a factor in selling price have supported these findings; some have even quantified the value added with specific percentages. A study of new home construction in Amherst, Massachusetts, showed that the cost of preserving residual trees on building lots was offset by "how fast the homes sold and higher prices for homes on wooded lots" (CTLA, 1992). The US Forest Service estimates that market values for homes increase at rates ranging from 7% to 20%, depending on the presence and extent of the trees (CTLA, 1992). The CTLA guidelines now suggest that well-maintained landscapes and mature, well-placed trees can increase property values by 20% to 30%.

8.2.2. Paired Sales Analysis

Another way to establish the market worth of plants on a particular property is by comparing selling prices of lots with trees to selling prices of lots devoid of trees or by looking at prices of comparable homes that differ primarily in the amount and quality of landscaping around them. Such paired sales analysis can help the appraiser present solid arguments for the value contributed by the amenities of trees and landscaping.

8.3. Arboricultural Appraisal Law and Property Value

In contrast to a traditional rural forest, where loss appraisals are usually based on timber value, appraisals of loss or damage in the urban forest are closely tied to property values. The courts have differentiated between the two types of forest properties in their attempts to resolve disputes in a fair and equitable manner. As a result, a body of case law has developed that addresses the unique situations caused by damages to trees in the urban setting (Merullo, 1992).

8.3.1. Measures of Damages

A number of legal cases have established precedents for measures of damages that may be available to urban forest property owners when unauthorized destruction occurs (CTLA, 1992). A summary of this case law by legal experts has concluded that "it is quite apparent from the courts' opinions that the prevailing view is to assess damage based on the resulting depreciation in the value of the land on which the trees stood" (Merullo, 1992). These legal conclusions have led to the growing recognition by tree industry professionals of the importance of real estate values as a key element to be considered in plant appraisal practices.

The same legal experts, however, have pointed out that the courts endorse the use of alternative methods to help in determining damages. These methods include the cost of replacing destroyed or injured trees (replacement cost method), the value of the injured or destroyed trees (trunk formula method), the cost of restoring the property on which they stood to its previous condition (cost of cure method), and the resulting loss of aesthetic value or the resulting deprivation of the comfort and convenience that the trees provided the landowner. In a case in Delaware, the court approved using both the costs of restoration and the before and after valuation of the land itself, because "allowing a jury to consider more than one measure of damages would permit flexibility and achieve just and reasonable results" (Merullo, 1992). The phrase "just and reasonable" is the recurring theme in tree appraisal, where the goal is to achieve a fair and equitable settlement for all parties concerned (ASCA, 1997).

8.3.2. Reasonableness Testing

One of the most common reasons for doing plant appraisals is that owners can receive partial or full remuneration for the value of trees that become casualties as a result of accidents or alterations to the landscape. The three methods for recovery of casualty losses include insurance coverage, income tax deductions, and civil damage claim. Where claims have resulted in litigation, the courts have established a concept called reasonableness testing to help determine values of landscapes in cases of loss or damage involving trees and other plants, as well as hardscape features such as stone walls and walkways. The concept of "reasonable and practical replacement cost" is based on two premises: (1) that the replacement should restore the property to approximately its former character and quality and (2) that the cost of such replacement should not be disproportionate to the end results of such restoration (CTLA, 1992). Cost of cure methodology closely mimics IRS qualifications for tax qualification purposes.

8.3.3. Appraisers as Expert Witnesses

In the past, monetary appraisals of trees and other plants most often have been tied to damage and litigation. The urban forest is by definition a forest within a populated area, and litigation relating to urban and community trees can be complex, involving private landowners, municipalities, public utility companies, private tree companies, practicing arborists, and highway agencies, among others (Merullo, 1992). Appraisers are often summoned to be expert witnesses in court cases and defend the methods and results of their appraisals. The need for credence and professionalism in court cases is a major reason why appraisers may choose to seek more training and credentials in the field of forensics, including professional certification by the Board of Forensic Examiners.

9. Summary: Role of the Appraiser and the Appraisal Process in Urban Forestry

A number of different methods are available to the appraiser in the quest to establish fair and reasonable values. There may be times when it will be advisable to use more than one method, in order to compare and test the validity of a resultant value. It is important, however, for the appraiser to know and remember that a reasonable value is also a real value, not an average. The results of each method used must stand alone and never be combined or averaged together in any way.

Appraisal values are only as good as the data on which they are based. The accuracy of the fact finding is one of two principal factors that will determine the quality and credibility of an appraiser's work. The other is the appraiser's ability to stand by those facts honestly to peers, property owners, and/or courts of law. The appraiser always should be aware of the background where a case originated, where it is at the time when the appraiser is consulted, and what end the client wants to accomplish. Awareness of these three items is necessary to enable the appraiser to weigh the ethical considerations carefully and maintain high-professional standards. Every situation is different, and a competent and qualified appraiser will continue to develop knowledge and skills with every experience.

The traditional approach to the tree appraisal process has been the reactive one, often precipitated by a damaging event. Fair and equitable values are more likely to result, however, if trees and landscape plants can be appraised while they are still whole, before they are damaged or destroyed. The recognition of value is a critical step not only in dispute resolution but also in convincing public and private owners to be responsible stewards of their urban forest resources.

The proactive approach can be a significant tool to help communities place value on their forest resources, to develop understanding and appreciation for the improved quality of life provided by the urban forest, and to educate citizens to be advocates for those valuable resources. When consultant arborists, foresters, or any other horticultural appraisers do appraisal work, they seek not only to achieve the goal of fairness and reasonableness in establishment of plant values; they also are making a contribution for protecting the equity of the community, the growing investment that is the urban forest.

References

American Society of Consulting Arborists (ASCA), 1997, *Tree Appraisal Workshop Material*, American Society of Consulting Arborists, Rockville, MD.

Bartlett Tree Experts, 1982, *Appraising Shade and Ornamental Trees*, Bartlett Tree Experts, Stamford, CT.

Clark, J. R., Matheny, N. P., Cross, G., and Wake, V., 1997, A model of urban forest sustainability, *J. Arboric.* **23**(1):17–30.

Council of Tree and Landscape Appraisers (CTLA), 1997, *Field Form Report for Cost of Cure*, International Society of Arboriculture, Savoy, IL.

Council of Tree and Landscape Appraisers (CTLA), 1995, *Field Report Guide for Trunk Formula*, International Society of Arboriculture, Savoy, IL.

Council of Tree and Landscape Appraisers (CTLA), 1992, *Guide for Plant Appraisal*, International Society of Arboriculture, Savoy, IL.

Council of Tree and Landscape Appraisers (CTLA), 1986, *Manual for Plant Appraisers*, Council of Tree and Landscape Appraisers, Washington, DC.

Fraedrich, B., 1993, *Hazardous Tree Evaluation and Management*, Bartlett Tree Research Laboratories, Charlotte, NC.

Harris, R. W., 1992, *Arboriculture—Integrated Management of Landscape Trees, Shrubs and Vines*, Prentice-Hall, Englewood Cliffs, NJ.

Ingram, J. B., 1993, *Landscape Insurance Study, Cape Cod, Massachusetts*, Bartlett Tree Experts, Osterville, MA.

Merullo, V., 1992, *Arboriculture and the Law*, International Society of Arboriculture, Savoy, IL.

Spicer, O. W., 1949, Evaluation of shade trees, *Arborist's News*, November, 137–142.

Chapter 20

Leaf Composting

Peter F. Strom[1] and Melvin S. Finstein[2]

1. Introduction

1.1. Background

Leaves fallen to the floor of a forest undergo a slow process of decomposition. This is brought about through the action of numerous organisms, with microorganisms playing a dominant role. Eventually, a thoroughly decomposed state is reached, the organic residue becomes part of the soil, and the nutrients may again be available to plants.

In contrast, leaves collected in developed areas represent a waste management problem. The traditional "solution" of open burning by residents was banned in the 1970s by many states or localities in the Northeast due to air pollution and safety concerns. This led to the further development of collection programs and shifted most of the burden for disposal to landfills.

Leaves differ from much other municipal solid waste (MSW) in the Northeast in that they occur seasonally in large volumes, and thus frequently are collected separately. In season, leaves may account for over half of the MSW collected and on a yearly basis may comprise 5% to 30% of the total MSW stream.

Disposal of leaves at landfills is also problematic and now banned in several states. The siting of new landfills has become difficult and landfill capacity has decreased sharply. Placing leaves in landfills uses up this resource for a relatively innocuous material. With fewer landfills, the hauling and waiting times and distances increase for many municipalities. Also, maintenance costs are high for trucks at landfills.

Peter F. Strom and Melvin S. Finstein [1]Professor and [2]Emeritus Professor Department of Environmental Science, ENRSB, Cook College and the New Jersey Agricultural Experiment Station, 14 College Farm Road, Rutgers University, New Brunswick, New Jersey 08901-8551.

Urban and Community Forestry in the Northeast, 2nd ed., edited by, J. E. Kuser.

Ironically, when leaves are placed in landfills their biodegradability, which makes them suitable for composting, adds to the gas, leachate and settling problems that landfills experience. Thus, landfilling of leaves is not only expensive but also environmentally unsound.

Leaves have also been banned from many incinerators. They are often too wet to burn well, reducing energy recovery. On the other hand, when leaves are dry, they tend to be light and fluffy, adding to the fly ash. In any case, incinerating leaves may be impractical because of their large volume and seasonal nature.

Like leaves in a forest, those collected from developed areas also can be decomposed microbially and the organic residue returned to the soil. However, the leaves must be processed in concentrated form, the decomposition accelerated, and the residue deliberately applied to soil. This can be accomplished economically by the composting process linked to a compost use program.

Other yard wastes in the Northeast include mainly grass clippings and woody materials such as tree trimmings. Grass clippings may account for 10% to 20% of the total MSW stream and woody materials perhaps half of that amount. In many cases, these historically have not been collected separately. Woody materials degrade too slowly to be amenable to composting, and grass clippings, although readily compostable, are problematic at leaf-composting facilities.

This chapter is designed to aid in the establishment and operation of leaf-composting facilities. Recycling of other yard wastes also will be discussed briefly. The best available scientific information is employed to find technically simple, cost-effective solutions that may be implemented by local personnel. Underlying principles are first explained so that the basis of the "how-to" recommendations may be understood. In this manner the composting operation may be flexibly adapted to meet site-specific needs.

1.2. Current Management Alternatives

In 1987, New Jersey became the first state to require that leaves (if collected) be source separated and recycled (NJSA 13:1E-99.11 *et seq.*). Several other states have now banned disposal of leaves, and in some cases have included grass clippings in the statutes.

Leaves and some other yard wastes can be managed by: (1) source reduction; (2) mixing (sometimes called "mulching") directly into agricultural soils; or (3) composting at public or commercial composting facilities. Several of these activities may be state or locally regulated, and could require permits.

1.2.1. Source Reduction

Leaf composting or mulching on a municipal or commercial scale requires separated leaf collection, as does the composting of other yard wastes. Source reduction, or waste minimization, is an alternative that has the benefit of avoiding collection and its associated costs. Several methods are possible.

1.2.1.1. Backyard Composting. Municipalities should encourage backyard composting as a part of their overall yard waste management program since all municipal collection, processing, and distribution costs are avoided for leaves composted by residents. Additionally, other materials, such as grass clippings and fruit and vegetable scraps, can be included in backyard composting, thus reducing handling of these wastes by the municipality as well. Numerous publications on backyard composting are available. Flower and Strom (1991) recommend a simple method, making it easier for residents to get started.

1.2.1.2. Grass: "Cut It and Leave It". Grass clippings can be easily eliminated from the MSW stream, while at the same time producing a healthier lawn. The clippings simply should be left on the lawn (Anonymous, 1993; Strom *et al.*, 1992).

1.2.1.3. Backyard Mulching. Another way of recycling leaves, grass clippings, and brush is to apply them to gardens or around shrubs and other plantings as a surface mulch. A 1-inch layer of mulch can help moderate soil temperatures and reduce weed growth, soil spattering and compaction, erosion, evaporation, and runoff. Brush needs to be chipped first, and leaves also make better mulch after shredding or composting. All of the materials help improve the organic content of soils, although with woody materials this will take a longer time.

1.2.2. Farmland "Mulching"

Collected leaves can be "mulched" on agricultural land. Leaves should be delivered unbagged; spread in a layer that is ≤6-inches deep (3 inches if preground, which may be beneficial) within a week; and incorporated into the soil by the next tillage season. Because of the high carbon-to-nitrogen ratio (C/N) of leaves, the need for additional fertilizer should be recognized.

1.2.3. Composting

For many municipalities, composting has proven to be the best and least expensive method for managing leaves. There are over 200 such facilities in New Jersey alone. Advantages include the cost savings and environmental benefits compared to many other alternatives, and the value of the compost produced. However, to fully realize these advantages and avoid the potential problems, care must be exercised in siting, designing, and operating the facility.

2. Process, Product, and Objectives

2.1. Composting (the Process)

When biodegradable organic materials containing sufficient moisture and inorganic nutrients are placed in a pile or windrow (elongated pile), a natural process known as self-heating typically occurs. Microorganisms, mainly bacteria and fungi, utilize the organics as a food source, growing rapidly and causing their decomposition. Because the microbes are not 100% efficient, some of the chemical energy stored in the organics is wasted and released as heat. A large enough pile will act as a thermal insulator, retaining heat and leading to an increase in temperature, which up to a point helps to speed up decomposition. Thus, the organic material "self-heats" through the intense metabolic activity of the microorganisms. Eventually the readily biodegradable food supply becomes exhausted, growth and heat generation slow down, and the pile cools.

The process which employs self-heating for waste treatment objectives is called *composting*. It has been used for many years for treatment of agricultural wastes and in more recent times for treating sewage sludge, some MSW fractions, certain industrial wastes, food scraps, and leaves and grass clippings.

2.2. Compost (the Product)

As composting progresses, the original material becomes less recognizable, although certain structures, such as twigs and the veins of oak leaves, persist longer than others. The material darkens, acquires a granular texture, increases in water-holding capacity, becomes neutral or slightly alkaline in pH, and eventually develops the pleasant odor characteristic of freshly turned soil. The stabilized finished product, *compost*, bears little resemblance to the original starting material.

2.3. Composting Objectives

The main objectives of leaf composting as a waste treatment process are: (1) a reduction in the mass and volume of the starting material and (2) the destruction of putrescible (odor-causing) substances. For compost production, reduction of the carbon-to-nitrogen ratio, and elimination of weed seeds and any plant pathogens, also are desirable. Cost-effectiveness, of course, is also a primary concern, and the economic and other advantages of composting leaves are dramatic when compared to many alternative strategies.

3. Underlying Scientific Principles

The composting system is an ecosystem and is dependent on the growth and metabolism of microorganisms. It is thus important to maintain conditions within a range that is suitable (although not necessarily always optimal) for microbial activity.

3.1. Temperature

At any given time, the temperature of the pile reflects the balance between microbial heat generation and the loss of heat to the surroundings. The rate of heat generation is a function of factors such as temperature, oxygen, water (moisture), nutrients, and the remaining concentration of easily biodegradable organic materials. The rate of heat loss is a function of factors such as ambient temperature, wind velocity, and pile size and shape.

Temperature is a powerful determinant of the rate of decomposition. Temperatures of less than 20°C (68°F) slow decomposition. Temperatures above 60°C (140°F), which is hotter than the setting of most home hot water heaters, are also unfavorable because they kill most of the desirable microorganisms. The range of favorable temperatures is thus approximately 20°C to 60°C. Precise control over temperature usually is not essential for leaf composting, but gross departure from the desired range should be avoided. Maintenance of the proper temperature, along with oxygenation, are the basic considerations underlying the recommendations for windrow size and turning operations.

Grass clippings are a more "energetic" material (generating more heat) than leaves and are produced and composted during the warmer part of the year. Overheating is thus more likely and low temperatures less likely than for leaves. This, along with the need for increased oxygen supply, is why a smaller pile size is recommended when grass clippings are included in a windrow.

3.2. Oxygen

Composting is mainly an aerobic process (requires oxygen), although anaerobic (without oxygen) activity also may occur. Most of the heat produced in composting results from the biodegradation of organic materials with consumption of oxygen and production of carbon dioxide and water. Thus, the pile must be sufficiently porous to allow oxygen (from the air) in and carbon dioxide out. For this reason, materials should be placed loosely in the piles and compaction avoided. Extensive anaerobic conditions can lead to odor production and slowed rates of decomposition.

3.3. Windrow Size and Turning

For leaves, control over process temperature and oxygen content can be exercised to a useful extent (though they are not optimized) through windrow size and turning. A basic problem is to reconcile the needs for oxygenation and heat conservation, which may conflict. The need for oxygenation favors small windrows to minimize the distance that air must penetrate within the pile. In contrast, the need for heat conservation, especially in the winter, argues for large windrows for greater insulation. Excessively large windrows, however, result in excessively high temperatures and anaerobic conditions. Specific recommendations are given later, but for almost all leaf composting, windrows should be no more than 6-feet high and 12 to 14-feet wide.

3.4. Water

Water is essential for biological functions in general, and composting is no exception. An initial moisture content of at least 50% (wet weight basis) is recommended. Depending on weather conditions prior to collection, the leaves might be sufficiently moist, but this cannot be relied upon. Adding water (when needed) at the start of composting is important to ensure adequate moisture throughout the pile. Once a pile is formed, the interior material is not easily wetted even by heavy rainfall or applying water to the surface. Such water penetrates only slowly, and unless a pile is turned during or shortly after wetting, much of it will simply evaporate to the air. Dryness is a common cause of slow leaf-composting rates, and as such should be prevented.

Leaves also can be excessively wet, slowing oxygen penetration. This condition is to some extent self-correcting, as excess water drains from the pile.

3.5. pH

Fresh leaves are nearly neutral (pH 7), which is desirable for rapid microbial activity. However, with the onset of decomposition even prior to composting, the production of organic acids causes the pH to decline to suboptimal levels, possibly below 4.5 if extensive anaerobic conditions develop. The pH subsequently recovers to a neutral or slightly alkaline level (perhaps pH 7.5) as the acids decompose in the presence of oxygen. Adjustment of the pH with additives is not ordinarily necessary. A persistently acidic pH is indicative of prolonged anaerobic conditions, while pHs above 8.5 occur when ammonia is released during composting of nitrogen-rich materials such as grass clippings.

3.6. Inorganic Nutrients

Leaves are deficient in nitrogen, and thus have a high carbon-to-nitrogen ratio ($C/N \approx 80$). This nutritional imbalance can slow microbial action early on, but corrects itself as carbon is lost in the form of carbon dioxide, while nitrogen is conserved within the system (final $C/N \approx 20$). Supplementation with nitrogen might accelerate decomposition but is not generally recommended because it could lead to other problems such as an increased need for oxygen supply. Unless this also is addressed, slow decomposition and odors might result. Appreciable deficiency of other nutrients, such as phosphorus, is not likely.

Grass clippings, on the other hand, contain excess nitrogen and thus have an undesirably low C/N. Unless sufficient available carbon (such as from leaves) is added, ammonia will be lost from the material, producing potential odor problems. The nitrogen also may contaminate ground or surface waters. Mixing of leaves and grass seems logical, but the potential problems are discussed later.

3.7. Microorganisms

Microorganisms found on leaves and yard waste are fully capable of starting the composting process and carrying it forward. A variety of commercial "inocula," "starters," and "bioaugmentation" products are offered for sale, and based on testimonials, are often claimed to be beneficial. However, there is no scientific support for these claims. Properly controlled experimentation indicates that inoculation has no useful effect on the process. Therefore, such products should not be purchased for leaf or yard waste composting operations.

3.8. Destruction of Pathogens and Weed Seeds

Pathogens and weed seeds are rapidly destroyed by the temperatures (Burge *et al.*, 1981) and vigorous microbial activity achieved during composting. Complete destruction is dependent on ensuring that all parts of the pile are exposed, by mixing, to these conditions.

3.9. Leaf Type

Maple leaves decompose more rapidly than oaks, and other leaf types doubtless differ as well. Mixtures would ordinarily be received at a composting facility, and no specific recommendation is made based solely on leaf type.

3.10. Pregrinding

Pregrinding or shredding of leaves make them more susceptible to microbial attack, potentially speeding up the composting process. This is *not* desirable in most cases, unless provision has been made for very frequent turning or blowers to supply the extra oxygen that will be needed, and remove the extra heat that will be generated. The guidelines given later assume no pregrinding; if any is done, smaller piles are recommended. The equipment typically used for the final shredding of finished compost usually is not suitable for shredding of leaves prior to composting.

4. Facility Siting and Initial Preparation

Site selection is an extremely important decision that should be made only after careful consideration, as each situation is unique. These deliberations should take into account proximity to residences and streams, prevailing winds, traffic patterns, travel distance and its effect on equipment and labor costs, and other factors such as local zoning requirements. Many of these are discussed in later sections, yet familiarity with local circumstances is essential.

4.1. Public Participation

When selecting a site, the importance of public participation must be stressed. Concerns raised may include odor, traffic, noise, litter, water pollution, and health issues. Surrounding property owners and the general public should be educated as to the benefits of composting, informed about the proposed facility, and assured that their concerns will be addressed. It is very important to have support within the community, and an informed public is less likely to oppose the siting and operation of a facility.

It may be necessary to deal with emotional reactions from some community members. Composting sites can be bad neighbors; first make sure this one will not be, then convince the public. An open dialogue should be maintained throughout the planning and operational phases of the project. Providing educational and informal discussion sessions during hours convenient to the public can be helpful. The county Cooperative Extension office may be able to provide special programs and expertise. Many sites offer the finished compost free of charge to residents, further increasing knowledge and support within the community.

The siting and operation of the facility may need to be approved by local, county, and state agencies. In some cases, specific permits may be required.

4.2. Area Requirement

A minimum of an acre per 3500 cubic yards of leaves collected usually is required for the actual composting operation. This assumes the use of the low-level technology described later, and is in addition to the requirement of a buffer zone (Table 1). Use of the intermediate level of technology may require additional space, since smaller windrows may be needed to accommodate a specific turning machine. A method of calculation of site capacity (Strom and Finstein, 1994) is shown in Table 2.

4.3. Buffer Zone

A buffer zone is required between the site activities and neighboring land use to minimize possible odor, noise, dust, health, and visual impacts. Other than "the larger the better," it is difficult to generalize buffer zone requirements for composting. It seems prudent to provide at least 50 feet between the composting operation and the property line, and at least 150 feet between composting activities and any sensitive neighboring land uses such as residential property lines. Additionally, at least a 250-foot buffer is recommended between composting activities and a place of human occupancy (house, school). If grass clippings will be brought to the site, at least 1000-foot buffer zones from the staging and grass clipping-handling areas are probably necessary.

Calculations of area requirements (Strom and Finstein, 1994) for buffer zones are shown in Table 3; note that they may increase site size considerably. The buffer zone may include a berm (often of finished compost) to serve as a visual barrier, help control vehicular access, and reduce noise levels off-site. A landscaping plan, including plantings, is recommended to enhance appearance.

Table 1. Appropriate Leaf Composting Technology

Level of technology	Capacity[a] (cubic yards/acre)	Buffer[b] (feet)	Time (month)	Relative cost	Use for grass clippings[c]
Minimal	4000	–/1000/–	36–60	Very low	No
Low	3500	50/150/250	16–18	Low	No
Intermediate	3000[d]	50/150/250	4–6	Low–moderate	Perhaps[e]
High	6000	50/–/–[f]	3–4	Moderate	Perhaps

[a]Based on fall collection of leaves in 1 year.
[b]Recommended distance from operations to: property line/sensitive neighboring land uses/place of human occupancy.
[c]At leaf composting site, with additional requirements.
[d]Assumes 5-foot pile height, 10-foot width, and 10-foot aisles.
[e]Buffer zone of 1000 feet from staging and grass clipping-handling areas to sensitive neighboring land uses.
[f]For totally enclosed system with odor control; otherwise, same as for intermediate.

Table 2. Maximum Initial Site Capacity (Cubic Yards per Acre of Windrowing Area)

Windrow height (feet)	Average aisle width (feet)							
	2	4	6	8	10	12	14	16
4	4055	3379	2896	2534	2253	2027	1843	1689
4.5	4665	3948	3421	3019	2701	2444	2231	2053
5	5280	4525	3960	3520	3168	2880	2640	2437
5.5	5897	5111	4509	4035	3650	3333	3066	2839
6	6517	5702	5068	4562	4147	3801	3509	3258
6.5	7138	6298	5635	5099	4655	4283	3966	3692
7	7761	6899	6209	5644	5174	4776	4435	4139

Resulting site capacities for various windrow and aisle sizes, considering the area used for windrowing only (including aisles between windrows, but not buffer zones, staging areas, and roads). For example, for 6-foot high windrows with average 14-foot wide aisles, 3509 cubic yards per acre can be composted.

Assumptions:

(1) windrow cross section is approximately a semicircle (assuming a triangular shape will underestimate windrow volume, while a rectangular shape will overestimate it; for some turning machines, using a trapezoidal shape may give a slightly better estimate);

(2) site is rectangular, and all windrows are the same size initially (for ease of calculation; does not directly affect the result).

For an individual windrow, let:

h = height, w = width = $2h$, l = length

A_c = cross-sectional area = $\frac{1}{2}\pi h^2$

A_b = area of base = $wl = 2hl$

V = volume = $A_c l = \frac{\pi h^2 l}{2}$

For the site (windrow area only, including aisles but not buffers, staging, etc.), let:

w_s = site width, a = aisle width (average if variable), l_s = site length = l (assuming that windrows run entire length of site), $A_s = w_s l_s$.

For a large site, with one aisle per windrow (although aisle width may vary):

N = number of windrows = $\frac{w_s}{2h + a}$

V_T = total volume of windrows = $NV = \frac{w_s}{2h+a}\left(\frac{\pi h^2 l}{2}\right) = \left(\frac{\pi h^2}{2h+a}\right) \times \frac{w_s l}{2}$

V_a = volume of windrows/acre = $\left(\frac{V_T}{A_s}\right)$ (4840 square yard/acre) = $2420\,\pi\left(\frac{h^2}{2h+a}\right)$ (square yard/acre)

4.4. Location

A centrally located facility is preferable to reduce transportation time and costs, although such sites are not often available or otherwise practical. Access is preferably over noncrowded, nonresidential, and hard surface roads. Avoidance of residential areas can minimize opposition to a facility.

4.5. Stream Encroachment and Water Pollution

Siting of a leaf composting facility in a flood plain normally is not allowed. During times of high water the windrows might impede water flow, and/or leaves and leachate might wash into the stream. Flooding of the site could pose serious operational difficulties, including problems with equipment access and operation. Flooding of the windrows also may lead to extensive anaerobic conditions and attendant problems of odor and lower decomposition rate.

Runoff into nearby streams or other surface waters is another concern because of the water pollution potential of leachate. If grass clippings are composted, nitrogen contamination of groundwater also must be considered.

Table 3. Buffer Zone Area Requirements (Additional Acres Needed for Specified Buffer Size)

Buffer zone width (feet)	Windrowing area required (acre)							
	1	2	3	5	10	20	30	50
50	1.2	1.6	1.9	2.4	3.3	4.5	5.5	7.0
100	2.8	3.6	4.2	5.2	7.0	9.5	11.4	14.5
150	4.9	6.1	7.0	8.5	11.2	14.9	17.8	22.4
200	7.5	9.1	10.3	12.2	15.8	20.8	24.7	30.8
250	10.5	12.5	14.0	16.5	20.9	27.2	32.0	39.6
500	32.5	36.5	39.6	44.4	53.3	65.8	75.4	90.7
1000	111.0	118.9	125.0	134.7	152.4	177.5	196.8	227.3

Assumes site is square; if length = $2x$ width, add 5%; if $3x$, add 10%. Table 3 can be used to determine the approximate additional acreage required to provide a buffer zone of a specified width on all sides of a site. For a given buffer size and windrowing area, find the additional acres needed from the table. For the example, if a 150-foot buffer zone is needed in addition to 5-acres needed for windrowing, this would require an additional 8.5 acres. The total acreage required would then be 5 + 8.5, as well as 1 or 2 acres for staging, roads, etc., or about 15 acres.

4.6. Slope and Grading

Steep slopes are unsatisfactory because of problems with erosion, vehicular access, and equipment operation. A gentle slope, however, is desirable to prevent ponding of runoff and leachate. The Soil Conservation Service allows a minimum slope of 0.5%, and a maximum of 10%, although 2% to 3% is preferable. Initial site preparation usually requires grading, and yearly maintenance should include regrading where necessary. Windrows should run up and down rather than across slopes to allow leachate and runoff to move between piles, rather than through them.

Drainage characteristics of a site can be determined from US Geological Survey topographic maps and a plot plan survey. The local Soil Conservation District should be contacted for information on permits or approvals that may be required for any soil disturbance in excess of 5000 square feet.

4.7. Percolation and Water Table

High-soil percolation rates are desirable so that excessive rainwater and leachate will not run off the site. Where drainage is poor, soil modification techniques may be used to improve it. With poor percolation, particular care must be taken to prevent ponding. While an impervious surface, such as pavement, may offer advantages for vehicle access and equipment operation, these usually are outweighed by the greater difficulties in leachate management.

The seasonal high-water table should be at least 2 feet below the surface. Wetlands and wetland buffer areas especially should be avoided. A high-water table is undesirable because it may lead to flooding of the site. Flooding makes operation more difficult and can lead to extensive anaerobic conditions. A high-water table also reduces the distance for percolation of leachate.

4.8. Water Supply

Water availability is important since it often is necessary to wet the incoming leaves. Water can best be supplied by using a hose from a fire hydrant or by pumping from a nearby pond, stream, or well. Use of a water truck usually is not practical because too much of water is needed. For the low-level technology approach, average water requirements are about 20 gallons per cubic yard of leaves (Strom and Finstein, 1994).

4.9. Security

Vehicular access to the site must be controlled to prevent illegal dumping. A gate across the entrance road is a minimum precaution. In some cases the entire site must be fenced, but usually preexisting features, such as streams, trees, and embankments, will provide partial security. A berm consisting of earth and finished compost often can serve in place of a fence at other points. Vandalism also may be of concern, especially if equipment is left on site.

4.10. On-Site Roads

Because of the heavy truck traffic during the collection period, a limited road network (possibly paved) within the site may be desirable to improve all-weather access. A circular traffic flow pattern is advantageous at heavily used sites. The purpose of the on-site roads is to help prevent trucks from getting stuck during muddy conditions; an extensive on-site road network is not needed.

4.11. Fire and Other Safety Considerations

Normally, windrowed leaves burn poorly, since the interior of the pile is wet. While vandals may be able to ignite the dry surface leaves, a major fire is unlikely. Fire safety is further accommodated by having a ready water supply and delivery capacity, wetting the leaves initially, and providing aisles between windrows as a fire break and for access.

Other safety precautions usual for any operation involving heavy machinery should be exercised. Public access should be restricted.

5. Appropriate Level of Technology

Four levels (minimal, low, intermediate, and high) of composting technology can be considered (Strom *et al.*, 1980, 1986, Strom and Finstein, 1994). The particular one that is most appropriate for a given application will depend mainly on the quantity and types of material accepted and the site selected, although the equipment and manpower available are also factors. Table 1 indicates that generally, the lower the level of technology, the greater the requirements for available space and composting time, but the lower the cost.

The level of technology refers to the extent to which the ideal conditions for composting are met. Minimal technology meets the conditions poorly, leading to slow processing and a strong odor potential. However, if site conditions permit this type of operation, it can be highly cost-effective. It might be considered for leaves only, or for some agricultural wastes.

The low-level technology provides somewhat better composting conditions, cutting the time requirements and odor potential. This is considered acceptable in many cases for smaller operations handling leaves only. It is discussed in the greatest detail below because of its widespread applicability and as a point of comparison for the other methods.

Intermediate-level technology invests more effort into speeding up the composting process. It is appropriate for larger sites, and is necessary where grass clippings are accepted at a site.

The high level of technology approaches optimum processing conditions. While not used for leaves, it is normally required to handle other MSW fractions, sludges, and other highly putrescible wastes and is mentioned for completeness.

5.1. Minimal Technology

If a large area that is well isolated from sensitive neighboring land uses is available, a very low-cost approach to leaf composting is possible. Leaves brought to the site are formed into large windrows (e.g., 12-feet high by 24-feet wide) using a front-end loader. Once every year the windrow is turned and reformed. Additional windrows are constructed with the new leaves each fall. After 3 to 5 years the material in a windrow is usually sufficiently well stabilized to be used as compost.

With this "minimal" technology the necessary conditions for rapid composting are not achieved. Much of the pile remains anaerobic for a year at a time between turnings. The center of the pile probably also reaches inhibitively high temperatures, especially the first year. However, the greatly reduced rate of activity is compensated for by providing a prolonged composting time.

Using this approach, odors can be expected for the first year, and high strength and offensive odors likely will be released during the first turning. Usually by the second turning, odors have diminished. Because of these odors, an extensive buffer zone is required. A quarter mile distance or more to sensitive neighboring land uses is recommended.

The obvious advantage of this approach is that it is very inexpensive. Only a few days per year of front-end loader operation is required. A second advantage is that relatively little space is needed for the composting itself. Even though the leaves must stay on site for several years, a 1-acre site (excluding buffer) is expected to be adequate for a yearly collection of 4000 cubic yards. Still, a large total area is required because of the large buffer zone. This might be useful in a wooded area, so that only a small clearing would be required, or at an isolated industrial site or public works yard.

5.2. Low-Level Technology

In the densely populated Northeast, siting of a minimal technology facility is rarely possible. Therefore, the necessary conditions for rapid and nuisance free composting have to be more nearly met. In particular this means that a better job must be done of ensuring adequate oxygenation, moisture content, and temperature control.

The simplest way to achieve the desired temperature range would seem to be to build piles large enough to conserve sufficient heat, but not so large as to overheat. On the other hand, adequate oxygenation by passive diffusion of air from the outside of the pile could be achieved if the piles were small enough. Unfortunately no single pile size completely reconciles these conflicting objectives. However, the desired conditions can be approached by starting with moderate size piles (6-feet high by 12 to 14-feet wide), then combining two piles after the first burst of microbial activity (which lasts approximately 1 month). Water addition at the outset is usually necessary to provide adequate moisture, and a few additional turnings are provided later.

Using this approach it is possible to produce a thoroughly decomposed (finished) compost in 16 to 18 months. The compost is ready for use in the spring, which is the time of peak demand for the product. Slight odors may be produced early in the composting cycle, but these usually are not detectable more than a few yards away from the windrows. After 10 to 11 months large curing piles are formed around the perimeter of the site, freeing the original area to accept the new leaf collection. Costs are still quite low, as only three to four operations with a front-end loader are required after initial windrow formation (one combining, one or two turnings, and one curing pile formation). Despite the fact that more space is required for the actual composting (roughly 1 acre per 3500 cubic yards of leaves) compared to the minimal technology, less total area is needed overall because of the reduced buffer requirement.

Unless otherwise indicated, the low-level technology is recommended for small to medium size sites composting leaves only. However, this technology generally is not appropriate at larger sites, or if grass clippings are accepted.

Table 4 summarizes the scheduling and estimates of manpower and equipment requirements for a moderate-sized leaf composting facility (15,000 cubic yards of leaves per year) employing the low-level technology. The individual steps are discussed in more detail below.

As indicated, a number of assumptions went into Table 4. The manpower and equipment time estimates, in particular, should be considered only as a general indication of the needs at a specific site, since they may be highly variable.

5.2.1. Site Preparation

Prior to each collection season the site must be readied to allow necessary truck access and front-end loader operation. The one part of the operation that has little scheduling flexibility is delivery of the collected leaves. Once leaves are collected, they must be promptly processed through the staging area and formed into windrows. It is critical, therefore, to prevent operational bottlenecks such as an area becoming so muddy that trucks get stuck trying to drop off their loads.

The yearly site preparation should include regrading and road and leachate system (if any) maintenance. All refuse and debris from the previous year's operation should be removed. Normally this step will require at most a few days work. It can be scheduled any time after the active site has been cleared of the leaves from the previous year (by formation of the curing piles), but before the new collection season begins.

Table 4. Generalized Schedule and Requirements for Moderate-Sized (15,000 cubic yard) Low-Level Technology Leaf Composting

Operation	Month	Flexibility in timing	Front loader (day)	Laborer (day)
Prepare site	September–October	Yes	2	2
Form windrows	November–December	No	30	30
Combine windrows	December–January	Yes	10	–
Turn windrows	March–April	Yes	5	–
Additional turnings	May–August	Yes	5 each	–
Form curing piles	August–September	Yes	5	–
Shredding/screening	March–May	Yes	20	20

General assumptions and notes:

(1) site has been prepared to allow all necessary truck access and loader operation under any expected weather conditions;
(2) leaves delivered in bulk or in paper bags (no plastic bags);
(3) adequate supply of water on site (average of 20 gallons water per cubic yard of leaves);
(4) daily supervision during periods of activity and regular checks at other times also needed;
(5) distribution of finished compost not included;
(6) other equipment, such as a grader, may be required;
(7) windrow size 6-feet high by 12 to 14-feet wide (avoid compaction);
(8) aisles 1 to 2 feet for pairs of windrows, 12 to 16 feet between pairs.

5.2.2. Delivery of Leaves/Staging Area

It is recommended that trucks dump their loads of leaves in a staging area, rather than trying to form windrows directly. Although a staging area involves additional labor, its use is justified for a number of reasons.

Leaves are normally collected and delivered in a variety of trucks, including garbage compactors, roll-offs, and vacuums. Windrows formed directly by dropping off loads from these trucks are highly nonuniform in size, moisture, and compaction, and hence decomposition rates. Wetting is virtually impossible in directly formed windrows since most of the water simply runs off the outside.

Use of a staging area leads to a more uniform windrow size and shape, giving both a better appearance and more efficient composting. Keeping trucks on the firmer surfaces, rather than backing into windrows, also decreases their chance of getting stuck and can otherwise speed the delivery process. It may be feasible to move the staging area periodically (weekly, for example) to minimize the distance to the active windrow forming area.

Supervision may be required to prevent dumping in undesired locations. Also, a record of the amount of leaves delivered should be kept. A daily tabulation of the number of loads for each individual truck of known capacity is an efficient estimation method, although commercial facilities may prefer to use scales.

The staging area also provides an opportunity to inspect loads as they are dropped off, and to remove incidental contamination from litter or other refuse. Heavily contaminated loads should be rejected, if possible, to avoid sorting and disposal costs, and to encourage better collection practices in the future. Plastic bags should not be accepted at the site, but paper bags represent no problem.

5.2.3. Wetting

Wetting of the leaves may be required during much of the collection season. Adequate wetting can only be achieved prior to or during windrow formation or turning. Most of the water applied to the outside of a windrow is simply shed by the leaves, or quickly reevaporated to the air. The water should be sprayed on the leaves with a firehouse as the loader breaks the masses apart in the staging area, or as they are placed in the windrows.

If the moisture content of the delivered leaves is known, the amount of water that must be added to get any particular desired moisture content can be calculated (Strom and Finstein, 1994). Informally, the rule of thumb is that it should be possible to squeeze a few drops of water from a fistful of the leaves.

5.2.4. Forming Windrows

Windrow formation must take place immediately after leaves are received. If freshly dumped leaves are allowed to sit for more than a day or so in the staging area, odor problems may develop.

Once the leaves have been dropped off, a front-end loader can be used to break apart and spread the compacted materials to facilitate wetting. Plastic, branches, curbing, and other incidental debris can be removed by hand. The front-end loader can then be used to place the uncompacted leaves in windrows.

The windrows initially should be 6-feet high by 12 to 14-feet wide. Any convenient length can be used. Two windrows can be formed side by side, with only 1 to 2 feet between, to conserve space. Sufficient aisle space between pairs of windrows (typically 12 to 16 feet) should normally be allowed for loader operation. Although in some cases it may be possible to have fewer aisles if space is limited, this may make turning operations awkward.

Neatly formed windrows with well-maintained aisles give the facility a professional appearance, while messy windrows give the impression of a "leaf dump." Care should be taken that equipment, especially the loader, does not ride up on the windrows, compacting them. Loosely piled leaves are required in order to ensure sufficient pore space for adequate air penetration into the windrows.

5.2.5. Combining Windrows

After approximately 1 month, much of the initial oxygen demand of the leaves has been exerted and the piles have been reduced to about half of their original size through decomposition and self-compaction. At this point, two windrows can be combined (Fig. 1) to form a single one that is still only about 6-feet high by 14-feet wide (about the same size as each of the initial windrows). Combining the windrows will help conserve heat during the colder weather. Portions of the center of the new, combined windrow may go anaerobic temporarily, but significant odors and acidification are not expected because much of the readily degradable material has already been consumed by the microorganisms.

FIGURE 1. Combining two small piles using front-end loader. (Photo by P. F. Strom.)

Combining should be done by moving and turning both piles, not by placing one on top of the other. The maximum degree of mixing and "fluffing" is desired.

To conserve space, combining may begin before leaf collection has been completed. In this way some of the space freed by combining windrows (formed with leaves collected in early November) can be used for new windrows made with leaves collected late in the season (mid-December).

5.2.6. Turning Windrows

As early as is practical in March or April, each windrow should be turned. Turning mixes the material, rewets the dry outer edges, reoxygenates the interior, and exposes the formerly cool edges to the hotter internal temperatures. The result is an increased rate of decomposition and improved destruction of any pathogens and weed seeds.

As with the prior combining operation, maximum mixing and "fluffing" is desired during turning. At this time additional water may be added if the material is too dry. However, every effort should be made to provide sufficient water initially.

Additional turnings throughout the summer would further enhance composting rate and product quality. At least one such turning is recommended in order to prevent any weed growth on the windrows from going to seed.

5.2.7. Curing

Using the low-level technology described here, much of the material will not be completely stabilized by the end of the summer, yet the composting area must be cleared to allow for site preparation for the next year's leaves. This does not represent a problem since the material is now moderately well decomposed, has little oxygen demand, and is unlikely to produce odors.

At this time, therefore, the material can be moved and formed into a large curing pile around the perimeter of the site. The curing pile may be made as large as 12-feet high by 24-feet wide to conserve space, but should not be compacted when formed. Moving the material also provides additional turning and mixing, while the large pile exposes a relatively small surface area to drying and freezing conditions. Additional weed and pathogen destruction is achieved at the temperatures reached within the large, well-insulated curing pile. This material is expected to be well stabilized by the following spring.

5.2.8. Shredding or Screening

Once composting is completed, shredding or screening is a final optional step to improve the physical quality and appearance of the finished compost, making it more acceptable for many home and commercial uses (Fig. 2). Depending on the equipment, this step breaks up clumps and separates out rejects consisting of uncomposted leaves, branches, rocks, plastic, and other extraneous materials. The "rejects" may be composted for an additional period, then reshredded or screened to minimize the amount requiring disposal.

This step is labor intensive. Leaf compost can only be processed at about half of the rated capacity of some of the equipment designed for soil, particularly if it is wet. Typically, a front-end loader is required for filling the hopper, and at least one person is required to operate the shredder/screener itself.

The major advantage of using a shredder or screener is that it yields a more uniform and debris-free final product. In some cases it also can be used to mix finished compost with soil. Disadvantages include the labor and equipment requirements, the need to dispose of rejects, and the capital cost of the specialized machine. For amending final landfill cover or sale to top soil companies (where it will be shredded during blending), shredding or screening is not needed.

FIGURE 2. Finished compost after shredding. (Photo by Franklin B. Flower.)

One way to reduce costs is to share a single unit among several sites or communities. Sharing is possible since the specialized equipment is only needed for a few weeks per year, and scheduling can be flexible.

5.3. Intermediate-Level Technology

More frequent turning of the windrows will speed the composting process through improved aeration and the physical mixing and grinding (particle-size reduction). Because of the increased rate of biological activity, turning must be continued regularly once it is started. If not, acid–anaerobic conditions and odors quickly develop, and the next turning releases the odor. During the first few weeks, two turnings per week may be required. This later can be reduced to one per week, then one every 2 weeks. Following this approach, finished compost can be produced in as little as 6 months or even less.

Except for very small sites, such frequent turning by front-end loader is impractical. The turning takes too much time, equipment and labor costs are too high, and the mixing and grinding is not very thorough. Also, compaction of the windrow is likely, and the soil at most sites would get too rutted or muddy. For these reasons, specialized turning machines are used.

Several commercially available turning machines are in use. Some are mounted on a tractor or front-end loader, and driven first along one side of the windrow and then the other, turning half of it at a time (Fig. 3). Others straddle the pile, turning the composting material all at once and displacing it backward. Another approach used is to lift the material and displace it to the side.

A major advantage of frequent turning may be the shorter composting period. However, this is really of benefit only if the site is required for another use (such as a beach parking lot) during the summer.

Perhaps surprisingly, turning machines may require more land. This is because in many cases windrow height is limited to 5 feet or less. Some larger models can

FIGURE 3. Example of a turning machine. (Photo by Franklin B. Flower.)

accommodate a 7-foot high windrow, but piles of this size are prone to odor generation and are not recommended except for isolated sites. Also, paired windrows cannot be used with types that are tractor or loader mounted because of the need to turn windrows from both sides. However, straddle type and side displacement turners may only require narrow aisles, saving space.

Contrary to what might be expected, turning may have the effect of reducing average overall oxygen levels within a windrow. Although the turning itself does incorporate additional oxygen, the higher rate of decomposition that results from the concurrent mixing and grinding can lead to rapid (a few hours) oxygen depletion and anaerobic conditions.

Specialized turning machinery may require a better graded surface for efficient operation. On the other hand, such equipment may create fewer ruts and less muddy conditions than front-end loaders.

Some turning equipment is very noisy. This should be mitigated to the maximum extent possible if residences or other sensitive land uses are located nearby. Noise codes typically specify that levels must not exceed 65 dB(A) at the property line. These considerations also apply to front-end loaders.

A staging area is not as important for an intermediate level of technology. Turning can be used to help reduce differences in initial windrow size, compactness, and composition. Frequent turning also makes initial water addition less critical. If inadequate moisture is present, turning during or immediately after it rains (or snows) can be used to incorporate water. This may make an on-site water source unnecessary (if fire officials approve).

Grass clippings (other than those incidental to fall leaf collection) should not be accepted at a site unless frequent turning will occur. Specialized turning equipment then is required.

Finally, the overall economic impact should be examined. The increased turning efficiency of the specialized equipment may justify the initial expense at larger sites even if the shortened-composting period is not a major factor. For sites of 10,000 cubic yards or less, turning machines may be uneconomical unless shared, but at sites of 30,000 cubic yards or more they probably are a necessity.

5.4. High-Level Technology

In order to approach a maximal rate of decomposition, near optimal levels of temperature and oxygenation are required. This can best be achieved by using an approach originally developed for sewage sludge composting, known as the Rutgers process control strategy (Finstein *et al.*, 1983, 1986, 1987). While this strategy has been successfully field tested for leaf composting, exact design and operation details for this application have not been developed.

Briefly, this approach consists of using forced pressure aeration of the composting pile, with the blower controlled by a temperature feedback system. When the temperature at a specific monitoring location within a pile exceeds a preset value, the blower automatically comes on to remove heat and water vapor and cool the pile. This ensures near optimum temperatures in the bulk of the material, and at the same time maintains a well-oxygenated condition. During the start-up period (and at other times, if needed), the blowers also come on under control of a timer (perhaps for 30 s every 15 min) to provide a minimal level of oxygen. After 2 to 10 weeks of composting, the aeration system would be removed, and the windrows turned periodically.

This degree of control is not needed for leaf composting. On the other hand, a high level of technology may be warranted for grass clippings, and is probably necessary for food wastes, sludges, and similar materials. It lends itself well to enclosed systems, and in fact some of the best composting "tunnel" systems developed in the mushroom industry are now being adapted for solid waste applications. These systems, in addition to employing temperature feedback control and being totally enclosed, are capable of recirculating a portion of the blower air stream. This makes odor control more efficient, and minimizes any potential problem with vermin.

6. Management of Other Yard Wastes

6.1. Grass Clippings

Grass clippings represent another significant seasonal solid waste. In some suburban communities they may account for one-third of the total MSW load during peak grass-growing periods. Although grass clippings are readily compostable, the odor problems they pose make this treatment option difficult to implement. Permitting requirements may be more stringent, particularly with respect to buffer zones, staging, leachate, and odor control, and collection costs may be substantial. As indicated in Section 1, the best alternative for grass clippings is not to collect them at all.

If grass clippings are to be composted, extra care must be taken to ensure that the windrows do not become anaerobic. Grass clippings are alive when first cut and are relatively high in nitrogen, moisture content, and readily degradable organics compared to the fallen leaves collected in autumn. For these reasons they decompose more rapidly, have a higher oxygen demand, and quickly go anaerobic. They are often highly odorous by the time they are delivered to a composting site. Therefore, it is especially important to properly implement odor control measures.

Additional precautions, such as expanding the buffer zone (at least 1000 feet from the grass-handling areas) and improved management of leachate, also will be necessary. Monitoring of nitrogen levels in leachate and groundwater, including background sampling both upgradient from the site and on-site prior to receiving materials, also may be necessary. *Please* consider carefully before bringing grass clippings onto a site. For more information, consult the Manual (Strom and Finstein, 1994).

If the grass clippings could be delivered to a composting site without causing odor problems, they should be incorporated (before the end of the day) into the partially composted leaf windrows. A ratio of no less than 3 volumes of partially composted leaves to 1 volume of grass clippings is recommended. Good mixing with a specialized windrow turning machine is essential.

Once the leaves and grass have been mixed in this way, no further odor problem is expected. The partially composted leaves act as a bulking agent to improve penetration of oxygen to the grass clippings, and as a sorbent to trap small amounts of odorous compounds. Because of their high C/N, the leaves also tie up ammonia as it is released from the decomposition of the clippings, minimizing both ammonia odors and the release of nitrogen in leachate.

Other bulking agents have been proposed for composting of grass clippings as a substitute for partially composted leaves. However, in addition to providing bulking for better aeration, any such materials also must supply sufficient available carbon to tie up ammonia as it is released from the clippings. If not, ammonia odors (even under aerobic conditions) and nitrogen contamination of water may occur. Woody materials, in particular, do not normally supply sufficient available carbon for this purpose, even though their C/N is high.

6.2. Woody Materials

Wood tends to decompose very slowly, making composting of woody materials impractical in most cases. Thus, woody materials should not be intentionally incorporated in leaf or leaf/grass composting windrows unless there is an end use for a mixed wood/compost product. (Separation by screening usually is too expensive.) Small amounts of incidentally included branches and twigs pose little problem.

For many woody materials, chipping, alone or followed by composting (with or without leaves or grass clippings), may produce a usable mulch. Direct incorporation of woodchips or other woody materials into the soil is not recommended because of the slow rate of decomposition and high C/N.

6.3. Other Organic Materials

Many other organics, such as most agricultural and food wastes, are potentially compostable. However, these materials may not be suitable for the composting technologies being used at yard waste composting facilities.

7. Potential Problems and Their Solutions

Table 5 summarizes the more common problems at leaf composting sites, their causes, and recommendations for their remedy. Most problems can be prevented by proper facility siting, design, operation, and maintenance. Grass clippings present additional concerns, which are also addressed later.

7.1. Odor

The major problem encountered even at leaf-only composting sites is odor. Those unfamiliar with handling large masses of leaves may be surprised at how serious a problem this can be. Starting with relatively innocuous leaves, it is possible to generate odors comparable to those of a barnyard or worse. Grass clippings greatly intensify both the odor strength and its unpleasantness.

In general, odor problems develop in four stages:

(1) odorous compounds must be present initially or be produced during processing;
(2) these odors must be released from the pile;

Table 5. Problems Encountered at Leaf Composting Sites

Problem	Cause	Recommendation
Odor	Piles too large	Initial windrows should be no larger than 6-feet high by 14-feet wide
	Leaf storage	Allow no more than 1 to 2 days between collection and windrow formation
	Leachate ponding	Eliminate ponding Add lime*
	Plastic bags	Do not accept plastic bags at the site
	Grass clippings	Do not accept grass clippings at the site Immediately incorporate into leaf windrows Mix at least 3 cubic yards partially decomposed leaves per 1 cubic yard grass
Leachate ponding	Inadequate slope, poor grading	Regrade site
	Improper windrow alignment	Run windrows down slope, not across it
Surface water pollution	Leachate discharge	Treat leachate before it leaves site by passing it through soil or sand
Groundwater pollution	Nitrogen leaching into groundwater	Limit grass clippings Mix at least 3 cubic yards partially decomposed leaves per 1 cubic yard grass Collect leachate for treatment
Slow composting	Material too dry	Add sufficient water initially, or as corrective measure during turning
	Acid anaerobic conditions (piles too large)	Make piles smaller, adding lime* if needed to raise pH and control odors
Illegal dumping	Uncontrolled access	Limit access to site (gate)

*Pulverized limestone, dry or as water slurry.

(3) the odors must travel off-site; and
(4) they must be detected by sensitive individuals (receptors).

An odor problem can be prevented by eliminating any stage.

With the minimal technology described previously, Stages 1 to 3 all occur, but since no receptors are present (Stage 4), there is not a problem. Except where very large buffer zones are present, however, this approach to odor "control" is not possible.

In most cases, prevention of odor problems can best be achieved by preventing odor formation in the first place (Stage 1). For leaf composting this means avoiding prolonged anaerobic conditions. Under anaerobic conditions, first volatile organic acids (which have vinegary, cheesy, goaty, and sour odors), alcohols, and esters (fruity and floral) are formed (acid fermentation). With extended anaerobiasis, amines and sulfur compounds (barnyard, fishy, and rotten odors) also can be produced. In contrast, with aerobic conditions only a mild earthy odor is expected. If excessive ammonia or urea-based fertilizer, grass clippings, or other high-nitrogen materials are added, an ammonia odor also may be produced even under aerobic

FIGURE 4. Steam released from an oversized pile. (Photo by Franklin B. Flower.)

conditions. Prevention of anaerobic conditions is virtually impossible with grass clippings alone.

The major cause of odor production at leaf composting sites is making the windrow too large (Fig. 4), especially when first assembled. Because of the initial high concentration of readily degradable material, there is a high demand for oxygen. If the piles are too large, sufficient oxygen cannot penetrate from the outside, and a large anaerobic core develops. Decomposition slows down, switching over to the odor-producing acid fermentation described above.

A second important source of odor production is failure to form windrows quickly enough once the leaves are collected. Leaves cannot be simply dropped at the site for later composting or collected and stored elsewhere. Although the intention might be to just hold them, temporary storage of leaves, unless they are very dry, can result in vigorous decomposition within 1 to 2 days, leading to anaerobic conditions and the production of offensive odors. Grass clippings, as discussed earlier, are almost always odorous already when they are delivered to the composting site.

If odors should be produced at a site, or if odorous materials are dropped off at the site (such as occurs with grass clippings or previously stored leaves), the second line of defense is to prevent their release (Stage 2). This might best be accomplished by leaving the odorous mass undisturbed until oxygen has penetrated sufficiently to destroy the odors. However, this may take several months or even years. Shaving off thin (1 to 2 foot) layers from the edges as they become aerobic may help speed this process.

If a long wait is not practical, another approach may be possible. Since many of the odorous compounds in leaf composting are acidic in nature, raising the pH (neutralizing the acids) will convert them to an ionized (negatively charged, dissociated) form. In this form they cannot be released to the air and will remain in the pile. For example, with the most commonly formed organic acid, acetic acid (vinegar), the reaction is:

$$CH_3\,COOH \Leftrightarrow CH_3\,COO^- + H^+$$

Application of pulverized limestone is probably the best way to raise the pH. Sprinkling the limestone in powdered form directly onto surfaces from which odors are escaping may be the simplest approach, although a liquid slurry of limestone in water is more effective. A layer (1 foot) of finished compost spread over the odorous material also helps to reduce odor release, serving as a biofilter to treat odors before they are released to the air.

The use of limestone may be ineffective with odors generated from grass clippings or other high-nitrogen wastes. Ammonia and amines are weak bases rather than acids, and raising the pH may therefore actually increase odor release:

$$NH_4^+ \Leftrightarrow NH_3 + H^+$$

If odors are still produced and released despite these precautions, it may still be possible to minimize their off-site impact (Stage 3). This approach relies on timing odor-releasing operations to coincide with favorable wind conditions. A wind sock should be installed at the site to determine wind direction, and odor-releasing operations performed only when the site is downwind of residences and other sensitive neighboring land uses. Also, higher winds are preferable to calm and light wind conditions because the higher the wind speed, the greater the dilution of any released odors. Unfortunately, winds often subside at dusk, and may shift, so that this method of control is unreliable.

Some commercially available products claim to mask or eliminate composting odors when sprayed onto windrows. Masking agents try to use another odor (such as lemon, pine, and roses) to hide the objectionable odors. To our knowledge, they have not been successful at composting sites. Odor elimination agents (with the exception of limestone) are also unsuccessful in our experience.

7.2. Leachate

Leachate is water that has come in contact with the composting materials and extracted some components from them. One way in which leachate may pose a problem is by forming small pools or "ponds" (Fig. 5). Ponding is a concern because it can create an odor problem (since anaerobic conditions are likely to develop both in the pool and in the base of any water saturated piles), serve as a place for mosquito breeding, and interfere with operations on the site (soft, muddy areas). Prevention, by properly grading the site, is the best remedy. Also, windrows should run down slopes rather than across, making it easier for the water to run off rather than accumulate between windrows. If ponding occurs and odors are released from the pools, adding pulverized limestone may be helpful.

Pollution of surface waters (lakes, streams) is the other major concern with leachate. While leachate from leaf composting is generally not toxic, it may deplete the dissolved oxygen in the water, possibly even to the point where fish kills could occur. It might also lead to discoloration of the water.

In order to prevent this potential pollution, leachate should not be allowed to enter surface waters without prior treatment. This treatment might consist of simple percolation down into or through the soil, or passage through a sand barrier constructed to

FIGURE 5. Leachate. (Photo by P. F. Strom.)

intercept any overland flow. In passing through the soil or sand, the leachate is both physically filtered and biologically degraded to remove a substantial portion of the pollutants. Contamination of ground water does not appear to be a problem associated with leaf composting.

With grass clippings, however, leachate may contain high levels of nitrogen. This may pose a problem of nitrogen contamination for both surface and groundwaters, and may not be adequately treated with simple soil or sand filters. Such contamination must be prevented either by limiting the nitrogen in the leachate (through control of the carbon-to-nitrogen ratio—by minimizing the amount of grass clippings, for example), or by more sophisticated (and expensive) leachate collection and treatment systems. Treatment of high-nitrogen leachate is not a simple matter.

7.3. Inadequate Composting Rate

Occasionally composting will progress too slowly in some windrows, usually because the material is too dry. Sufficient water should be added initially, either before or as the windrows are being formed. Other opportunities to add water are during the combining operation (low-level technology) and scheduled or extra turnings. Adequate wetting usually cannot be accomplished simply by spraying water on the outside of the piles. Similarly, rainfall is not effective unless it is followed almost immediately by turning.

Another cause of slow composting is piles that are too large. Once acidic anaerobic conditions occur, the material tends to be preserved ("pickled") rather than decomposed.

7.4. Bags

Some communities have found bagging of leaves and grass clippings to be a convenient collection method. Plastic bags, however, pose a major problem at composting sites because they must be removed from the compostable material. Even so-called "degradable" plastic bags are incompatible with composting because they degrade far too slowly. If plastic bags are to be used, it is very strongly recommended that they be removed prior to delivery of the material to the composting site (e.g., by opening and dumping them curbside, at the point of collection).

An acceptable alternative is the use of paper bags. These do not have to be removed, since they decompose at about the same rate as the composting material. Plastic bags have a lower initial unit cost than paper bags, but the expense of removing them may exceed the price differential.

7.5. *Aspergillus fumigatus*

One concern with leaf composting is the release of spores of *A. fumigatus*. This is a common, widespread, and naturally occurring fungus found in soil and on vegetative materials. Its airborne spores may produce an allergic response in some people, and in a few cases they are capable of causing infection in individuals with a compromised (weakened) immune system. Because these spores have been of some limited concern in sludge composting, research has been conducted to examine their importance in yard waste composting.

On the basis of these findings, *A. fumigatus* appears to pose little risk to neighbors of composting sites with adequate buffer zones. However, workers at a site may receive high exposures, and therefore some precautions seem warranted. Potential workers at the composting site should be screened for conditions that might predispose them to infection or allergic response. Such conditions include asthma, a history of allergic responses, a weakened immune system, the taking of antibiotics or adrenal cortical hormones, and a punctured eardrum. Workers having such conditions should not be assigned to the composting operation (nor other tasks putting them at similar elevated risk) unless a health specialist is consulted.

Additionally, wearing an approved dust mask during leaf drop off, windrow formation and turning, screening, and similar dust-generating operations is recommended. Air conditioner filters in loaders and turning machines should be cleaned frequently. It also is expected that adequate wetting of the windrows, as recommended here, will help to reduce potential exposure.

7.6. Toxic Contaminants

Leaves as collected contain low levels of toxic materials. Lead, for example, is present because of its former use in gasoline and paint. Limited testing, however, has found only low levels, and these appear to be declining as use of lead decreases. Lead levels in leachate typically meet drinking water standards.

Some pesticides also may be present, particularly in grass clippings, but again levels ordinarily are expected to be too low to be of concern. In samples from six sites in New Jersey only one pesticide, chlordane, was detected (Strom, 2000). Since chlordane is no longer used and it is not taken up by plants, it is believed that this came from the residential soil mixed with the yard waste during raking.

Overall, yard waste compost is considered safe for residential use without specific testing. An exception to this would be for composts containing materials from golf courses or orchards, where more intensive use of more toxic and persistent compounds is common. Such materials may require testing for specific metals and pesticides before general use.

7.7. Other Potential Problems

Noise may be a concern, depending on siting, buffers, and the equipment used. Noisy operations should be performed as far from sensitive receptors as possible and at the least objectionable times. Berms can help reduce off-site noise.

Dust from the windrows can be minimized by proper wetting. Dust from the roads and aisles, however, may be a problem during dry weather.

Traffic must be considered during original siting of the facility. Limiting hours of operation may be necessary in some cases.

8. Use of Leaf and Yard Waste Compost

Early in the development of a composting facility, it is desirable to plan for distribution of the end product. While the nutrient content of leaf compost is too low for it to be considered a fertilizer, it does act as a soil conditioner and organic amendment, improving the physical, chemical, and biological properties of the soil (Flannery and Flower, 1991). Most soils benefit considerably from the increase in organic matter content that leaf compost can provide.

Most municipal leaf composting facility managers like to make a portion of their finished compost available to individual users in the community. Some allow public access to the site itself, while others prefer to make the compost available at another location such as the public works yard or recycling center.

Some state and local recycling initiatives require that government agencies must give consideration and preference to the use of compost material in all land maintenance activities that are paid for with public funds. The municipality, particularly the parks and roads departments, may use the compost in place of purchased organic soil amendments. The compost also may be blended with poor soils to produce a good quality topsoil.

Other bulk users may include farmers, nurseries, landscapers, builders, top soil companies, and landfills (for amendment of final cover). The compost may be offered to such users at no cost, but often a modest charge is imposed.

Acknowledgments

This chapter draws heavily (with copyright permission) upon "New Jersey's Manual on Composting Leaves & Management of other Yard Trimmings" and its predecessor, "Leaf Composting Manual for New Jersey Municipalities," both authored by P. F. Strom and M. S. Finstein and distributed by the New Jersey Department of Environmental Protection (NJ DEP). We would therefore like to acknowledge the following people who contributed to those publications: Richard M. Abramowitz, Tim Bartle, Jeffrey Callahan, Jean Clark, Jae-Chun Chung, Donn A. Derr, Mark DiDomenico, Steven W. Fass, Patricia Ferriola, Franklin B. Flower, Jonathan H. Forsell, Ellen McShane Fox, Roger M. Guttentag, Don Hansen, John A. Hogan, Pegi Ballister-Howells, Helen Kushner, Alan C. Little, Ming-Huei (Phillip) Liu, Roy Meyer, Frederick C. Miller, Harry Motto, Keun Chan Oh, Clarence "Pete" Peterson, Brian Petitt, Paul Petto, Ned Scannel, William Schulz, Mary T. Sheil, Aletha Spang, James J. Stefel, Daniel Stein, Mary Sue Topper, Matt Vastano, Vivette S. Walker, Michael Winka, and Eric Zwerling.

Funding for the field and laboratory studies that provided much of the information incorporated in the Manuals was provided by the NJ DEP Division of Solid Waste Management, the Montclair Organizations for Conservation, Essex County, and the New Jersey Agricultural Experiment Station, supported by State funds.

References

Anonymous, 1993, *Grass: Cut It and Leave It*, New Jersey Department of Environmental Protection, Division of Solid Waste Management (NJ DEP, DSWM), Trenton, 8pp.

Burge, W. D., Colacicco, D., and Cramer, W. N., 1981, Criteria for achieving pathogen destruction during composting, *J. Water Pollution Control Fed.* **53:**1683–1690.

Finstein, M. S., Miller, F. C., Strom, P. F., MacGregor, S. T., and Psarianos, K. M., 1983, Composting ecosystem management for waste treatment, *Bio/technology* **1:**347–353.

Finstein, M. S., Miller, F. C., and Strom, P. F., 1986, Waste treatment composting as a controlled system, Chapter 10 in *Biotechnology,* Vol. 8: *Microbial Degradations*, Verlag Chemie (German Chemical Society), pp. 363–398.

Finstein, M. S., Miller, F. C., Hogan, J. A., and Strom, P. F., 1987, Analysis of EPA guidance on composting sludge: Part I—Biological heat generation and temperature, *BioCycle* **28**(1):20–26; Part II—Biological process control, *BioCycle* **28**(2):42–47; Part III—Oxygen, moisture, odor, pathogens, *BioCycle* **28**(3):38–44; Part IV—decomposition rate and facility design and operation, *BioCycle* **28**(4):56–61.

Flannery, R. L., and Flower, F. B., 1991, *Using Leaf Compost*, Rutgers Cooperative Extension FS117, New Brunswick, NJ, 2pp. (available free on-line at http://www.rcre.rutgers.edu/pubs/publication.asp?pid=FS117)

Flower, F. B., and Strom, P. F., 1991, *Backyard Leaf Composting*, Rutgers Cooperative Extension FS074, New Brunswick, NJ, 2pp. (available free on-line at http://www.rcre.rutgers.edu/pubs/publication.asp?pid=FS074)

Strom, P. F., 2000, Pesticides in yard waste composting, *Compost Sci. Utilization* **8:**54–60.

Strom, P. F., Morris, M. L., and Finstein, M. S., 1980, Leaf composting through appropriate, low-level, technology, *Compost Sci.* **21**(6):44–48.

Strom, P. F., Flower, F. B., Liu, M. H. P., and Finstein, M. S., 1986, Municipal leaf composting: recommended methods, *BioCycle* **27**(9):48–52.

Strom, P. F., Murphy, J. A., and Indyk, H. W., 1992, *Minimizing Waste Disposal: Grass Clippings*, Rutgers Cooperative Extension FS389, New Brunswick, NJ, 2pp. (available free on-line at http://www.rcre.rutgers.edu/pubs/publication.asp?pid=FS389)

Strom, P. F., and Finstein, M. S., 1994, *New Jersey's Manual on Composting Leaves and Management of other Yard Trimmings*, NJ DEP, DSWM, Trenton, NJ, 81pp. (available free on-line at http://www.state.nj.us/dep/dshw/rrtp/compost/front.htm)

Chapter 21

Urban Tree Removals

Donald F. Blair

1. Introduction

PEGSMOR is an acronym for the life of a tree: planting, establishment, growth, structure, maturity, overmaturity, removal. Since this text addresses PEGSMO, it is appropriate to include a chapter on the "R" factor. Tree removals are an essential arboricultural responsibility.

In 1993, The National Arborist Association and the US Forest Service conducted a survey to determine the number of trees administered to by arborists in the United States and what those treatments were. The survey concluded that 25,575,000 were pruned, fertilized, cabled, treated for insects and disease, removed, and/or planted. Of these, 1,663,235 trees were removed. Ninety percent of the work surveyed was done by commercial tree care firms, the balance by municipal agencies.

The two most expensive events in the life of an urban tree are its planting and its removal. In a talk given in 1992, the Director of Forestry for the City of Savannah, Dr Don Gardner, stated that his average tree maintenance costs worked out to:

1. First year: purchase, planting, and maintenance: $250.00
2. Annual average cost of pruning: $10.00
3. Removal: $350.00

Obviously, the more years of productive health a tree can enjoy between numbers 1 and 3 (or P and R), the more cost-effective that tree is going to be to the landscape.

Donald F. Blair Sierra Moreno Mercantile Company, Big Pool, Maryland 21711
Urban and Community Forestry in the Northeast, 2nd ed., edited by, J. E. Kuser.

1.1. Reasons for Removal

Urban trees will have to be removed for a variety of reasons. Among them are the following:

1. Overmaturity. This could mean that the trees are either in decline as a result of their age or are too large for their planting area. Many communities have planted thousands of flowering trees such as plum, cherry, or pear, only to have them decline within three decades.
2. Hazardous structure. Trees with codominant stems, including bark or other structural defects may have to be removed as part of a risk management program.
3. Infrastructure. Tree removal made necessary as a result of road widening or development. Perfectly healthy trees may have to be removed because they have outgrown their planting space, causing upheaval of sidewalks or interference with vehicular traffic or utility space. Tree-caused obstruction of traffic signals and stop signs and tripping hazards caused by uplifted sidewalks are a major source of litigation against municipalities. Trees have been removed because they interfered with air traffic control or radar tracking of shipping in a bay, harbor, or waterway.
4. Storm damage. Hurricanes, thunderstorms, ice storms, snow storms, and flooding all take their toll on the urban forest in the Northeast.

2. Understanding Tree Failure

Since tree failure is going to be the cause of many tree removals, it is useful to understand that there are two distinct categories of tree failure: biological and mechanical.

2.1. Biological Failure

Biological failure results in a standing dead tree. As a result of advanced age or disease or insect attack, the tree has died. Environmental changes, such as soil compaction or changes in grade, can lead to biological failure. Weather events, such as drought or flooding, can kill trees.

In 1972 and again in the late 1980s, hard freezes in the San Francisco Bay Area caused severe dieback in eucalyptus, carob, mimosa, and grevillia by the thousands. Although many trees were not totally killed, the amount of fuel wood created by dead limbs caused great concern over the potential fire hazard. In 1991, fueled by eucalyptus, Monterey pine, and the homes themselves, a firestorm in the East Bay Hills above Oakland, California, destroyed hundreds of homes and thousands of trees. In the aftermath of the fire, helicopter logging techniques had to be employed to clear steep slopes and hillsides of hazardous trees.

2.2. Mechanical Failure

The collapse or uprooting of a tree is a mechanical failure. Uprooting may occur as a result of poor root stock, disease, or improper root pruning. Limb and leader failure may be the result of poor genetic structure such as weak attachments or codominant stems.

A common cause of mechanical failure is improper pruning. Topping and flush cuts are often the reason why large columns of decay or cavities develop in trees, weakening them to the point of collapse or complete failure.

Mechanical failure will occur when loading overcomes resistance. Limb breakage and total tree loss as a result of snow storms, ice storms, hurricanes, and thunderstorms are an expensive and serious problem in the Northeast.

3. Removal Options

There are almost as many ways to remove a tree as there are reasons to take one down. In an effort to bring some semblance of order to the methods of removal, the options have been divided into two main headings: mechanical and manual.

3.1. Mechanical Options

3.1.1. Bulldozer

Just shove the tree over and get on with life. Although this method works well in right-of-way work and lot clearing, it is certainly not an option downtown or in a residential neighborhood. However, if your site makes it feasible to use heavy equipment, the time saved can be spectacular when compared to more conventional methods.

Years ago, at Stanford University, Palo Alto, California, I watched in awe as experienced equipment operators using four-in-one loaders put in a "face cut" excavation and pushed huge 150-foot eucalyptus over. Two loaders and a chainsaw bucker, working as a team, pulled the leaders apart, bucked them to length, and loaded them into tractor trailers. We had bid against them, planning on conventional felling, brush chipping, and stump-grinding techniques. They were tens of thousands of dollars lower than my bid, and I am sure they made more money with less effort than I would have. I learned that the low bid does not always mean the lowest profit; in this case, my competitor's experience, equipment, and perspective were completely different from mine, and in this case more suited to the job at hand.

3.1.2. Crane

Compared to manual methods and technical rigging with rope, all things being equal, a crane may be your most economical option. My personal business philosophy about crane rigging was that I could rent much larger than I could ever afford to own. Over the years I have developed the following guidelines for working with large cranes. Many public agencies as well as tree companies make good use of truck-mounted cranes that are not large enough to qualify for these recommendations.

The following recommendations are just good, basic, sound advice; apply them as applicable to your particular situation:

1. Work to develop a relationship with a crane company so that you know what their equipment is capable of and which operator has the best feel for working with a climber.
2. Before bidding a job, contact the crane company and show the job to their estimator. That person can check the site for access and recommend the best crane for the job.
3. Keep in mind that the distance the boom has to extend to reach the load has a tremendous bearing on the actual weight the crane will be able to handle. This is why you might need to use a 150-ton crane to do the job that a 25-ton unit could do if it was right next to the tree.
4. Big cranes are heavy. In parks and along secondary roads, you might have to factor in repair of pavement or curbing as part of the cost of the job.
5. Big cranes are large. Make sure that there is more than enough clearance from energized conductors and other installations.
6. Crane rigging is a highly specialized skill. The climber and the crane operator have to work in perfect harmony or risk injury, death, or property damage.
7. Tree removal by crane must be done in compliance with ANSI Z133.1 (American National Standards Institute, 1994) (current standard at time of publication; see Appendix B).

A potential downside of a successful crane job can be the amount of wood that piles up in an amazingly short period of time. You either have to plan a staging area or have enough trucks available to keep the flow of the job going. An excellent strategy for those who have a smaller crane in-house is to use the big crane to get the tree(s) on the ground and to use your crane to load the wood at your pace. In this way, you might be able to reduce your large crane rental from a full day to a half day with a net increase in efficiency and economy. Or, if you are working on a large project, you can send the large crane to the next job, following behind with the cleanup phase.

Rental agreements should include a clause that suspends charges in the event of mechanical problems with the crane. If the crane is down, make sure the clock stops on the rental until it is serviced, repaired, or working again.

3.1.3. Auxiliary Equipment

Most removals are going to make use of a truck and chipper. Many removals in the Northeast utilize an aerial lift for worker safety and convenience. As noted, cranes in a variety of sizes are often employed. Beyond that, any piece of equipment that helps with handling the wood and brush is welcome. Skid loaders with log forks on the bucket can be very useful in some situations, while knuckle-boom loaders can be very helpful in others. When it comes to auxiliary equipment, the specific situation, the job budget, your experience, and the availability of a piece of equipment are your best guides to selection.

3.2. Manual Methods

Manual removal refers to any method that requires extensive chainsaw work and/or roping to accomplish the task. Manual methods can range from ground felling to advanced technical rigging. Since an aerial lift is nothing more than a platform to access the work from, I have not included them in either list. An aerial lift can find gainful employment in crane work as well as manual methods. Manual methods include but are not limited to the methods subsequently described.

3.2.1. Felling from the Ground

Wedges, fiber rope tag lines, winches, hydraulic jacks, and block and tackle may all be used to guide the tree down. In some cases, it is necessary to top the tree or limb a portion of it to get it to fit into the drop zone. As access becomes more and more restricted, the amount of rigging necessary to bring the tree down safely increases.

3.2.2. Dismantling

Piecing down, chunking down, whatever you want to call it, someone has to get in the tree, brush it out, and get the limb wood and base wood to the ground. This work can be done by a rope and saddle climber or an aerial lift operator or a combination of the two.

3.2.3. Rigging

The primary function of rigging is worker safety, when the tree cannot be felled or dismantled without causing collateral damage to adjacent structures, utilities, or sensitive landscaping. When a tree is out of reach of a crane, rigging may become necessary. Rigging may range from simple to technical:

1. Simple rigging makes use of lowering lines and guide ropes called *tag lines*, but relies on natural crotches and trunk wraps for rope placement and control.
2. Technical arboricultural rigging involves the use of ropes, pulley blocks, lowering devices, slings, and false crotches to create a system that provides the greatest degree of precision, accuracy, and control.
3. More important than the equipment are the knowledge and skills necessary to make it all work safely and efficiently.

It is not within the scope of this chapter to go into technical "how to" instruction on techniques and methods for removal. Information of that nature is best reserved for a very large volume devoted entirely to the subject.

3.3. Safety in Tree Removal

The leading causes of death and serious injury to arborists are falls, being hit by falling objects, and electrocution. In addition to these causes, chainsaw cuts and back strains account for a significant number of painful and debilitating injuries. Tree removal, by its very nature, carries with it a high potential for injury.

- All tree maintenance and removal operations shall be performed in full compliance with all applicable American National Standards Institute (ANSI) and Occupational Safety and Health Administration (OSHA) regulations (see Appendix 1).

- Always observe the general condition of the tree to be removed.
- Locate all conductors and utility equipment.

Be aware of:

- Ivy obscuring the root collar, trunk, and/or limb structure from view
- Major deadwood
- Broken, hanging limbs
- Cables and bracing rods
- Split crotches
- Bird or other nests
- Cavities
- Bee, wasp, or hornet activity
- Decay fruiting bodies
- Poison oak, ivy, or other allergy-inducing plants
- Ground-level obstructions such as irrigation heads, holes, and stakes. Such tripping hazards may be particularly dangerous if obscured by low-growing ground cover (Blair, 1995)

3.3.1. Site Security

The following removal site hazards must be managed.

3.3.1.1. Traffic. Many urban removal sites are going to be along roadways, park paths, or utility corridors. Provisions must be made to secure the work site. Road cones, flagging, and lane closings may all be necessary to protect the work crew and the public. From experience, I can tell you that the public-at-large, whether they are motoring, bicycling, or on foot, are not a particularly observant lot. Extraordinary measures must be taken sometimes to protect them from themselves and your workers. Scopes, signs, and cones must comply with local standards and regulations. Residential streets have different requirements from major throughways.

3.3.1.2. Pedestrian. Many pedestrians stroll through life so totally absorbed in themselves that they are deaf to chainsaws, blind to road cones, and oblivious to tree work in progress. I have been witness to unbelievable acts of ignorance by pedestrians. Do not trust them any more than you would a crazed motorist. Do not trust them to notice your work or to heed normal traffic signs. Sometimes you have to literally fence an area off or post a full-time guard to control the movements of passersby.

3.3.1.3. Site. You are on the job to remove a tree, not to destroy the surrounding area. Some sites are an arborist's dream: open, flat, no competing obstructions. Other sites are what we call a "challenge." In developing the work plan, find out what is expendable and what has to be protected. Plan accordingly. Simple measures like scaffolding and plywood can protect understory vegetation from branch drop. We have been known to drape buildings with plastic sheeting to protect the walls from being stained by sawdust and chain oil. Plywood is less expensive than plate glass. Consider protecting expensive windows with plywood *before* they get broken. Advanced rigging techniques have evolved to enable arborists to cope with site security while accomplishing an efficient removal.

4. Planning a Removal Operation

Tree removal calls for far more consideration than the choice between a hinge cut and a jump cut or which brand of chainsaw is most desirable. Some municipalities have in-house crews, some contract all of their work, and many use a combination of both. When drafting specifications for competitive bids, be sure to cover as many factors of the removal as necessary to ensure that the work is done safely, efficiently, and with adequate safety measures for protection of property and passersby.

Safe and efficient removal begins with a work plan. In the planning stages, it is far more important to know what your access is than whether you will be using a bowline or a clove hitch to tie off limbs. The following work plan assessment forms have been developed to enable a person to develop a clear picture of how to cope with any removal ranging from a single tree to a major project (Figs. 1 and 2). By working through the questions on the assessment form, the project planner will create a clear picture of what it is going to be encountered and what will be required to accomplish the task.

The sample assessment forms in Figs. 1 and 2 have sample answers in italics to show the reader how a completed form might work. There are blank forms in Appendix C that may be copied for actual field use.

5. Thoughts on Tree Removal

There are commercial arborists who specialize in large-scale tree removal. There are others who have built up successful practices that do very little, if any, tree removal. With an obligation to public safety, municipalities do not enjoy the luxury of deciding whether they want to do removals. If it is a public tree and it is a hazard, it has to be removed.

5.1. Changing Cultural Practices May Reduce the Need for Future Removal

Sooner or later, every tree is going to die and require removal. The unfortunate fact with too many municipal trees is that they succumb "sooner" rather than "later." Premature removals could be greatly reduced if the following commonsense practices were followed more often and with greater diligence.

5.1.1. Species to Site Match

Too often the wrong tree gets planted in the wrong place. If a selected tree is unsuitable to the site it may die prematurely or outgrow the site. Example: Over the vehement objection of the arborist, a city once planted tulip poplar as the shade tree in its downtown area. Each tree went into a square hole in the sidewalk. The trees grew quickly and large. The leaf drop increased the street sweeping burden. The annual infestation of aphids and sooty mold made the sidewalks and vehicles parked near them black and sticky. The vigorous root systems buckled the sidewalks. Within

A. Tree Removal Work Plan Assessment Form

1. **Site Location of Tree(s) to be removed** *509 Bailey Avenue, Mountain View*
2. **Inventory Number/GPS location** *CL9672210*
3. **Number of Tree(s) to be removed** *One*
4. **Species of Tree(s) to be removed** *Quercus macrocarpa. Bur Oak*
5. **Reason for removal** *Clogged drain system created wet conditions favoring water mold fungi. Tree is dying and needs to be removed while it is still safe to work in.*
6. **Heritage Tree removal permit required? Yes** ***XXX*** **No** _____ **Permit on file** ***XXX***
7. **Diameter of Tree to be removed (answer only if single tree)** *38-inches @ 4.5ft above ground*
8. **Height of Tree to be removed (answer only if single tree)** *76-feet*
9. **Range of diameter of trees to be removed** *NA*
10. **Range of height of trees to be removed** *NA*
11. **Access**
    A. _____ **Excellent. Accessible to any motorized equipment**
    B. _____ **Limited. Describe** ____________________
    C. ***XXX*** **Restricted. Describe** *Inner courtyard of Children's Museum*
12. **Debris Disposal Method** *Chip and haul to mulch pile*
    A. **Estimated Quantity of brush and/or chips** *30 cu. Yds. Of chips*
    B. **Estimated quantity of logs/wood** *8 cords of firewood*
13. **Stump removal** Yes ***XXX*** No _____
    A. **Number of stumps to grind** *Just the one*
    B. **Estimated total inches/feet of stump to grind** *78-inches (large butt swell at ground)*
    C. **Stump grinder needed** *Small, self-propelled unit*

B. Job Site Safety

1. **Visible Tree Defects/Hazards Present No** _____ **Yes** ***XXX***
   Describe *Die-back in crown. Extensive signs of Armillaria mellea at root crown. Large pruning wound cavity in NE leader. 3-span cabling system*.
   Plan: *Anchoring root system still sound enough to support climbing and rigging. Do not tie-in to NE leader. Do not cut cables until the leaders have been lightened*.
2. **Utility Installations Above ground** *None* **Below ground** *Conduit for courtyard lighting located. Not in conflict with job*.
   Recommended Course of Action *None needed at this site*
3. **Traffic**
   A. **Vehicular** *Busy street, library traffic*
   B. **Pedestrian/bicyclist** *On street, and at entrance, okay in courtyard*
   C. **Other (equestrian, air, aquatic, etc.)** *NA*
   D. **Traffic Security Plan** *Cordon off staging area in parking lot. Close courtyard to public. Do work in stages 6:00 a.m. to 10:00 a.m. to avoid disruption*
4. **Job Site Concerns**
   A. **Structures** *Glass walls of library, slate roof*
   Plan: *Build a plywood barricade in front of east wall to protect glass. Hang black plastic sheeting on other walls to prevent chain oil and sawdust staining of adjacent walls. Put plywood on roof where necessary to stand. Use blower on roof to remove sawdust and fine debris in clean-up phase*.
   B. **Landscape** *Fish pond, fern garden, flintstone paving*
   Plan: *Cover pond. Put a scaffold over the ferns. Do not service saws on stone paving. Do not drop sections of wood onto paving. Lay plywood down over most heavily worked areas*.
   C. **Noise (hospital zone, noise ordinances, etc.)** *Library, shhhhhh!*
   Plan: *Schedule work in phases from 6:00 a.m. to 10:00 a.m. to avoid disruption*.
   D. **Other** ____________________
   Primary Challenge: *Crane is not an option. Out of reach. Can't afford to tip a crane over on the structure*

FIGURE 1. Sample work plan for single tree. This sample shows how a work plan can be developed from the data provided to plan for the removal of a single tree.

Removal Plan

Personnel Required _Standard team._

Phase One—*Speedline brush over roof to staging area. Use traveling block and "crossline" set-up with 4:1 yachting blocks to manage tension and control descent.*

Phase Two—*Using Hobbs Lowering Device™ and Rope Brake™, tip-tie and butt hitch scaffold limbs, lower them into courtyard and bring wood out through library with hand truck and four-wheel rubber-tired dolly.*

4. Phase Three—*Grind-out stump, clean-up, remove tarps, scaffolding and plywood.*

Removal Equipment Needed:

Full range of chain saws, Hobbs Lowering Device™, Rope Brake™, single and double braided lowering lines, "Crossline" SPEEDLINE gear, rigging bag with slings, shackles, carabiners, rescue pulleys, plywood, scaffolding, tarps, hand truck, ARBOR CART™, wedges, can't hook

FIGURE 1. *(continued)*

10 years, they were all removed and replaced with locust. An appallingly poor species to site match resulted in controversy, unnecessary maintenance, and premature removal.

Planting a tall tree beneath a power line ensures that the tree will have to be severely headed back periodically to keep the utility space clear of obstruction. Periodic severe heading of such species as maple or yellow poplar practically guarantees that cavities and decay will develop, dramatically reducing the effective life of the tree.

5.1.2. Improper Pruning

There is a long list of problems related directly to incorrect pruning. Among them, heading cuts (topping), flush cuts, and overthinning can be the cause of serious wounds, cavities, and defects that can lead directly to premature mechanical failure.

Proper pruning in the first 5 years of a tree's life is essential to helping the tree develop the proper structure. In the formative years, a hand pruner can remove many potential problems that, if left unattended, can grow into structural defects that cannot be resolved without removal.

5.1.3. Good Plant Material

Even if you select a suitable species for the site and have made provisions for establishment pruning and after planting care, the tree has to have a good root system and a decent structure to begin with to ensure any measure of long-term success. Take an interest in what gets purchased, how it is transported and planted. Remember that the birth of a baby is when the care begins. The planting of a tree is when the real commitment begins. People feel good about planting trees, but I think many forget that they need to be irrigated, trained, and pruned for years to come. Make sure that you have detailed planting specifications appropriate to the site and species drawn up for reference and review.

A. Tree Removal Work Plan Assessment Form

1. **Site Location of Tree(s) to be removed** El Camino Real
2. **Inventory number/GPS location** _All trees between Matthew Dr. and Mackenzie Terrace._
3. **Number of Tree(s) to be removed** _350_
4. **Species of Tree(s) to be removed** _Ulmus americana_
5. **Reason(s) for removal** _road improvement, trees are also overmature and hazardous_
6. **Heritage Tree removal permit required? Yes** _XXX_ **No**_____ **Permit on file** _XXX_
7. **Diameter of Tree to be removed (answer only if single tree)** _NA_
8. **Height of Tree to be removed (answer only if single tree)** _NA_
9. **Range of diameter of trees to be removed (in inches)** _18 to 84_
10. **Range of height of trees to be removed** _40 to 125_
11. **Access**
    A. _XXX_ **Excellent. Accessible to any motorized equipment**
    B. _____ **Limited. Describe** ____________________
    C. _____ **Restricted. Describe** ____________________
12. **Debris Disposal Method** _Load as large as possible, haul to disposal site and process_
    A. **Estimated Quantity of brush and/or chips** _4200 cu. Yds of chips_
    B. **Estimated quantity of logs/wood 150,000 cu. ft.** (1171 cords)
13. **Stump removal**
    A. **Yes** _XXX_
    B. **No** _____
    C. **Number of stumps to grind** _350_
    D. **Estimated total inches/feet of stump to grind** _1,000 feet_
    E. **Stump grinder needed** _Track-Mounted, self propelled, largest unit available, sub-contract work to specialist_.

B. Job Site Safety

1. **Visible Tree Defects/Hazards Present No** _____ **Yes** _XXX_
   Describe _Too many to note herein_.
   Plan: _Mark trees with hazards or defect and brief crew before commencing work_.
2. **Utility Installations**
   A. **Above ground** _Yes, high voltage to 12kv_
   B. **Below ground** _Yes, natural gas_
   C. **Recommended Course of Action** _Have utility company top trees below the power lines, work with gas company to locate and work around gas lines during stump removal_.
3. **Traffic**
   A. **Vehicular** _Main thoroughfare, 33,000 vehicle daily count_.
   B. **Pedestrian/bicyclist** _On street, and sidewalk_
   C. **Other (equestrian, air, aquatic, etc.)** _NA_
   D. **Traffic Security Plan** _close inside lane and sidewalk traffic on work side of road_.
4. **Job Site Concerns**
   A. **Structures** _Trees overhang private yards and some residential structures_
   Plan: _rig as necessary to protect property, use crane on largest trees and most hazardous_.
   B. **Landscape** _privately owned lawns, shrubs and shade trees in front yards_
   Plan: _rig and rope as necessary to protect property. Use crane on largest trees_.
   C. **Noise (hospital zone, noise ordinances, etc.)** _residential neighborhood_.
   Plan: _Schedule work from 7:30 a.m. to 5:00 p.m. to avoid disruption_.
   D. **Other** ____________________
   Primary Challenge: _A whole lot of big trees. Lots of brush, wood and traffic_.
   Personnel Required _Standard team plus flagging crew_.

FIGURE 2. Sample work plan for large project. This sample shows how a work plan can be developed from the data provided to plan for the removal of a large quantity of trees.

C. Removal Plan

1. Have contract line clearance contractor come in and top all trees below the power lines.
2. send ground falling crew down the line, felling those that can be dropped in one piece.
3. send aerial lift and climbing crew down the line, removing and rigging what they can get.
4. send crane crew down the line. Recommend leaving large trunks upright and "on the stump", where practicable for crane to "pick and load" in one piece. Action will reduce damage to sidewalks that aren't being removed in the widening and reduce chance of damage to private lawns.
5. Grind stumps and final clean-up.

Notes: With the volume of brush, logs and wood coming off this job, I recommend using a grapple to load as much material as possible for transport to a biomass recycling center for high-volume processing.

Removal Equipment Needed: 28-inch road cones, lane barricades and 48-inch traffic warning signs in sufficient quantity to comply with D.O.T. regulations. Aerial lift, self-loading log trucks, grapple-mounted skid loader, debris box trucks, Hobbs Lowering Device™, Rope Brake™, single and double braided lowering lines, "rigging bag". Plywood and tarps for site protection as needed.

FIGURE 2. *(continued)*

Most removals in the urban forest may be classified as either being: maintenance or emergency:

1. Maintenance removals would be the more or less scheduled removal of those trees that have died or need to be taken down for some other reason.
2. Emergency removals are those necessitated by storm emergencies.

Each type of removal requires planning and preparation.

Maintenance removals require the sort of planning that accompanies the process of filling out the work plan assessment form. With maintenance removals, you have the luxury of being able to plan for the removal of a tree or trees under more-or-less controlled conditions.

Emergency removals require planning in advance to be able to cope with the chaos that comes with a hurricane, ice storm, or other natural disaster. Every municipality should have a storm emergency contingency plan worked out in advance that addresses tree removal. The first thing that happens after a natural disaster like Hurricane Hugo, the great ice storm of January 1998, or countless other storm emergencies is a period of chaos, confusion, and isolation immediately following the aftermath. Untold miles of roads are going to be clogged with trees, power lines, and poles. Everybody will demand they be taken care of first. Have you made plans to cope with the situation?

During hurricane or the winter season, watch the weather reports carefully. Keep the trucks in good repair, the tanks filled with fuel, and the chip boxes empty. Have plenty of spare saw chains, spark plugs, saw mix and oil, and essential parts.

What roads are you going to clear first? Do you have contracts in place with commercial tree care firms to provide back-up emergency help? Are all your equipment eggs parked in one basket, or do you have them placed in strategic locations throughout the city?

If you have not been through a natural disaster of even a small scale, you cannot imagine the impact that small, usually insignificant details can have if overlooked.

Not usually accustomed to pruning trees at 2:00 a.m., will your crews be able to see in the dark? Caver's headlights on the hardhats are an invaluable aid to the tree workers. Portable lighting can be essential for safe storm emergency work. Have you made provisions to give them a place to rest between shifts if they cannot go home? The danger of accidents during storm emergency work cannot be overemphasized.

In California, our celebrated Pacific Storms can pack inches of rain for days, driven by 80 mph winds, and knocking down trees by the thousands. But the weather was always tolerable. My heart goes out to anyone who has to cope with the immediate aftermath of an ice storm while the weather is still brutal.

Fatigue coupled with heavy work under slippery conditions is a recipe for disaster. Make sure that your crews are properly clothed and equipped for the conditions. A worker in the best of shape can only push just so far and just so hard before he/she will break down. Make sure that your people get enough to eat and are required to take a reasonable break for sleep. You will get even less done if your people collapse or end up in the ER as a result of fatigue-induced injury.

Without trying to provide instruction in removal, the following outline details a few of the tricks, procedures, and lessons that I have learned over years of maintenance and emergency removal:

1. Chainsaws. Learn all you can about maintenance, safe use, and sharpening. An executive from Husqvarna once told me that they spend millions of dollars trying to shave a few ounces of weight off a saw or wring out a few more rpms, when the operator can gain an immediate increase in power of 20% by merely seeing that the chain is correctly sharpened.

Buy a dealer and not necessarily a brand. Make sure that the saws you buy can be serviced and supplied with parts in a timely manner. A saw waiting for parts is a saw wasting time.

Make sure you have plenty of saws on the job. If one breaks down, you should be able to grab another and keep on working. Saw maintenance should be done in the shop, at the workbench, not on the stump.

That said, I also recommend making up a field kit that has a spare plug, replacement nuts and bolts that are prone to loss, some loops of sharp chain sized to the saws issued, and the tools needed to tighten or assemble.

2. Rope. Use ropes recommended by the manufacturer as suitable for arborist use. Know what their tensile strengths are and how to calculate a safe working load. Adopt an inspection protocol that incorporates the following three elements:

a. Initial. Make sure the rope is what you ordered and in good condition when you first receive it.
b. Frequent. Train field personnel in daily inspection for cuts, abrasion, heat fusion,
c. Periodic. On a scheduled basis, a person in authority inspects the rope as well as other safety and production gear and passes it or fails it for use until the next inspection cycle.

The most commonly offered cut lengths by vendors are 120- and 150-foot lengths. Buy your rope in the lengths that serve your needs best. Always remember that the working height is half the length of the rope. A 150-foot rope is going to be too short to work in a 60-foot tree if you consider wraps and the need to provide the line handler with enough rope to keep him or her away from the drop zone.

Take good care of your ropes. Keep them clean and coil and hang them or bag them when not in use. Let wet ropes dry out whenever possible.

3. Backsavers. Tree removal is tough enough without making it impossible. One tool that I do not see often enough on the job is a cant hook. Few hand tools are more useful for rolling logs than a cant hook. Every crew geared up for removal should have at least one cant hook as standard equipment. Plastic falling wedges and a logger's falling axe are also essential tools in my removal kit.

6. Bid Specifications

You are the city arborist. Your small tree maintenance crew is hopelessly backlogged with routine pruning and planting work. The finance department finally approves your budget request for funds to hire outside contractors. Now the trick is to get the most work from the best contractor with the least amount of hassle. The key to success is in your bid specifications.

As a commercial contractor who has spent years reading good and bad bid specifications, when applicable, I like to have the following questions answered in the specifications before I have to spend time trying to track down the answers on my own:

1. Be specific about the location, species, and size of the tree(s) to be removed.
2. Mark the trees or tag them so that there can be no doubt about which trees are to be removed.

   ANECDOTE: We once took a job removing most but not all street trees in a downtown area. Upset that his tree obscured his storefront, a merchant took me to task when I felled trees on either side of his store, leaving his tree healthy and full of leaf. I told him that without its fluorescent orange death sentence, I could not remove it. About an hour later, the merchant told me I had made a mistake, that I had failed to notice the orange paint on the tree trunk. He had failed to notice the orange paint on his hands! The tree stayed.
3. Let us know if there are any time restrictions or noise ordinances that we will have to operate in compliance with.
4. Let us know if there is a completion date. Let us know if there are liquidated damages, what they are, and if there are allowances made for bad weather.
5. Be specific about debris disposal. Some municipalities will issue dump tickets to handle the disposal of debris, but many do not. Some cities have a wood yard, but many do not. Some cities will allow firewood to be left on site, but many do not.
6. On large projects, it can be helpful to all parties to provide the contractor with secure parking for equipment at the corporation yard.
7. Always require that all work be done in compliance with current and applicable ANSI (1994) Z133.1 standards as well as other appropriate municipal, state, and federal standards.

8. Be specific about stump removal. If you want the stumps removed, say so; do not try to convince the contractor that "you meant the stumps" after you put the job to bid. If you want them ground, specify the width and depth of grind out that you want and whether the chips can be left on site or cleaned up and hauled away.
9. On large projects, it is always wise to plan a bid tour and orientation. I have asked so many questions on some bid tours that the exasperated tour guide had to grant an extension to the submission date in order to answer our concerns. From my perspective as a contractor, I am not willing to assume the risk of an open-ended project that leaves issues unresolved or open to interpretation that could cost me profit or my professional reputation.
10. Be clear about who assumes the responsibility for traffic control and lane closure, if necessary.
11. Establish a rate for add-ons either through unit pricing or an established labor and equipment rate.
12. When appropriate to the job site or situation, make provisions to shield the contractor from the public by making it clear that any questions are to be answered by the appropriate agency and not by the contractor.
13. For everyone's protection, make sure that the preexisting condition of curbs, side-walks, structures, and so on is established and documented before the contractor begins work.
14. Many municipalities require removal permits or environmental impact reports. Make the job easier and less expensive for the contractor by taking care of the red tape in the course of preparing the bid specifications.
15. In jobs requiring backfill, specify what is acceptable. Show samples at the bid tour.
16. Be wary of having tree work buried deep in a large project such as a downtown redevelopment.

Case History: In 1979, we ended up as a sub-to-a-sub-to-a-general contractor! The general contractor had bid the total job, subbing out the tree removal to the excavating contractor who had the job of tearing up the street. We in turn, subbed the stump grinding to our stump grinding specialist. We had bid $80.00 per tree, removed and ground out. Later the "stump guy" asked for another $5.00 per tree because it did not turn out that we could do all six blocks at once.

Our boss (the sub-to-the general) had no problem with an extra charge for a change in the original agreement. About halfway through the project, the city arborist called me up and asked what my bid was. I told him about the $80.00 plus $5.00 bid. The arborist thought our price fair enough, but on principle, he did not like getting charged $280.00 per tree by the general contractor! I did not much appreciate the thought that someone I had never met was making 2.5 times the amount we were without getting wet, dirty, and covered in sawdust. The contract had been let for one side of the street at a time, so the arborist eliminated tree removal from phase two of the project and did the work with his own crew.

Through no fault of our own, we lost out on phase two of a nice job. The city lost out because it probably cost them more per tree than our bid to do the work in-house (but a lot less than the general contractor's bid). While occupied with Main Street, other tree work was not getting done. From that job forward, I have always been careful about getting buried too deep in a bid.

7. Conclusion

Tree removal is an unavoidable aspect of urban forestry in the Northeast and anywhere else for that matter. Tree removal may become necessary for reasons as varied as biological failure, storm damage, or urban renewal. Tree loss through hurricanes and ice storms is more or less unavoidable, but premature failure as a result of poor selection, inadequate establishment practices, or improper pruning is very avoidable. A much better use of taxpayer revenue and arborist skills is in quality tree maintenance as opposed to removals that could have been prevented or postponed for decades in the life of a tree.

Large-scale tree removal is a highly specialized, potentially dangerous arborist skill. Personnel must be properly trained, equipped, and supervised in such diverse skills as chain-saw use, rope and saddle climbing, felling, limbing, bucking, knots ropes, and rigging (see Appendix D). Work that is put out to bid should be as clear and concise in intent and objective as possible.

For all its blood, sweat, tears, and aggravation, few assignments in arboriculture are more challenging or rewarding than a large-scale removal well-planned and skillfully executed. I will tell you something that is, however, tree preservation. For me, nothing is more nourishing to my soul than the practice of the preservation arts, for example, pruning, cabling, soil manipulation, and diagnosis. The hope and promise that my grandchildren may someday be able to enjoy the shade of a favorite tree that their grandpa labored to preserve when he was a young man is the essence of this profession.

Appendix A

Cordage Manufacturers

Some of the arborist suppliers and most of the cordage manufacturers are beginning to publish reference material along with product information.

The American Group/Samson Division
2090 Thornton Street
Ferndale, WA 98248
(360) 384–4669

New England Ropes, Inc.
848 Airport Road
Fall River, MA 02720
(800) 333–6679

Yale Cordage, Inc.
PO Box 3820
Portland, ME 04101
(207) 282–3396

Chainsaw Manufacturers

The major chainsaw manufacturers such as Husqvarna, Stihl, and Shindaiwa publish good, basic safety, and operational information regarding chainsaw use and safety.

Husqvarna Forest and Garden Company
9006-J Perimeter Woods Drive
Charlotte, NC 28216
(800) 438–7297

Shindaiwa, Inc.
11975 SW Herman Road
Tualatin, OR 97062
(800) 521–7733

Stihl, Inc.
PO Box 2015
536 Viking Drive
Virginia Beach, VA 23450–2015
(757) 486–9100

Associations

American National Standards Institute
11 W 42nd Street
New York, NY 10036

The MF Blair Institute of Arboriculture
PO Box 292
Big Pool, MD 21711
(301) 842–2544
Donald F. Blair, Director
Arborist skills training, consultation.

The International Society of Arboriculture
PO Box 3129
Champaign, IL 61826
(217) 355–9411

The ISA has over 10,000 members, worldwide, organized into state and/or regional chapters. The ISA has an arborist certification program, as well as publications and videos on a wide selection of safety and performance-related topics. Membership is open to anyone with an interest in the advancement of arboriculture.

The National Arborist Association
PO Box 1094
Amherst, NH 03031-1094
(603) 673–8952

Membership is limited to the owners and principals of contracting firms. Training seminars offered at various times and locations throughout the country. Publications and videos on a wide selection of safety and performance-related topics are available. The National Arborist Association is working with this author, Ken Johnson and Robert Phillips to develop a training program on rigging for removal that will include a videotape, manual, and continuing education credits.

Arborist Supplies

Karl Kuemmerling, Inc.
129 Edgewater Avenue NW
Massillon, OH 44646
(216) 477–3457

Sherrill Arborist Equipment and Supply
3101 Cedar Park Rd.
Greensboro, NC 27405-9657
(800) 525–8873

The Sierra Moreno Mercantile Company
PO Box 292
Big Pool, MD 21711
(800) 262–0800
(301) 842–2544
Donald F. Blair, founder

Tools for the tree health professional—work-proven solutions worldwide since 1975. Sierra Moreno Mercantile Company is the innovative firm responsible for introduction of the Hobbs lowering devices, Hobbs rigging blocks, rope brakes, custom climbing gear, and a comprehensive stock of arborist tools, supplies, and equipment. Available direct and through authorized dealers worldwide.

Appendix B

Applicable American National Standards

- Z133.1, Tree care operations: safety
- A300, Standard practices for trees, shrubs and other woody plants: pruning, cabling, etc.
- A10.14, Requirements for safety belts, harnesses, lanyards, lifelines, etc. for industrial use
- A14.1, Ladders; portable wood; safety requirements
- A14.2, Ladders; portable metal; safety requirements
- A92.2, Vehicle-mounted elevating and rotating aerial devices
- B175.1, Gasoline powered chain saws; safety requirements
- Z87.1, Practice for occupational and educational eye and face protection
- Z89.1, Personnel protection; protective headgear for industrial workers; requirements

Applicable Federal Regulations

US Department of Labor, Occupational Safety and Health Administration:

- CFR 29 1910.000, General industry
- CFR 29 1910.67, Vehicle-mounted elevating and rotating work platforms
- CFR 29 1910.95, Occupational noise exposure
- CFR 29 1910.151, Medical services and first aid
- CFR 29 1910.268, Telecommunication
- CFR 29 1910.269, Electric power generation, transmission, and distribution
- CFR 29 1910.331.335, Safety; electric-related work practices
- CFR 29 1910.1200, Hazard communication
- 49 CFR (transportation regulations)

Appendix C

Tree Removal Work Plan Assessment Form

1. Site location of tree(s) to be removed ______________________________
2. Inventory number/GPS location ______________________________
3. Number of tree(s) to be removed ______________________________
4. Species of tree(s) to be removed ______________________________
5. Reason for removal ______________________________

6. Heritage tree removal permit required? Yes _____ No _____ Permit on file _____
7. Diameter of tree to be removed (answer only if single tree) _____________
8. Height of tree to be removed (answer only if single tree) _______________
9. Range of diameter of trees to be removed ______________________________
10. Range of height of trees to be removed ________________________________
11. Access _____________
    A. _____ Excellent. Accessible to any motorized equipment
    B. _____ Limited. Describe
    C. _____ Restricted. Describe
12. Debris disposal method _____
    A. Estimated quantity of brush and/or chips _____
    B. Estimated quantity of logs/wood _____
13. Stump removal
    A. Yes _____
    B. No _____
    C. Number of stumps to grind
    D. Estimated total inches/feet of stump to grind
    E. Stump grinder needed

Notes: ____

____

Job Site Safety

1. Visible tree defects/hazards present No _____ Yes _____
   Describe ____
   Recommended course of action ____
2. Utility installations
   A. Aboveground ____
   B. Belowground ____
   C. Recommended course of action ______________________________________
3. Traffic
   A. Vehicular ____
   B. Pedestrian/bicyclist ____
   C. Other (equestrian, air, aquatic, etc.) _____________________________
   D. Traffic security plan ____
4. Job site concerns
   A. Structures ____
   Plan: ____
   B. Landscape ____
   Plan: ____
   C. Noise (hospital zone, noise ordinances, etc.) _______________________
   Plan: ____
   D. Other ____

Personnel required ____

Removal Plan

Phase one ____________________

____

Phase two ____________________

____

Phase three ____________________

____

Removal equipment needed:

____

____

____

Notes:

____

____

____

____

____

____

____

____

____

____

____

Appendix D

Recommended Guidelines for Safety Training and Standard Performance for Qualified Personnel

For a person to be considered qualified to do tree removal as well as general tree maintenance, they should first receive training and instruction from competent persons in the following areas.

General

- Job description appropriate to job assignment
- Introduction to immediate supervision and crew members
- Instruction in selection, care, use, and maintenance of personal protective equipment
- Familiarization with equipment
- Introduction to company policies and procedures
- Safe work practices as related to job assignments

Tree Knowledge Appropriate to Job Assignments

- Education and training in accordance with prevailing national standards for tree removal and/or general tree maintenance.
- Education and training in accordance with local, state, or regional standards for removal and/or general tree maintenance as well as those specified by contract.

Tree Knowledge for Removal and/or General Tree Maintenance Appropriate to Job Assignments

- Provide education and training relative to predominant tree species within geographic area, such as identification, growth habits, structure, and wood strength.

General Safety

- OSHA standards
- ANSI standards
- Public safety and traffic control
- Electrical hazards
- Emergency conditions
- Job site briefings
- Lifting
- Load handling
- Direct supervision
- Noise level compliance

Equipment Safety

- Mobile equipment and aerial lifts
- Aerial equipment and electrical hazards
- Chain saw, power tool, and hand tool use and safety
- Climbing equipment

Operational Safety

- Climbing techniques
- Rigging and tree removal
- Hazard communications

Personal Safety

- Personal protective equipment
- Emergency response procedures
- Back and other accidental injury prevention
- Poison plant/animal identification and avoidance

Note: A training program that does not document the employees' progress and participation is an inadequate program. Make sure you keep records of your safety meetings, seminars, and field exercises and have all attending sign in. Keep your records for at least 5 years.

References

I have been lecturing, demonstrating, and writing about arborist skills since 1972. I have welcomed this opportunity to assemble many of these thoughts and experiences for the first time in published format. In addition to common sense, experience, and my personal notes, I have referred to the following published data in researching this chapter:

American National Standards Institute (ANSI), ANSI Standard Z133.1, *Safety Requirements for Pruning, Trimming, Repairing, Maintaining and Removing Trees for Cutting Brush*, American National Standards Institute, New York.

Blair, D. F., 1995, *Arborist Equipment*, International Society of Arboriculture, Champaign, IL.

Blair, D. F., 1986, *Catalog of Arborist Supplies*, Sierra Moreno Mercantile Company, Big Pool, MD.

At present, there is no comprehensive manual or guide for tree removal. In addition to books used in the bibliography, the following are excellent references in their fields of specialization:

Beranek, 1996, *The Fundamentals of General Tree Work*, Beranek Publications, PO Box 251, Fort Bragg, CA 95437

Jepson, J. L., 1997, *The Tree Climber's Companion*, Jeff L. Jepson, Longview, MN.

Chapter 22

Recycling Urban Tree Removals

Edward A. Lempicki and Edward Cesa

The purpose of this chapter is to provide information and guidance for utilizing street tree removals; a "recycling" strategy that can potentially turn a cost-burden scenario into an income-generating opportunity. The strategy involves merchandising sawmill-size logs from street tree removals to sawmills or other companies that have unique uses for street tree logs. Recycling municipal trees by converting street tree removals to valuable sawlogs can generate income as well as reduce the amount of time and labor costs involved in processing this material into firewood.

1. Introduction

Many municipalities and local governments are currently experiencing budgeting problems in meeting community needs. Street tree management and maintenance budgets are among those becoming strained. As a result, the quality of our street trees cannot help but suffer as economic considerations continue to reduce tree management budgets.

Currently, much of the wood generated from street tree removals brings little economic return to tree management budgets. Because of this, most tree management and maintenance programs are being run as a cost burden to municipal budgets. Although most tree management crews are hardworking and efficient, the products rendered from street tree removals are usually low value, which returns little money to municipalities. In fact, in New Jersey, it is estimated that more than 50% of an average municipality's tree management budget is spent on the cost of tree removals alone (NJ Division of Parks and Forestry, 1996).

Edward A. Lempicki New Jersey Forestry Services, Trenton, New Jersey 08625.
Edward Cesa US Department of Agriculture, Forest Service, Morgantown, West Virginia 26505.
Urban and Community Forestry in the Northeast, 2nd ed., edited by, J. E. Kuser.

Tree mortality from natural occurrences, such as insects, diseases, and storms, plus a myriad of man-made circumstances, such as roadway widening, right-of-way maintenance, and utility construction activities, takes a huge toll on street trees. This results in a continuing need for tree maintenance on a municipal level. Much of this harvested wood, if produced and marketed effectively, can generate income for municipalities to support tree management and maintenance programs.

This chapter is designed as a guide for those who deal with disposal of street tree removals. Ideas and suggestions here are offered as potential alternatives for current practices. Many sawmill-size logs from municipal tree removals are marketable. Advantages of merchandising such sawlogs include:

- Potential income generated from selling logs or developing barter arrangements
- Reduction in labor cost by reducing the amount of time work crews need to process logs into firewood
- Reduction in amount of woody material going to landfills
- Reduction in landfill costs for disposal of material
- Reduction in volume of firewood material that must be stored in municipal maintenance yards until it is sold
- Conservation of forest land resources by generating sawlogs from street trees that must be removed anyway

Many municipalities, particularly the ones located in urban and suburban areas, are already recycling and marketing street tree removals to some degree. In most cases, the use of tree removals involves providing products directly to residents in the municipality. For example, leaves and twigs are composted and used by residents as fertilizer. Small branches are chipped and used residentially or municipally as mulch, and large branches and tree trunks are cut into firewood.

Firewood is the most common product produced from street tree removals. Usually, this material is left at the roadside for local residents to pick up, or it gets transported to a central yard or storage site where it is piled for future processing. Firewood production generally involves tree material from every size class, quality type, and species group. Whatever there is to be removed, whether it is oak, sycamore, ash, maple, pine, or spruce, usually goes into the firewood pile. In addition, many municipalities split the larger pieces, thus further adding to the labor and costs of removing street trees.

The premise of this chapter is that there must be a better way; a better way for street tree management than maximum cost–minimum return. The removal work itself must be done, but there is a potential opportunity for changing this cost-burden scenario into one that is more cost-efficient by exploring the ideas in this chapter. Instead of cutting a good log into firewood, leave it "as is"; a readily marketable commodity, one that can be marketed to sawmilling industries that process logs into a broad range of wood products. The dollar return potential of selling the log for lumber products exceeds the return potential of selling the log for firewood by at least two to four times.

2. The Market

The concept of utilizing street trees in sawmills is not new. Some sawmills have been sawing products from street trees for many years because they have found a unique niche for using street tree sawlogs. These mills process both hardwoods (trees that lose their leaves every year, i.e., oaks, maples) and softwoods (trees with needle-like leaves that do not fall off in the dormant season, i.e., pine, spruce).

From street tree logs, sawmills can manufacture products such as pallets and pallet stock, landscape ties, truck bed stock, fencing, heavy timbers, construction lumber, posts, bridges, and park benches. Furniture-grade lumber also can be produced from these logs, which then can be processed after kiln drying into a very broad range of finished wood products such as furniture, moldings, and flooring. Sawmills are a major market and opportunity to which street tree logs can be merchandised.

Some of the wood generated from municipal trees holds special potential for unique and figurative characteristics. One example is spalted wood, which results when logs are invaded by certain fungi. The fungi produce a highly unique coloration and pattern in the wood that is very appealing and special. Burlwood and crotchwood also have highly unique and figurative characteristics. These types of wood are valuable and sometimes found in municipal trees that have been traditionally turned into firewood.

The retail price for some of these figurative woods can be as much as four times the retail price for standard lumber used to produce the same product. Crafters seek these types of wood because of the many special effects they give their finished products. Generally, the mills using street trees are not typical high-production operations. They are smaller in size and may have special markets and product lines compared to standard high production-oriented sawmills. The keys to marketing street tree logs are:

1. Locating local sawmills
2. Working with mills to learn sawlog requirements
3. Making sure that the merchantable sawlogs are free of metal and other foreign material
4. Storing sawlogs until a salable quantity is accumulated
5. Being flexible and persistent enough to try this concept

3. Products and Specifications

Street trees that are at least 12 inches in diameter at breast height (4.5 feet from the ground) and have a log of at least 6 feet in length have sawlog potential. Normally, the most valuable part of the tree is the first 8 to 16 feet closest to the ground. This is where the greatest volume of wood is located. It is also where the highest quality and most valuable wood is found. The first log cut closest to the ground is called the *butt log*. A sawmill's raw material requirements are directly influenced by its markets. Consequently, the demand and price for your potential sawlogs depend on this relationship. Knowing what a sawmill requires is an important first step for successfully merchandising sawlogs. Listed below are general sawlog specifications that should assist you in determining what a sawmill requires. These are general specifications and could vary considerably in your local area.

For example, persimmon is listed as "fair" in species desirability because most Northeast sawmills do not have a high demand for these sawlogs. However, in Tennessee there is a large market for persimmon, which centers on its use in manufacturing golf club heads. This drives up the price and demand for persimmon in that region. Similar examples exist for other species such as Osage orange and mulberry. You must keep in mind that special markets dictate higher values for particular species, depending on local market conditions.

3.1. General Sawlog Specifications

1. General species desirability
   - Best: Walnut, butternut, ash, oaks (except pin oak), cherry, paulownia
   - Good: Maples, elms, most fruitwoods, basswood, sycamore, cedar, poplar
   - Fair: White pine and other softwoods, mulberry, Osage orange, persimmon, beech
   - Poor: Gum, ailanthus, pin oak
2. General size requirements:
   - Best: 16 inches or larger diameter at small end of log 8 feet or longer in length
   - Good: 14 inches or larger diameter at small end of log 8 feet or longer in length
   - Fair: 12 inches or larger diameter at small end of log 6 feet or longer in length (NOTE: Logs should be generally sound, i.e., relatively free of rot, decay, and holes.)
   - Poor: Small in length and diameter; or large in length and diameter with many knots, branches, holes, rot, or cracks; or with large or numerous metal objects

3.2. Metal and Other Foreign Material in Street Trees

One of the primary reasons why demand for street tree sawlogs has been low in the past is the presumption that metal and other foreign objects are often associated with urban-source logs. The reputation of these logs having metal in them (i.e., nails, wire, spikes, or even car parts) is common among sawmillers.

Metal can become a serious problem during log sawing because it dulls and/or damages saw blades and sawmill equipment. It also can be a safety hazard for workers in a mill because of flying debris when a blade hits large metal objects. These considerations dictate that metal is a problem that must be ensured against by taking appropriate actions.

The best way to correct this problem is to scan logs for metal before they go through the sawing process. Standard metal detectors are normally adequate. If metal is discovered, it must be removed. If large quantities of metal are detected in a log, it should not be sold as a sawlog. If a metal-laden log is shipped as part of a load to a sawmill, it will probably be the last load you ever sell to that particular mill.

Normally, most metal is located within the first 4 to 6 feet of a street tree. This is the section of the tree that people use for hanging signs and securing fencing for yards or pastures. This also is the section that children like to pound nails into. Consequently, butt logs need to be screened more carefully than logs that come from higher up in the tree.

Typical metal detection techniques include a visual inspection of the log surface for metal objects, such as wire and protruding nails, as well as any discoloration that

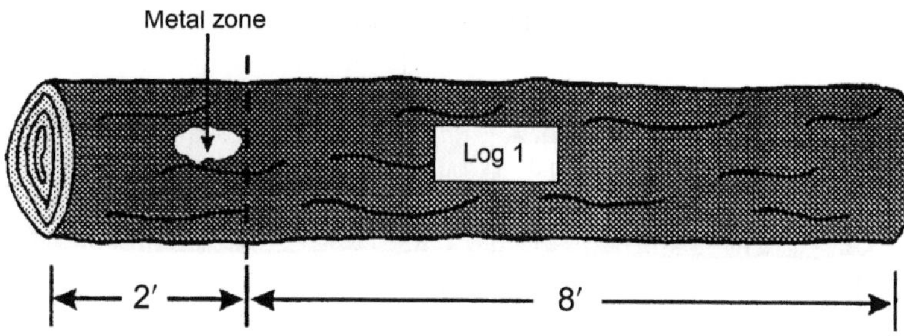

FIGURE 1. Remove the end section of the log that contains metal. You do not want to remove too much "good" wood beyond the metal, yet at the same time you do not want to hit the metal with the chainsaw. If you are uncomfortable removing the metal, let your log buyer show you how.

normally appears as a black–blue stain on the end of the log. Following a thorough visual inspection, a careful scan with a metal detector is needed.

When metal is seen or detected, appropriate steps are needed to remove it. If metal is located at the end of the log, that part can be sawn off (Fig. 1). If metal is detected toward the middle of the log near the surface, then the section containing the metal can be removed (Fig. 2) (Cesa *et al.*, 1994). If you do not feel comfortable removing the metal, then mark the area with paint and let the log buyer know it contains metal.

Depending on where the metal is located and how deep it is in the log, a determination must be made as to whether it is worth removing (Fig. 3). If it is not, the log should be classified unusable as a merchantable sawlog. It may be best, for the first couple of sales, to discuss these marginal logs with the log buyer.

Metal detectors vary in size, cost, and capability. Some detectors tell you not only where the metal is but also its depth. If a municipality wants to test the sawlog market initially, a metal detector could be borrowed for log scanning. Basic metal detectors range in price from $250 to $500+. Listed below are some metal detector manufacturers that can be contacted for specific details:

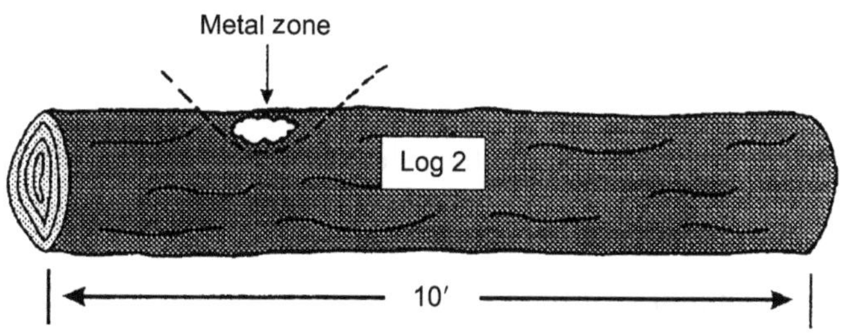

FIGURE 2. Remove metal in this area provided that it is not too deep in the log. If metal is located in the first several inches of the log, removal of it will not affect the volume or quality of lumber produced from the log because this section is usually cut off in the sawing process.

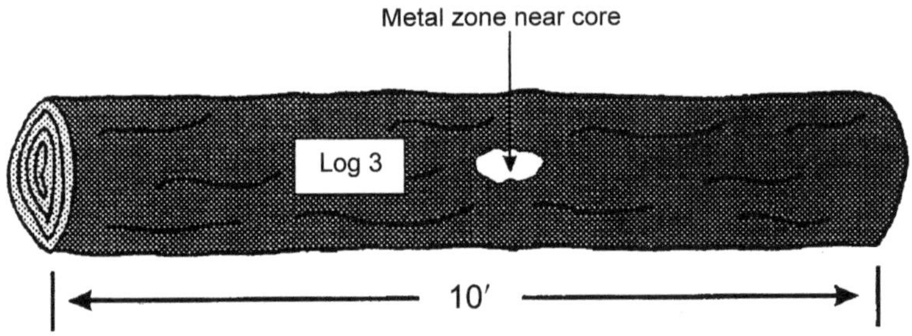

FIGURE 3. When metal is located near the core of the log, removing it would drastically reduce the volume of lumber that could be sawn from the log. Leaving it could cause considerable damage to sawmill equipment. These types of logs having metal should be clearly marked and discussed with the log buyer or processed into firewood.

Fisher Research Laboratory
200 W. Willmott Road
Los Banos, CA 93635
Phone: 209–826–3292
Fax: 209–826–0416

Garrett Electronics, Inc.
1881 W. State Street
Garland, TX 75042–6761
Phone: 972–494–6151
Fax: 972–494–1881

Rens Manufacturing Co.
83868 N. Pacific Hwy
Creswell, OR 97426
Phone: 541–895–2172
Fax: 541–895–3580

Tesoro Electronics, Inc.
715 White Spar Road
Prescott, AZ 86303
Phone: 520–771–2646
Fax: 520–771–0326

White's Electronics, Inc.
1011 Pleasant Valley Road
Sweet Home, OR 97386
Phone: 541–367–6121
Fax: 541–367–2968

Other foreign materials sometimes found in street tree logs are cement and car parts. Any nonwood material within a log poses serious problems to sawmill operators and equipment in the sawmilling process; therefore, every precaution needs to be taken to ensure that saw logs are free of foreign materials. This effort alone could make or break the concept of a municipality merchandising logs to a sawmill.

Once a working relationship has been established with a sawmill and the mill manager realizes that you are taking steps to locate and remove metal and other foreign material from your logs, the mill then should become a consistent outlet for your merchantable sawlogs.

3.3. Proper Log Manufacturing

In order to maximize the dollar value of street tree sawlogs, it is imperative that they are properly manufactured. The definition of "properly manufactured logs" depends on the requirements specified by the sawmill. However, the following list and Fig. 4 provide some general guidelines for proper log manufacturing (Stump, 1967):

1. Safety always comes first
2. Follow the sawmill's log specifications correctly
3. Keep log ends straight
4. *Cut logs for highest quality*. Group defects when possible, preferably near the ends of the logs manufactured
5. Leave trim allowance on all logs in accordance with the sawmill's specifications. Normally, trim allowance is an additional 2 inches; in other words, a 10-foot log is actually 10 feet 2 inches in length
6. Trim limbs close to the log

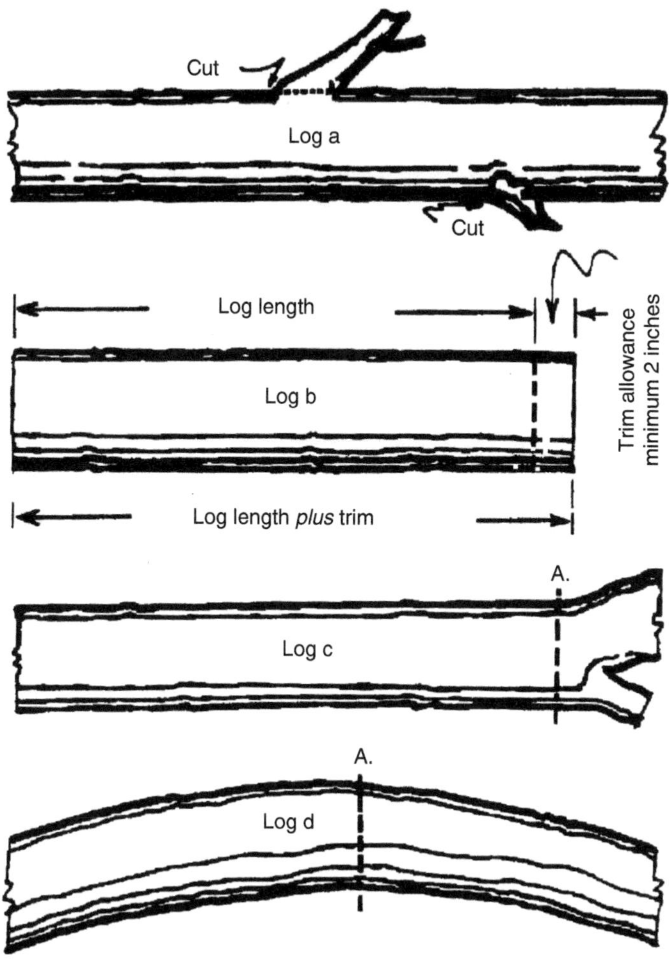

FIGURE 4. Proper log manufacturing (Stump, 1967). (a) Logs with handles do not roll. Cut limbs off close to log for easier handling. (b) Always cut logs at least 2 inches longer. This is called trim allowance. Check with buyer before cutting logs to be sure of required trim allowance. (c) Log should be cut in back of crotch at A. (d) Cut at A to reduce or eliminate sweep in log. (e) Make two logs. Cut at A and B if each log can be made 8 feet or longer. (f) Cut at A to avoid loss of footage. (g) Keep logs as free from defects as possible. Do not cut at A. Cut at B and C to make two clear logs. (h) Cutting at A instead of B yielded an additional 50 board feet of lumber.

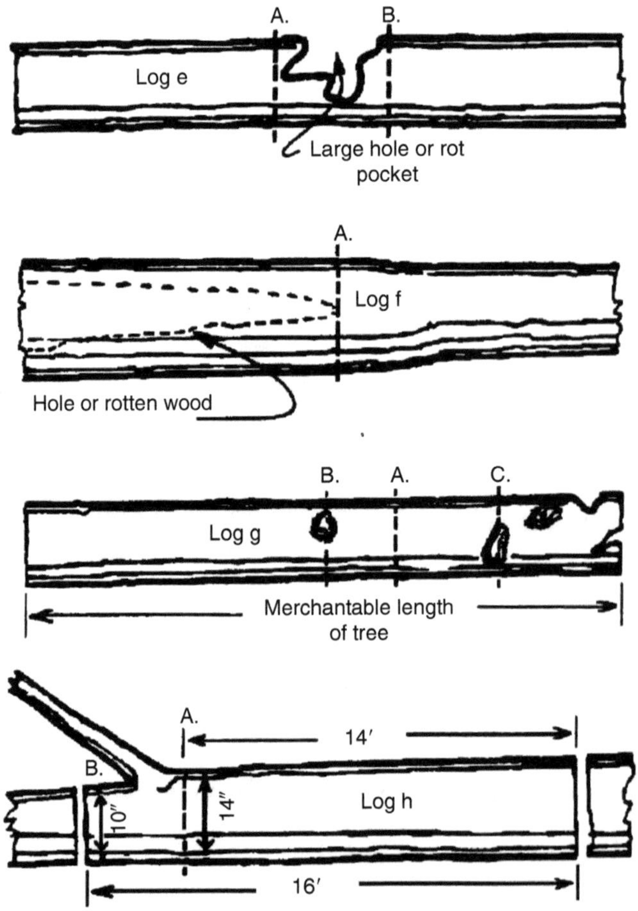

FIGURE 4. *(continued)*

7. If the base of a tree is hollow, taking a short log first may make the next log higher in quality. If a large hole is present at the base of the tree, cutting a long log would result in a loss of quality, footage, and money
8. Standard log lengths are 8, 10, 12, 14, and 16 feet for softwoods and both even and odd lengths between 8 and 16 feet for hardwoods. After a few logs are cut, have the log buyer from the sawmill come to check your logs and show you the mill's method of determining volume

Figure 4a-h show methods for properly manufacturing sawlogs that will be processed into standard lumber products.

3.4. Log Loading and Unloading Safety Procedures

Safety should always be the number one priority when dealing with heavy, movable materials such as logs. In 1989, the Occupational Safety and Health Administration

(OSHA) (OSHA, 1989) listed proposed regulations pertaining to safety in logging operations. Within these regulations, log loading and unloading procedures are identified. The final ruling on the proposed regulations has not been made as of the date of this publication. However, it is believed these will not be changed. "Safety First" is the number one priority when doing tree removal work and processing the resulting material into usable products. These OSHA safety procedures for chainsaw operation as well as log loading and unloading should always be followed.

For additional information on chainsaw safety procedures and log loading and unloading safety procedures contact:

American Forest and Paper Association
1111 19th Street NW
Suite 800
Washington, DC 20036
Phone: 202–463–2766
Fax: 202–463–2791

3.5. Log Quantities and Volume Determination

The sawmill will normally require a certain quantity of logs before a log truck is sent to your yard to pick them up. A standard log truck can legally transport about 2500 to 3500 board feet of logs measured by the international 1/4 inch log scale rule. One board foot is a piece of wood that is 12-inches wide by 12-inches long by 1-inch thick.

The international 1/4 inch log scale rule is one of several log scales used in the United States to estimate the amount of lumber on a board foot basis that can be sawn from logs. To determine what log scale is used in your area, contact a local sawmill. For the purpose of this publication, we have assumed it is the international 1/4 inch log scale rule.

The easiest way to estimate board foot volumes of logs is to measure them with a scaling stick. Although scaling logs might sound complicated, it is quite easy. In order to scale logs, you need to know several things:

1. Inside bark diameter of small end of log
2. Length of log to nearest foot
3. Estimated amount of nonsolid wood in logs (i.e., rot, decay, crookedness, and so on)

A scaling stick is a tool that enables you to determine the board foot volume of a log. It lists volumes based on the inside bark diameter of the small end of the log and its length. Thus, if a log measures 18 inches in diameter at the small end and it is 12 feet in length, you can refer to the scaling stick to determine that the board foot volume of the log is 170 board feet. If a scaling stick is not available, volume can still be determined by referring to a log volume table (see Table 1).

Both the scaling stick and table volumes assume that the log is solid and has no defects. If defects are present or a section of the log is missing, a volume reduction is necessary. To estimate this reduction, ask the log buyer to explain how to estimate

Table 1. International 1/4 Inch Log Scale Table[a]

Small end diameter (Inches)	Length of log (feet)					
	6	8	10	12	14	16
	Contents in board feet					
6	5	10	10	15	15	20
7	10	10	15	20	25	30
8	10	15	20	25	35	40
9	15	20	30	35	45	50
10	20	30	35	45	55	65
11	25	35	45	55	70	80
12	30	45	55	70	85	95
13	40	55	70	85	100	115
14	45	65	80	100	115	135
15	55	76	95	115	135	160
16	60	85	110	130	155	180
17	70	95	125	150	180	205
18	80	110	140	170	200	230
19	90	125	155	190	225	260
20	100	135	175	210	250	290
21	115	155	195	235	280	320
22	125	170	215	260	305	355
23	140	185	235	285	335	390
24	150	205	255	310	370	425
25	165	220	280	340	400	460
26	180	240	305	370	435	500
27	195	260	330	400	470	540
28	210	280	355	430	510	585
29	225	305	385	465	545	630
30	245	325	410	495	585	675

[a]USDA Forest Service, 1989.

volume reductions or refer to the United States Department of Agriculture (USDA) Forest Service publication on grading hardwood logs (Kenna, 1981). This publication describes the process of log grading, log scaling, and methods for determining scaling reductions. Copies may be obtained by contacting the USDA Forest Service:

USDA Forest Service
1720 Peachtree Street, NW
Atlanta, GA 30367
Phone: 404–347–7206 or 4243
Fax: 404–347–2776

Remember, you will not be measuring logs as accurately as a sawmill. The purpose of your log scaling is to give you an idea of how much log volume you have.

Log scaling sticks and tally sheets (forms for recording log volumes) can be purchased from the following companies:

Ben Meadows Company	Forestry Suppliers, Inc.
3581 Broad Street	205 W. Rankin Street
PO Box 80549	PO Box 8397
Atlanta, GA 30366	Jackson, MS 39284–8397
Phone: 770–455–0907	Phone: 601–354–3565
Fax: 770–457–1841	Fax: 601–355–5126

3.6. Storage Requirements

As stated earlier, in order for a sawmill to pick up a load of sawlogs, the mill will need a predetermined quantity of logs. This quantity will depend on the size of the truck. Therefore, the sawmill will give you an estimated volume figure that their log truck can legally handle. For example, if a sawmill's truck can safely transport about 3000 board feet of oak logs at one time, and one 18 inch × 12-foot log is 170 board feet, then about 18 oak logs of this size would be the approximate load for this truck.

Maximum load limits are based on weight. Consequently, the quantity of logs shipped on a log truck will vary depending on the species mix, moisture content, log size, and so on. For example, red oak logs are heavier than basswood; therefore, fewer red oak logs of the same size can be transported than the same size basswood logs. Keep in mind that you will need a storage area for the logs somewhere in the municipality. Depending on topography, one-quarter acre of flatland could be sufficient, provided that a log truck can easily park next to the logs.

4. Marketing Merchantable Sawlogs

4.1. Commercial Sawmill Operations

Although typical production-oriented sawmills shy away from street tree logs, some, particularly the small mills (such as a mill that saws less than 1.5 million board feet per year), may purchase them. Normally, sawmills located near urban and suburban areas are relatively small and may be best suited for your needs. In fact, some urban and suburban mills actively buy street tree logs. There are many ways to locate potential sawmills interested in buying street tree logs. The easiest ways include contacting the following:

- State forestry offices
  - Sawmill directories
  - Marketing bulletins
- Consulting foresters
- Forestry/wood products extension offices
- Resource conservation and development (RC&D) program

4.1.1. State Forestry Offices

Most state forestry offices have staff specialists who work closely with the forest products industries of the state. These individuals may be able to direct you to sawmills that buy street tree logs. When contacting a state forestry office, ask to talk with a forest product utilization and marketing specialist. These specialists generally are the best source of assistance in helping you make contact with sawmills.

A service that most state forestry agencies provide is the publication of a statewide sawmill directory. Normally, these directories list sawmills by county and include helpful information about each. Consulting this type of directory is a good way to locate sawmills in your area. Contact your state forestry office for a copy (see Appendix A).

Another service that some state forestry agencies offer is a monthly or quarterly marketing bulletin. State marketing bulletins serve as an avenue that connects suppliers and buyers of wood products with each other. Advertisements about your available logs can be placed in your state's marketing bulletin. Normally, there is no charge for advertising.

4.1.2. Consulting Foresters

Consulting foresters provide forestry services and expertise to landowners for a fee agreed to by the landowner and forester. Consulting foresters are usually familiar with sawmills in their work area. They may be able to direct you to potential sawmills interested in purchasing your logs. Many state forestry offices maintain listings of consulting foresters in their individual states. Also, the Association of Consulting Foresters of America maintains a detailed directory of its members' services. A copy can be purchased by contacting them at the following address:

The Association of Consulting Foresters of America, Inc.
1403 King Street
Alexandria, VA 22314–2714
Phone: 703–548–0990
Fax: 703–548–6395

4.1.3. Forestry/Wood Products Extension Offices

Many land-grant universities employ forestry/wood product extension specialists as part of their forestry/wood technology programs. These professionals can provide information pertaining to sawmills located in your area.

4.1.4. Resource Conservation and Development Program

The RC&D program is a national program that assists rural communities in improving their economies through wise use and development of natural resources. Each RC&D area has a full-time coordinator who oversees the daily program operations. Many coordinators are familiar with sawmills in their area (which usually consists of several counties). They may be able to provide you with names and addresses of sawmills that might purchase your logs.

4.2. Local Crafters/Hobbyists/Custom Sawmillers

Beyond commercial sawmills, many hobbyists and crafters are interested in purchasing unique logs for their hobbies and/or profession. Those who have portable bandmills are often interested in purchasing small quantities of sawlogs to cut into lumber. For example, spalted logs are highly desirable for their beauty and income potential once turned into finished products. Consequently, hobbyists or crafters seeking this type of material can become a good market for your logs. Some ways of locating these people are:

1. State forestry offices
2. Sawmill equipment manufacturers
3. Craft organizations

4.2.1. State Forestry Offices

Your state forestry agency's forest products utilization and marketing specialist may be able to direct you to potential crafters/hobbyists/custom sawmillers who are interested in street tree logs.

4.2.2. Sawmill Equipment Manufacturers

Portable or small-sized sawmill manufacturers usually maintain lists of customers that have bought their mills. Often, they are willing to share these lists with potential customers of their mills. If you contact these manufacturers and explain that you want to market sawlogs and are looking for potential buyers, they may share their lists with you. Some manufacturers of portable sawmills are:

Breezewood, Inc.
PO Box 266
Reynoldsville, PA 15851
Phone: 814–653–9500
Fax: 814–653–8895

Canadian Board Master
RR 2
Marmora, Ontario
Canada KOK2MO
Phone: 613–472–2122
Fax: 613–472–5499

Hurdle Machine Works, Inc.
16195 Highway 57
Moscow, TN 38057
Phone: 901–877–6251
Fax: 901–877–6260

Meadow Mobil Sawmill
Meadow Mill Company
PO Box 1288
North Wilkesboro, NC 28659
Phone: 910–838–2282
Fax: 910–667–6501

Mighty-Mite Industries
3931 NE Columbia Boulevard
Portland, OR 97220
Phone: 503–288–5923
Fax: 503–288–5582

Mobile Dimension
Mobile Manufacturing Co.
798 NW Dunbar Ave.
PO Box 250
Troutdale, OR 97060

Phone: 503–666–5593
Fax: 503–661–7548

Timber Harvester
1310 Wat-Gen Road
Waterloo, NY 13165
Phone: 315–539–5095
Fax: 315–539–8000

Timberking
1331 North Topping Ave.
Kansas City, MO 64120
Phone: 816–483–1007
Fax: 816–483–7203

Woodland Portable Sawmills
PO Box 903
Poplar Bluff, MO 63902
Phone: 573–785–3810
Fax: 573–785–0962

Wood-Mizer Products, Inc.
8180 W. 10th Street
Indianapolis, IN 46214
Phone: 317–271–1542
Fax: 317–273–1011

4.2.3. Craft Organizations

Most states have some type of statewide craft organization. Usually, these organizations include crafters who produce specialty wood products. Some of these individuals produce lumber for their own needs. If not, they purchase it. Regardless, these individuals may be able to direct you to small-scale sawmill operators who are interested in purchasing street tree logs.

4.3. In-house Markets: Viable Options for Municipalities

Today's portable sawmill technology allows the option for municipalities to produce products from street tree removals for their own use. These portable mills are relatively easy to operate and maintain and are usually priced in $12,000 to $20,000+ range. These mills can produce landscape ties, truck bed material, fence posts, fencing, timbers, tree stakes, and a broad range of lumber products to satisfy municipal needs. The advantages of utilizing this option can be quite considerable, depending on the municipalities' annual need for the kinds of raw products that can be manufactured from the tree resources available. Direct cost purchases of the host of primary products then can be largely avoided with the purchase and use of portable sawmill technology. In order to properly evaluate this option, several factors must be analyzed first. (1) Annual municipal demand for raw products that can be produced as well as the regular costs of those products; (2) thorough knowledge and understanding of portable sawmilling technology as well as purchase, operational, and maintenance costs of its use; and (3) present costs of existing log use or nonuse. The portable sawmill manufacturers can provide detailed information about costs, production rates, maintenance, and use of these machines.

Many of today's portable sawmills are small-scale bandmills. Detailed information on portable sawmills, such as descriptions, costs, and operation and production rates, are available directly from portable sawmill manufacturers. Those manufacturers are listed under sawmill equipment manufacturers in this chapter.

Appendix A: State Forestry Offices

Most state forestry agencies have staff personnel with expertise in forest product utilization and marketing. These specialists will be able to assist you in locating sawmills in your area. Furthermore, most state forestry agencies publish some type of forest product directory(s) that could be of assistance to you. These directories have traditionally been oriented toward sawmills and primary processors of wood products. Listed below are the names of the directories that include sawmills and manufacturers of primary wood products. To speak with a forest product utilization and marketing specialist or to obtain copies of directory(s), contact the appropriate agency.

Alabama
Directory of Alabama's
Forest Industry
Forestry Commission
6650 Old Highway 31 N.
Gardendale, AL 35071
Phone: 205-631-2552
Fax: 205-631-9128

Alaska
Forest Products Manufacturers Directory
Division of Forestry
PO Box 7-005
Anchorage, AK 99510
Phone: 907-762-2117
Fax: 907-561-6659

Arizona
Wood Products Trade Directory
State Land Department
Forestry Division
3650 Lake Mary Road
Flagstaff, AZ 86001
Phone: 520-774-1425
Fax: 520-779-2143

Arkansas
Forest Industry Directory
Arkansas Forestry Commission
PO Box 3821
Rosevette Road
Little Rock, AR 72204
Phone: 501-664-2531
Fax: 501-664-5906

California
Forest Products Marketing Directory
Department of Forestry and Fire Protection
PO Box 944246
Sacramento, CA 94244-2460
Phone: 916-653-9449
Fax: 916-653-8957

Colorado
Forest Products Directory
State Forest Service
203 Forestry Building
Colorado State University
Fort Collins, CO 80523
Phone: 970-491-6303
Fax: 970-491-7736

Connecticut
Primary Processing Directory
Department of Environmental Protection
Division of Forestry
State Office Building, Room 260
79 Elm Street
Hartford, CT 06102-5066
Phone: 860-566-5348
Fax: 860-566-7921

Delaware
Primary Processors Directory
Department of Agriculture
Forestry Section
2320 South Dupont Highway
Dover, DE 19901
Phone: 302-739-4811
Fax: 302-697-6287

Florida
Wood Using Industry Directory
Department of Agriculture and Consumer Services
Division of Forestry
3125 Conner Boulevard
Tallahassee, FL 32399-1650
Phone: 850-488-6000
Fax: 850-488-0863

Georgia
Wood Using Industry Directory
Forestry Commission
Forest Products U, M & D
PO Box 819
Macon, GA 31298-4599
Phone: 912-751-3521
Fax: 912-751-3522

Hawaii
Division of Forestry and Wildlife
1151 Punchbowl Street
Honolulu, HI 96813
Phone: 808-587-0166
Fax: 808-587-0160

Idaho
Directory of Idaho Wood Products Manufacturers
Forest Products Department
University of Idaho
Moscow, ID 83843-1132
Phone: 208-885-6126
Fax: 208-885-6226

Illinois
Sawmill Directory
Department of Conservation
Division of Forest Resources
600 North Grand Avenue
Springfield, IL, 62706
Phone: 217-782-2361
Fax: 217-785-8405

Indiana
Primary Forest Products Industries
Department of Natural Resources
Division of Forestry
PO Box 283
Connersville, IN 47331
Phone: 317-825-6769
Fax: 317-825-6769

Iowa
State Forestry Nursery
2404 South Duff Street
Ames, IA 50010
Phone: 515-233-1161
Fax: 515-233-1131

Kansas
Primary Processing Directory
State and Extension Forestry
Harold G. Gallaher Building
2610 Clafin Road
Manhattan, KS 66502
Phone: 785-537-7050
Fax: 785-539-9584

Kentucky
Primary Wood Industry
Directory
Division of Forestry
627 Comanche Trail
Frankfort, KY 40601
Phone: 502-564-4496
Fax: 502-564-6553

Louisiana
Directory of Louisiana's Primary Forest Industries
Department of Agriculture and Forestry Officer
of Forestry
PO 1628
Baton Rouge, LA 70821-1628
Phone: 225-925-4500
Fax: 225-922-1356

-or-
Department of Agriculture and Forestry
Office of Marketing
PO Box 3334
Baton Rouge, LA 70821-3334
Phone: 225-922-1280
Fax: 225-922-1289

Maine
Primary Processor Mill List
Department of Conservation Bureau of Forestry
State House Station #22
Augusta, ME 04333
Phone: 207-289-4995
Fax: 207-289-2400

Directory of Forest
Products Industries
College of Forest Resources
201 A Nutting Hall
University of Maine
Orono, ME 04469
Phone: 207-581-2857
Fax: 207-581-2858

Maryland
Primary Directory
Forest, Park, and Wildlife Service
Tawes State Office Building
580 Taylor Avenue
Annapolis, MD 21401
Phone: 410-974-3776
Fax: 410-974-5550

Massachusetts
Department of Environmental Management
Division of Forests and Parks
100 Cambridge Street
Suite 1900
Boston, MA 02202
Phone: 617-727-3180
Fax: 617-727-9402

Michigan
Forest Products Producers, Truckers, Brokers, and Dealers
Wood Products in Michigan: A Directory of Mills and Manufacturers
Department of Natural Resources
Forest Management Division
Stevens T. Mason Building
PO Box 30028
Lansing, M1 48909
Phone: 517-373-1275
Fax: 517-373-2443

Minnesota
Forest Products Directory
Department of Natural Resources
Division of Forestry
DNR Bldg., Box 44
500 Lafayette Road
St. Paul, MN 55155-4044
Phone: 651-296-6491
Fax: 651-296-5954

Mississippi
Directory of Primary Industry
Forestry Commission
Suite 300
301 N. Lamar Street
Jackson, MS 39201
Phone: 601-359-1386
Fax: 601-359-1349

Missouri
Directory of Primary Wood Processors
Forestry Division
2901 West Truman Boulevard
PO Box 180
Jefferson City, MO 65102
Phone: 573-751-4115
Fax: 573-893-6079

Montana
Forest Products Directory
Department of State Lands
Division of Forestry
2705 Spurgin Road
Missoula, MT 59801
Phone: 406-542-4300
Fax: 406-542-4217

Nebraska
Primary Processors Directory
Department of Forestry, Fisheries, and Wildlife
101 Plant Industry Building
Lincoln, NE 68503
Phone: 402-472-3645
Fax: 402-472-2964

Nevada
Division of Forestry
123 West Nye Lane
Carson City, NV 89710
Phone: 775-687-4350
Fax: 775-687-4244

New Hampshire
Directory of Sawmills
UNH Cooperative Extension
Pettee Hall
University of New Hampshire
Durham, NH 03824
Phone: 603-862-1096
Fax: 603-862-1585

New Jersey
Primary Processing Directory
Division of Parks and Forestry
Forestry Services
PO Box 404
501 East State St.
Trenton, NJ 08625
Phone: 609-292-2531
Fax: 609-984-0378

New Mexico
Wood Industry Directory
Forestry and Resource Conservation Division
PO Box 1948
Santa Fe, NM 87504
Phone: 505-827-5835
Fax: 505-827-3903

New York
Directory of Primary Wood-Using Industries
Bureau of Land Resources
Environmental Conservation Department
50 Wolf Road, Room 404
Albany, NY 12233-4252
Phone: 518-457-7431
Fax: 518-457-1088

North Carolina
Buyers of Forest Products
Division of Forest Resources
PO Box 27687
Raleigh, NC 27611
Phone: 919-733-2162
Fax: 919-733-0138

North Dakota
Sawmill Directory
Forest Service
PO Box 604
Lisbon, ND 58054
Phone: 701-683-4323
Fax: 701-683-5895

Ohio
Sawmill Directory
Department of Natural Resources
Division of Forestry
Fountain Square
Columbus, OH 43224
Phone: 614-265-6703
Fax: 614-265-6709

Oklahoma
Forest Industry Directory
Department of Agriculture
Forestry Division
2800 N. Lincoln Boulevard
Oklahoma City, OK 73105
Phone: 405-521-3864
Fax: 405-521-4912

Oregon
Department of Forestry
2600 State Street
Salem, OR 97310
Phone: 503-945-7392
Fax: 503-945-7212

Pennsylvania
Sawmill Directory
Department of Environmental Resources
Bureau of Forestry
PO Box 8552
Harrisburg, PA 17105-8552
Phone: 717-787-2105
Fax: 717-783-5109

Rhode Island
Primary Producers Directory
Division of Forest Environment
Arcadia Headquarters
260 Arcadia Road
Hope Valley, RI 02832
Phone: 401-539-2356

South Carolina
Directory of Primary Forest Industries
Forestry Commission
PO Box 21707
Columbia, SC 29221
Phone: 803-896-8860
Fax: 803-798-8097

South Dakota
Directory of Wood Products Industries
Department of Agriculture
Division of Forestry
523 E. Capitol Ave.
Pierre, SD 57501-3182
Phone: 605-773-4260
Fax: 605-773-5926

Tennessee
Directory of Forest Industries
Department of Agriculture
Division of Forestry
Ellington Agricultural Center
Box 40627, Mel Rose Station
Nashville, TN 37204
Phone: 615-360-0733
Fax: 615-360-0756

Texas
Forest Products Industries in Texas
Texas Forest Service
Texas Forest Products Laboratory
PO Box 310
Lufkin, TX 75902-0310
Phone: 409-639-8180
Fax: 409-639-8185

Utah
Division of State Lands and Forestry
355 W. North Temple
3 Triad Center, Suite 400
Salt Lake City, UT 84180-1204
Phone: 801-538-5508
Fax: 801-355-0922

Vermont
Directory of Sawmills & Veneer Mills
Department of Forests, Parks, and Recreation
Agency of Environmental Conservation
103 South Main Street
Waterbury, VT 05676
Phone: 802-244-8716
Fax: 802-244-1481

Virginia
Forest Products Industry Directory
Department of Forestry
McCormick Road
PO Box 3758
Charlottesville, VA 22903
Phone: 804-977-6555
Fax: 804-296-2369

Washington
Forest Industry Mill Directory
Department of Natural Resources
Forest Landowner Assistance
PO Box 407046
Olympia, WA 98504-7046
Phone: 360-902-1650
Fax: 360-902-1788

West Virginia
The Forest Industry
Department of Commerce, Labor, and Environmental Resources
Division of Forestry
1900 Kanawha Boulevard East
Charleston, WV 25305
Phone: 304-558-2788
Fax: 304-558-0143

Wisconsin
Primary Directory
Department of Natural Resources
3911 Fish Hatchery Road
Route 4
Madison, WI 53711
Phone: 608-275-3276
Fax: 608-275-3338

Wyoming
Timber Industry Directory
State Forestry Division
1100 W. 22nd Street Cheyenne, WY 82002
Phone: 307-777-7586
Fax: 307-637-8726

References

Cesa, E. T., Lempicki, E. A., and Knotts, J. H., 1994, *Recycling Municipal Trees, A Guide for Marketing Sawlogs from Street Tree Removals in Municipalities*, NA-TP-02-94, USDA Forest Service, NA S&PF, Morgantown, WV, p. 52.

Kenna, K., 1981, *Grading Hardwood Logs for Standard Lumber*, USDA Forest Service, Southern Region, Atlanta, GA, p. 19.

NJ Division of Parks and Forestry, 1996, *Trees in Crisis, A Statewide Assessment of New Jersey's Street Trees* NJ Forest Service.

Occupational Safety and Health Administration (OSHA), 1989, *Federal Register* **54**(83): 19.

Stump, W. G., 1967, *More Money for Your Logs When Properly Cut*, USDA Forest Service, NAS&PF, Upper Darby, PA, p. 14.

United States Department of Agriculture (USDA) Forest Service, 1989, *Reference Handbook for Foresters*, NA S&PF, NA-FR-15, Morgantown, WV, p. 35.

Chapter 23

Tree City USA

John Rosenow and Mary Yager

1. Introduction

Trees make a world of difference in communities: the difference between sandy, dirty lots and shaded parks for baseball, picnics, and quiet walks; and between steamy, sun-baked streets, and friendly, shady neighborhoods. Trees help conserve energy, give wildlife a home, and increase property values. They also clean the air we breathe, hold the topsoil, and keep rivers running clear. But trees do not just happen. Communities need to plant trees and provide for their care.

The Tree City USA program provides incentive to communities to plant and care for trees. Tree City USA can provide benefit to a community regardless of its size and the current status of its community forestry program.

2. History

Tree City USA was created in 1976 as a joint bicentennial project of the National Arbor Day Foundation (NADF), the US Department of Agriculture (USDA) Forest Service, and the National Association of State Foresters. One of the first programs of the then-fledgling Arbor Day Foundation, Tree City USA helped to expand the concept of Arbor Day from tree planting to total tree care. The foundation, Forest Service, and state foresters recognized a serious need in America's towns and cities—a need for more and better tree care. Together, they designed and launched a community improvement program to meet that need and improve the level of tree care in communities. The program combined the strengths of the three organizations in technical assistance, information exchange, and public education expertise. In the program's first year, 42 communities in 16 states received Tree City USA designation and helped to usher in a new era of community tree care (NADF, 1996).

John Rosenow and Mary Yager National Arbor Day Foundation, Nebraska City, Nebraska 68410.
Urban and Community Forestry in the Northeast, 2nd ed., edited by, J. E. Kuser.

Steady growth and a number of milestones marked the 28 years that followed the introduction of the Tree City USA program. Deneke and Grey (1986, p. 216) wrote in the second edition of their classic text, *Urban Forestry*, "Its (Tree City USA) impact as an incentive for development of local municipal forestry programs has been significant." Tree City USA indeed had grown in significance in those 10 years to include 679 communities in 43 participating states. Nineteen of the forty-two charter communities had recertified and maintained their Tree City USA status for each of the 10 years and were honored in 1986 for their accomplishment.

As the program grew, it became apparent that there was a need for easily digestible tree care information for busy community forestry volunteers and city officials. In 1988, a new education component was introduced—the *Tree City USA Bulletin*. The bimonthly bulletin, launched with a seed money grant from the Forest Service, features important issues in community forestry and tree care, including tree planting, pruning, and preservation tactics (NADF, 1995a).

Near the 15th year of the program, frequent requests from both longtime participating communities and state forestry personnel for a new component to stimulate continued interest spurred the foundation to develop a companion program. In 1991, the Tree City USA Growth Award was introduced to encourage communities to go beyond the basics and continually strive to improve their tree programs. Communities are recognized in four areas: education and public relations, partnerships, planning and management, and tree planting and maintenance. Participation in the Growth program has been strong with an average of 448 Tree City USAs qualifying annually (NADF, 1992, 1993c, 1994b, 1995, 1996, 1997, 1998, 1999, 2000, 2001, 2002, 2003.)

In 1994, another education component was introduced: the Tree City USA National Conference. Held biennially at Arbor Day Farm's Lied Conference Center in Nebraska City, the conference features exchange of ideas among Tree City USA foresters, tree board members, and city officials (NADF, 1996).

As more communities became active in tree planting and care and community groups developed to support local programs, the crucial role of utilities came to the forefront. Utilities can have a greater impact on community trees than any other single element. Clearing lines for safe and dependable utility transmission often has resulted in tree mutilation by topping. The Tree Line USA program was launched in 1992 in cooperation with the National Association of State Foresters to foster better tree care by utilities and promote cooperation between utilities and the communities they serve (NADF, no date-b).

Twenty-eight years into the program, more than 110 million Americans live in Tree City USAs; 3,007 towns, villages, townships, and cities were recognized as Tree City USAs for 2003. The four standards were purposely set to be obtainable by a community of any size. Communities of all sizes in all 50 states are involved in the program today. From Calvin, North Dakota, with a population of 26 to Los Angeles with its 3.7 million citizens, Tree City USA is attainable by all (NADF, no date-c).

3. Benefits of Tree City USA

Why have communities committed the resources necessary to achieve and maintain Tree City USA status? First, because they recognize the value of trees to the quality of life in their communities. Second, because they value the Tree City USA designation itself.

Trees in our communities beautify streets, homes, and parks. They improve air quality, reduce noise, and moderate temperatures (see Chapter 2, this volume). Urban trees attract songbirds (Chapter 25, this volume) and create natural elements in the environment. Trees help maintain soil moisture and mitigate storm water runoff by allowing rain to more slowly percolate into the soil. Trees add to the quality of life in communities, which in turn increases property values up to 15% in areas with well-tended trees. In many communities, tree programs have been an important stimulus for economic development. Towns and cities beautifully planted with trees are more likely to attract new businesses and are more appealing to tourists (NADF, no date-e).

Citizens often experience the benefits provided by trees without consciously recognizing trees as the source. They often fail to value the community forestry program and the dedicated community employees, officials, and volunteers who help make sure the trees receive the care they need.

The Tree City USA name and logo remind citizens of the value of trees and open the door to educate the public to the benefits of trees and how essential their care is to the quality of life in the community. The Tree City USA signs at community entrances tell visitors that the community cares about its environment. It is also an indication to prospective businesses that the quality of life may be better here. Tree City signs and the use of the logo (Fig. 1) on town stationery, trucks, annual reports, visitor information, and banners all contribute to a greater understanding by citizens and visitors.

Gaining and retaining Tree City USA recognition is a reward to the tree workers, managers, volunteers, tree board members, and others who work on behalf of better care of a community's trees. The general public, too, often shares a sense of pride that theirs is a Tree City USA. This may translate to better care of trees on private property and a willingness to support the program in the future.

In addition to helping educate the public, Tree City USA provides valuable tools for continuing the education of employees and volunteers. Resources from state forestry offices and state urban forestry councils include publications, conferences, and workshops. The foundation provides national conferences and the *Tree City USA Bulletins* (NADF, 1996).

The publicity and educational opportunities surrounding the Tree City USA award can help gain the support of city officials and the public. This support can be critical when the time comes to defend annual tree care budgets.

Tree City USA communities have successfully used their track record and knowledge to more successfully apply for grant money and emergency funds in the wake of natural disasters. Tree Cities can easily demonstrate the value of their tree resource and ability to manage funds for the good of the community.

4. How to Become a Tree City USA

To become a Tree City USA a community must meet four standards. The standards were designed to be flexible and allow individual communities to address the requirements in a way that fits local needs, while providing a general framework for a

FIGURE 1. Tree City USA logos.

strong community forestry program (NADF, no date-e). These standards were established to ensure that every qualifying community would have a viable tree management plan and program. Importantly, they were also designed so that no community would be excluded because of size.

3.1. Standard 1: Tree Board or Department

Someone must be legally responsible for the management of the community's public trees. This may be a professional forester or arborist, a forestry department, or a volunteer tree board. A tree board or commission is a group of concerned volunteer citizens charged by ordinance with developing and administering a comprehensive tree management program. A volunteer board is often the practical choice for smaller communities. *A Handbook for Tree Board Members* (Grey 1993) is a valuable resource available from the NADF.

A professional city forester or arborist is ideal for larger communities and may be supported by a volunteer advisory board. Smaller communities may find it feasible and desirable to partner with other communities to share a forester or contract with a forester from a large community for part-time assistance.

3.2. Standard 2: Tree Care Ordinance

The tree ordinance designates the tree board, forester/arborist, or forestry department and gives the designee the responsibility for writing and implementing an annual community forestry work plan. The ordinance determines public tree care policies for planting, care, and removal of street, park, and other public trees. Resources are available to assist communities in writing or improving a tree ordinance (NADF, 1991, 1993a).

3.3. Standard 3: Community Forestry Program with Annual Budget of at Least $2 per Capita

The essential, ongoing activity for the care of trees along streets, in parks, and in other public areas is the community forestry program. An annual work plan developed by the tree board or department outlines planting, pruning, insect and disease control, and removal work to be done. The work is carried out by city employees or commercial contractors. In small towns the tree board might actually implement the program. Funding and planning guidance is available in *Tree City USA Bulletins 29* and *34* (NADF, 1993b, 1994a).

3.4. Standard 4: Arbor Day Proclamation and Observance

This is the easiest and probably the most enjoyable standard to accomplish. An Arbor Day observance can be a simple tree-planting event or an all-day or all-week celebration complete with educational activities, contests, and awards. Obtain a *Celebrate Arbor Day* packet from the foundation for a sample proclamation and celebration ideas (NADF, no date-d).

The Tree City USA application process may be started by an interested citizen, city employee, or public official. Wherever it begins, support should be gained from city council, city staff, and citizens. Resources are available from state forestry and the Arbor Day Foundation to help gain widespread support (NADF, 1987, no date-e).

Help is also available in meeting the four standards. The urban and community forestry coordinator in your state forester's office is happy to work with communities in taking the first steps toward better community forestry.

A complete application should be submitted to the state forester by December 31st. Following certification by staff in the state forester's office, the application is forwarded to the foundation for final approval. Communities are notified by letter, An Arbor Day Foundation release announcing the award is made public. State foresters present the Tree City USA signs for community entrances, a flag, a plaque, and other awards at an Arbor Day ceremony in the community or a statewide awards program or conference. Following certification, communities are reminded annually of the need for recertification and the opportunity for a Growth Award.

5. Tree City USA Growth Award

Effective tree care is an ongoing process of growth and renewal, a program of planting and care that continues through the years. Tree City USA provides a solid foundation for that process. But community leaders realize that there is also a need to do more and a better way.

The Growth Award was created in 1991 to recognize environmental improvement and higher levels of tree care in Tree City USAs. It was designed to both recognize communities and provide ideas to help communities plan improvements to their forestry programs in the areas of planning and management, tree planting and maintenance, education and public relations, and partnerships.

Two initial requirements must be met before a Growth application can be considered. First, the community must be a recertifying Tree City USA. That is, it must have been a Tree City USA for at least the previous year. Second, the municipality's expenditures for its forestry program must be no less than those of the year before.

To receive a Growth Award, a recertifying Tree City must complete several activities in one or more of the four categories. Each activity is assigned a number of points from 2 to 10 based on degree of difficulty. A community must earn at least 10 points in a given year to receive the award (NADF, no date-a).

Growth activities encourage communities to adopt a systematic approach to tree care with points for implementing a computerized tree management system, a long-range management plan, a scheduled tree pruning program, or hazard tree program. Interpretive programs, tree care workshops, and youth education inspire a more supportive public. And partnerships with city departments, green industry, and utilities lead to cooperation. Qualifying communities receive Growth Award additions for their Tree City USA signs and a certificate of achievement enumerating their award-winning activities.

6. Tree Line USA

Partnership with utilities in the interest of better tree care was deemed so critical that it inspired a program of its own. Tree Line USA promotes the dual goal of dependable utility service and abundant healthy street trees. On the basis of the success of Tree City USA, three requirements were designed to be met by utilities wishing to be designated a Tree Line USA by the foundation. The three requirements were designed to give maximum benefits to both the utility and the communities it serves.

To qualify for Tree Line USA, a utility must formally adopt work practices for proper pruning (Shigo, 1990). These practices avoid topping, tipping, removing branch collars, and leaving long stubs. Each worker who performs line clearance, including contractors, must read and understand these practices and have a copy at every work site for quick reference.

Work practices also must be recommended for trenching and tunneling near trees to reduce the destruction of roots and injury to trees (Fazio, 1992). Each worker who performs trenching or tunneling for the utility, including contractors, must receive information about these practices.

When trenching and tunneling for the utility is provided by the developer, the utility must educate the developer on proper trenching and tunneling and include in the packet of utility instructions a copy of the *Trenching and Tunneling* pocket guide and the sample contract provisions available from the foundation (NADF, no date-b).

Annual documented training on following the work practices is required for all employees, contractors, and supervisors who do pruning or trenching/tunneling work for the utility. An arborist, forester, or other trained utility employee is designated by the utility to ensure that the training takes place and the work practices are followed.

An ongoing community tree-planting program must be sponsored by the utility. Utility employees may plant the tree, or the utility may fund tree planting by municipalities, volunteer groups, or homeowners. It is suggested that the utility should work toward an annual expenditure of at least 10 cents per customer.

One or more mailings that include education information about trees are made annually to all homeowner customers. These mailings can be in the form of bill inserts or customer newsletters and cover topics such as appropriate tree species for planting near utility lines, energy-efficient landscaping to reduce cooling and heating loads, and tips on how to prune trees safely (NADF, no date-f). Finally, the utility must sponsor or participate in an annual Arbor Day event.

The utility benefits from lower line clearance costs resulting from long-term savings from proper pruning versus topping and more low-growing trees being planted near utility lines. Other benefits include improved public relations, reduced tree mortality resulting from proper trenching/tunneling practices, and increased reliability of service because healthier trees mean fewer downed lines during storms. Better placement of trees helps moderate temperature and lowers peak air conditioning and heating demand, and more trees help to mitigate carbon dioxide produced by power plants that burn fossil fuels. The community benefits from all of this, as well as healthier and more abundant community forests and utility support for community tree planting, Arbor Day, and public education.

7. Conclusion

Tree City USA has been a long-term success (Fig. 2)—a success because of the strength and simplicity of the program and program's relationship-building quality. Communication among city foresters, tree board members, utility, and state foresters

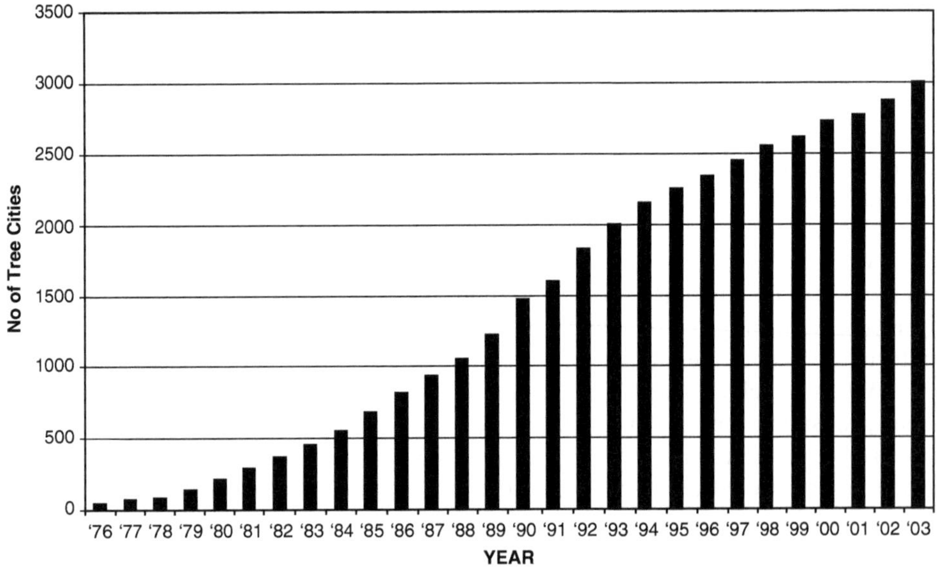

FIGURE 2. Tree City USA: Tree cities 1976–2003.

has led to spread of technical and educational information and exchange of ideas. This communicative relationship has also led to extensions and improvements in the program over the past years and promise of continued success in the future.

References

Deneke, F., and Grey, G., 1986, *Urban Forestry*, Wiley, New York, p. 216.

Fazio, J., 1992, *Trenching and Tunneling Near Trees: A Field Pocket Guide for Qualified Utility Workers*, National Arbor Day Foundation, Nebraska City.

Grey, G., 1993, *A Handbook for Tree Board Members*, National Arbor Day Foundation, Nebraska City.

National Arbor Day Foundation (NADF), 1987, *A Special Kind of Care, Audio-Visual Presentation*, National Arbor Day Foundation, Nebraska City.

National Arbor Day Foundation (NADF), 1991, How to write a municipal tree ordinance, *Tree City USA Bulletin* No. 9, National Arbor Day Foundation, Nebraska City.

National Arbor Day Foundation (NADF), 1992, *Annual Report and Directory*, September 1992, National Arbor Day Foundation, Nebraska City.

National Arbor Day Foundation (NADF), 1993a, Tree protection ordinances, *Tree City USA Bulletin* No. 31, National Arbor Day Foundation, Nebraska City.

National Arbor Day Foundation (NADF), 1993b, How to plan for management, *Tree City USA Bulletin* No. 29, National Arbor Day Foundation, Nebraska City.

National Arbor Day Foundation (NADF), 1993c, *Annual Report and Directory*, National Arbor Day Foundation, Nebraska City.

National Arbor Day Foundation (NADF), 1994a, How to fund community forestry, *Tree City USA Bulletin* No. 34, National Arbor Day Foundation, Nebraska City.

National Arbor Day Foundation (NADF), 1994b, *Annual Report and Directory*, National Arbor Day Foundation, Nebraska City.

National Arbor Day Foundation (NADF), 1995a, *Tree City USA Bulletins*, 40 Bulletin Set, National Arbor Day Foundation, Nebraska City.

National Arbor Day Foundation (NADF), 1995b, *Tree City USA Annual Report*, National Arbor Day Foundation, Nebraska City.
National Arbor Day Foundation (NADF), 1996, *Tree City USA Annual Report*, September 1996, National Arbor Day Foundation, Nebraska City.
National Arbor Day Foundation (NADF), No date(a), *Tree City USA Growth Award Application*, National Arbor Day Foundation, Nebraska City.
National Arbor Day Foundation (NADF), No date-b, *Tree Line USA Application*, National Arbor Day Foundation, Nebraska City.
National Arbor Day Foundation (NADF), No date (c), *Tree City USA Application*, National Arbor Day Foundation, Nebraska City.
National Arbor Day Foundation (NADF), No date (d), *Celebrate Arbor Day*, National Arbor Day Foundation, Nebraska City.
National Arbor Day Foundation (NADF), No date (e), *Keep A Great Thing Growing America, Tree City USA*, National Arbor Day Foundation, Nebraska City.
National Arbor Day Foundation (NADF), No date (f), *Plant the Right Tree in the Right Place* (*bill insert*), National Arbor Day Foundation, Nebraska City.
National Arbor Day Foundation (NADF), 1997, *Tree City USA Annual Report*, September 1997, National Arbor Day Foundation, Nebraska City.
National Arbor Day Foundation (NADF), 1998, *Tree City USA Annual Report*, September 1998, National Arbor Day Foundation, Nebraska City.
National Arbor Day Foundation (NADF), 1999, *Tree City USA Annual Report*, National Arbor Day Foundation, Nebraska City.
National Arbor Day Foundation (NADF), 2000, *Tree City USA Annual Report*, National Arbor Day Foundation, Nebraska City.
National Arbor Day Foundation (NADF), 2001, *Tree City USA Annual Report*, National Arbor Day Foundation, Nebraska City.
National Arbor Day Foundation (NADF), 2002, *Tree City USA Annual Report*, National Arbor Day Foundation, Nebraska City.
National Arbor Day Foundation (NADF), 2003, *Tree City USA Annual Report*, National Arbor Day Foundation, Nebraska City.
National Arbor Day Foundation (NADF), 2004, *Tree City USA Annual Report*, National Arbor Day Foundation, Nebraska City.
Shigo, A., 1990, *Pruning Trees Near Electric Utility Lines: A Field Pocket Guide for Qualified Line-Clearance Tree Workers*, Shigo & Trees, Associates, Durham, New Hampshire.

Chapter 24

A Case Study of Paramus, New Jersey: A "Tree City USA" Community

William R. Comery

1. Introduction

Paramus, New Jersey was there at the birth of the "Tree City USA" award and has been part of this national recognition program ever since. For over 30 years, "Tree City USA" signs, plaques, and flags have been prominently displayed throughout the borough. Official stationary, entrance signs to the community, public buildings and schools, and municipal vehicles display the "Tree City USA" identification. This award has caught the attention of the American people and brought trees to the forefront of public awareness. In Paramus, "Tree City USA" is a household phrase because people appreciate the importance and beauty of trees.

2. History of Paramus, New Jersey

Paramus, New Jersey is a suburban community, 8 miles west of New York City via the George Washington Bridge. For most of its life small farms dotted the landscape. From 1713 to 1930 the major crops grown in Paramus ranged from celery to rhubarb to raspberries. The population was Dutch, English, and later German immigrants at the end of the 19th century. George Washington slept here.

In 1931, the New Jersey countryside became accessible to the people of New York City with the opening of the George Washington Bridge. This was the beginning of change in Paramus. Ten years earlier Paramus was incorporated as an independent borough. The 1200 residents farmed the land and after the bridge opened could easily sell their produce at the New York City market. The farms continued to prosper

William R. Comery, C.L.A. Shade Tree and Parks Commission, Borough of Paramus, Paramus, New Jersey 08625.
Urban and Community Forestry in the Northeast, 2nd ed., edited by, J. E. Kuser.

FIGURE 1. The Shade Tree and Parks Commission logo welcomes all visitors to Paramus. Photo, Williams R. Comery.

and survived the war years. Returning GI's sought out affordable tract housing in Paramus for their young families. The population exploded to 3000 residents as per the 1950 census.

As early as 1948, tree-preservation buffer zones were established to protect the residential community by setting apart commercial properties. Farm land abutting the two major bisecting highways (State Highway Rt. 4 and Rt. 17) were sold and developed for commercial use. Other farms were gradually sold and subdivided for residential use. *Forty percent* of Paramus had no trees because of the farms. The only significant trees to be found were those next to the farmhouse or in hedge rows. The remainder of the land had stands of native low land trees. There were small pockets of hardwood trees scattered throughout the borough but no major hardwood forest.

In the early 1950s two large tracts of land were sold to developers who built the earliest regional shopping centers in the country. This changed the destiny of Paramus forever. Today, Paramus is known as one of the top two retail centers in the United States rivaled only by Fifth Avenue in New York City. As those two parcels of land were sold to developers the Paramus government saw a need to further protect and enhance the integrity of the community.

3. Shade Tree Commission Established

The Paramus Shade Tree Commission was established in 1951 by ordinance of the Mayor and Council, to protect and preserve existing trees and establish new tree plantings in the community. This three-member board of commissioners were the "sole and

FIGURE 2. Paramus has planted and maintains over 600 street trees on State Highway #4 and State Highway #17, Photo, William R. Comery.

exclusive" authority over trees in the municipal right-of-way as defined in New Jersey State Statue N.J.R.S. 40:64-14. One of its first official acts was to require a uniform street tree planting on all new streets. Planting and species selection were controlled by the commission from the beginning. This insured quality trees and proper planting standards. Contracts were let for tree planting until 1957 when the commission hired its first arborist and crew. By 1965, the population of Paramus grew to almost 23,000 residents. The commission was now planting over 1000 new street trees a year. In order to maintain that pace and contain costs, it established its own tree farm in the early 1960s. As the new trees matured, a comprehensive maintenance program was developed. This included fertilization, insect and disease control, trimming, and routine inspection. Diseased and/or hazardous trees along the streets were removed and replaced.

The commission's responsibilities were expanded to include municipal parks and public properties in the late 1960s. With these new responsibilities, the board of commissioners was enlarged to five members, and the name was changed to Shade Tree and Parks Commission. Also, additional employees were hired to handle this added work. Today, the commission maintains over 20,000 street trees, 12 Municipal Parks with a land area of 160+ acres, a 1200 tree municipal nursery, and two leaf and organic material composting sites. With Paramus 95% developed, over 650 new street trees were planted in 1996 by the Shade Tree and Parks Commission.

"Keep America Beautiful" was a national recognition program during the Lyndon B. Johnson administration. From 1964 to 1968, Ladybird Johnson administered "Keep America Beautiful" and acknowledged Paramus for its outstanding beautification achievements. This was the first national recognition award received by

FIGURE 3. Paramus trims over 2,500 street each year. There are over 32,000 street along Paramus streets. Photo, William R. Comery.

the borough for the work of its Shade Tree and Parks Commission. Projects established during this period continue to flourish and have established Paramus and its Shade Tree and Parks Commission as a paramount leader in beautification and environmental awareness.

4. "Tree City USA" Award

In 1976, Americans celebrated their 200th birthday, and these year-long festivities emphasized all that is good in America. The National Arbor Day Foundation celebrated the Bicentennial with the inaugural granting of the "Tree City USA" award. Paramus, New Jersey was one of the 19 communities nationwide to receive this award. The State Forester's office administered the award in New Jersey for the Foundation. From its inception, the State Foresters office and the borough treated this recognition as a prestigious award. This second taste of national recognition continues to stimulate and motivate our efforts so that new and creative projects are undertaken.

When the State Foresters office notified Paramus of the National Arbor Day Foundation "Tree City USA" award, all Paramus had to do was fill out the application. This community had already exceeded the eligibility requirements. The commission, its ordinance, regulations, budget, and Arbor Day observance predated the award requirements for two decades. The State Forester's office had no trouble certifying Paramus as a "Tree City USA."

Recognition as an original "Tree City USA" recipient made us a role model for other communities. As more and more neighboring towns became "Tree City USA" Paramus developed new and innovative programs, especially Arbor Day projects to keep ahead of our fellow "Tree City USA" communities. Each year Paramus included more groups and organizations to help celebrate Arbor Day. Cosponsors for Arbor Day have included Paramus Public Schools, private schools, Bergen Community College, service organizations such as Rotary, Lions, Elks, Kiwanis, Garden Clubs, Woman's Clubs, PTA, Unico, Knights of Columbus, Veterans organizations, and others. The business community has also been active in all our programs, & notably Arbor Day. The Shade Tree and Parks Commission has recognized the business community by giving landscape excellence awards, certificates of appreciation, and assisting them with their own beautification projects. The result has been a total community effort in the advancement of trees for Paramus and the environment.

Flying the "Tree City USA" flag at town hall means "TREE POWER." Elected officials are always eager to boast that their community is a "Tree City USA." They

FIGURE 4. Arbor Day is celebrated each year in Paramus. All schools participate in planning trees. Photo, William R. Comery.

become willing participants in the promotion of arboriculture in order to continue to receive the "Tree City USA" designation. The positive effect of this award on community leaders turned them into tree advocates. "TREE POWER" in Paramus has resulted in increased public awareness, funding, tree care, and appreciation of community trees. This has been demonstrated specifically in Planning and Zoning Ordinances protecting trees. Trees on undeveloped properties must be reviewed by the Shade Tree and Parks Commission before the Planning Board or Zoning Board of Adjustment finalizes site development. The opinion of the commission is held in highest regard and evident by the conditions contained in all site-plan approvals with regard to trees. Site landscaping is also reviewed and approved by the commission. The approved landscape plan is a site condition and must be maintained in perpetuity. The commission and/or its superintendent is required by ordinance to sign all commercial certificates of occupancy for landscaping and tree preservation.

Deposits are required for street tree planting at all new and renovated residential and commercial sites. The commission determines the amount, quantity of trees and species selection. All plantings are done by the commission utilizing their own crews. This assures quality trees, properly planted at a fair cost. All of this is made possible and easier because Paramus is "Tree City USA."

5. "Tree City USA" Growth Award

The "Tree City USA" Growth Award is provided by the National Arbor Day Foundation, in cooperation with the National Association of State Foresters and the USDA Forest Service, to recognize environmental improvement and encourage higher levels of tree care throughout America. This award is designed not only to recognize achievement but also to communicate new ideas and help the leaders of all "Tree City USA's" plan for improving community tree care. A community may be eligible for the "Tree City USA" Growth Award if it is "Tree City USA" for at least the second consecutive year and has spent at least as much on its community forestry program this year as it did last year.

"TREE POWER" enabled Paramus to do new things. This expanded leadership role elevated Paramus to where its Shade Tree and Parks Commission was sought out for advise and guidance by residents, public officials at home, and in neighboring communities. This took the form of public awareness programs, joint meetings, and working with groups attempting to organize or improve local tree care. Because of the interaction with other communities, organizations, and agencies new recognition on the regional, state, and national level was received. To name a new: (1) The Bergen County Board of Chosen Freeholders for innovative Arbor Day and tree planting programs; (2) The State of New Jersey Community Forestery Council Award for Outstanding Leadership in promoting tree care in New Jersey; (3) New Jersey State Legislature Resolutions for Arbor Day and Tree Planting; (4) International Society of Arboriculture "Gold Leaf Award" for Outstanding Arbor Day Programs; and (5) The American Forestry Association prestigious "Sylvia" award for community forestry efforts.

To elevate Paramus to a premier "Tree City USA," the Shade Tree and Parks Commission introduced new and innovative projects. When the National Arbor Day

Foundation began the "Tree City USA" Growth Award this community was prepared to meet this new challenge. The criteria and how this community complies is outlined as follows:

5.1. Category A: Education and Public Relations

The citation for some of the public relation efforts achieved by Paramus are Youth Education, Community-Wide Tree Event, Local Awards Program, and Literature Distribution.

The citation for some Education efforts achieved by Paramus are Tree-Care Workshops and Tree-Worker Safety program.

5.2 Category B: Partnerships

The citation for some of the partnerships are Utility Partnership, Green Industry Partnership, Cooperative Purchasing/Contracting, and Land Use Planning Coordination.

5.3 Category C: Planning and Management

The citation for some of the planning and management are Sidewalk/Curb Policy, Improved Ordinance, Standards and Specifications, Park Open Space, and Disaster Plan.

FIGURE 5. Educational programs help bring "Tree Awareness" to all Paramus students. Photo, William R. Comery.

5.4 Category D: Tree Planting and Maintenance

The citation for some of the tree planting and maintenance are Special Tree Planting Project, Public Utility Tree Care, Recycling, Street Tree Planting, Integrated Pest Management, Plans for Donations, and Street Tree Pruning.

Leaf composting in Paramus began in 1971. In 1985 we refined the raw compost into a valuable asset. Since that time we have improved upon this program and made it a self-sustaining operation. This successful operation has permitted us to expand and improve so that no top soil has to be purchased for any of our programs including tree planting.

The driving force provided by the "Tree City USA" awards caused Paramus to seek and obtain additional funding for expanded programs. The Small Business Administration Grant Program planted trees on streets constructed before the adoption of a uniform street tree planting. The State of New Jersey Green Communities Challenge Grants allowed Paramus to publish and distribute to every property owner a "Paramus Tree City USA Bulletin" giving essential information on trees and tree care. The latest grant from the state of New Jersey is for a public information and training course on proper tree and shrub care and trimming.

Our gift to the people of Paramus is a community of tree-lined streets, excellent parks, with residential and commercial areas that are environmentally compatible. As early as the incorporation of the borough in 1922 there was a plan. That plan has been refined to meet the needs of a changing and growing community. The early farmers in Paramus knew the importance of the land. Their legacy to the future is felt in every decision made to adjust to new circumstances. A "Special Thanks" goes to the residents of Paramus who served on the Mayor and Council, Board of Shade Tree and Parks Commissioners, and other local boards especially Mark S. Hillman for his assistance and help with this chapter. They insisted on the highest level of tree care for Paramus, because "they care about trees".

Chapter 25

Attracting and Managing for Wildlife*

Kathleen E. Clark

1. Introduction

The urban and suburban forests of the Northeast fulfill many functions that people value, not the least of which is providing habitat for wildlife. Forests, woodlots, parks, and backyard trees all support wildlife and provide the basis for wildlife management.

The principles of wildlife management may be applied at the backyard, local, and community levels to enrich the quality of our daily living environment, as well as to enhance regional wildlife habitat. Living near and viewing wildlife enhances quality of life in often subtle but real ways, providing a sense of connection with nature and respite from everyday life. Research shows that nearly one third of residents in the Northeast participated in watching wildlife in 1996 (US Fish and Wildlife Service, 1997). Most citizens watch wildlife in their own yards, as well as traveling to view wildlife. An estimated $6.4 billion was spent in 1996 on wildlife feeding, photography, and traveling to view wildlife by Northeast residents alone (US Fish and Wildlife Service, 1997). Many homeowners go further than feeding birds by improving the habitat of their yards for wildlife (National Wildlife Federation, 1997). The idea of managing urban and suburban communities for wildlife also has gained popularity, but is usually a side effect of management with other goals, such as watershed protection and greenways.

While local wildlife management benefits enthusiasts, it also helps sustain regional populations for many species, particularly migratory birds. Migratory birds are especially capable of exploiting patchy habitats, fragmented and smaller natural areas, which are typical of urban–suburban habitats. Here they seek food and cover to carry them through the stressful migration period. In a study of passerine (songbird)

---

*This Chapter is dedicated to my father.

Kathleen E. Clark Endangered and Nongame Species Program, New Jersey Division of Fish and Wildlife, Woodbine, New Jersey 08270.

Urban and Community Forestry in the Northeast, 2nd ed., edited by, J. E. Kuser.

migration along the Atlantic flyway, McCann *et al.* (1993) concluded that patchy habitats in suburban and coastal plain communities can help support the migration. They stressed, however, that conservation of the migration requires a mosaic of habitats and conserved natural areas, in addition to enhanced "backyard" habitat management.

2. Attitudes Toward Wildlife

The concept of managing vegetation to attract wildlife assumes that people in the community value wildlife as part of their environment. Conversely, the need to control or discourage problem wildlife also suggests that there are negative associations. In general, it is recognized that wildlife is considered an important element of people's quality of life and valued as part of their living environment.

In the Northeast, more than elsewhere in the United States, wildlife has been the focus of both good and bad perceptions. Here the interface of human and natural environments is large in impact and geography, and interactions with wildlife happen on both positive and negative levels. This is especially evident in suburban residential areas where species like white-tailed deer find refuge and food and in the process exceed the normal "carrying capacity" of the area. While many community residents value seeing deer, others suffer from increased incidence of deer–car collisions and lose valuable trees and shrubs to deer foraging. How residents value deer in this situation will tend to change as they personally experience the negative impacts of too many deer.

There are many positive experiences with wildlife in Northeast communities, as evidenced by widespread bird feeding, wildlife photography, and the increasing popularity of wildlife landscaping. In general, popular species are those that pose no threat to people's homes and yards and are not usually found in overabundance. Even lesser-recognized wildlife, such as frogs, salamanders, butterflies, and dragonflies, is being popularized by identification and management guides (e.g., Stokes and Stokes, 1991; Glassberg, 1993; Stein, 1997). Still, birds tend to reign in popularity because they are highly visible, colorful, and occur in great variety, especially during migration.

3. Elements of Wildlife Habitat

Wildlife habitat can be broken down to its components of cover, food, and water, and can be managed correspondingly. The best-quality habitat has all elements in proper proportions and proximity to each other. The design of the habitat, that is, the type and placement of vegetation, thereby serves to encourage some species and discourage others, providing the basis of management for wildlife.

3.1. Cover

Cover is vegetation that supplies shelter from weather, protection from predators, and places for nesting. The structure of vegetation, its height, density, and texture is often more important than species composition in providing cover and overall habitat for wildlife (DeGraff and Witman, 1979; Ehrlich *et al.*, 1988: 541).

3.1.1. Shelter and Escape Cover

All wildlife needs cover in which to sleep and escape from predators. Good cover also reduces thermal stress during harsh winter weather. Vertical structure of vegetation provides a variety of locations for different species. For many birds, cover is provided by evergreen trees and shrubs or thick deciduous vegetation that also serves to camouflage their body outlines. For mammals, such as raccoons and opossums, shelter is found in tree cavities or stumps, while small mammals (mice and voles) find cover in tall and dense grasses. White-tailed deer often rely on dense conifer stands for both shelter and food.

3.1.2. Nesting/Breeding Cover

Cover for nesting and breeding differs among species. Birds use a great variety of nest sites, ranging from conifers, tall deciduous trees, and shrubs, to ground nests found in heavy and light understory. In the urban–suburban regions, common species are well served with a range of evergreen and deciduous trees and shrubs. Developed in clumps, they offer cover and some protection from ground predators such as cats. Birds that are cavity nesters require older trees, or can be accommodated by nest boxes placed to specifications (Hassinger, 1997). Mammals, such as raccoons, tend to find breeding cover in cavities formed in older trees and stumps and in man-made structures like barns. Deer use cover provided in woodlots and often use high grass to hide new fawns.

3.2. Food

Wildlife food sources are as varied as animal species themselves. Native wildlife have evolved with native trees and shrubs, which provide a seasonal variety of food. Food plants supply fruits, seeds, and nectar as well as foliage in the form of trees, shrubs, vines, and herbaceous plants. Many birds, bats, and small mammals rely on insects, which in turn may be associated with specific host plants and trees. A variety of plants producing food in all seasons will support the greatest variety of wildlife, although winter and early spring is usually the time when food is most limited.

3.3. Water

Water is essential to all wildlife and can be the limiting factor in the distribution of wildlife in an area. Many species require a source of open water on a daily basis. In urban–suburban habitats that are often geographically patchy, access to water can be limiting. In managing to attract wildlife, water can be simple to provide. The types of water systems range from a bird bath to ponds, circulating pools, and misters. Resident wildlife learn the locations of water sources, but migrating birds are often attracted by the sound of moving water. Ponds and artificial pools are essential to attract water-dependent wildlife such as amphibians (frogs, toads, salamanders) and dragonflies.

3.4. Arrangement of Elements in the Landscape

The proximity of cover to food and water is key to high-quality wildlife habitat for nearly all species. This is most evident in animals having small territories, or home ranges, such as amphibians and reptiles, and "neighborhood" birds like robins and blue jays. For animals with small home ranges, the elements of cover, food, and water must be found in a relatively small area. Most amphibians are limited by availability of water and do not occur more than 1 km from it. For more mobile species, like deer and many birds, the elements may be farther apart, but their home range is consequently larger to meet their needs. For example, deer will forage on grains in agricultural fields, yet seek shelter in woodlots usually adjacent to those fields. While there is no specific maximum distance between food, water, and cover for a group of species, the closer the elements are, the greater the diversity of wildlife tends to be.

4. Wildlife in Urban and Suburban Landscapes

A key to attracting and managing for wildlife in urban and suburban areas is the concept of scale. While the size of the yard, woodlot, or park will always be fundamental to the type of management, so is its place within the landscape. The yard or neighborhood may be too small to support white-tailed deer, but deer may visit, for foraging, from the larger landscape area. If the landscape does not support the population, the animals may not be available to the smaller parcels. Managing habitat for wildlife works at the backyard or local level; the types and variety of wildlife that subsequently use the habitat are determined by the surrounding habitats, movement corridors, and animals' mobility.

In the face of continued habitat loss, the larger forests and parks of the northeast are essential to support wildlife populations on a larger scale, especially for area-sensitive species (those with large size requirements). These larger forests tend to be strongholds for "neotropical" songbirds—those that breed in North America and winter in Central and South America. Many species of this class of birds are in serious population declines (Robbins *et al.*, 1986). These and other songbirds may use the larger forests for nesting, but they rely on good quality habitat for food and cover during migration. The proximity of large forested areas to population centers of Northeast cities also makes these forests popular for wildlife-oriented recreation.

4.1. Typical Species

In the Northeast, most wildlife in the urban–suburban landscape are birds, whose mobility allows them to access good quality habitat "patches," and mammals that have adapted to human structures and habitats. The urban–suburban habitats, however, have a heightened importance to migratory birds. Highly mobile and dependent on food and cover all along the migration pathway, migrating birds will take seasonal advantage of good habitat wherever it occurs.

4.1.1. Urban Landscape

In most Northeast cities, large parks form the reservoir of wildlife species to be found, except perhaps for rock doves (or pigeons) (see Table 1 for species' names). Typical species tend to be those most adapted to the human environment, including

Table 1. Species Names Referenced in Text and Typical Habitats[a]

Common name	Latin name	Habitat type
Birds		
Red-tailed hawk	*Buteo jamaicensis*	Woodlands, fields
Northern bobwhite	*Colinus virginianus*	Pastures, fields
Rock dove	*Columba livia*	Cities, suburbs, farms
Mourning dove	*Zenaida macroura*	Woodland and suburban edges
Great horned owl	*Bubo virginianus*	Woodlands, field edges
Northern flicker	*Colaptes auratus*	Wood edges, suburbia
Downy woodpecker	*Picoides pubescens*	Open woodland, wood edges
Tree swallow	*Tachycineta bicolor*	Fields, wood edges
Blue jay	*Cyanocitta cristata*	Woods, edges, suburbs
Black-capped chickadee	*Parus atricapillus*	Woods, edges, parks
Tufted titmouse	*Parus bicolor*	Woodland, parks
House wren	*Troglodytes adeon*	Woods, thickets
Eastern bluebird	*Sialia sialis*	Forest edge, field with scattered trees
Northern mockingbird	*Mimus poyglottus*	Parks, suburbs, shrub–field
Gray catbird	*Dumetella carolinensis*	Thickets, hedges
American robin	*Turdus migratorius*	Parks, suburbs, woodland
European starling	*Sturnus vulgaris*	Cities, suburbs, farms
Eastern meadowlark	*Sturnella magna*	Pastures, hayfields
Northern oriole	*Icterus galbula*	Open woodlands, edges
Common grackle	*Quiscalus quiscula*	Parks, shade trees
Red-winged blackbird	*Agelaius phoeniceus*	Marshes, fields
Brown-headed cowbird	*Molothrus ater*	Open woods, edges
Northern cardinal	*Cardinalis cardinalis*	Thickets, suburbia
Purple finch	*Carpodacus purpureus*	Coniferous woods, edges
House finch	*Carpodacus mexicanus*	Cities, towns, farms
American goldfinch	*Carduelis tristis*	Weedy fields, open woodland
Chipping sparrow	*Spizella passerina*	Pine woods, edges
Field sparrow	*Spizella pusilla*	Brushy fields, edges
House sparrow	*Passer domesticus*	Cities, suburbs, farms
Song sparrow	*Melospiza melodia*	Thickets, shrubbery
Mammals		
White-tailed deer	*Odocoileus virginianus*	Woodlands and fields
Woodchuck, groundhog	*Marmota mona*	Upland woods, fields
Cottontail rabbit	*Sylvilagus floridanus*	Farmland, thickets
Raccoon	*Procyon lotor*	Woods, swamps, suburbs
Opossum	*Didelphis virginiana*	Woodlands

(*continued*)

Table 1. (*Continued*)

Common name	Latin name	Habitat type
Gray squirrel	*Sciurus carolinensis*	Woodlands, suburbs
Flying squirrel	*Glaucomys volans*	Woodlands
Eastern chipmunk	*Tamius striatus*	Hardwood forests, rock walls
Eastern mole	*Scalopus aquaticus*	Fields, gardens, lawns
White-footed mouse	*Peromyscus leucopus*	Woodlands, thickets
Meadow vole	*Microtus pennsylvanicus*	Fields
Reptiles		
Black rat snake	*Elaphe obsoleta*	Field, edges, open woods
Eastern Garter snake	*Thamnophis sirtalis*	Fields, roadsides, gardens
Box turtle	*Terrapene carolina*	Open woodlands, fields

[a]Harper and Row (1981), Conant (1991), Ehrlich *et al.* (1988).

rock doves, house sparrows, starlings, cowbirds, house finches, and mammals such as raccoons, opossums, and white-footed mice. The more mobile birds tend to move daily or seasonally into smaller lots in search of water or food from seed- or fruit-bearing herbaceous, shrub, or tree vegetation. It is not uncommon for migrating birds to use urban parks, lots, and even window box feeders during spring and fall migration periods.

4.1.2. Suburban–Residential Landscape

The suburban–residential landscape is common in the Northeast, created by a predominance of single-family houses on parcels less than 1 to 2 acres. When these housing developments are created, most existing native vegetation is removed. New trees and shrubs may be placed by homeowners, but the landscape generally is changed significantly. The process of revegetation may resemble natural succession, growing from simple herbaceous plant communities to more complex communities of a varied structure and species composition. Over time, these landscapes become valuable to wildlife once more, as trees mature and shrubs develop into understory providing cover.

Typical wildlife species found in this landscape will also change if the vegetation is allowed to mature. In early succession types of neighborhoods, with grass lawns and some trees, common species may be American robins, blue jays, house sparrows, house wrens, Northern mockingbird, mourning doves, cardinals, and gray squirrels. As trees and shrubs develop to create a varied-structure of cover and food, a greater variety of species may occur, including Northern flicker, downy woodpecker, black-capped chickadee, tufted tit-mouse, gray catbird, Northern oriole, cottontail rabbit, raccoon, opossum, and chipmunk. Other ground dwellers like small snakes and salamanders occur with the addition of brush piles, old logs, and other ground cover. In general, as vegetation grows in complexity of species and structure across neighborhoods, a greater variety of wildlife will be supported and attracted from adjacent habitats.

4.1.3. Suburban–Farmland Landscape

The suburban–farmland landscape is composed of the intersection of suburban–residential "habitat" and adjacent agricultural lands. This type of landscape has been a common element in recent years partly because farmland is aesthetically attractive and easily developed for housing. The agricultural landscape thus has been declining, giving way to suburban residential development. Still, where it exists near suburban areas, farmland is a rich reservoir of wildlife. Smaller farms, more common in the Northeast than elsewhere, tend to have a variety of cover types—field, hedgerow, woodlot—in close proximity, which is beneficial to many species. Suburban habitats, however, must be developed for wildlife to maximize the benefit of adjacent farmland. Typical species that may "spill over" between agricultural and residential areas are white-tailed deer, woodchuck, cottontail rabbits, a number of voles, shrews, and mice, and resident and migratory bats. Among the birds, red-winged blackbird, seed-eating sparrows (song, chipping, field) and finches (goldfinch, purple finch), Eastern meadowlark, tree swallow, bluebird, red-tailed hawk, great horned owl, and even bobwhite might be added to the more common birds of suburban areas. Further, a large variety of migratory birds, including warblers, sparrows, and hawks, will frequent this landscape.

4.2. Benefits

A rich variety of wildlife near one's residence provides benefits that are difficult to quantify. The sight of a deer or unusual bird often alerts us to the outside natural world and refocuses our attention away from the human-structured environment. This might be described as an aesthetic or spiritual value of wildlife. This ethereal value, however, may often translate to an economic level. Homes in urban–suburban neighborhoods may achieve higher market values when they incorporate trees (Ebenreck, 1989) or are integrated into natural areas that support wildlife (Leedy *et al.*, 1978).

Areas successfully managed for wildlife usually have the effect of enhancing the regional ecology. Good habitat promotes development of more complex food "webs," including beneficial insects, less visible reptiles and amphibians, and small mammals. As the system becomes more complex in diversity of species, certain pest problems tend to take care of themselves. Infestations of Japanese beetles or tent caterpillars can be solved by birds or other insects in an ecological balancing act. A single yard managed to promote wildlife can be an island, but an entire neighborhood can work as an ecosystem. In a healthy ecosystem (or neighborhood) the need for costly chemicals and maintenance is reduced.

4.3. Problems and Nuisance Wildlife

Some wildlife species can become pests when they occur in some places or in over-abundance. In Northeast urban and suburban areas, typical problems include the following:

- White-tailed deer involved in deer–car collisions
- White-tailed deer foraging on residential shrubbery

- Black bears raiding bird feeders, beehives, or garage containers
- Interaction of rabid wildlife with people or domestic animals
- Canada geese overpopulating suburban areas
- Raccoons inhabiting yards, feeding on garbage
- Beavers causing flooding of properties
- Squirrels or bats inhabiting attics
- Moles causing damage to lawns
- Woodpeckers affecting houses
- Snakes in backyards

Many of these problems arise when high populations of wildlife occur in suburban–open space interfaces, often where wildlife find refuge from hunting pressure. This is often the case with white-tailed deer and Canada geese, which are two of the major problem species in the Northeast. The incidence of rabies in wildlife is a concern, but usually perceived to be a greater threat than it actually is. County or municipal health departments usually address local rabies occurrence with emphasis on pet vaccinations and educating the public. Deer are part of the life cycle of the spirochete that causes Lyme disease. While there is not a clear cause-and-effect relationship between high suburban deer populations and incidence of Lyme disease in people, this may be the public's perception. Other problems, such as woodpeckers, squirrels, bats, and moles, may be occasional and temporary inconveniences, treatable with the help or advice of professionals. Some problems may be more a matter of perception than real problems: snakes occur widely and do not normally pose a threat to people, yet they are not well tolerated when observed in yards.

5. Managing for Wildlife

Managing habitat for wildlife centers on creating cover, food, and water in close proximity, to meet the daily and seasonal needs of desired wildlife. In the urban–suburban landscape, small-scale habitats can be enhanced using trees and shrubs that provide superior cover and food (Fig. 1). In addition, an examination of the larger landscape may reveal elements that are missing or in short supply; by providing the elements in short supply, we meet wildlife needs for the area. Local and regional planners can enhance habitat by promoting conservation of native vegetation in construction projects and along streams and wetlands, and developing open spaces and parks with wildlife habitat in mind (Leedy and Adams, 1984).

5.1. Attracting Birds

Birds are among the easiest wildlife to attract into small-scale habitats like backyards; birds thereby can serve as "umbrella" species. Managing successfully for birds tends to promote other wildlife as well. The mobility of birds allows them to exploit good habitat, sometimes regardless of size, to find seasonal foods. Bird species most likely to find a new site may depend on the type and size of local natural habitats: forests and fields will support populations that can colonize smaller habitats. Nesting

FIGURE 1. Development of habitat from simple, individual trees (left) to complex vegetative structure that benefits wildlife (right).

birds must find all elements for survival within a nesting territory and defend it against others. Major migrations of songbirds and hawks occur in the Northeast in April–May (northbound) and in September–November (southbound). Migrating birds are not constrained by territorial behavior and have acute needs for quality foods and escape cover. Birds need food in the form of seeds, nuts, insects, catkins, nectar, and fruit. The greater variety of food-producing shrubs and trees in the area, the more birds it may support.

5.1.1. Landscape Elements and Tree–Shrub Species

An urban or suburban habitat for birds should include tall trees, both deciduous and evergreen, for cover and nesting. Excellent tree species are Eastern red cedar, white pine, black cherry, oaks, maples, sweet gum, and black gum, among others. Deciduous and evergreen shrubs, as understory and massed with trees, help create a multilayered or tiered structure excellent for cover and necessary for some species' nesting. Recommended shrubs include bayberry, viburnum, sumac, elderberry, sweet pepperbush, butterfly bush, cotoneasters, and blueberry (Table 2), but native shrubs and trees will volunteer or seed naturally in many cases. In most urban–suburban neighborhoods, tree–shrub structure is usually limited, so its addition provides the greatest early benefit to birds. Grass lawns provide almost no benefit to wildlife but can serve as backdrop for bird activity along native shrub borders or hedgerows. In otherwise suitable habitat, nest boxes can substitute for natural tree cavities to accommodate bird and squirrel nesting and bat roosting (Hassinger, 1997).

Table 2. Some Trees, Shrubs, and Vines Beneficial to Wildlife[a]

Species	Latin name	Benefit
Trees		
American holly	*Ilex opaca*	Food (winter berries), cover
Ash	*Fraxinus* spp.	Food (seeds)
Beech	*Fagus* spp.	Food (nuts)
Birch	*Betula* spp.	Food (catkins, insects)
Black gum	*Nyssa sylvatica*	Food (fruits)
Black cherry	*Prunus serotina*	Food (fruits)
Chokecherry	*Prunus virginiana*	Food (fruits, insects)
Crab apple	*Malus* spp.	Food (fruits)
Eastern hackberry	*Celtis occidentalis*	Food (berries, insects)
Eastern hemlock	*Tsuga canadensis*	Cover, nesting, food
Eastern red cedar	*Juniperus virginiana*	Cover, nesting, food
Flowering dogwood	*Cornus florida*	Food (fruit)
Hawthorn	*Crataegus* spp.	Food (winter berries)
Maple	*Acer* spp.	Food (seeds), cover, nesting
Mulberry	*Morus* spp.	Food (berries), cover
Oak	*Quercus* spp.	Food (nuts), cover, nesting
Persimmon	*Diospyros virginiana*	Food (fruit, insects)
Sassafras	*Sassafras albidum*	Food (seeds, insects)
Serviceberry (shadbush)	*Amelanchier laevis*	Food (fruit)
Shagbark hickory	*Carya ovata*	Food (nuts), nesting
Spruce	*Picea* spp.	Cover, food
Sweet gum	*Liquidambar styraciflua*	Food (fruit)
Tulip tree (tulip poplar)	*Liriodendron tulipifera*	Food (seeds)
White pine	*Pinus strobus*	Cover, food (seed)
Shrubs		
Alders	*Alnus* spp.	Food (seeds)
Bayberry	*Myrica pensylvanica*	Food (fruit), cover
Butterfly bush	*Buddleia davidii*	Food (nectar)
Common spicebush	*Lindera benzoin*	Food (fruit)
Cotoneaster	*Cotoneaster* spp.	Food (winter fruit), cover
Elderberry	*Sambucus canadensis*	Food (fruit, insects)
Firethorn	*Pyracantha coccinea*	Food (winter berries), cover
Juniper	*Juniperus* spp.	Food (berries), cover
Red osier dogwood	*Cornus stolonifera*	Food (berries)
Sumac	*Rhus spp.*	Food (winter berries)
Sweet pepperbush	*Clethra alnifolia*	Food (fruit)
Tartarian honeysuckle	*Lonicera* spp.	Food (nectar), cover
Viburnum	*Viburnum* spp	Food (fruit), cover
Vines		
English ivy	*Hedera helix*	Cover, food (insects)
Grape	*Vitis* spp.	Food (fruits), cover
Trumpet vine	*Campsis radicans*	Food (nectar)
Virginia creeper	*Parthenocissus quinquefolia*	Food (berries, foliage), cover

[a]National Wildlife Federation (1974), Degraff and Witman (1979), Harrison (1979), Ernst (1987), Sutton (1989), Cox (1991).

Native vegetation on site should be encouraged or incorporated into landscape design, with additions of food-producing shrubs and trees (Table 2). Shrubs should be selected to provide food year round ideally, requiring some evaluation of what is currently available in the area as well as plants' seasonality. Trees also provide food for birds in seeds, catkins, and the insects they support. Older trees and dead limbs should be left, where feasible, since they provide for woodpeckers and cavity-nesting birds.

The simplest method of providing water for birds is a bird bath, about 1m high and 1 to 2 m from shrubs. More complex water sources, including ponds and circulating pools, will attract birds as well as other wildlife. Retention and detention ponds provide water for wildlife and can be improved for this purpose by allowing native vegetation to grow around their banks and avoiding use of fertilizers.

5.2. Attracting Other Wildlife

Managing for species such as mammals, butterflies, dragonflies, amphibians, or reptiles follows the same principles of providing the animal's requirements. Large mammals are generally limited by the size of habitat and availability of wooded corridors for moving between habitats. In the urban–suburban area, large- and medium-size mammals, such as deer and raccoons, if present at all, may easily become pests. They have adapted to human structures and food sources, making them difficult to control, however, smaller wildlife can be promoted in small habitats with less likelihood for problems.

5.2.1. Landscape Elements and Tree–Shrub Species

Small mammals, such as chipmunks and squirrels, may inhabit small habitats that have some connection to population centers in larger habitats. Both eat seeds and nuts available from oaks, maples, hickories, and grasses and other herbs. Gray squirrels require tall deciduous trees for breeding and cover, flying squirrels require tree cavities, while chipmunks are ground dwellers, using fallen logs, rock walls, and other ground cover. Other small mammals, primarily mice and meadow voles, inhabit dense herbaceous vegetation that serves both as cover and their seed-food source. Many small mammals obtain water through their food or moist vegetation.

Reptiles, including garter snakes, black rat snakes, and box turtles, can be common but unseen in suburban habitats. They do best as part of a healthy ecosystem, relying on insects and/or fruits found in varied habitats of trees, shrubs, and grasses. Managing for the larger area, following suggestions for bird management, usually meets their needs.

Amphibians are an interesting group of species united by their dependence on a reliable source of water. They include a wide variety of frogs, toads, and salamanders, some of which can inhabit smaller habitats that include water (at ground level) and cover. A shallow pond that does not support fish is necessary, along with ground cover in which they can move. Cover varies by species: treefrogs require trees near ponds, while frogs and salamanders do well with forest floor vegetation like leaf layers and decomposing logs. Within live (usually herbaceous) and decomposing vegetation they forage on insects. It is best to allow these species to colonize a pond naturally rather than introduce them; the species for which the habitat is suitable will find it.

5.3. Managing to Discourage "Problem" Wildlife

Wildlife can become out of balance with natural systems when they adapt well to the human landscape and learn to find food and cover in residential areas. Human food sources or refuge provided by suburban areas can artificially elevate the natural carrying capacity, creating an imbalance. Problems occur when these animals take up residence in homes, attics, or garages, interact with domestic pets, or damage landscaped yards (Cummings, 1979). Some problems can be diminished with habitat management that brings about greater ecological balance, but some are related to high regional populations and require solutions at the population level (within the jurisdiction of state wildlife agencies). It is important to understand, however, that wildlife live where there is suitable habitat: snakes may be perceived as problems in some cases, but actually are part of a system involving food (insects, rodents) and cover. Insect problems may be resolved using integrated pest management (IPM) (see Chapter 18, this volume), available through agricultural extension services. Successful wildlife management helps restore a system of multiple ecological levels.

5.3.1. Landscape Design

The arrangement of trees and shrubs can be managed to reduce wildlife problems. Most wildlife require cover in which to move about or seek out specific foods. By reducing the connectiveness of vegetative corridors, some species, such as rodents that are reluctant to cross open areas, will be reduced or limited in range. In cases of small mammals entering houses, foundation plantings might be changed to woody shrubs that reduce ground cover and seed production. Food left outside for dogs and cats will attract nuisance animals such as raccoons, skunks, and opossums and therefore should be avoided. Squirrels and small mammals but not birds are deterred from bird feeders when capsaicinoid pepper is mixed with the seed.

Animals with known food preferences may be discouraged by limiting preferred foods. White-tailed deer have a wide diet of herbaceous plants and woody "browse," including many trees in the sapling stage. In many areas of the Northeast, deer have a strong influence on habitat: they can reduce shrub and tree understory and prevent forest regeneration. Where deer are abundant, local habitat development must employ tree and shrub species that deer usually do not eat (e.g., box elder, cotoneaster, English ivy, holly; consult an agricultural extension agent). Canada geese problems flourish at sites with short grass adjacent to ponds or lakes and are easily solved by replacing grass with wildflower meadow or long grass. Long grass management, mowing only once per year, is recommended for deterring Canada geese and gulls at airports (Blokpoel, 1976).

5.3.2. Mechanical and Chemical Controls

Wildlife problems can be prevented or reduced using physical means that are preferable to chemical methods when promoting wildlife habitat. Physically sealing holes and cracks in foundations and screening attic ventilation holes will exclude animals like raccoons, squirrels, and bats that seek nest or roost sites. Animals that do gain entry can be trapped and removed by the homeowner or a pest control company; bats will leave after dusk, then holes can be sealed to prevent their reentry (one should be sure there are not breeding bats with young). Heavy-duty screening can be used on chimneys to prevent access by raccoons, squirrels, and birds, and tree branches should be at least 2 m away from roofs to prevent animals from reaching the house. If animals take up residence under out-buildings or in crawl spaces, ammonia-soaked rags often discourage them. Fencing around vegetable gardens will prevent foraging by deer, rabbits, and woodchucks. Woodchucks will also burrow, so submerged galvanized wire fence will prevent them from residing under decks. Valuable shrubbery can be protected from browsing deer by covering with nylon netting, especially during winter and early spring when deer forage is low. The lower (0.5 m) stems of small trees can be wrapped with screening to prevent mice and rabbits from damaging bark. Woodpeckers pecking on houses can be a sign of insect infestation, which should be investigated first; otherwise, pecking can be discouraged by covering particular areas with wire mesh (US Department of interior, 1978) or hanging Mylar balloons or tape; some pecking is done as part of territorial advertising and is unrelated to food. Trash cans and bird and grass seed should be stored securely to prevent access by bears, raccoons, and mice. Most wildlife are protected from lethal controls without proper permits; state wildlife agencies should be consulted for regulations applicable to nuisance wildlife.

Chemicals can be used to interrupt wildlife problems and provide short-term fixes, but are usually not a long-term solution. Chemical repellents (e.g., Hinder—Shield-Brite, Inc., Magic Circle—J. C. Ehrlich) can be effective on small areas in reducing damage to shrubs and trees caused by deer and rabbits and is available from garden supply stores. Gas cartridges are available in garden centers to kill moles in lawns, but will not prevent other moles from moving in where food (especially grubs) remains available. Woodpeckers can be discouraged from wood-sided houses by application of wood preservative to the siding; wood should be free of insects before application. Most chemical controls must be used repetitively unless other steps are taken to eliminate food or cover resources locally.

References

Blokpoel, H., 1976, *Bird Hazards to Aircraft*, Books Canada, Buffalo, NY.

Conant, R., 1991, *A Field Guide to Reptiles and Amphibians of Eastern and Central North America*, 3rd ed., Houghton Mifflin, New York.

Cox, J., 1991, *Landscaping with Nature*, Rodale Press, Emmaus, PA.

Cummings, M. W., 1979, Wildlife damage problems, in *Wildlife Conservation: Principals and Practices* (R. D. Teague, and E. Decker, eds.), The Wildlife Society, Washington, DC, pp. 167–171.

DeGraff, R. M., and Witman, G. M., 1979, *Trees, Shrubs and Vines for Attracting Birds, A Manual for the Northeast*, University of Massachusetts Press, Amherst.

Dennis, J. V., 1988, *The Wildlife Gardener*, Alfred A. Knopf, New York.

Ebenreck, S., 1989, The values of trees, in *Shading Our Cities: A Resource Guide for Urban and Community Forests* (G. Moll, and S. Ebenreck, eds.), Island Press, Washington, DC, pp. 49–57.

Ehrlich, P. R., Dobkin, D. S., and Wheye, D., 1988, *The Birder's Handbook*, Simon and Schuster, New York.

Ernst, R. S., 1987, *The Naturalist's Garden*, Rodale Press, Emmaus, PA.

Glassberg, J., 1993, *Butterflies through Binoculars*, Oxford University Press, New York.
Harper and Row, 1981, *Complete Field Guide to North American Wildlife*, Harper and Row, New York.
Harrison, G. H., 1979, *The Backyard Bird Watcher*, Simon and Schuster, New York.
Hassinger, J., 1997, *Woodcrafting for Wildlife: Homes for Birds and Mammals*, Pennsylvania Game Commission, Harrisburg.
Leedy, D. L., and Adams, L. W., 1984, *A Guide to Urban Wildlife Management*, National Institute for Urban Wildlife, Columbia, MD.
Leedy, D. L., Maestro, R. M., and Franklin, T. M., 1978, *Planning for Wildlife in Cities and Suburbs*, American Society of Planning Officials, Report No. 331, Washington, DC.
McCann, J. M., Mabey, S. E., Niles, L. J., Bartlett, C., and Kerlinger, P., 1993, A regional study of coastal migratory stopover habitat for neotropical migrant songbirds: Land management implications, *Trans. 58th North Am. Wildlife Natural Res. Converence* **58:**398–407.
National Wildlife Federation, 1974, *Gardening with Wildlife*, National Wildlife Federation, Washington, DC.
National Wildlife Federation, 1997, *Backyard Wildlife Habitat Program*, National Wildlife Federation, Vienna, VA.
Robbins, C. S., Bystrak, D., and Geissler, P. H., 1986, *The Breeding Bird Survey: Its First Fifteen Years, 1965–1979*, US Department of Agriculture, Fish and Wildlife Service Resource Publication 157, Washington, DC.
Stein, S., 1997, *Planting Noah's Garden*, Houghton-Mifflin, New York.
Sternberg, G., and Wilson, J., 1995, *Landscaping with Native Trees*, Chapters Publishing, Shelburne, VT.
Stokes, D., and Stokes, L., 1991, *The Butterfly Book*, Little, Brown, Boston.
Sutton, P., 1989, *Backyard Habitat for Birds*, New Jersey Audubon Society, Cape May Point.
US Department of Interior, 1978, *Controlling Woodpeckers*, Animal Damage Control 101, Washington, DC.
US Fish and Wildlife Service, 1997, *1996 National Survey of Fishing, Hunting, and Wildlife-Associated Recreation: Preliminary Findings*, Washington, DC.

Further information can be obtained from:

1. State wildlife agencies.
2. Cooperative wildlife research units associated with state universities.
3. National Institute for Urban Wildlife, 10921 Trotting Ridge Way, Columbia, MD 21044.
4. National Wildlife Federation, Backyard Wildlife Habitat Program, 8925 Leesburg Pike, Vienna, VA 22184-0001.
5. US Department of Agriculture, Animal and Plant Health Inspection Service-Animal Damage Control, National Wildlife Research Center, 1201 Oak Ridge Drive, Fort Collins, CO 80225.

Chapter 26

Managing Urban Ecosystems: A Look to the Future of Urban Forestry

L. Robert Neville

1. Introduction

Near the end of the 19th century, the American conservation community, led by Gifford Pinchot, was discussing the need for scientific management of forests and related natural resources. The primary concerns at that time were destructive and wasteful timber-harvesting practices and the need for a sustained yield policy (Pinchot, 1947). Urbanization was not an issue. Nineteenth-century cities were very compact in form due to the constraints of transportation and access. It is understandable, therefore, that urban growth was not perceived by Pinchot as a threat to the long-term viability and productivity of forests. Now, in retrospect, as we approach a new millennium, it has become obvious that a century of unparalleled population growth and urbanization has had an extremely detrimental impact on the natural systems and processes that sustain life on this planet. Forests in the heavily populated regions of the Northeast have been decimated and in many places can no longer maintain their functional efficiency in stabilizing soil, purifying water and air, and sustaining biological diversity.

One of the reasons for the demise of forests and associated natural systems is the role that foresters have assumed in urban areas. For most of the 20th-century the urban forestry profession concentrated on the care, management, and replacement of trees in the public rights-of-way and other public property. The emphasis was on the trees themselves with little or no consideration for the role of vegetation, in a broader context, as it contributes to greater societal needs such as clean water and air. Any agency of government did not assume the responsibility for protecting and conserving the functional effectiveness of natural systems in developing areas. Consequently,

L. Robert Neville Wolfe Mason Associates, Hampton, New Hampshire 03842.
Urban and Community Forestry in the Northeast, 2nd ed., edited by, J. E. Kuser.

flooding is a recurring problem, storm water infrastructure is inefficient and costly to maintain, summer temperatures in urban areas are higher than in surrounding forested areas by as much as 10°C, energy consumption is excessive, open space is limited, and biological diversity has been significantly reduced.

In looking at the future of urban forestry it is apparent that a new focus must emerge. Restoring the role of forests and related natural resources to the extent possible in existing developed areas and ensuring their proper place in allocating future land use will provide new challenges for the profession in the next century.

1.1. Sustainable Development

1.1.1. Effects of Urban Growth on Natural Systems and Processes

Ignoring natural systems and processes in urban and urbanizing areas can have costly long-term consequences for community sustainability. This has been demonstrated by older cultures in Europe and Asia that have long histories of dealing with the effects of increasing population concentrations on a limited resource base. Centuries of human occupation and intensifying land use in the Netherlands, for example, have reshaped the natural environment there into what has been described as "an almost completely cultural landscape." Ongoing land and resource exploitation is effecting the transfer of genetic information through loss of habitat and species diversity while stressing natural regulatory functions, resulting in excessive air and water pollution (Vos and Zonneveld, 1993). In response, there is a movement in the Netherlands to reintroduce natural areas into the landscape continuum to restore critical ecosystem function (Harms *et al.*, 1993). The Dutch government has prepared a nature policy plan to compensate for the deterioration of natural systems by setting aside 50,000 hectares (123, 550 acres) of land as part of its national physical planning process (Harms *et al.*, 1993). One of the principal strategies to implement the plan is to establish a landscape-level framework that includes a pattern of interconnected zones in which long-term sustainable conditions for nature development and water supply are provided. This system proposes to restore and protect upland groundwater recharge areas in addition to wetlands and riparian zones through reforestation. The process is also expected to restore and enhance biodiversity through the development and dispersal of related plant and animal species (Van Buuren and Kerkstra, 1993).

Fortunately, conditions in the United States have not deteriorated to the extent they have in the Netherlands, since there is 15 times the land area per capita (Vos and Zonneveld, 1993). Signs of environmental stress are beginning to surface in specific locations, however, especially urban areas, owing to the heavy subsidies of energy and materials that must be imported from other ecosystems in order to maintain them (Ecological Society of America, 1995) and the huge quantities of waste products that must be disposed of. The situation is not limited to cities only; environmental issues associated with land use change at the urban–rural fringe are surfacing, as indicated by the New York–New Jersey Highlands regional study (US Department of Agriculture Forest Service, 1992). The Highlands study illustrates the encroachment of urban development into previously rural watersheds, which threatens to degrade the water supply for half of the population of the state of New Jersey.

Chesapeake Bay is one example where human activity has caused a deterioration in water quality resulting in a dramatic decline in aquatic species composition and abundance that has exceeded a threshold level of sustainability (Brush, 1995; US Environmental Protection Agency, 1983). From a reconstructed paleoecological record of indicator organisms and materials preserved in Chesapeake Bay sediments, Brush (1995) concluded that the transformation from a forested to a non-forested landscape was responsible for converting a diverse benthic estuary in the Chesapeake Bay into one dominated by plankton. The benthic resource has high potential for productivity in shallow water systems when light can penetrate the water column. However, accelerated sedimentation and nutrient enrichment caused by changing land use within the surrounding watershed was so intense that it altered the flux of light and energy within the water column, causing a permanent change in aquatic habitat (Brush, 1995; Chesapeake Bay Monitoring Subcommittee, 1989). This situation manifested itself in a significant decline in the relative abundance and quality of various finfish and shellfish. Oyster harvesting, for example, which at one time yielded as much as 15 million bushels annually from the Chesapeake Bay, has been reduced to less than 1% of historical levels (Chesapeake Bay Monitoring Subcommittee, 1989).

Population growth and related land use change, especially over the second half of the 20th-century, have been linked directly to the decline in water quality and the living resources of the Chesapeake Bay (US Environmental Protection Agency, 1983). Nearly six million people live in the Baltimore–Washington–Annapolis metropolitan area, which ranks fourth in the nation behind New York, Los Angeles, and Chicago. Since 1950, total population within the Chesapeake Bay basin has increased more than 60%, from about 8.3 million to 13.6 million (Year 2020 Panel, 1988). This represents an increase in the land consumption rate of from 0.07 to 0.26 ha per person (0.18 to 0.65 acre per person) (Swanson, 1992). Population is expected to reach 16.2 million by the year 2020 which, at the present consumption rate will convert another 687,980 hectares (1.7 million acres) from forests and agriculture to urban use. Most of this growth is expected to occur in counties adjacent to metropolitan areas where 75% to 100% increases are expected. Furthermore, a disproportionate amount of this development is expected to take place within 300m (1000 fect) of the Bay (Year 2020 Panel, 1988), where maximum impact to natural patterns and processes can be anticipated. The implications of this growth and change are enormous, since this situation is not limited to the Chesapeake Bay. Development patterns are similar all along the Atlantic coastal zone.

1.1.2. Effects of Environmental, Social, and Economic Systems on Urban Decay

Conditions of poverty, social injustice, and environmental contamination in many inner-city areas like Boston, New York, Chicago, Detroit, and Baltimore are testimony to the lack of attention that is being given to the sustainability of urban areas. It is becoming increasingly apparent that all three components of a sustainable ecological system, that is, social stability, economic vitality, and environmental quality, are mutually interdependent like three legs of a stool. If one component is lacking, the system collapses.

For many years, social and economic concerns have driven land use and resource allocation decisions in populated areas, which according to the three-legged stool analogy are not sustainable. As Allen and Hoekstra (1993) have indicated, in order for disturbed ecosystems to function there must be management intervention to substitute the structural integrity that is lacking. This concept seems to be exaggerated in urban environments. For example, little, if any, attention is paid to natural systems in poor neighborhoods where residents (the social component) are concentrating on subsistence issues (the economic component). Therefore, in order to maintain the stability of the system, policy makers, through resource managers, must intervene to support the biophysical component. A clean and well-maintained urban environment is an asset that supports a stable economic base and contributes to stable social systems and institutions.

1.2. Existing Management Responsibilities for Urban Natural Systems

The principle agencies at the federal level that address environmental issues in urban areas are:

- US Forest Service
- US Environmental Protection Agency
- US Fish and Wildlife Service
- Natural Resource Conservation Service

Each agency has programs that provide either financial and technical assistance to states and communities or environmental review of proposed actions or even site-specific remediation. Neither of these agencies, however, has legislative authority to assume natural resource management responsibility on nonfederal land. Nor do state agencies maintain any systematic record of environmental conditions at a specific enough scale that the information can be used to coordinate the efforts of various jurisdictions or individual landowners. Occasionally there is a watershed compact or similar regional entity that reviews individual project proposals in the watershed context. It is rare, however, that regional compacts include multijurisdictional coordination with regard to natural systems and processes. Occasionally, there is a local unit of government where natural resource management is integrated with community services. Unfortunately, these communities are seldom linked to others in the same watershed or ecosystem where collective decisions can be made about natural resource allocations that will assure the sustainability of critical natural systems.

The unfortunate fact is that no agency of government at the federal, state, or local level is responsible for the collective environmental impacts of uncoordinated land and resource allocation decisions in urban areas.

1.3. The Need for a Comprehensive Management Approach

Mitigating the negative effects of urbanization that include excessive loss of liquid water and essential nutrients, soil erosion, and reduction of biological diversity to create sustainable communities will require multiple ecological, social, and economic strategies. Sustainable communities can be described as those that maintain a balance between ecological, social, and economic systems in order to satisfy present needs while protecting the options for future generations to meet their needs as well. One potentially cost-effective strategy that has been overlooked is the aggressive conservation and management of vegetation as a natural element within the built environment.

2. Establishing a Context for Urban Forest Resource Management

2.1. The Urban–Rural Continuum

Urban expansion is primarily a 20th-century phenomenon. Previously, topography and mobility were the major constraints to the growth of cities. Topography was a factor because it required so much more energy to build on steep slopes, and mobility was limited to a distance that people could reasonably travel on foot. By the late 19th century widespread distribution of electricity allowed for the introduction of mass transit systems that extended the city boundary into the urban fringe to create adjacent suburbs. Urban form remained relatively compact around transit stops however, since development was still constrained by a reasonable walking distance (Lazaro, 1979).

It was the automobile that fundamentally changed urban form and blurred the boundary between the city and the surrounding rural lands. Early 20th-century suburbs gradually filled the area between radial mass transit lines while being extended to include a reasonable commuting distance by automobile for the elite at the city's edge (Laurie, 1986). Following World War II, the economic expansion made automobile ownership possible for nearly every working family. Congestion increased; and to meet the need for more efficient movement of traffic as well as improve military mobility, the interstate highway transportation system was initiated. With personal transportation and easy access, workers were liberated from having to live and work in the same location, and the suburban lifestyle became the norm (Lazaro, 1979). A new model is emerging at the end of the 20th-century based on the technological revolution where advanced communication allows workers to live wherever they want.

Rather than providing an escape from the congestion, traffic, and irritation of city living, the suburbs became recognized as an extension of the same urban lifestyle except that the deleterious impact on the land base increased considerably. The Regional Plan Association of New York has estimated that the tristate metropolitan area of New York City, which includes adjacent counties in New Jersey and Connecticut, experienced a 60% increase in urban land use between 1968 and 1990, despite a population growth of only 6% (US Department of Agriculture, Forest Service, 1992). By 1984, 80% of New Jersey's population lived in the suburbs and 84% of the state's labor force was employed there. Corresponding national statistics of 48% and 45%, respectively, indicate that the older more densely populated Northeast has decentralized at a much more rapid rate than the rest of the country (Hamill *et al.*, 1989). Environmental consequences of this urban decentralization include social and economic destruction of inner-city neighborhoods and significant deterioration of water availability and quality in watersheds surrounding metropolitan areas due to the removal of forest cover (Chavooshian *et al.*, 1987; Delleur, 1982). Related problems include the loss of wildlife habitat and functional open space.

2.2. Ecosystem Classification: Natural to Human-Dominated

Urbanization as a process that has substantial environmental implications has been overlooked as a significant source of ecological inquiry according to McDonnell and Pickett (1990). In its recent white paper, *The Scientific Basis for Ecosystem Management*, the Ecological Society of America (1995) essentially ignores urbanization as a long-term sustainability issue. McDonnell and Pickett (1990) have suggested that: "The study of the metropolis as an ecosystem, including its human inhabitants and institutions, would be a radical expansion of ecology" (p. 1232). Furthermore, they propose a framework to help guide the design and integration of a variety of research activities that accounts for the factors that constitute urbanization, that is, the effects of urbanization on the biota and the physical environment and the resultant effects on ecosystems.

The current US Department of Agriculture (USDA) Forest Service process of mapping ecosystems across a range of scales is intended to stratify the earth into progressively smaller areas of land with increasingly uniform ecological potential. It includes eight ecological sub-regions ranging in scale from the domain at the global level down to the land type phase that can be less than 10 acres in size (ECOMAP, 1993). This ecological classification will eventually assist land managers with developing goals and objectives to provide context for sustainable management from the global to the local level that links ecosystem science to natural resource management actions (Ecological Society of America, 1995; Kaufmann *et al.*, 1994; ECOMAP, 1993). Unfortunately, unlike European classification systems (Haber, 1990), this process does not account for the change in ecosystem structure and function that has been brought about by intensive human activity.

After many centuries of economic growth and change, European ecologists are more advanced in their consideration of urban structure and function (Vos and Opdam, 1993). Haber (1990) characterizes ecosystems according to decreasing naturalness ranging from bioecosystems, dominated by natural components and biological processes, to technoecosystems, which are totally dependent on human control and external sources of matter and energy. This view of ecological classification is shaped by the European experience where population densities in relation to the United States range from a fourfold increase in Denmark to a 15-fold increase in the Netherlands (Vos and Zonneveld, 1993). Although these countries have substantially higher population densities overall, there are extensive urban concentrations within the United States, particularly along the East and West coasts, with equivalent densities to that of the Netherlands. The State of New Jersey, for example, which is the most densely populated state in the United States, has almost twice as many people per unit area as the Netherlands (New Jersey State Planning Commission, 1992).

The national hierarchical framework of ecological units (ECOMAP, 1993) contains no recognition of anthropogenic factors. This oversight is unfortunate since it ignores the obstacles to long-term sustainability identified by the Ecological Society of America (1995), namely, the adequacy of information on the biological diversity of environments and widespread ignorance of the function and dynamics of ecosystems. Whereas the framework (ECOMAP, 1993) is based solely on biophysical criteria from the domain (global) to the land type phase (site), urban areas are dominated by man-made structures and materials that completely alter ecosystem structure and function.

Various frameworks have been developed for quantifying naturalness that describe the variability of external energy sources required to maintain a system in its present state. For example, a natural forest stand is self-maintaining (Ecological Society of America, 1995), whereas urban–industrial land uses (technoecosystems) have little or no self-maintaining capacity (Haber, 1990). Allen and Hoekstra (1993) advocate a strategy to address ecosystem management in human-dominated systems. It is based on ecological context (which implies self-maintaining capacity) that provides stability and the ability to recover from repeated disturbances. Accordingly, the most effective management will recognize the extent to which the context is missing, identify the resulting lost functional capacity, and subsidize the managed area to the extent possible. By following the hierarchical approach expressed in ECOMAP (1993), the often confusing issue of scale is dealt with systematically (Table 1). The hierarchy accounts for scale at each level from the continent to the site. Ecological units at the various levels in the hierarchy are identified based on associations of biotic and other environmental factors that directly affect or indirectly express moisture, energy, and nutrient inputs that regulate the structure and function of ecosystems. However, because human activity is a major contributing factor that can have a significant effect on ecosystem structure and function, it too needs to be included in the hierarchy. The issue of scale is important here and must be considered when establishing the point where human influences become dominant elements.

Sanders and Rowntree (1983) concluded that the most appropriate framework within which to classify urban vegetation is the standard metropolitan statistical area

Table 1. Ecological Subregions of the United States[a]

Planning and analysis scale	Ecological units	Purpose, objectives, and general use	General size range
Ecoregion			
Global	Domain	Broad applicability for modeling and sampling, strategic planning and assessment	Million to tens of thousands of square miles
Continental	Division		
Regional	Province		
Subregion	Section	Strategic, statewide multi-agency analysis and assessment	Thousands to tens of square miles
	Subsection		
Landscape	Land type assoc.	Multiple community planning and watershed	Thousands to hundreds of acres
Land unit	Land type Land type phase	Management area and project planning and analysis	Hundreds to less than ten acres

[a]Adapted from ECOMAP (1993).

(SMSA), a designation used by the Bureau of the Census. Although the primary objective is to classify vegetation they needed to first identify those areas that can be characterized as urban. The SMSAs account for approximately 20% of the nation's land area. Each SMSA contains at least one central city with a population of no less than 50,000 and includes the economically integrated surrounding counties (Sanders and Rowntree, 1983). These are areas where natural systems and processes are already substantially affected by human activity or they soon will be.

Statewide multiagency analysis and assessment is the stated purpose at the subregional scale in the adjusted national hierarchy (Table 1). Separate guidelines for urban classification and analysis can be prepared at this level that are directed toward ecosystem restoration, management, and long-term sustainability based on existing and projected land use. Land use is an excellent indicator of relative change from natural landscape conditions and used as an indicator of energy, moisture, and nutrient inputs and outputs in urban hydrology and water quality analysis (Lazaro, 1979; Chavooshian *et al.*, 1987).

Within the SMSAs, management emphasis would be on the health, safety, and well-being of the public as opposed to commodity production, which characterizes rural management strategies, all driven, of course, by the fundamental precepts of ecosystem management, which are:

> (1) long-term sustainability as a fundamental value, (2) clear operational goals, (3) sound ecological models and understanding, (4) understanding complexity and connectedness, (5) recognition of the dynamic character of ecosystems, (6) attention to context and scale, (7) acknowledgement of humans as ecosystem components, and (8) commitment to adaptability and accountability (Ecological Society of America, 1995).

2.3. Linking Ecosystem Structure and Function

Structure and function are extremely important concepts when considering an ecological approach to management. In a woodland or rural setting, ecosystem structure begins with the soil that is the foundation for the various layers of the biotic community including mosses, forbs, and grasses, shrubs, subcanopy, and canopy vegetation. Each structural layer is unique in its ability to carry out various functional activities. For example, in processing the energy and matter associated with rainfall, there is a considerable difference between bare soil and a multilayered forest canopy. Rainfall striking bare soil dislodges particles that are carried downslope by sheet flow, which rapidly concentrates into rills and gullies, further eroding the soil due to the concentration of energy and contributing to rapid surface water runoff and downstream flooding.

In contrast, a multilayered forest canopy intercepts the rainfall and distributes its energy throughout the leaves of the canopy as it falls through the boughs to the ground where forest litter facilitates its infiltration into the soil. Infiltrated water is gradually released to the stream channel to maintain stream flow and aquatic life forms. Ecosystem structure in this case significantly alters functional response.

In managing ecosystems' it is important to know the response that is desired and the optimal structure to achieve that response. This is more feasible in rural settings than in urban settings due to the complexity of the urban ecosystem structure, which includes a wide variety of man-made elements including buildings of various sizes and impervious surfaces. Optimal structure to achieve a desired function, such as creation and maintenance of migratory bird habitat, is seldom possible under urban conditions. However, when managing urban ecosystems it is possible to maximize opportunity through vegetation management, given the structural conditions that exist and recognizing that compromise is essential.

2.4. The Multifunctional Role of Vegetation in Urban Environments

Urban forests have historically been managed solely for the purpose of maintaining trees for their own sake and not to achieve any additional societal benefits. This contrasts sharply with the original idea of Pinchot (1947) who defined the concept of conservation, that is the greatest good for the greatest number, that linked forestry to the stewardship of all natural resources. In this age of declining municipal budgets the first programs to be trimmed (no pun intended) are those that have minimal economic value to the community. Trees are pretty and people like them, but in almost every case they will lose out to police, fire, and public works when budgets are tight.

The alternative is to manage forests and related natural resources in populated areas to serve the needs of the people for a clean and healthy environment and demonstrate the ecological, social, and economic benefits that will accrue. Research is currently being conducted that quantifies the value of trees and other vegetation in such areas as carbon sequestration, reducing air and water pollution, heat island mitigation, energy conservation, and storm water management (McPherson *et al.*, 1994; Neville, 1996; see also Chapter 2, this volume). These are issues of value to society that can be addressed in a very cost-effective way by analyzing the functions that the urban forest is to perform and the optimal structure to accomplish it. Management plans provide the blueprint to achieve the desired structure over time.

3. Expanding the Scope of Urban Forestry

3.1. Linking Green Infrastructure to Economic Development

Review the table of contents for this volume or almost any text on urban forestry for that matter. Techniques for planting, pruning, fertilizing, maintaining, and removing trees as well as conflicts between trees and utilities are often discussed along with root and sidewalk and sewer problems. Some recent texts, like this one, include community involvement to plan for and plant trees in their neighborhood, and techniques for inventories are always discussed. The focus is typically on the trees themselves and what can be done to better manage the individuals without knowledge of the population (the forest) and the social, economic, and ecological forces that affect its condition and long-term viability.

Traditional urban forestry programs deal primarily with the physical improvements in the landscape; those components that were introduced once the natural environment was altered and specifically the tools and techniques that are needed to maintain them. There is seldom a question about the significance of functioning the natural system in populated areas and its importance to maintain the integrity to the extent possible. There is a place for the what and the how of maintaining the urban forest, but to address the increasingly important issue of sustainability (the why), the green infrastructure—the ecological component in populated areas—must be linked to economic development and re-development.

The why will vary from one location to another. Managing tree canopy cover (structure) to reduce summer energy demand (function) while restoring neighborhoods and providing summer jobs and life skills for youth might be a viable option in the city of Philadelphia. In the outlying counties, however, where intensive new development is taking place, environmental and social concerns may require strategies to maintain riparian vegetation and tree cover on upland groundwater recharge areas to reduce downstream flooding and maintain water quality. Issues vary but natural systems can and should be an integral part of the decision-making process that relates to economic viability. This process must lead to the development of comprehensive natural resource management plans to guide land use decisions and operational activities on the ground that will lead to community sustainability.

3.2. Eleven Steps Toward Managing Urban Ecosystems

Effectively managing ecosystems in urban areas requires a systematic process that can ensure the essential information for good decision-making is accounted for. The complexity of ecosystems dictates that specific ecological issues are identified within a study location in order that structural and functional information can be acquired, which addresses those, which issues. This helps to avoid the often repeated mistake of acquiring all the resource data that can possibly be acquired about a particular area only to find that most of them are irrelevant to the intended solution. Although this might appear obvious, it is a common problem. How many communities have developed a "natural resource inventory" that sits on a shelf because no one identified how the data would be used prior to their acquisition.

Through the efforts of the USDA Forest Service Northeastern Area office, in cooperation with the Center for Urban Forestry at Morris Arboretum, an 11-step process, developed by the Revitalizing Baltimore project, was adapted to provide a framework for an Ecosystem Managers Workbook (Morris Arboretum, 1994). Figure 1 illustrates the cyclical nature of this process, which is necessary to ensure that accomplishments continuously contribute toward achieving the vision and goals set forth for the study location. There are a few key points that must be considered when proceeding through the various steps.

The process is driven by a central issue or series of issues. For example, nonpoint source contamination from runoff over paved surfaces is causing algal blooms and advanced eutrophication in creeks and water bodies in the community. With consensus this issue can develop into a vision for clean water with specific goals for aggressive management of critical watersheds, riparian areas, and vegetative cover throughout the community to reduce storm flow and maximize nutrient uptake. This vision and associated goals provide focus for all project activities and must be periodically revisited to ensure that compliance is maintained.

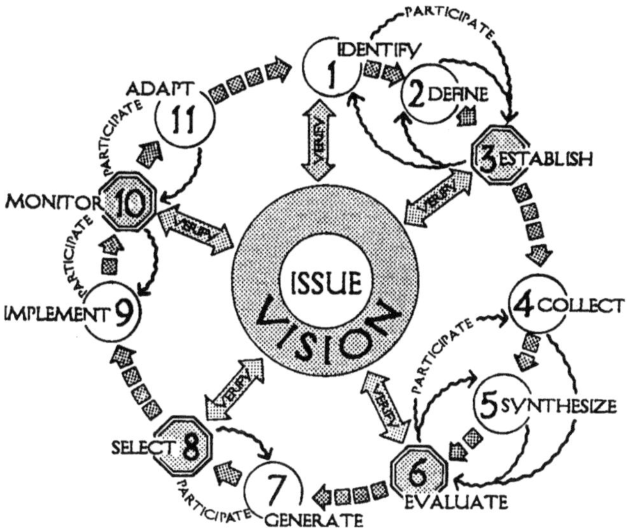

FIGURE 1. Steps in the process. (1) Identify the issues and opportunities to be addressed. (2) Form the project team and define the project scope. (3) Establish a vision and goals through public participation. (4) Identify and collect data to respond to the project goals. (5) Synthesize and analyze data. (6) Evaluate direction and verify goals. (7) Develop alternatives for action. (8) Select the preferred alternative. (9) Implement the preferred alternative. (10) Monitor and evaluate the program. (11) Adapt management strategies to changing conditions.

The process is cyclical, interactive, and adaptive. The stop signs at Steps 3, 6, 8, and 10 serve as reminders for review, testing, and approval. For example, the vision and goals for the project are established through public input in Step 3. This sets the stage for specific data gathering and analysis to determine existing conditions and the effects of alternative future conditions. Step 6 provides the opportunity to reevaluate the vision and goals based on the results of the analysis. Alternative courses of action are then identified and again verified in Step 8 according to the stated vision and goals. Once the project is implemented, a process of monitoring and evaluation will allow for periodic verification that the vision and goals are being achieved. What is learned from project implementation can be used at this point to adapt the management strategy or adjust the vision and goals for the project and continue through the cycle once more.

This process is a tool that can be used at any scale for any project, from restoring a wetland to managing a metropolitan landscape. These is seldom any process simple enough to be linear as is indicated by the 11 sequential steps in the model, especially when volunteers are involved. Decisions frequently have to be made before data are available and analysis is complete. These are the times when it is important to verify that actions to be taken are compatible with the vision and goals for the project.

Disciplines involved and numbers of participants will vary by project, along with complexity of data required, roles in decision-making, and implementation strategies; however, the 11 steps assist greatly in maintaining control of a complex process.

4. Conclusion

4.1. Sustainability as a Central Issue in the 21st Century

The concept of sustainability, that is, meeting current societal needs without compromising the needs of future generations, will continue to remain controversial into the next century because of the abstract nature of the definition. Borman and Likens (1979) relate sustainability to ecosystem stability, which explains the concept in more practical terms. Accordingly, "Every ecosystem is subject to an array of external energy inputs: radiant energy, wind, water, and gravity. All these represent potential destabilizing forces that may destroy or diminish ecosystem organization or sweep away the substance of the ecosystem" (Borman and Likens, 1979:5)." For an ecosystem to grow or even maintain itself (i.e. to remain sustainable) it must be able to channel or meet these potentially destabilizing energetic forces in such a way that their full destructive potential is not achieved within the ecosystem. To the forces mentioned here must be added those introduced through human caused change, as in the Chesapeake Bay example, in what Haber (1990) refers to as the *anthrosphere*. This phenomenon has resulted in transforming the natural landscape to a cultural landscape, which in general obfuscates the natural ecological productivity. Whereas the biosphere relies on cyclical production, consumption, and decomposition, the human-dominated system produces and consumes *technomas* that requires a supply of energy and matter from external sources and in relative terms is not biodegradable. The result is a growing discrepancy between natural orderliness and accelerated human disorder as reflected in environmental degradation (Vos and Zonneveld, 1993; Haber, 1990).

It is in the process of change from natural- to human-dominated systems that the potentially destabilizing forces undermining basic ecological function, and therefore long-term sustainability, must be controlled (Vos and Zonneveld, 1993). The rapid pace of development in this country and around the world will ensure that sustainability will increase in significance in the next century. By protecting and conserving vegetation as a primary ecosystem component, the natural processes of matter and energy transformation can be used to reduce the total impact of human caused change (Brush, 1995). According to Borman and Lukens (1979): "The success of an ecosystem in resisting destabilization may be judged by its ability to minimize the loss of liquid water and nutrients and control erosion" (p. 6). If this is true, urban forestry can play a key role in maintaining a sustainable society.

4.2. The Importance of Context in Natural Resource Decision-Making

It is given that all ecosystems have been impacted to varying degrees either directly or indirectly by human activity; furthermore, ecosystems are constantly in transition and will continue to be (Kaufmann *et al.*, 1994). Therefore, an essential step in the process of ecosystem management is to establish existing conditions that can be related to historical conditions for reference in terms of the change that has taken place with regard to structure and its effect on ecological function. From this, existing and potential productive capacity can be determined for use in developing sustainable future options for ecosystem management that can meet societal demands (Kaufmann *et al.*, 1994).

Clearly, human-dominated ecosystems must link to the national hierarchical frame-work (ECOMAP, 1993) in order to deal with long-term ecological and socioeconomic sustainability. The continuous exchange of energy, materials, and waste products across the urban–rural continuum must become a part of a comprehensive management framework (Vos and Zonneveld, 1993; Girardet, 1992; Haber, 1990). Context in the urban setting is especially critical since cumulative effects can be severe in a metropolitan region where hundreds of local jurisdictions are making independent decisions about the environment with no ecological framework (US Department of Agriculture Forest Service, 1992).

The Highlands area in Northwestern New Jersey and Southeastern New York States is an example of the need for context in natural resource decision-making. Ninety-two separate municipalities with "home rule" authority granted by the states are making independent, uncoordinated land and resource use decisions that threaten biological diversity and habitat, as well as the water supply for over three million people (US Department of Agriculture Forest Service, 1992). Obviously, there is a need for context in this instance that encompasses both ecological and hydrological boundaries and relates the existing condition of the resource to long-term societal needs. Furthermore, information about the resource base needs to be continually updated and made available to municipalities and the community at large through an institution of government that has some regulatory authority. Unless there is context for natural resource decision making in the Highlands and elsewhere, it is unlikely that sustainability of the resource can be maintained under heavy development pressure.

4.3. Managing Urban Ecosystems: An Expanded Role for Urban Foresters

Presently there are millions of acres of forests in various land use categories in populated areas throughout the United States. These forests, for the most part, are not managed except for the municipal trees on public lands and rights-of-way in the more environmentally conscious communities. Decisions continue to be made daily about the disposition of forested lands to other uses without the benefit of information regarding the impacts to natural systems and the long-term costs associated with such essential societal needs as clean air and water, storm water management, and energy consumption. Lands set aside as open space remain unmanaged and are seldom viewed in an ecological context where they serve to provide diversity and connectivity for plants and animals and related functional uses such as high-quality lumber production or needed outdoor recreational use.

Comprehensive management of the urban forest will not just happen, and ironically it is where the natural system has been substantially altered that management is most needed (Allen and Hoekstra, 1993). This is an unfilled niche that creates a significant opportunity for natural resource professionals who understand ecological concepts and are equally comfortable dealing with social and economic issues. It is not necessary for us to repeat the mistakes of more mature cultures like the Netherlands before recognizing the need for conserving and protecting natural resources in populated areas. By expanding their professional role, urban foresters can fill this niche and significantly contribute to long-term community sustainability through ecosystem-based management.

References

Allen, T. F. H., and Hoekstra, T. W., 1993, *Toward a Unified Ecology*, Columbia University Press, New York.

Borman, F. H., and Likens, G., 1979, *Pattern and Process in a Forested Ecosystem*, Springer-Verlag, New York.

Brush, G. S., 1995, *History and Impact of Human Activities on Chesapeake Bay*, unpublished manuscript prepared for the Renewable Natural Resources Foundation, Annapolis, MD.

Chesapeake Bay Monitoring Subcommittee, 1989, *The State of the Chesapeake Bay: Third Biennial Monitoring Report—1989*, Chesapeake Bay Program, Annapolis, MD.

Delleur, J. W., 1982, Introduction to urban hydrology and stormwater management, in *Urban Stormwater Hydrology*, Water resources monograph 7, American Geophysical Union, Washington, DC.

Ecological Society of America, Ad Hoc Committee on Ecosystem Management, 1995, *The Scientific Basis for Ecosystem Management: An Assessment by the Ecological Society of America*, Ecological Society of America, Washington, DC, prepublication copy.

ECOMAP, 1993, *National Hierarchical Framework of Ecological Units*, US Department of Agriculture Forest Service, Washington, DC.

Girardet, H., 1992, *The Gaia Atlas of Cities: New Directions for Sustainable Urban Living*, Anchor Books, New York.

Haber, W., 1990, Basic concepts of landscape ecology and their application in land management, *Physiol. Ecol. Japan* **27** (Special Number):131–146.

Hamill, Jr., S. M., Keene, J. C., Kinsey, D. N., and Lewis, R. K., 1989, *The Growth Management Handbook: A Primer for Citizen and Government Planners*, Middlesex, Somerset, Mercer Regional Council, Princeton, NJ.

Harms, B., Knappen, J. P., and Rademakers, J. G., 1993, Landscape planning for nature restoration: Comparing regional scenarios, in *Landscape Ecology of a Stressed Environment* (C. C. Vos, and P. Opdam, eds.), Chapman and Hall, London, pp. 197–217.

Kaufmann, M. R., Graham, R. T., Boyce, Jr., D. A. *et al.*, 1994, *An Ecological Basis for Ecosystem Management*, Gen. Tech. Rep. RM-246, US Department of Agriculture Forest Service, Fort Collins, CO.

Laurie, M., 1986, *An Introduction to Landscape Architecture*, 2nd ed., Elsevier, New York.

Lazaro, T. R., 1979, *Urban Hydrology*, Ann Arbor Science, Ann Arbor, MI.

McDonnell, M. J., and Pickett, S. T. A., 1990, Ecosystem structure and function along urban–rural gradients: An unexploited opportunity for ecology, *Ecology* **71**:1232–1237.

McPherson, E. G., Nowak, D. J., and Rowntree, R. A., 1994, *Chicago's Urban Forest Ecosystem: Results of the Chicago Urban Forest Climate Project*, Gen. Tech. Rep. NE-186, US Department of Agriculture Forest Service, Northeastern Forest Experiment Station, Radnor, PA.

Morris Arboretum, 1994, *An Ecosystem-Based Approach to Urban and Community Forestry: An Ecosystem Manager's Workbook*, unpublished manuscript, Philadelphia.

Neville, L. R., 1996, *Urban Watershed Management: The Role of Vegetation*, Ph.D. Dissertation, State University of New York, College of Environmental Science and Forestry, Syracuse.

New Jersey State Planning Committee, 1992, *The New Jersey State Development and Redevelopment Plan*, Trenton.

Pinchot, G., 1947, *Breaking New Ground*, Island Press, Washington, DC.

Saunders, R. A., and Rowntree, R. A., 1983, *Classification of American Metropolitan Areas by Ecoregion and Potential Natural Vegetation*, US Department of Agriculture Forest Service Research Paper, NE–516, Broomall, Pennsylvania.

Swanson, A. P., 1992, *Growth Management Efforts in the Chesapeake Bay Region*, unpublished paper, Chesapeake Executive Council, Annapolis, MD.

US Department of Agriculture Forest Service, 1992, *New York–New Jersey Highlands Regional Study*, Government Printing Office, Washington, DC.

US Environmental Protection Agency, 1983, *Chesapeake Bay: A Framework for Action*, Region 3, Philadelphia.

Van Buuren, M., and Kerkstra, K., 1993, The framework concept and the hydrological landscape structure: A new perspective in the design of multifunctional landscapes, in *Landscape Ecology of a Stressed Environment*. (C. C. Vos, and P. Opdam eds.), Chapman and Hall, London, pp. 219–240.

Vos, C. C., and Opdam, P. (eds.), 1993, *Landscape Ecology of a Stressed Environment*, Chapman and Hall, London.

Vos, C. C., and Zonneveld, J. I. S., 1993, Patterns and processes in a landscape under stress: The study area, in *Landscape Ecology of a Stressed Environment* (C. C. Vos, and P. Opdam, eds.), Chapman and Hall, London, pp. 1–25.

Walton, G. F., 1987, *Watershed Management Strategies for the State of New Jersey*, New Jersey Agricultural Experiment Station publication, H–17505–1–87, New Brunswick, NJ.

Year 2020 Panel, 1988, *Population Growth and Development in the Chesapeake Bay Watershed to the Year 2020*, Report to the Chesapeake Executive Council, Chesapeake Bay Program, Annapolis, MD.

Index

Abies species, 194
Acidity of soil; see Soil reaction/pH,
Adelges tsugae (hemlock woolly adelgid), 319
Aeration, soil, 170
Aesculus, 281, 289
Aesculus carnea briotti (ruby red horse chestnut), 319
Aesculus pavia (red buckeye), 195
Aesthetic pruning, 269
Aesthetics
  cost-benefit considerations, 36
  funding fluctuations and, 141
  planning, design, and management considerations, 41
  pruning for, 239
Agricultural lands, 89, 90, 93, 116, 351, 443
Ailanthus altissima (tree of heaven), 9, 196
Air pollution, 27–29, 349
  benefits, 26
  costs, 26
  removal of pollutants, 39
  VOCs and maintenance activity emissions, 31
Air quality, 26, 27, 29, 32, 36, 41, 101, 116, 222, 223, 421
  cost-benefit considerations, 36
  planting site selection, 233
Albedo, 28
Alkalinity of soil; *see* Soil reaction/pH
Alée, 5, 114
Amelanchier (serviceberry, shadbush), 193, 195, 196, 446
Amelanchier X *grandiflora* (shadblow), 193, 258
Amendments, soil, 178, 230, 375
American Standards for Nursery Stock, 204, 215
American urban development, 26
America the Beautiful grants, 138
Amsterdam, 5
Amur cork tree (*Phellodendron amurense*), 193
Annual reports, 55, 56, 64, 148, 235, 421
Anthrosphere, 462
Antiguity, 14
Apiognomonia veneta (sycamore anthracnose), 310
Apple (*Malus*) species and cultivars, 197
  crab apple, 189, 190, 194, 258, 446
    disease resistant, 304
    eastern tent caterpillar, 311
    pest management, 295
    susceptibility to pests (key plant), 302
    wildlife benefit, 446
  eastern tent caterpillar, 311
  as low VOC emitters, 32
  pest management, 295
  wildlife benefit, 446
Appraisal, tree, 323
  cost of repair and cost of cure methods, 342
  methods, 326
  procedures, 325
    instruments and tools, 325
    qualifications, 325
  property value, 343–346
  replacement cost
    compounded, 342
    method, 326
  role in urban forestry, 346
  trunk formula method, factors affecting value, 327–338
    condition, 327
    location, 330
    replacement cost and basic price, 339
    size, 337
    species, 336
    summary of method, 340, 341

Arbor Day, 12, 153, 162, 257, 419, 420, 423–425, 433, 434
Arboreta, 9, 101, 187, 323
Arboriculture
  development of, 12
  legal issues; *see* Laws and ordinances
  standards, local ordinance, 209
  tree selection, 188, 189
Arboricultural practices, care, and maintenance
  crew training in, 145
  newley planted trees, 224, 227, 234, 235
  nursery production practices affecting survival and growth, 205–208
  planting, 235
  prevention of premature tree removal, 383, 385
Arborvitae (*Thuja occidentalis*), 194, 197
Arborvitae, giant (*Thuja plicata*), 194
Ash (*Fraxinus*) species and cultivars, 196, 262, 281
  as low VOC emitters, 32
  green (*Fraxinus pennsylvanica*), 192, 195, 271, 317, 446
  lightning susceptibility, 281
  mountain (*Sorbus alnifolia*), 193
  root systems, 262
  sawlog desirability, 402
  white (*Fraxinus americana*), 125, 191, 197, 310
  wildlife benefit, 446
Aspen (*Populus grandidentat, Populus tremuloides*), 197
Aspergillus fumigatus allergenicity, 374
Atlanta, 29–31
Avenues, 5–7, 70, 74, 76, 259, 321, 410

Bacillus thuringiensis (BT), 298, 307, 312, 315
Backyard composting, 350, 351
Backyard mulching, 351
Bald cypress (*Taxodium distinchum*), 192, 195
Balled and burlapped planting, 230
Baltimore, 30, 31, 33, 453, 460
Banded ash clearwing borer (*Podosesia aureocincta*), 317
Bare root planting, 226, 230
Bare-root plants
  nursery stock, 226
  protection from desiccation, 226
Bark
  decay and, 284
  hazard tree features, 284
  pruning technique, 239
  root decay symptoms, 285
  staked and guyed trees, 233
Bark damage 330
  herbicides, 202
  during planting, 202
Baroque boulevard, 6
Barriers, root, 273–276
Basal pruning, 241
Basic price, tree appraisal value, 327
Basswood, sawlog desirability, 402
Bayberry (*Myrica pennsylvanica*), 446
Beech (*Fagus*) species, 281, 402
  lightning susceptibility, 281
  sawlog desirability, 402
  wildlife benefit, 446
Beneficial insects 296, 304, 315, 319, 443
  hemlock woolly adelgid control, 303
Benefits
  natural environment, planning considerations, 79
  systematic management, 120
  volunteers, 159
  wildlife, 446
Benefit and costs
  funding considerations, budget allocations, 134
  sawlogs, recycling removed trees, 399–418
    see also Sawlogs, street tree recycling as
Benefits and costs of urban forest ecosystems, 25–46
  benefit-cost analysis, 40
  implications for planning, design, and management, 41
  physical-biological, 26–36
    atmosphere, 27–33
    hydrology, 33, 34
    noise, 34
    wildlife and biodiversity, 34
  social-economic, 36
    communities, 37
    individuals, 36
    miscellaneous, 39
    real estate values, 38
    tree value formulas, 39
Bid specifications
  nursery stock, 218
    realities, 217
    resources, 218
  tree removal, 390
Biological control, 12, 185, 297, 306, 307, 311, 312, 315–317, 319, 321
  chestnut blight, 321
  eastern tent caterpillar, 311
  gypsy moth, 306, 315
  hemlock woolly adelgid, 307, 319
Biological factors, planting site selection, 224
Biological failure, 378, 391

Biological properties of soils, 166, 173, 174, 375
Birch (*Betula*) species 75, 191, 197, 196, 281, 289, 305, 446
  decay susceptibility, 289
  lightning susceptibility, 280, 281
    disease resistance, 314
    Heritage, 305
  white (*Betulla alba*), 302, 305
  wildlife benefit, 446
Birds, attracting, 444
Black gum (*Nyssa sylvatica*), 192, 195, 196, 445, 446
Block grants, 138, 139
Blueways, planning and design, 103, 104, 106
Boston
  air pollutant removal figures, 30
  Louisburg Square, 112
  pollution removal figures, 30
Boston Common, 6, 104
Botanical gardens, 8, 9, 101, 323
  Middle Ages, 4
Botanic gardens, 9
Boulevard, baroque, 6
Boulevards, 6, 15, 240, 317
Bounday layer heights, 28, 29
Branches
  scaffold, 238, 241
  temporary, 242
Branching habit, tree selection considerations, 189
Broussonetia papyrifera (paper mulberry), 190
Buckeye (*Aesculus*), 289
Buckeye (*Aesculus carnea briotti*), 258
Buckeye (*Aesculus pavia*), 195
Budgeting; *see* Appraisal, tree; Financing and budgeting
Buffer zone, composting facility, 374
Buildings
  energy conservation, costs and benefits, 27
  site conditions, tree selection considerations, 186
  site selection for planting, 222
Bulldozer, 71, 343, 379
Burl, decay and, 285
Burlap, plastic, 203, 207, 208
Burlapped planting, 230, 231
Burlap wrapped soil balls, 204, 205
Business area landscaping, 38

Caliper/trunk diameter
  appraisal of tree value, 323, 325–327
  contract specifications, 208
  minimum thickness formula, wood decay threshold, 289
  nursery stock, 201, 205
  sawlog potential, 401
callery pear, 193, 225, 241
Cankers
  and decay, 285, 288
  herbicides and, 206
  nursery stock, 202
Canopy
  appraisal of tree value, 327
  design considerations, 107
  nursery stock, 202
  temperature increase, conditions causing, 27, 28
Cant hook, 389
Carbon dioxide levels, 26
Care and maintenance of plants, *see* Arboricultural practices, care, and maintenance
Carpinus betulus (European hornbeam), 193, 271
Carpinus caroliniana (American hornbeam), 195, 258, 271
Catalpa (*Catalpa*), 196, 223, 281, 300
Cation exchange capacity, soil, 171, 172
Cedar
  Atlantic white (*Chamaecyparis thyoides*), 195
  Cedrus species, 194
  eastern red (*Juniperus virginiana*), 195, 196, 305, 335, 445, 446
  Japanese (*Cryptomeria japonica*), 196
  (*Juniperus virginiana*) cedar apple rust, 305
  northern white (*Thuja occidentalis*), 194
  Port Oxford (*Chamaecyparis lawsoniana*), 194
  sawlog desirability, 402
  western red (*Thuja plicata*), 194
Cedrus (cedar) species, 194
Celtis (hackberry), 75, 191, 196, 446
Central Park, 7, 104
Cercis (redbud), 75, 195, 258
Chainsaws, 379–383, 388, 392, 403, 407
Chamaecyparis (false cedar), 194, 195
Chemical contaminants, site conditions, 188
Chemical control methods, 317, 448 *see also* Pesticides
Chemical properties of soils, 167, 171, 173, 188
Cherry (*Prunus*) species and cultivars, 194
  black (*Prunus serotina*), 196, 446
  chokecherry (*Prunus virginiana*), 258, 446
  eastern tent caterpillar, 311
  as low VOC emitters, 32
  oriental (*Prunus serrulata*), 75, 258
  sawlog desirability, 402
  wildlife benefit, 443, 446
Chestnut, American (*Castanea dentata*), 321
Chestnut, Chinese (*Castanea mollissima*), 321

Chestnut blight (*Cryphonectria parasitica*), 313, 321
Chicago, 9, 29, 31, 35, 37, 88, 100, 453
Chilocorus (lady beetle), 306
Chokecherry (*Prunus virginiana*), 258, 446
Citizen planning commissions, 84
Civic facilities, planning and design, 101
Civic organizations
  public relations, 152, 163
  volunteer searches and recruitment, 153, 163
Climate, tree selection considerations, 189
Cold storage for nursery stock, 226
Comment forms, 151
Commercial areas, landscaping, 38
Commissions, shade tree, 52
Common law rule, 49
Commons,
  Boston, 6, 104
Community
  benefits and costs of urban forests, 36
  definitions and attributes, 80, 81
  gateways, 113, 114
Community development, 81
  block grants, 138, 139
Community involvement, 147–164
  public awareness, 147
    annual reports, 148
    comment forms, 151
    image analysis survey, 152
    leaflets, 148
    lectures, 150
    news releases, 149
    television, 151
    videotapes, 151
  relations with local officials, 152
  volunteers, involving, 153
    advertising, 155
    applications for screening, 155
    benefits, 157
    challenges, 159
    financing and locating teachers, 159
    selection, 155
    targeting needs, 155
    training, 155–159
Community planning
  community, definitions and attributes, 81
  community development, 81
  definitions, 80
  ecosystem approach, 95
  elements of
    comprehensive plan, 83, 85
    natural resource inventory, 88
    zoning and subdivision ordinances, 87
  environmental ethic, 96
  leadership, 95
  legal basis, 80, 81
  linking green infrastructure to economic development, 459
  managers and planners working together, 94
  natural environment, benefits of, 79
  problems, 88, 89
  strategies for planning and land use regulation, 85–90
  strengthening, 95
Compaction, soil; *see* Soils, compaction
Compensatory damages, 51
Composting; *see* Leaf composting
Comprehensive plan, 59, 79, 81–87, 91, 95
Compressed air, soil fracturing, 178
Computers
  Internet plant locator resources, 218
  selection of trees, 198
  shade tree inventory, 122, 127–129
Condition, appraisal, 327
Confined urban space, 271, 272
Connecticut statutes, 48–51, 53
Conservation areas, planning and design, 101
Conservation subdivisions, 92, 93, 101
Conservation zoning, 92
Consultants, 126, 140, 209, 213, 324, 325
  for bid specifications, 213
  grant writing, 140
Consulting foresters, 323, 324, 409, 410
Container planting, 231
  nursery stock, 205, 216
  soil volume, 178
Contaminants, specification of, 180
Context
  importance in decision making, 462
  map, community planning and land use regulation strategies, 92
  urban forest resource management, 455–458
    ecosystem classification, natural to human-dominated, 456, 457
    linking ecosystem structure and function, 458
    multifunctional role of vegetation in urban environment, 459
  urban-rural continuum, 455
Contoured sidewalks, 274–276
Contract, nursery stock, 211–218
  business section details, 212, 213
  general form, 211
  project section details, 214
  results, 211–217
Cooperative Assistance Program, 138
Cooperative Extension Office, 18, 151
Coppice growth, 282
Cordage, 392
Cork tree, Amur (*Phellodendron amurense*), 193

Cornus; see Dogwood
Cost of repair and cost of cure methods, 342
Costs
  and benefits; *see* Benefits and costs
  budgeting considerations, 375
  leaf composting facility, minimal technology, 375
  of program; *see* Financing and budgeting
  tree removals, 377, 379, 380
  tree size and, 225
Course, 12, 14, 19, 82, 83, 85, 100, 149, 159, 163, 176, 185, 190, 238, 277, 280, 352, 375, 384, 386, 390, 395, 436, 458, 461
Cover, wildlife needs, 438–440
Crab apple; *see* Apple species and cultivars
Crane, removal methods, 387
Crataegus (hawthorn), 193, 194, 446
Crews
  training, 145
  tree removal, safety training and performance standards, 393
  volunteer; *see* Volunteers
  work assignments, inventory utilization for, 145
Criminal penalties for tree damage, 51
Crown
  and root decay, 281
  cleaning, 242, 244
  hazard trees, 247, 256
  microclimate modification, 12
  nursery stock, identification of, 205
  pruning, 238–246
  reduction, 242, 243, 245
  site selection for planting, 224
  tree selection considerations, 189, 190
Cryphonectria parasitica (chestnut blight), 321
Cryptomeria japonica (Japanese cedar), 196
Cultural control, pest management, 297, 303
Cultural practice; *see* Arboricultural practices, care, and maintenance
Cypress
  bald cypress (*Taxodium distichum*), 192, 195
  Lawson (*Chamaecyparis lawsoniana*), 194

Damages, legal liability, 47
Dark Ages, 4, 10
Data collection, shade tree inventory, 123, 124
Datascape Guide to Commercial Nomenclature, 210
Deadmen, 233
Dead trees, management guidelines, 288
Debris removal, tree removal projects, 384, 386
Decay
  hazard trees, 277, 287
  minimum thickness formula, 289
  root failure, 285
  stem, strength loss thresholds, 288
Decision making; *see also* Selection of trees
  inventory and, 144, 145
  shade tree inventory and, 119, 120
Delaware statutes, 49, 51
Delaware & Raritan Canal, 106
Desert regions, 34
Design; *see* Planning and design, urban/community forestry
Detroit, 15, 453
Development; *see* Community planning
Development rights, 84, 93, 94
Diameter, tree; *see* Caliper/thrunk diameter
Diospyros virginiana (persimmon), 401, 402, 446
Disease; *see also* Integrated pest management
  alteration of native vegetation structure and, 27
  and decay, 284
  appraisal of tree value, 327
  budgeting priorities, 142
  contract specifications, 216
  hazard tree features, 287
  nineteenth and early twentieth centuries, 19, 20
  nursery stock, 208
  pruning wound care, 246
  root decay symptoms, 285
  site conditions, tree selection considerations, 186
  tree selection considerations, 186
Disposal of trees, recycling as sawlogs, 399–417; *see also* Sawlogs, street tree recycling as
Dogwood, 162, 193, 195, 241, 258, 302, 317, 335
  flowering (*Cornus florida*), 75, 195, 446
  hybrids, 193, 195
  kousa (*Cornus kousa*), 75, 193
  red osier (*Cornus stolonifera*), 446
Double damages, 51
Drainage, soil; *see* Soils, drainage
Drip line, 177–179
Drop crotch pruning, 254
Dry climates, root system growth, 264
Dry sites, tree selection, 195, 196
Dutch elm disease (*Ophiostoma ulmi*), 13, 14, 120, 142, 188, 189, 304, 313

Easement, planting, 269–271
Eastern tent caterpillar (*Malacosoma americanum*), 305, 311, 312
Economic development
  linking green infrastructure to, 459
  planning, *see* Community planning

Economics; *see also* Appraisal, tree; Financing and budgeting
benefits and costs of urban forests, 35
communities, 37, 38
individuals, 36, 37
miscellaneous, 39
real estate values, 38
tree value formulas, 39
failure to plan and, 83
urban decay, 452
Ecosystem
alteration of native vegetation structure and, 27
linking structure and function, 458
natural to human-dominated, 456
urban forester role, 459
urban-rural continuum, 455
Ecosystem approach, 95
Education benefits, 36
Elder, box, 289, 448
Elderberry, 445, 446
Elm bark beetle, 13, 304
Elm leaf beetle, 300, 305, 307, 314
Bacillus thuringiensis and, 307
pest management, 305, 307
Elm (*Ulmus*) species and cultivars
American (*Ulmus americana*), 75, 192, 195, 271, 313
as low VOC emitters, 32
Chinese (*Ulmus parvifolia*), 192, 193, 196, 314
Dutch (*Ulmus X hollandica*), 196
hybrid, 271
lightning susceptibility, 281
root systems, 262
sawlog desirability, 402
Siberian (*Ulmus pumila*), 192, 314
"Urban" (*Ulmus glabrata*), 314
Wilson (*Ulmus wilsoniana*), 314
Emergency conditions
tree removal, 387, 397
utility-municipality cooperation, 257
Emissions from trees, 27, 32
Encroachment, 50, 357, 452
Endangered species, 35
Endowments, 134, 137, 138
Energy conservation, costs versus benefits, 27, 32, 33
Energy reserves, nursery stock, 203
Entomophaga maimaiga, 315
Environmental education, 40, 102, 115
Environmental ethic, community planning, 96
Equipment
budgeting, 144, 145
tree appraisal, 325
tree removal, 379
Eucalyptus, 32, 378
Euonumus (key plant), 302
Europe, history of urban forestry, 17
Evergreens,
for parks, windbreaks, and landscapes, 194
pruning, 226, 240, 246
Exotic species
early horticulture in US, 18, 19
insect predators, 306
problems with, 35
Expert witnesses, tree appraisers as, 131, 346

Fagus; *see* Beech species
Failure profiles, 278
Falling axe, 389
Falling wedges, 389
Farmland mulching, 351
Farmland suburban landscapes, 443
Federal agencies, environmental issues, 140
Fencing, prevention of soil compaction, 176
Fertilization
budgeting, 142
foliage growth and insect pests, 303
nursery practices, 205
pest and disease susceptibility, 336
planting, 229
root-building, 232
soil chemical property modification, 173
Field form reports for plant appraisers, 326, 343
Financing and budgeting, 133
budgeting, 140
allocation of funds, 141
competition for funds, 134
establishment of priorities, 142
competition for funds, 134
history and overview, 133
inventories and role of, 142
justification for budget funds, 135
sources, 134
federal and state government, 137
property tax, 136
sources of funding, 134, 135
volunteer training, 157
Fines or jail sentences, 52
Fir (*Abies*) species, 194, 289
Fire hazard, leaf composting facility, 378
Firethorn, 446
Firewood, 2, 328, 384, 389, 390, 400, 401, 404
Flare, trunk, 283, 339
Flooding, 27, 33, 188, 222, 329, 357, 358, 378, 444, 452, 458, 460
Fly ash, soil amendment with, 178
Foliage; *see also* Crown
appraisal of tree value, 330
nursery stock, indicators of health, 203
tree selection considerations, 192

Food, wildlife, 439
Forensics, 346
Foresters, consulting, 323–325, 338, 409, 410
Forestry committee, 56
Form, pruning for, 238
Fracturing, soil, 178
Fraxinus; *see* Ash species and cultivars
Frontage tax, 137
Fruit properties, tree selection considerations, 192
Fruitwoods, sawlog desirability, 402
Full-fee acquisition, 93
Funding; *see also* Financing and budgeting
  community planning, 94
  inventory and, 143
  public relations, relationships with local officials, 152
  shade tree inventory, 119
  state legislation, 53
  volunteer training program, 159
Future prospects, 451–453
  comprehensive management, 454
    eleven steps, 460
    linking green infrastructure to economic development, 459
    need for, 454
  context, importance in decision making, 462
  context for urban forest resource management, 455
    ecosystem classification, natural to human-dominated, 456
    linking ecosystem structure and function, 458
    multifunctional role of vegetation in urban environment, 459
    urban-rural continuum, 456
  existing responsibilities for urban natural systems, 454
  planning and design, 115
  role of urban forester in managing urban ecosystems, 463
  sustainability as central issue, 462
  sustainable development, 452
  urban decay, economic system effects on, 453

Garden allée, 5
Gas tax, 136–138
Gateway planning and design, 113, 114
General funds, 134–138, 145
Ginkgo, 9, 75, 192, 195, 271, 303
Girdling, rope, 142, 205, 208, 225, 231, 233, 234, 329, 332
Gleditsia triacanthos (honey locust), 75, 192, 196, 271
Goldenrain tree (*Koelreuteria paniculata*), 193, 258
Governmental immunity, municipality liability, 50
Government funding, 134–136
Government policy; *see also* Laws and ordinances; Legislation; State government
  overregulation, 84
  planning considerations, 100
  regulations
    tree removal, 394
    utility wires, 250
Grade cut, 283
Grants, 18, 19, 52–54, 60, 65, 95, 134, 136–140, 159, 436
Grant writing, 139
Grape, 446
Grass clippings, 350–354, 356, 357, 360, 361, 367–374
Greenbelts, 89, 90, 103, 116
Greenways, 86, 91, 93, 101, 103, 104, 106, 437
Gro-bags, 204, 205, 218
Ground failure, 280
Growing conditions; *see also* Arboricultural practices, care, and maintenance
  appraisal of tree value, 327
  pest and disease susceptibility, 336
Growth, community planning
  community options, 80
  economic development versus development of community, 81
  zoning and subdivision ordinances, 87
Growth habit, site selection for planting, 225
Growth history, nursery stock, 201
Guarantee, bid/contract specifications, 217
Guide for Plant Appraisal, 39, 326, 334, 338
Gum, 191, 223
  black (*Nyssa sylvatica*), 192, 195, 196, 402, 445, 496
  sawlog desirability, 402
  sweet (*Liquidambar styraciflua*), 223, 445, 446
  wildlife benefit, 446
Guying, 228, 231, 233
Gymnocladus dioicus (Kentucky coffee tree), 196
Gypsy moth (*Porthetria dispar*), 12–14, 300, 305–307, 311, 315, 316

Habitat, wildlife, 34, 67, 89–91, 102, 437, 438, 440, 448
Hackberry (*Celtis*), 75, 191, 196, 446
Halesia carolina (Carolina silverbell), 75, 193, 194, 258
Handling
  nursery stock, contract specification, 208
  planting process, 225, 226

Hardiness, tree selection considerations, 187, 269
Hawaii rule, 49, 50
Hawthorn (*Crataegus*), 194, 200, 258, 302, 311, 446
Hazard reduction pruning, 243
Hazard trees, 277–293
  identification of, 278
    decay, 284
    location, 279
    problem conditions and locations, 280
    root defects, 285
    symptoms of root problems, 285
    tree defects, 281
  management, 287
    decision making, 287
    storms, 292
    strength loss through stem decay, 288
    utility line clearance, 291
  management program, 142
  methods, 278
  removal, reasons for, 378; *see also* Removals
Heading back (pruning method), 242, 342
Health, tree; *see also* Integrated pest management
  appraisal of tree value, 327–330
  certification, 216
  contract specifications, 214
  electricity and, 250
  funding fluctuations and, 140
  hazard trees; *see* Hazard trees
  newly planted trees, 235
  nursery stock, 205–207
  pest and design considerations, 336
  pruning for, 238
  root cutting considerations, 111–115
Health problems, human, 36
Height, nursery stock, 201–203, 207, 209, 215
Hemlock (*Tsuga*), 281, 319
  integrated pest management, 302
  lightning susceptibility, 281
  susceptibility to pests (key plant), 301
  wildlife benefit, 446
Hemlock woolley adelgid (*Adelges tsugae*), 302, 303, 307, 319
Herbicide damage, nursery stock, 206
Hickory, 223, 269, 310, 446
Highway interchange planning and design, 115, 116
Highway planning and design, 107
Hillier's manual of Trees and Shrubs, 210
Hilling, nursery stock, 205
History and origins of urban forestry, 1–20
  agricultural revolution, 3
  America, 12
  American urban development, 6–8
  arboricultural practices 10
  community trees, 19
  design of urban landscapes, 3
  development of urban areas and vegetation use, 3
  Europe, 10
  legislation and scientific treatment of trees, 11–14
  organizations, 18
  urban forest management, 14
Holly (*Ilex*), 32, 75, 194, 196, 281, 335, 446
Honeylocust (*Gleditsia triacanthios*), 75, 192, 196
Honeysuckle, tartarian (*Lonicera*), 335, 446
Hornbeam
  American (*Carpinus caroliniana*), 193, 195
  European (*Carpinus betulus*), 193, 271
  hop (*Ostrya virginiana*), 75
Horse chestnut (*Aesculus*), 196, 258, 281, 293
Hortus Third, A Concise Dictionary of Plants Cultivated in the United States and Canada, 24
Human-dominated ecosystems, 456, 457, 463
Humidity, 28, 304; *see also* Water
  pest and disease susceptibility, 298
  root system growth, 262, 263
Hydrology, 33, 458; *see also* Water

Idaho laws, 52
Ilex (holly), 32, 75, 194, 196, 281, 446
Illinois statutes, 49, 51
Image analysis survey, 151, 152
Incentives, community planning, 95
Income tax, municipal, 136, 137
Indiana statutes, 49, 52, 53, 60
Industrial Revolution and romanticism, 3
Infrastructure repair, 27; *see also* Sidewalks and sewers; Utility lines
Inner city decay, 453, 455
Insecticides
  banded ash clearwing borer control, 317
  hemlock woolley adelgid control, 319
Insect pests; *see also* Integrated pest management,
  alteration of native vegetation structure and, 27
  appraisal of tree value, 327
  contract specifications, 213
  nineteenth and early twentieth century, 14
  nursery stock, 202, 203
  root protection after planning, 241
  site conditions, tree selection considerations, 189
  tree selection considerations, 189
  tree-species to site matches, 383

Insects, beneficial
hemlock woolley adelgid control, 307
pesticides and, 296
pest management, 304
Inspection
hazard trees, 277–293; *see also* Hazard trees
transportation, care, and handling, 230
Institutional facilities, planning and design, 101
Integrated pest management
case studies, 310–322
banded ash clearwing borer (*Podosesia aureocincta*), 317
chestnut blight (*Cryphonectria parasitica*), 321
Dutch elm disease (*Ophiostoma ulmi*), 313
eastern tent caterpillar (*Malacosoma americanum*), 311
gypsy moth (*Porthetria dispar*), 315
hemlock woolley adelgid (*Adelges tsugae*), 319
sycamore anthracnose (*Apiognomonia veneta*), 310
monitoring, 298–301
key plants and key pests, 301
thresholds, 301
pest control strategies, 303–308
biological control, 306
cultural control, 303
mechanical control, 305
microbial and viral control, 307
pesticides and biorational pesticides, 307
resistance to pesticides, 290
resurgence of pests, 290
secondary ourbreaks, 290
Interior avenues, 5
International Code of Nomenclature for Cultivated Plants, 210
Internet plant locator resources, 218
Introduced species; *see* Exotic species
Inventory, shade tree, 120
automation, 127
benefits of, 120
budgeting and funding considerations, 140–142
case study, 126
computer system
automation, 127
hardware, software, and training materials, 127
record-keeping system, 122
data collection, 124
mechanics of conducting, 123
recourse assessment, 128
systematic management, 120
training issues, 120
types of, 121
work plan
development, 129
evaluation of, 130
review and implementation, 130
Iowa statutes, 49, 51
Irrigation
nursery stock, 207, 217
pest and disease susceptibility, 298
Ivancic v. Olmstead, 47, 61

Jail sentences, 52
Japanese beetles, 189, 224, 296, 300, 306, 307, 443
Jay v. Norwalk Lodge No. 730 BPOE, 48, 61
Job training partnership act, 138
Joint ventures, utilities and municipalities, 257, 258
Juniperus virginiana (Eastern red cedar), 195, 196, 305

Katsura tree (*Cercidiphyllum japonicum*), 75, 195, 271
Kentucky coffee tree (*Gymnocladus dioicus*), 75, 192, 196
Key plants and key pests, 301–303
Koelreuteria paniculata (goldenrain tree), 193, 258

Lady beetle (*Chilocorus kuwanae*), 306
Lady beetle (*Pseudoscymnus*), Asian, 307
Lady beetle (*Pseudoscymnus tsugae*), 319
Land acquisition, community, 93, 94
Landowner liability, *see* Laws and ordinances
Landscape, tree selection considerations, 183–185
Landscape Institute, table of landscape trees and shrubs, 197
Landscape Plants for Eastern North America, 211
Landscape Specification Guidelines, 210
Landscaping, 3, 10, 38, 111, 267, 324, 344, 345, 356, 381, 425, 434, 438
Land-use planning, 17, 80, 84, 88, 91, 95, 96, 435
Land-use regulations, 84–86, 91, 94
Larch (*Larix*), 195, 315
Large tree transplanting, 232
Larix (larch), 195, 315
Lawn area, 272, 333
Laws and ordinances 47, 63
appraisal, 345, 346
arboriculture law in Northeast, 47
general tree law, 47

Laws and ordinances (*Continued*)
municipal commissions, enabling statutes, 52–60
community planning, 83, 85, 90, 95
sensitive area ordinances, 92
zoning and subdivision ordinances, 87
local/municipal, 63–77
drafting, 65
examples, 70–77
functions, 68
legal authority, 64
planning, 64
sections, 68
specifications and standards, 69
Leachate, leaf composting facility, 357, 361, 372
Leadership, planning, 90
Leaf composting, 349–376
current management alternatives, 350
facility siting and preparation
area requirement, 356
buffer zone, 356
fire hazards and safety considerations, 359
location, 357
percolation and water table, 358
public participation, 355
roads, 359
security, 359
slope and grading, 358
stream encroachment and water pollution, 357
water supply, 359
grass clippings and other yard wastes, 368
principles, 352–355
problems, 369–375
Aspergillus fumigatus allergenicity, 374
illegal dumping, 370
leachate, 372
odor, 369
plastic bags, 373
slow composting, 370
toxic contaminants, 374
water pollution, 376
process, product, and objectives, 351
technology, 360–366
high-level, 367
intermediate-level, 366
low-level, 360–366
use of, 375
Leaflets, 148, 159
Leaf type, composting considerations, 355
Leaning tree hazards, 290, 291
Leases, rolling, 94
Lectures, public relations, 149, 151
Legal issues, *see also* Appraisal, tree; Laws and ordinances; Liability
community planning, authority for, 83, 85
sidewalk accidents, 266
tree appraisers as expert witnesses, 346
Legislation
Illinois, 51, 53, 60
New Jersey, 49, 52, 265
urban forestry, 18
Liability, *see also* Appraisal, tree; Laws and ordinances
contractors, bid specifics, 211
cost-benefit considerations, 49
hazard trees; *see also* Hazard trees
New Jersey legislation, 265
shade tree in inventory benefits, 121
sidewalk accidents, 266
sidewalk and sewer problems, 266
types of damages, 51
utility-municipality cooperation, 254
volunteers, 154
Life cycle
pruning, 209, 209
of tree; *see* Removals
Light
planting site selection, 224
site conditions, tree selection considerations, 186
tree selection considerations, 185
urban conditions, pruning considerations, 237
Lightning, 222, 280–282, 288
Lilac, tree (*Syringa amurensis*), 75, 194, 258
Linden (*Tilia*) species, 192, 196
Amsterdam canals, 5
decay susceptibility, 289
'Greenspire' (*Tilia cordata*), 75, 192, 196, 269, 271, 289
tree selection considerations, 192
Liquidambar styraciflua (sweet gum), 75, 192, 195, 446
Liriodendron tulipifera; *see* Tulip tree
Litigation; *see* Laws and ordinances; Legal issues; Liability
Local government; *see* Community planning; Municipalities
Location, appraisal considerations, 327, 331
Locust, black (*Robinia pseudoacacia*), 196, 281
Locust, honey (*Gleditsia triacanthos*), 75, 192, 271, 333
Locust species, 75
Loggia v. Grobe, 50, 61
London, planetrees, 108, 192, 195, 310
Longevity, tree, 225
Lonicera, 446

Magnolia, 192–196
Mailing list, public relations, 148
Maine statutes, 49, 51, 53, 57, 58
Maintenance; *see also* Arboricultural practices, care, and maintenance: Leaf composting
  and air pollution, 27
  budgeting for, 135, 138, 141–145
  challenges to tree management, 135
  cost-benefit considerations, 27, 31, 32, 38–41
  grants for, 137
  management program, 138
  planning, design, and management considerations, 41
  pruning, 137
  shade tree inventory and, 124
  tree removal, 387
  tree selection considerations, 188, 189
Malacosoma americanum (eastern tent caterpillar), 311
Malls, 5
Malus; *see* Apple species and cultivars
Management; *see also* Inventory, shade tree; Leaf composting
  benefit and cost considerations in, 25
  budgeting, inventory use to create management program, 142
  comprehensive, 460, 463
    eleven steps, 460
    linking green infrastructure to economic development, 459
    need for, 454
  cost-benefit considerations, 26
  crew training, 149
  grants for, 137
  hazard trees, guidelines for, 288
  site conditions, tree selection considerations, 186
Managers, planning, 94
Manual for Plant Appraisal, 343
Manual of Cultivated Conifers, 210
Manual tree removal methods, 377, 379, 381, 383
Maple (*Acer*) species and cultivars, 193, 195
  Amur (*Acer ginnala*) 193, 258, 262
  'Armstrong' freeman (*Acer x freemanii*), 191, 195, 271
  decay susceptibility, 289
  frost cracking, 230
  hedge (*Acer campestre*), 75, 193, 258, 271
  Japanese (*Acer palmatum*), 195
  lightning susceptibility, 289
  low-growing species and varieties, 258
  Norway (*Acer platanoides*), 35, 75, 191, 194–196, 225, 262, 269, 271
    'Emerald Queen', 191, 269
    fastigate, 271
    invasion by, 35
    root system, 262
  red (*Acer rubrum*), 124, 180, 191, 195, 200, 269, 271
    'Armstrong', 269
    nursery stock, selection of variety, 200
    'October Glory', 191, 200, 271
    'Red Sunset', 191, 200, 271
  root systems, 262
  sawlog desirability, 402
  silver (*Acer saccharinum*), 195
  sugar (*Acer saccharum*), 190, 191, 233, 269, 289
    'Green Mountain', 191, 192, 195, 200, 269, 271
    selection considerations, 190, 199
  sycamore (*Acer pseudoplatanus*), 191, 196, 289
  Tatarian (*Acer tataricum*), 193
  tree-species to site matches, 383
  Trident (*Acer buergeranum*), 193
  wildlife benefit, 446
Mapping
  natural resources, community planning, 93
  pest monitoring, 298
Market evidence, plant appraisal procedure, 344
Maryland statutes, 49, 51, 53, 59
Massachusetts rule, 15, 49, 52
Massachusetts standards, 53
Mass transit, 455
Mechanical breakup of soil, compaction remediation, 177
Mechanical digging, 227, 230
Mechanical failure, 378, 379, 385
Mechanical stresses and injury
  appraisal of tree value, 327
  hazard trees, 290
  site selection for planting, 222
Mechanical tree removal methods, 380
Mechanical tree spades, 206, 232
Medieval times, 6
Metal detectors, 402, 403
Metasequoia glyptostroboides (dawn redwood), 192, 271
Michigan, 1, 15, 49, 52
Microorganisms
  composting, 351, 354
  damage to root collar tissue, 227, 234
  microbial pesticides, 307
  root-building products, 232
  soils, 173
Middle Ages, 4
Minnesota, 49, 51

Missouri, 49, 51
Mites, predaceous, 307
Miticides, 296, 307
Monitoring
  newly planted trees, 235
  pest management, 235, 295
    key plants and key pests, 301
    thresholds, 301
Morus; *see* Mulberry species
Mounded trees, hazard tree features 283
Mulberry (*Morus*) species, 34, 196, 197, 335, 402, 446
  sawlog desirability, 402
  wildlife benefit, 446
Mulberry, paper (*Broussonetia papyrifera*), 196
Mulching
  backyard, 351
  damage to root collar tissue, 227
  pest and disease susceptibility, 336
  planting, 233
  prevention of compaction, 175
  soil compaction remediation, 176
Munich, Germany, pedestrian zone plantings, 107
Municipal budget; *see* Financing and budgeting
Municipal income tax, 136
Municipalities; *see also* Community planning: Utilities, and municipalities
  codes and ordinances, 100; *see also* laws and ordinances
  cost-benefit considerations, 41
  liability, governmental immunity, 50
  management of trees in public areas, 41
  planning and land use, 83
  reuse of street trees as lumber, 412
  state law, 47, 49, 51, 52–54
  Tree City USA; *see* Tree City USA
Myrica pennsylvanica (bayberry), 446

National Arbor Day Foundation, 18, 19, 419, 434, 435; *see also* Tree City USA
Natural boundaries, 90
Natural environment
  community planning, 79, 82
    problems, 88, 89
    strategies, 90
  planning considerations, 83, 108
  shift to human-dominated ecosystem, 462
  urban systems, existing responsibilities for, 454
Natural factors, planting site selection, 222
Natural pruning, 254
Natural resource inventory, 85, 88, 91, 96
Natural resource managers, community planning, 94
Natural resource mapping, 93
Nematodes, 173, 224, 227, 306, 317, 329
New Brunswick, NJ gateway, 114
New England, historical development, 6
New Hampshire, 25, 49, 52, 53, 56
New Jersey
  Delaware & Raritan Canal, 106
  shade tree inventory, 121
  shade tree legislation, 52, 53
  statutes, 49, 52, 53
News releases, public relations, 149, 150, 161
New York City, 13, 31, 32, 87, 225, 429, 430, 455
  Central Park, 7, 104
  pollution removal figures for, 29
New York State statutes, 49, 51, 53
Noise, 27, 34, 66, 337, 355, 356, 367, 375, 384, 386, 389, 394, 395, 397, 421
Nomenclature, 199, 210, 218
Norway maple; *see* maple species and cultivars, Norway
Nuclear polyhedrosis virus, 307, 315
Nuisance wildlife, 443, 449
Nursery maintenance practices, crew training in, 145
Nursery stock, 199–208
  bid specification writing, 208
    realities, 208
    resources, 209
  Internet plant locator resources, 218
  selecting quality trees, 199–208
    aboveground quality, 201
    belowground quality, 203
    nursery sources, 200
    plant nomenclature, 199
    production practices affecting survival and growth, 205
  specifying quality (contract), 211
    business section details, 214
    general form, 211
    project section details, 214
    results, 211
Nutrients, soil, 89, 188
  composting, 354
  root system growth, 263
  specification of, 180
  urban soils, 174
Nutrition; *see also* Fertilization
  root-building products, 232
  root system growth, 263
Nylon twine, 208
Nyssa sylvatica (black gum, tupelo), 192, 195, 196, 271, 446

Oak (*Quercus*) species and cultivars, 7, 195–197, 269, 271, 281, 310
as VOC emitter, high-level, 32
English (*Quercus robur*), 192, 271
lightning susceptibility, 289
pin (*Quercus palustris*), 75, 125, 192, 195, 271
red (*Quercus rubra*), 75, 125, 162, 192
root systems, 262
sawlog desirability, 402
street space, 106
wildlife benefit, 446
Odor, leaf composting problems, 352–357, 360, 361, 363, 364, 366–372
Ohio statutes, 49, 51–53
Open space, planning, 41, 99, 101, 102, 111
problems, 88
regional planning and zoning, 90, 91
strategies, 90
subdivision ordinance, 87
Ophiostoma ulmi (Dutch elm disease), 313
Ordinances; *see* Laws and ordinances
Organic matter; *see also* Mulching
microorganisms and, 173
specification of, 180
Osage orange, 402
Ostryia virginiana (hop hornbeam), 75, 193
Overmaturity, 377, 378, 386
Ownership, planning considerations, 100

Paired sales analysis, 344, 345
Pall mall, 5
Paramus, NJ, 429–436
awards, 432–436
history, 429–430
shade tree commission, establishment of, 430–432
Parasitoids, pest management, 298, 306, 312, 315, 316
Paris squares and plazas, 112
Parking lots, planning and design, 111
Parks, 7, 10, 11, 13, 15, 16, 36, 38, 39, 55, 57, 58, 67, 70, 72, 74, 79, 85, 92, 93, 100, 101, 103, 116, 127, 136, 140, 161, 194, 278, 290, 323, 375, 380
planning and design, 100, 103, 111
tree selection, 185
US cities, 7
Pathogens; *see also* Disease; Integrated pest management
composting and, 352, 354
Paulownia, 402
Pear (*Pyrus*) species, 75, 193, 202, 225, 241, 378
PEGSMOR; *see* Removals
Pennsylvania statutes, 49, 51, 52, 54
Pepperbush, sweet, 445, 446
Persimmon (*Diospyros virginiana*), 401, 402, 446
Personality issues, volunteer program, 160
Pesticides; *see also* Integrated pest management
contaminant specification, 188
tree selection considerations, 188, 189
Pesticides and biorational pesticides, 307
Pest management; *see also* Integrated pest management
budgeting, 141
newly planted trees, 235
nineteenth and early twentieth century, 14
Pests; *see also* Disease; Insect pests
nursery contract specifications, 199
site conditions, tree selection considerations, 188, 189
soil-borne, 227
pH
composting considerations, 352, 354
soil; *see* Soil reaction/pH
Phellodendron amurense (Amur cork tree), 193
Philadelphia, 1–3, 6–9, 15, 28, 30, 31, 460
Phloem necrosis, 313
Physical-biological benefits and costs of urban forests, 25–35
atmosphere, 27–35
hydrology, 33, 34
noise, 34
wildlife and biodiversity, 34, 35
Physical factors
planting site selection, 224
soil properties, 165, 166
Picea (spruce), 194, 196, 197, 281, 446
Austrian (key plant), 302
lightning susceptibility, 289
sawlog desirability, 402
white (*Pinus strobus*)
sawlog desirability, 402
wildlife benefit, 446
Place des Vosges, 112
Placement, and tree value, 327–336
Place (squares), 6
Plane tree (*Platanus*), 8, 32, 191, 192, 195, 196, 223, 310
Planners, cooperation with managers, 94
Planning; *see also* Community planning
cost-benefit determination for, 26
environmental impact mitigation by trees, 25
local laws and ordinances, 63
selection of trees, 184
shade tree inventory and, 123, 126
sidewalk and sewer problem considerations, 261, 459

Planning; *see also* Community planning (*Continued*)
  siting of plants for optimal health and pest resistance, 303
  tree removal, 383, 385–387
Planning and design, urban/community forestry, 99–118
  benefits and costs of urban forests, 36
  future opportunities, 115
    highway interchanges, 115
    productive landscapes, 115
    vacant and underutilized land and brown fields, 115
  links, 102
    blueways, 104
    greenways, 103
    highways and rail corridors, 107
    streets, 102
    visual, 107
  siting of plants for optimal health and pest resistance, 304
  tree functions, selection of trees, 184
  urban forest, 99
    objectives, 100
    operational definition, 99
    responsibility, 100
  urban forest elements, 101
    civic, private, and institutional facilities, 101
    conservation and natural areas, 101
    parks and open spaces, 101
    plazas and squares, 101
    streets, 108–110
Plant health; *see* Health, tree
Planting, 221–236
  annual reports, 148
  arboricultural practices, tree selection considerations, 188
  budgeting, 142–144
  care and maintenance, 235
  contract specifications, 216
  cost-benefit considerations, 36
  grants for, 111
  mulching, 233
  municipality-utility cooperation, 259
  pruning of newly transplanted trees, 241
  site selection, 222–224
    air quality, 223
    biological factors, 224
    disease control and pest management considerations, 310
    light, 223
    natural and urban factors, 224
    physical and morphological factors, 224
    soil conditions, 224
    temperature and hardiness, 222
    water, 223
  site-specific factors, 224
  staking and guying, 233
  timing and techniques, 226–232
    balled and burlapped planting, 230
    bare-root planting, 230
    container planting, 231
    fertilizer, 229
    large tree transplanting, 232
    planting hole, 227
    pruning, 229
    soils, 227
  transportation, care, and handling, 225
  Tree City USA; *see* Tree City USA
  watering, 234
Planting depth, root crown inspection, 208
Planting easement, 269–272
Planting programs, 37, 142, 145, 208, 209, 425, 434; *see also* Tree City USA
Plastic bags, leaf composting problems, 362
Plastic burlap, 203, 207, 208
Plantanus (planetree), 8, 32, 192, 195, 196, 223, 281, 310
Plazas, planning and design, 101, 111
Plowing, compaction remediation, 178
Podosesia aureocincta (banded ash clearwing borer), 317
Police power, defined, 84, 87
Politics; *see also* Community planning
  planning considerations, 100
  relations with local officials, 152
Pollarding, 229, 240, 243
Pollen, 27, 37
Poplar (*Populus*) species, 281, 315, 383
  as VOC emitter, high-level, 281
  ligthning susceptibility, 281
  Lombardy, 7, 9
  sawlog desirability, 402
  tulip, 383
    ligthning susceptibility, 281
    wildlife benefit, 446
  white (*Populus alba*), 196
  yellow (*Liriodendron tulipifera*); *see* Tulip tree
Population growth, 81, 451, 453, 455
Populus species; *see* Aspen; Poplar species
Portable sawmills, 411, 412
Porthetria dispar (gypsy moth), 12–14, 300, 305–307, 311, 315, 316
Pot-in-pot system (PI), 205, 231
Power lines; *see* Utilities, and municipalities; Utility lines
Preservation zoning, 87
Private facilities, planning and design, 105
Problem and nuisance wildlife, 443, 448, 449

Production, pruning for, 237, 240, 246
Professional associations, 390
Promenades, 5, 6
Property owners, shade tree inventory takers, 127
Property rights, community planning and land use regulation, 89
Property taxes, 38, 77, 94, 134–137, 394
Property values,
  and tree value, 25
  benefits and costs of urban forests, 25
  cost-benefit considerations, 26
  sidewalk problems and, 231
  tree appraisal, 288
Pruning, 237–247
  budgeting process, 134, 135, 140
  categories, 242
  hazard tree features, 283, 284
  in nature, 237
  liability of public utility companies, 51
  maintenance, 244
  methods, 244
  nursery practices, 201, 203
  pest management, cultural control, 336
  planting, 219
  premature mechanical failure, 383, 385
  proper cuts, 242
  reasons for, 238
  seasonal timing, 240
  site conditions, tree selection considerations, 184
  Tree Line USA, 436
  tree selection considerations, 183
  utility, 246; *see also* Utilities, and municipalities
  utility company-municipality cooperation, 257, 258
  wound care, 246
Prunus species, 32, 193, 194, 196, 258, 446; *see also* Cherry species and cultivars
Pseudoscymnus, 319
Pseudoscymnus tsugae, 319
Public; *see* Community involvement
Public areas, management of trees in, 41
Public forum, community planning, 94
Public Gardens, Boston, 104
Public policy; *see* Government policy; Laws and ordinances
Public-private partnerships, community planning, 94
Public relations, 147
  annual reports, 48
  challenges to tree management, 135
  comment forms, 151
  composting facility site selection, 354
  image analysis survey, 152
  leaflets, 148
  lectures, 150
  news releases, 149
  television, 151
  videotapes, 151
  wildlife, 435
Public safety, budgeting priorities, 140
Public use, hazard trees, 70
Public utilities; *see* Utilities
Punitive damages, 51
Pyrus species, 32, 75, 193

Quercus; *see* Oak species and cultivars

Radial trenching, 177
Rail corridors, 107
Raised planting beds, 177, 179
Random sampling, shade tree inventory, 121
RC&D program, 409, 410
Real estate value; *see* Property values
"Reasonable and necessary" standard, 51
Reasonableness testing, tree appraisal, 345, 347
Record keeping
  appraisal of tree value, 331, 332
  pest management, monitoring forms, 298, 299
  systematic management, 122, 123
Recreation areas, 40, 41, 86, 278
Recycling
  composting leaves; *see* Leaf composting
  street trees as sawlogs, 402; *see also* Sawlogs, street tree recycling
Redbud (*Cercis*), 75, 195, 241, 258, 310
Redwood, dawn (*Metasequoia glyptostroboides*), 192, 271, 321, 339
References
  bidding contractors, 218
  nursery, 201, 210–212
Regional planning and zoning, 91
Regulations; *see* Government policy
Relative humidity, 28
Removals
  bid specifications, 389
  budgeting, 140
  cost-benefit considerations, 28
  costs, 381
  cultural practice to prevent premature removal, 383
  emergency, 377
  hazard trees, root decay and, 290
  maintenance, 377, 385, 387
  methods,

Removals (*Continued*)
manual, 381
mechanical, 379
planning, 383–390
professional associations, 392
reasons for, 378
recycling, 399–418; *see also* Sawlogs, street tree recycling
safety, 382
American national standards, 394
job site, 395
training and performance guidelines for personnel, 396
sidewalk damage and, 273
supplies and equipment
arborist, 393
manufacturers, 392
tree failure, 378
Renaissance, 3, 4
Replacement, types of damages, 51
Replacement cost
appraisal, 326
tree appraisal method, 326, 327
tree appraisal value, 327, 339
tree value formulas, 50
Replacement planting
bid/contract specifications, 218
municipality-utility cooperation, 249
sidewalk and sewer problems, 273
Report, comprehensive community plan, 83
Residential suburban landscapes, 442
Resistance to diseases and pathogens, 310, 313, 314
Resistance to pesticides, 296
Restatement rule, 49, 50
Restricted suburban space, 271, 272
Resurgence of pests, 296
Rhamnus species, 32
Rhode Island statutes, 49, 51, 52, 54, 56, 57
Rhododendron (key plant) 302, 335
Rigging
cranes, 379, 380
manual tree dismantling, 381
Rights-of-way planting, 40, 41, 451. 463
Roads and streets; *see* Streets and roads
Robinia pseudoacacia (black locust), 196, 281
Rodents, 227, 230, 233, 234, 448
Rolling leases, 94
Romantic movement, 7
Roman writers on trees, 10
Root collar
appraisal of tree value, 329–332
hazard tree features, 281, 283, 286
Root crown inspection, 231
Root cutting, 265, 273, 274, 280
Root decay, 229, 280, 283, 285, 286, 290, 291
Root failure, 256, 285
Root flare, 215, 227, 230, 234
Roots, 165–179; *see also* Soils
appraisal of tree value, 325, 328–332
hazard trees, 283
defects, 281
evaluation of defects, 278
symptoms of problems, 285
larger tree transplantation, 232
mulch and, 233, 234
nursery stock, 203–208
circling, 205
contract specifications, 213
hilling and, 205
specification of handling procedures, 213
planting
development, 227, 230–232
protection of, 233
site selection, 222
restriction, soil volume and, 176
sidewalk and sewer problems, 261
cost-benefit considerations, 34
growth of roots, 262
identification of, 262
impeding growth of, 263
responsibility for sidewalk and sewer problems, 266
sewer problems, 265
sidewalk problems, 264
site conditions, tree selection considerations, 186
soil properties, 165, 166
Rope
girdling by, 225
removal equipment, 381, 388, 389, 391, 394
Runoff, 27
costs versus benefits, 33, 34
leaf composting facility, 358
parking lot design, 113
salt accumulation in soil, 180
Rutgers process control strategy, 367

Safety; *see also* Hazard trees
leaf composting facility, 349, 359
pruning for, 239
root cutting considerations, 273
sawlogs, street tree recycling as
log loading and unloading, 406
metals and foreign material hazards, 402
tree removal, 377, 399
American national standards, 394
assessment form, 394
emergency removals, 385, 387, 388
guidelines, 396
job site, 395

training and performance guidelines for personnel, 396
Salix species, 32, 195, 197, 289
Salts
accumulation in soil, 174
specification of, 180
Sampling, shade tree inventory, 121
Sanitation
elm bark beetle, 313
pest management, 305
sycamore anthracnose, 310
Sapwood decay, 290
Sassafras, 196, 446
Sawlogs, street tree recycling as, 399–418
markets, 400
commercial operations, 409
in-house, 412
local crafters/hobbyists/custom sawmillers, 411
products and specifications, 401
metal and foreign material hazards, 402
quantity and volume determination, 407
safety, log loading and unloading, 406
specifications, 402
storage requirements, 409
state forestry offices, 413
Saws, 246, 384, 385, 388, 394
Scaffold branches, 238, 241, 242, 282, 331, 332
appraisal of tree value, 332
hazard trees, 282
Scholar tree (*Styphnolobium japonicum*), 192, 196, 271
Scientific Basis for Ecosystem Management, 456
Scientific treatment of trees, historical perspectives, 20
Seasonal timing
pesticide applications, 296
pruning, 240
tree selection considerations, 186
Secondary pest outbreaks, 296
Security
leaf composting facility, 350, 357, 361
tree removal safety, 381
Selection of trees, 183
arboricultural practices, 188
computer methods, 198
criteria, 189, 190
criteria, lists tree species matching, 191–197
dry sites, 195
parks, windbreaks, and landscapes, 194
street trees, 191–194
wet sites, 195
for sidewalk planting, 269
construction practices, 272
planning for future, 271
planting easement, 271
planting location, 271
repair options, 273
species, 269
nursery stock; *see* Nursery stock
process, 184
purpose of tree in landscape, defining, 184
site conditions, evaluating, 186
species to site matches, 383
Self-help rule, 49, 50, 383
Sensitive area ordinances, 42
Serviceberry (*Amelanchier*), 193, 195, 196, 258, 446
Sewer damage; *see* Sidewalks and sewers
Shadblow (*Amelanchier X grandiflora*), 258
Shadbush (*Amelanchier*), 195, 196, 446
Shade tree commissions, 8, 16, 52–55, 60, 61, 65, 67, 69, 257, 287, 430
Shade tree inventory; *see* Inventory, shade tree
Shade tree program, municipality-utility cooperation, 256
Shaping around utility lines, 255
Shearing, 240–242
Shopping centers, 38, 430
Short v. Kerr, 48, 62
Shovels, mechanical, 77, 177, 226, 325, 326
Side trimming for utility lines, 254, 255, 326
Side trim shelf, 256
Sidewalk planting, tree selection for, 269–273
Sidewalks and sewers, 261–276
cost-benefit considerations, 40
hazard trees, 278
root systems, 262
growth of roots, 263
identification of, 262
impeding growth of, 263
responsibility for sidewalk and sewer problems, 266
sewer problems, 265
sidewalk problems, 264
selection of trees for sidewalk planting, 273
construction practices, 272
planning for future, 271
planting easement, 271
planting location, 272
repair options, 273
species, 269
Silverbell, Carolina (*Halesia carolina*), 75, 193, 194, 258
Site
and plant health, 297
appraisal of tree value, 327
disease control and pest management considerations, 295, 303
evaluating for tree selection, 189, 190
planting, 221–224

Site (*Continued*)
air quality, 223
biological factors, 224
light, 223
physical and morphological factors, 224
site-specific factors, 224
soil conditions, 224
temperature and hardiness, 222
water, 223
Size of trees
appraisal of tree value, 326, 337, 338
contract specifications, 216
diameter; *see* Caliper/trunk diameter
mature dimensions, tree selection considerations, 186, 187, 191
municipality-utility cooperation, 260
nursery stock, 207
tree appraisal value, 337, 338
Slow composting, leaf composting problems, 370, 373
Small Business Association, tree-planting grants, 138, 436
Socio-economic benefits and costs of urban forests, 36
communities, 37
individuals, 36
miscellaneous, 39
real estate values, 38
tree value formulas, 39
Softwoods, sawlog desirability, 402
Soil balls, nursery stock, 201, 203–208
contract specification, 216
digging with mechanical tree spades, 296
plastic burlap, 207
Soil cracks, and root decay, 285
Soil failure, hazard trees, 283, 285, 290
Soil reaction/pH, 172, 173, 175, 180
root system growth, 263
specification of, 180
tree selection considerations, 189
urban soil, 174
Soils, 165–182
compaction, 176
and root system growth, 263
particle size, and, 176
prevention, 175
remediation, 176
site conditions, tree selection considerations, 186
soil volume, 178
urban soils, 174
drainage, 170
hazard tree conditions, 280
urban soils, 174
hazard tree conditions, 280
nursery stock, 207
planning and design process, 110, 111
planting considerations, 223, 224
larger tree transplantation, 232
site selection, 222
properties, 166
biological, 173
chemical, 171
physical, 166
root decay symptoms, 285
root interactions, 34, 166, 173
structure and texture and, 263
site conditions, tree selection considerations, 186
specification summary, 180
structure, 168
microorganisms and, 173
root system growth, 263
urban soils, destruction of, 174
texture, 166, 167
and root system growth, 263
compaction remediation, 176
prevention of compaction, 175
specification of, 180
tree selection considerations, 186
urban versus natural, 174
aeration and drainage, 170
pH, elevated, 175
structure, 168
volume, 178
watering considerations, 234
Solar radiation, 27
Source reduction, 350
Spades, mechanical, 206, 232
Spruce (*Picea*), 195
Squares (place), 6, 76, 101, 111, 112, 116
Staking, 17, 161, 189, 228, 231, 233, 303
Standardized Plant Names, 210
Standard metropolitan statistical area (SMSA), 57
Standard Practices for Trees, Shrubs, and Other Woody Plant Maintenance, 210
Standards
arboricultural practices, local ordinance, 63, 64
for bid specifications, 211
local laws and ordinances, 63, 64
plant health certification, 216
source for contractor, 214
tree removal, 387
projects, 389
utility line clearance contractor certification, 252
State forestry offices, 18, 409–411, 413
State government
arboriculture law in Northeast, 47

community planning legislation, 95
funding sources, 134
power over municipalities, 83
State planning, and land use regulation, 91
Stem decay, hazard trees, 288
Sticky traps, 300, 306
Storage, sawlogs, street tree recycling as, 409
Storms; *see also* Runoff
causes of mechanical failure, 379, 385
hazard tree management, 292
utility-municipality cooperation, 257
Storm sewers, 265, 268
Storm water treatment costs, 40
Streets and roads; *see also* Hazard trees; Sidewalks and sewerts; Traffic
historical development, 3–9, 12, 13, 15–17
liability for trees, 47
planning and design, 41, 99–103, 108, 109
traffic hazards, 280, 359
tree selection, 184, 185, 186, 188
Strength, pruning for, 238
Strengthening, 81, 95
Strength loss thresholds, from stem decay, 288
Stressors
appraisal of tree value, 328, 329, 333, 340
site conditions, tree selection considerations, 188
Structural defects, 278, 288, 378, 385
Structure, soil; *see* Soils, structure
Structures; *see* Buildings
Stubbing, 243, 255
Stump hazards, 283
Stump removal, 141, 384, 386, 390, 395
Styphnolobium japonicum (scholar tree), 192, 196
Subdivision ordinances, 84, 86–88, 90, 95
Suburbanization, 89
Suburbs
restricted space, 272
wildlife habitats, 437–441, 443–445, 447, 448
Surface roots, 190, 262, 263
Sustainable development, 452
Sweethbay, 195
Sweet gum *(Liquidambar styraciflua)*, 72, 192, 195
Sycamore anthracnose (*Apiognomonia veneta*), 300, 302, 310
Sycamore (*Platanus*), 4, 8, 191, 195, 196
frost cracking, 230
key plant, 302
Syringa amurensis (tree lilac), 75, 258, 227, 230, 269, 281, 289, 300, 302, 310, 317, 400, 402
Syringa (tree lilac), 75, 194, 258
Systematic technology, community planning process, 83
Tap roots, 262
Taxes, 39, 60, 86, 94, 134–136, 145
Taxodium distichum (bald cypress), 192, 195
Technomas, 462
Television, public relations, 155
Temperature
and VOC emissions by trees, 31
composting, 351–353, 360, 361, 364, 365, 367, 368
cost-benefit considerations, 27
costs versus benefits, 27–29
pest and disease susceptibility, 301
planting considerations, 222
tree selection considerations, 188
Ten percent sample, shade tree inventory, 121
Texture, soil; *see* Soils, texture
Thinning out (pruning method), 240
Thuja occidentalis (arborvitae), 194, 195, 197
Thuja occidentalis (Northern white cedar), 194
Thuja plicata (giant arborvitae), 194
Thuja plicata (Western red cedar), 194
Tilia; *see* Linden species
Tilling, compaction remediation, 178
Timing; *see* Seasonal timing
Toomey v. State of Connecticut, 48, 50, 62
Topography, 194, 409, 455
Topping, 69, 76, 243, 254, 281, 282, 379
Town commons, 2
Toxic contaminants, leaf composting problems, 374
Tracing, 231
Traffic
compaction remediation, 176
planning and design, 103, 109, 111
tree damage, 280–283
tree removal projects, 378, 382, 384, 386
Traffic hazards, 280, 287, 378
Training
crews, 144, 145, 148
shade tree inventory takers, 126–128
volunteers, 157–160
Transferable development rights, 84, 93
Transportation
contract specification, 212, 213
planting, 225
Tree appraisal, 323–347; *see also* Appraisal, tree
Tree area, 272; *see also* Size of trees
Tree City USA, 419–435
benefits, 420
eligibility standards, 423
growth award, 424
history, 419
Paramus, NJ, 429
awards, 432
history, 429

Tree City USA (*Continued*)
shade tree commission, establishment of, 430
Tree Line USA, 424
Tree failure, 37, 277, 285, 287, 290, 292, 293, 324, 378
Tree inventory; *see* Inventory, shade tree
Tree lilac (*Syringa*), 75, 194, 258
Tree Line USA, 109, 420, 424, 425
Tree of heaven (*Ailanthus altissima*), 9, 196
Tree removals; *see* Removals
Tree replacement; *see* Replacement
Tree selection; *see* Selection of trees
Tree spades, mechanical, 206, 232
Tree value, 39, 290, 324, 237, 330, 331, 339, 340; *see also* Appraisal, tree
Tree wardens, 12, 15, 52, 53, 55–59, 61
Trenching, radial, 177
Trichogramma wasp, 306
Trimming
budgeting process, 13
liability of public utility companies, 47, 51
shade tree inventory data, 124, 125
sidewalk damage prevention, 273, 274
utility, 251–257, 259
Triple damages, 51
Trunk
appraisal of tree value, 327, 330, 331, 339, 340
hazard trees, 287
mulch and, 234
protection, staked and guyed trees, 233
Trunk decay, management guidelines, 288
Trunk diameter; *see* Caliper/trunk diameter
Trunk flare, hazard tree features, 283
Trunk formula appraisal method, factors affecting value, 32, 340–345
Trunk formula method, factors affecting value, 326–340
condition, 327
location, 331
replacement cost and basic price, 339
size, 337
species, 336
Trusts, community planning, 90, 91
Tsuga, *see* Hemlock
Tucson, AZ, 34
Tulip tree (*Liriodendron tulipifera*), 289, 315, 446
decay susceptibility, 289
gypsy moth repellant, 315
tree-species to site matches, 383
wildlife benefit, 446
Tupelo (*Nyssa sylvatica*), 192, 195, 196, 271
Turbulence, 28
Twine, nursery stock, 204, 208

Ulmus species; *see* Elm species and cultivars
Ultraviolet radiation, 28, 36
University extension specialists, 410
University of Toronto, urban forestry concept, 19
Urban decay, 453
Urban factors, planting site selection, 222
Urban forest effects model, pollution removal, 29
Urban-rural continuum, 455
Urban space, confined, 271, 272
Urban stresses
site conditions, tree selection considerations, 186
tree selection considerations, 190
Urban wildlife habitats, 440–443
Utilities
hazard trees, 278
liability of, 51
and municipalities, 249–260
avoiding conflicts, 256
background, 249
cooperation, 254
joint ventures, 257
line clearance programs, 252
regulations, 251
summary, 259
wire conflict, 250
planning, design, and management considerations, 41
pruning, 246
sewers; *see* Sidewalks and sewers
site selection for planting, 222
Tree Line USA, 424
tree-species to site matches, 383
Utility lines, 12, 71, 109, 145
clearance programs, 252
conflict, 250
hazard tree management, 277
hazard trees, 277
planting site selection, 222
site conditions, tree selection considerations, 186
Tree Line USA, 425
Utility pruning, 246

Value of tree
appraisal; *see* Appraisal, tree
formulas, 29
hazard trees, 289, 290
Vandalism
peat/bark grown nursery plants, 204, 205
site conditions, tree selection considerations, 188
V-crotches, 238, 282, 286, 288, 290, 328
Vegetative growth
and frost cracking, 230

canopy size; *see* Canopy
vehicles, 5, 13, 32, 72, 176, 185, 186, 188, 223, 239, 283, 301, 330, 358, 383, 386, 394, 429; *see also* Traffic
leaf composting facility access, 356, 358
soil compaction remediation, 177
Vermont, 25, 49
Videotapes, 151, 159, 393
Virginia, 49, 52, 315, 317, 319
Vista pruning, 245
Visual linkages, planning and design, 103, 107
Volatile organic compound emissions, 27, 31, 32
Volume, soil, 111, 167, 174, 178, 179, 221, 224, 229, 329
Volunteer planning commissions, 84
Volunteers, 153–160
advertising, 155
applications for screening, 155
benefits, 157
challenges, 159
financing and locating teachers, 159
monitoring of newly planted trees, 235
shade tree inventory, 127
training, 159

Wall promenade, 5
Washington, DC, 7, 9, 30, 32, 138
Waste disposal, 27
Waste management; *see* Leaf composting
Water; *see also* Runoff
benefit and costs of urban forests, 27, 28, 33, 34
composting, 353
hazard tree conditions, 280
leaf composting facility, 350
planting site selection, 222
soil properties
aeration and drainage, 170, 171, 174
compaction, 176
fracturing, 178
texture and, 166, 168, 175
volume, 179
tree selection, 221–224
wet sites, tree selection, 221–224
wildlife needs, 447
Water deficits, 179
Water gels, root protection, 226, 232, 234
Watering
peat/bark grown nursery plants, 205
pest and disease susceptibility, 289
planting, 234
Water quality
costs versus benefits, 33
leaf composting facility and, 353–355, 357–359, 361–364
Watersheds, decentralization and, 455
Waterside promenade, 5
Water stress, 179
Weather conditions, root system growth, 264, 274, 292, 300, 307, 353, 362
Weed management
bark damage by herbicides, 202
composting and, 350
nursery stock, 202
pest and disease susceptibility, 301
West Virginia statutes, 50, 51
Wet climates, *see* Humidity
Wildlife, 437–450
attitudes toward, 438
benefits, 443
benefits and costs of urban forests, 26
decentralization and, 455
elements of habitat, 438
managing for, 444
attracting birds, 444
attracting other wildlife, 447
discouraging problem wildlife, 448
problem and nuisance species, 443
urban and suburban landscapes, 440
Willow (*Salix*) species, 75, 192, 195, 197, 289, 311
Wind
microclimate modification, 27
site conditions, tree selection considerations, 186
Windbreaks, tree selection, 194
Windrow size and turning, composting, 353
Windthrow (soil failure)
hazard trees, 285, 290
root cutting considerations, 273
Winter
frost cracking, 230
root freezing during, 227, 230
site selection for planting, 224
tree selection considerations, 189
Wire baskets, nursery stock, 201, 204, 208
Wires
guying, 233
power lines; *see* Utilities; Utility lines
Wisconsin statutes, 49, 51, 52, 54
Wood decay; *see* Decay
Wood products; *see* Sawlogs, street tree recycling as
Woody materials, composting, 350, 351, 369
Woolly adelgid (*Adelges tsugae*), 302, 303, 307, 319
Work plan, tree removal, 381–384, 386, 387, 394
Wound care, pruning, 238, 242, 246

Zelkova (*Zelkova serrata*), 75, 193, 196
Zoning
local ordinances, 87, 90, 91
regional, 90, 91

MIX
Papier aus verantwortungsvollen Quellen
Paper from responsible sources
FSC® C105338

If you have any concerns about our products,
you can contact us on
ProductSafety@springernature.com

In case Publisher is established outside the EU,
the EU authorized representative is:
Springer Nature Customer Service Center GmbH
Europaplatz 3, 69115 Heidelberg, Germany

Printed by Libri Plureos GmbH
in Hamburg, Germany